FUNDAMENTALS OF POLYMER ENGINEERING

PLASTICS ENGINEERING

Founding Editor

Donald E. Hudgin

Professor
Clemson University
Clemson, South Carolina

FUNDAMENTALS OF POLYMER ENGINEERING

Second Edition
Revised and Expanded

Anil Kumar
Indian Institute of Technology
Kanpur, India

Rakesh K. Gupta
West Virginia University
Morgantown, West Virginia, U.S.A.

MARCEL DEKKER, INC.   NEW YORK · BASEL

Library of Congress Cataloging-in-Publication Data
A catalog record for this book is available from the Library of Congress.

ISBN: 0-8247-0867-9

The first edition was published as *Fundamentals of Polymers* by McGraw-Hill, 1997.

This book is printed on acid-free paper.

Headquarters
Marcel Dekker, Inc.
270 Madison Avenue, New York, NY 10016
tel: 212-696-9000; fax: 212-685-4540

Eastern Hemisphere Distribution
Marcel Dekker AG
Hutgasse 4, Postfach 812, CH-4001 Basel, Switzerland
tel: 41-61-260-6300; fax: 41-61-260-6333

World Wide Web
http://www.dekker.com

The publisher offers discounts on this book when ordered in bulk quantities. For more information, write to Special Sales/Professional Marketing at the headquarters address above.

Current printing (last digit):
10 9 8 7 6 5 4 3 2 1

PRINTED IN THE UNITED STATES OF AMERICA

To the memory of my father.

Anil Kumar

To the memory of my father.

Rakesh Gupta

Preface to the Second Edition

The objectives and organization of the second edition remain essentially unchanged. The major difference from the first edition is the inclusion of new material on topics such as dendrimers, polymer recycling, Hansen solubility parameters, nanocomposites, creep in glassy polymers, and twin-screw extrusion. New examples have been introduced throughout the book, additional problems appear at the end of each chapter, and references to the literature have been updated. Additional text and figures have also been added.

The first edition has been successfully used in universities around the world, and we have received many encouraging comments. We hope the second edition will also find favor with our colleagues, and be useful to future generations of students of polymer science and engineering.

Anil Kumar
Rakesh K. Gupta

Preface to the First Edition

Synthetic polymers have considerable commercial importance and are known by several common names, such as plastics, macromolecules, and resins. These materials have become such an integral part of our daily existence that an introductory polymer course is now included in the curriculum of most students of science and engineering. We have written this book as the main text for an introductory course on polymers for advanced undergraduates and graduate students. The intent is to provide a systematic coverage of the essentials of polymers.

After an introduction to polymers as materials in the first two chapters, the mechanisms of polymerization and their effect on the engineering design of reactors are elucidated. The succeeding chapters consider polymer characterization, polymer thermodynamics, and the behavior of polymers as melts, solutions, and solids both above and below the glass transition temperature. Also examined are crystallization, diffusion of and through polymers, and polymer processing. Each chapter can, for the most part, be read independently of the others, and this should allow an instructor to design the course to his or her own liking. Note that the problems given at the end of each chapter also serve to complement the main text. Some of these problems cite references to the literature where alternative viewpoints are introduced. We have been teaching polymer science for a long time, and we have changed the course content from year to year by adopting and expanding on ideas of the kind embodied in these problems.

Since polymer science is an extremely vast area, the decision to include or exclude a given subject matter in the text has been a difficult one. In this endeavor, although our own biases will show in places, we have been guided by how indispensable a particular topic is to proper understanding. We have attempted to keep the treatment simple without losing the essential features; for depth of coverage, the reader is referred to the pertinent technical literature. Keeping the student in mind, we have provided intermediate steps in most derivations. For the instructor, lecturing becomes easy since all that is contained in the book can be put on the board. The future will tell to what extent we have succeeded in our chosen objectives.

We have benefited from the comments of several friends and colleagues who read different parts of the book in draft form. Our special thanks go to Ashok Khanna, Raj Chhabra, Deepak Doraiswamy, Hota V. S. GangaRao, Dave Kofke, Mike Ryan, and Joe Shaeiwitz. Professor Khanna has used the problem sets of the first seven chapters in his class for several years.

After finishing my Ph.D. from Carnegie-Mellon University, I (Anil Kumar) joined the Department of Chemical Engineering at the Indian Institute of Technology, Kanpur, India, in 1972. My experience at this place has been rich and complete, and I decided to stay here for the rest of my life. I am fortunate to have a good set of students from year to year with whom I have been able to experiment in teaching various facets of polymer science and modify portions of this book continuously.

Rakesh Gupta would like to thank Professor Santosh Gupta for introducing polymer science to him when he was an undergraduate student. This interest in polymers was nurtured by Professor Art Metzner and Dr. K. F. Wissbrun, who were his Ph.D. thesis advisors. Rakesh learned even more from the many graduate students who chose to work with him, and their contributions to this book are obvious. Kurt Wissbrun reviewed the entire manuscript and provided invaluable help and encouragement during the final phases of writing. Progress on the book was also aided by the enthusiastic support of Gene Cilento, the Department Chairman at West Virginia University. Rakesh adds that these efforts would have come to nought without the determined help of his wife, Gunjan, who guarded his spare time and allowed him to devote it entirely to this project. According to Rakesh, "She believed me when I told her it would take two years; seven years later she still believes me!"

I doubt that this book would ever have been completed without the constant support of my wife, Renu. During this time there have been several anxious moments, primarily because our children, Chetna and Pushkar, were trying to choose their careers and settle down. In taking care of them, my role was merely helping her, and she allowed me to divide my attention between home and work. Thank you, Renu.

Anil Kumar
Rakesh Gupta

Contents

Contents

1

Introduction

1.1 DEFINING POLYMERS

Polymers are materials of very high molecular weight that are found to have multifarious applications in our modern society. They usually consist of several structural units bound together by covalent bonds [1,2]. For example, polyethylene is a long-chain polymer and is represented by

$$-CH_2CH_2CH_2- \quad \text{or} \quad [-CH_2CH_2-]_n \qquad (1.1.1)$$

where the structural (or repeat) unit is $-CH_2-CH_2-$ and n represents the chain length of the polymer.

Polymers are obtained through the chemical reaction of small molecular compounds called monomers. For example, polyethylene in Eq. (1.1.1) is formed from the monomer ethylene. In order to form polymers, monomers either have reactive functional groups or double (or triple) bonds whose reaction provides the necessary linkages between repeat units. Polymeric materials usually have high strength, possess a glass transition temperature, exhibit rubber elasticity, and have high viscosity as melts and solutions.

In fact, exploitation of many of these unique properties has made polymers extremely useful to mankind. They are used extensively in food packaging, clothing, home furnishing, transportation, medical devices, information technology, and so forth. Natural fibers such as silk, wool, and cotton are polymers and

TABLE 1.1 Some Common Polymers

Commodity thermoplastics	
Polyethylene	$-[CH_2-CH_2]-$
Polystyrene	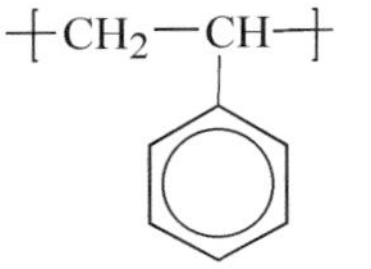
Polypropylene	$-[CH_2-CH(CH_3)]-$
Polyvinyl chloride	$-[CH_2-CH(Cl)]-$
Polymers in electronic applications	
Polyacetylene	$-[CH=CH]_n-$
Poly(*p*-phenylene vinylene)	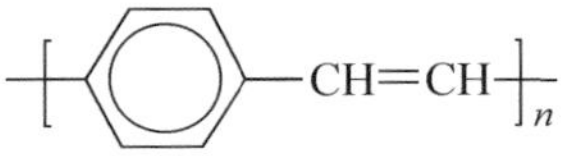
Polythiophene	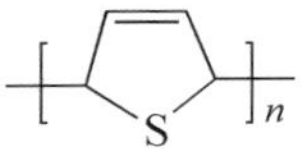
Polyphenylene sulfide	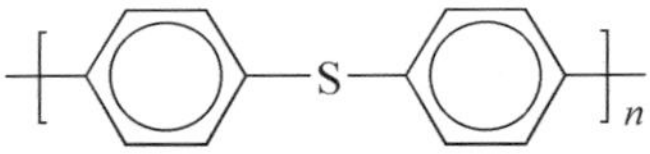
Polyanilines	$-[CH_2-CH(NH_2)]_n-$
Biomedical applications	
Polycarbonate (diphenyl carbonate)	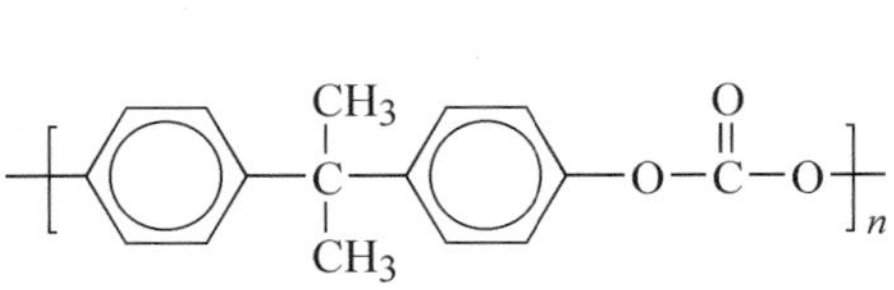
Polymethyl methacrylate	$-[CH_2-C(CH_3)(COOCH_3)]_n-$
Silicone polymers	$-[Si(CH_3)_2-O]_n-$

Specialty polymers

Polyvinylidene chloride

$$\left[CH_2-\underset{\underset{Cl}{|}}{\overset{\overset{Cl}{|}}{C}}\right]$$

Polyindene

$$\left[-CH-CH-\right]_n$$

with benzene ring and CH_2

Polyvinyl pyrrolidone

$$\left[CH_2-CH_2\right]$$
$$N$$
$$CH_2 \quad C=O$$
$$CH_2-CH_2$$

Coumarone polymer

$$\left[-CH-CH-\right]$$
with benzene ring and O

---

have been used for thousands of years. Within this century, they have been supplemented and, in some instances, replaced by synthetic fibers such as rayon, nylon, and acrylics. Indeed, rayon itself is a modification of a naturally occurring polymer, cellulose, which in other modified forms have served for years as commercial plastics and films. Synthetic polymers (some common ones are listed in Table 1.1) such as polyolefins, polyesters, acrylics, nylons, and epoxy resins find extensive applications as plastics, films, adhesives, and protective coatings. It may be added that biological materials such as proteins, deoxyribonucleic acid (DNA), and mucopolysaccharides are also polymers. Polymers are worth studying because their behavior as materials is different from that of metals and other low-molecular-weight materials. As a result, a large percentage of chemists and engineers are engaged in work involving polymers, which necessitates a formal course in polymer science.

Biomaterials [3] are defined as materials used within human bodies either as artificial organs, bone cements, dental cements, ligaments, pacemakers, or contact lenses. The human body consists of biological tissues (e.g., blood, cell, proteins, etc.) and they have the ability to reject materials which are "incompatible" either with the blood or with the tissues. For such applications, polymeric materials, which are derived from animals or plants, are natural candidates and some of these are cellulosics, chitin (or chitosan), dextran, agarose, and collagen. Among synthetic materials, polysiloxane, polyurethane, polymethyl methacrylate, polyacrylamide, polyester, and polyethylene oxides are commonly employed because they are inert within the body. Sometimes, due to the requirements of mechanical strength, selective permeation, adhesion, and/or degradation, even noncompatible polymeric materials have been put to use, but before they are utilized, they are surface modified by biological molecules (such as, heparin, biological receptors, enzymes, and so forth). Some of these concepts will be developed in this and subsequent chapters.

This chapter will mainly focus on the classification of polymers; subsequent chapters deal with engineering problems of manufacturing, characterization, and the behavior of polymer solutions, melts, and solids.

1.2 CLASSIFICATION OF POLYMERS AND SOME FUNDAMENTAL CONCEPTS

One of the oldest ways of classifying polymers is based on their response to heat. In this system, there are two types of polymers: thermoplastics and thermosets. In the former, polymers "melt" on heating and solidify on cooling. The heating and cooling cycles can be applied several times without affecting the properties. Thermoset polymers, on the other hand, melt only the first time they are heated. During the initial heating, the polymer is "cured"; thereafter, it does not melt on reheating, but degrades.

A more important classification of polymers is based on molecular structure. According to this system, the polymer could be one of the following:

1. Linear-chain polymer
2. Branched-chain polymer
3. Network or gel polymer

It has already been observed that, in order to form polymers, monomers must have reactive functional groups, or double or triple bonds. The functionality of a given monomer is defined to be the number of these functional groups; double bonds are regarded as equivalent to a functionality of 2, whereas a triple bond has a functionality of 4. In order to form a polymer, the monomer must be at least bifunctional; when it is bifunctional, the polymer chains are always linear. It is pointed out that all thermoplastic polymers are essentially linear molecules, which can be understood as follows.

In linear chains, the repeat units are held by strong covalent bonds, while different molecules are held together by weaker secondary forces. When thermal energy is supplied to the polymer, it increases the random motion of the molecules, which tries to overcome the secondary forces. When all forces are overcome, the molecules become free to move around and the polymer melts, which explains the thermoplastic nature of polymers.

Branched polymers contain molecules having a linear backbone with branches emanating randomly from it. In order to form this class of material, the monomer must have a capability of growing in more than two directions, which implies that the starting monomer must have a functionality greater than 2. For example, consider the polymerization of phthalic anhydride with glycerol, where the latter is tri-functional:

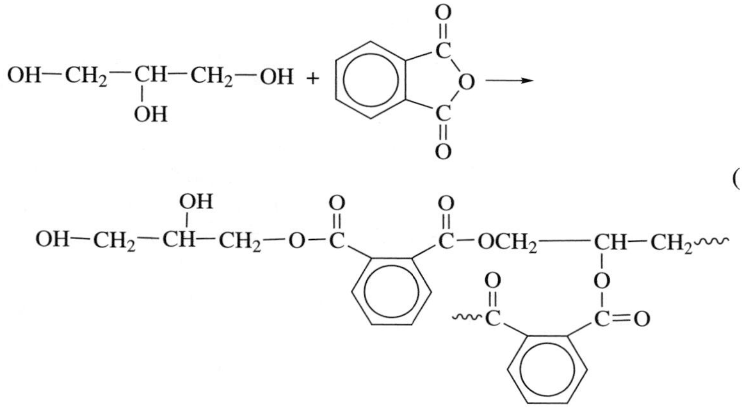

$$OH-CH_2-CH-CH_2-OH \;+\; \text{(phthalic anhydride)} \longrightarrow$$
$$\underset{OH}{|}$$

(1.2.1)

$$OH-CH_2-\underset{OH}{\underset{|}{CH}}-CH_2-O-\overset{O}{\overset{\|}{C}}\text{(benzene ring)}\overset{O}{\overset{\|}{C}}-OCH_2-CH-CH_2\sim$$

The branched chains shown are formed only for low conversions of monomers. This implies that the polymer formed in Eq. (1.2.1) is definitely of low molecular weight. In order to form branched polymers of high molecular weight, we must use special techniques, which will be discussed later. If allowed to react up to large conversions in Eq. (1.2.1), the polymer becomes a three-dimensional network called a *gel*, as follows:

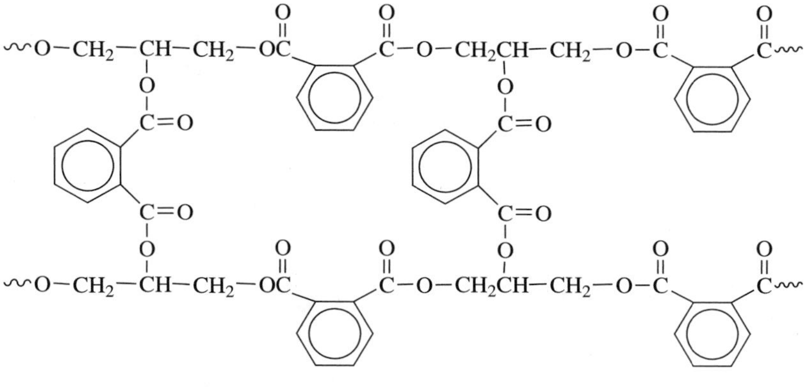

(1.2.2)

In fact, whenever a multifunctional monomer is polymerized, the polymer evolves through a collection of linear chains to a collection of branched chains, which ultimately forms a network (or a gel) polymer. Evidently, the gel polymer does not dissolve in any solvent, but it swells by incorporating molecules of the solvent into its own matrix.

Generally, any chemical process can be subdivided into three stages [viz. chemical reaction, separation (or purification) and identification]. Among the three stages, the most difficult in terms of time and resources is separation. We will discuss in Section 1.7 that polymer gels have gained considerable importance in heterogeneous catalysis because it does not dissolve in any medium and the separation step reduces to the simple removal of various reacting fluids. In recent times, a new phase called the *fluorous phase*, has been discovered which is immiscible to both organic and aqueous phases [4,5]. However, due to the high costs of their synthesis, they are, at present, only a laboratory curiosity. This approach is conceptually similar to solid-phase separation, except that fluorous materials are in liquid state.

In dendrimer separation, the substrates are chemically attached to the branches of the hyper branched polymer (called dendrimers). In these polymers,

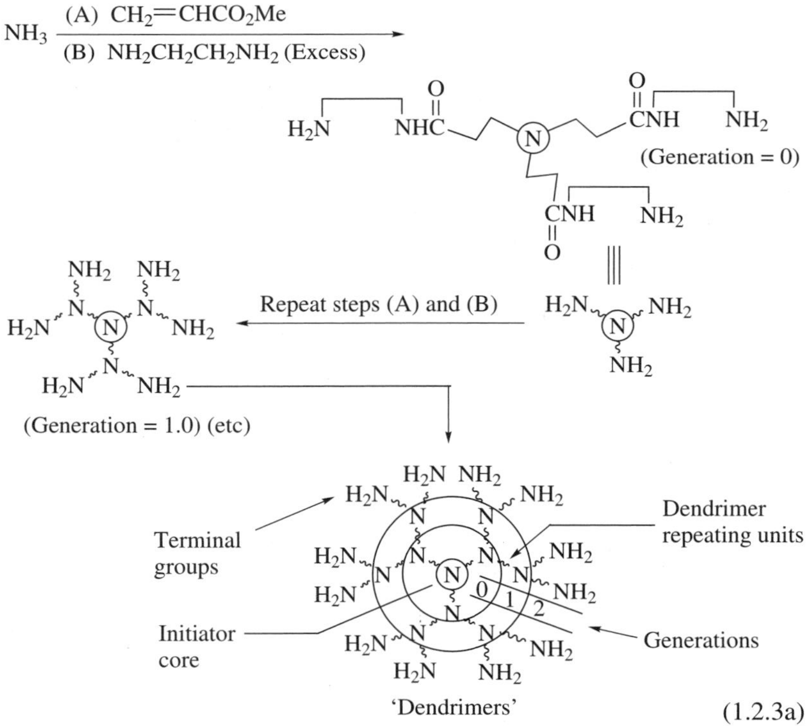

'Dendrimers'

(1.2.3a)

the extent of branching is controlled to make them barely soluble in the reaction medium. Dendrimers [6] possess a globular structure characterized by a central core, branching units, and terminal units. They are prepared by repetitive reaction steps from a central initiator core, with each subsequent growth creating a new generation of polymers. Synthesis of polyamidoamine (PAMAM) dendrimers are done by reacting acrylamide with core ammonia in the presence of excess ethylene diamine.

Dendrimers have a hollow interior and densely packed surfaces. They have a high degree of molecular uniformity and shape. These have been used as membrane materials and as filters for calibrating analytical instruments, and newer paints based on it give better bonding capacity and wear resistance. Its sticking nature has given rise to newer adhesives and they have been used as catalysts for rate enhancement. Environmental pollution control is the other field in which dendrimers have found utility. A new class of chemical sensors based on these molecules have been developed for detection of a variety of volatile organic pollutants.

In all cases, when the polymer is examined at the molecular level, it is found to consist of covalently bonded chains made up of one or more repeat units. The name given to any polymer species usually depends on the chemical structure of the repeating groups and does not reflect the details of structure (i.e., linear molecule, gel, etc.). For example, polystyrene is formed from chains of the repeat unit:

$$CH_2-CH$$

(1.2.3b)

Such a polymer derives it name from the monomer from which it is usually manufactured. An idealized sample of polymer would consist of chains all having identical molecular weight. Such systems are called *monodisperse polymers*. In practice, however, all polymers are made up of molecules with molecular weights that vary over a range of values (i.e., have a distribution of molecular weights) and are said to be *polydisperse*. Whether monodisperse or polydisperse, the chemical formula of the polymer remains the same. For example, if the polymer is polystyrene, it would continue to be represented by

$$X-CH_2-CH-\left[CH_2-CH\right]_n-CH_2-CH-Y$$

(1.2.4)

For a monodisperse sample, n has a single value for all molecules in the system, whereas for a polydisperse sample, n would be characterized by distribution of values. The end chemical groups X and Y could be the same or different, and what they are depends on the chemical reactions initiating the polymer formation.

Up to this point, it has been assumed that all of the repeat units that make up the body of the polymer (linear, branched, or completely cross-linked network molecules) are all the same. However, if two or more different repeat units make up this chainlike structure, it is known as a copolymer. If the various repeat units occur randomly along the chainlike structure, the polymer is called a random copolymer. When repeat units of each kind appear in blocks, it is called a block copolymer. For example, if linear chains are synthesized from repeat units A and B, a polymer in which A and B are arranged as

$$-AAAAAAAABBBBBBBBBB- \qquad (1.2.5)$$

is called an AB block copolymer, and one of the type

$$-AAAAAAAABBBBBBBH_2BH_2AAAAA- \qquad (1.2.6)$$

is called an ABA block copolymer. This type of notation is used regardless of the molecular-weight distribution of the A and B blocks [7].

The synthesis of block copolymers can be easily carried out if functional groups such as acid chloride (~COCl), amines (~NH$_2$), or alcohols (~OH) are present at chain ends. This way, a polymer of one kind (say, polystyrene or polybutadiene) with dicarboxylic acid chloride (ClCO~COCl) terminal groups can react with a hydroxy-terminated polymer (OH~OH) of the other kind (say, polybutadiene or polystyrene), resulting in an AB type block copolymer, as follows:

$$\underset{ClC\sim CCl}{\overset{O\ \ \ O}{\overset{\|\ \ \|}{}}} + OH\sim OH \longrightarrow Cl\left[\underset{-C\sim C}{\overset{O\ \ \ O}{\overset{\|\ \ \|}{}}} -O\sim O \right]_n -H \qquad (1.2.7)$$

In Chapter 2, we will discuss in more detail the different techniques of producing functional groups. Another common way of preparing block copolymers is to utilize organolithium initiators. As an example, *sec*-butyl chloride with lithium gives rise to the butyl lithium complex,

$$CH_3-\underset{CH_3}{\underset{|}{CH}}-CH_2Cl + Li \longrightarrow CH_3\underset{CH_3}{\underset{|}{CH}}-CH_2Li^+ \cdots Cl^- \qquad (1.2.8)$$

which reacts quickly with a suitable monomer (say, styrene) to give the following polystyryl anion:

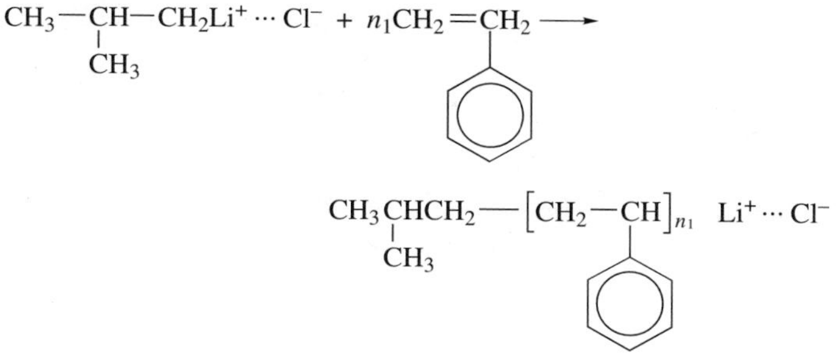

$$CH_3-CH-CH_2Li^+ \cdots Cl^- + n_1CH_2=CH_2 \longrightarrow$$

$$\underset{CH_3}{|}$$

(1.2.9)

$$CH_3\underset{|}{CHCH_2}-\left[CH_2-CH\right]_{n_1} Li^+ \cdots Cl^-$$

$$\underset{CH_3}{|}$$

This is relatively stable and maintains its activity throughout the polymerization. Because of this activity, the polystyryl anion is sometimes called a living anion; it will polymerize with another monomer (say, butadiene) after all of the styrene is exhausted:

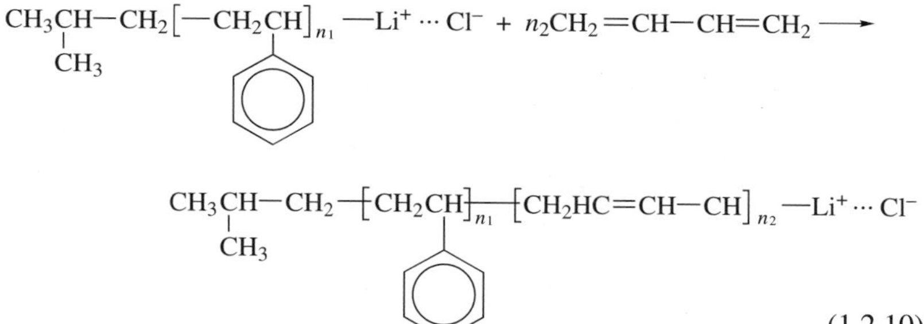

$$CH_3\underset{|}{CH}-CH_2\left[-CH_2CH\right]_{n_1}-Li^+ \cdots Cl^- + n_2CH_2=CH-CH=CH_2 \longrightarrow$$

$$\underset{CH_3}{}$$

$$CH_3\underset{|}{CH}-CH_2-\left[CH_2CH\right]_{n_1}\left[CH_2HC=CH-CH\right]_{n_2}-Li^+ \cdots Cl^-$$

$$\underset{CH_3}{}$$

(1.2.10)

In this way, we can conveniently form an AB-type copolymer. In fact, this technique of polymerizing with a living anion lays the foundation for modifying molecular structure.

Graft copolymers are formed when chains of one kind are attached to the backbone of a different polymer. A graft copolymer has the following general structure:

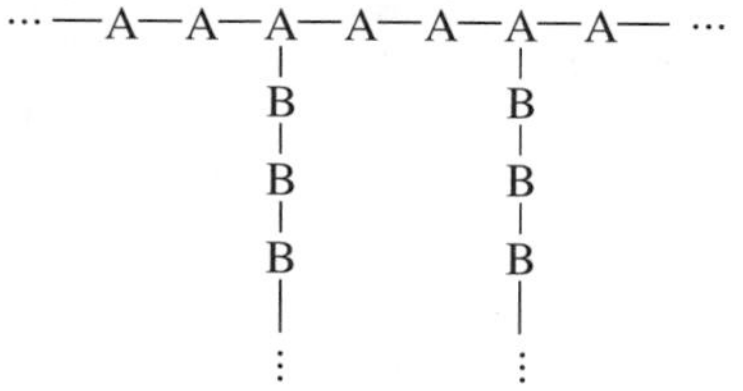

$$\cdots -A-A-A-A-A-A-A- \cdots$$
$$\qquad \quad | \qquad \qquad \quad |$$
$$\qquad \quad B \qquad \qquad \quad B$$
$$\qquad \quad | \qquad \qquad \quad |$$
$$\qquad \quad B \qquad \qquad \quad B \qquad\qquad (1.2.11)$$
$$\qquad \quad | \qquad \qquad \quad |$$
$$\qquad \quad B \qquad \qquad \quad B$$
$$\qquad \quad \vdots \qquad \qquad \quad \vdots$$

Here $-(A)_n$ constitutes the backbone molecule, whereas polymer $(B)_n$ is randomly distributed on it. Graft copolymers are normally named poly(A)-g-poly(B), and the properties of the resultant material are normally extremely different from those of the constituent polymers. Graft copolymers can be generally synthesized by one of the following schemes [1]:

The "grafting-from" technique. In this scheme, a polymer carrying active sites is used to initiate the polymerization of a second monomer. Depending on the nature of the initiator, the sites created on the backbone can be free-radical, anion, or Ziegler–Natta type. The method of grafting-from relies heavily on the fact that the backbone is made first and the grafts are created on it in a second polymerization step, as follows:

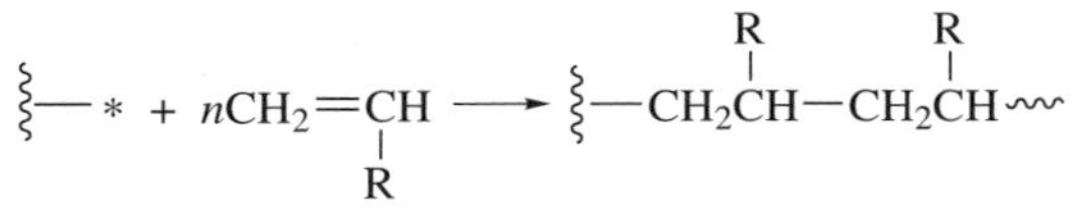

$$ \text{\textbf{---}} * + n\text{CH}_2\text{=CH} \longrightarrow \text{---CH}_2\text{CH--CH}_2\text{CH} \qquad (1.2.12) $$

This process is efficient, but it has the disadvantage that it is usually not possible to predict the molecular structure of the graft copolymer and the number of grafts formed. In addition, the length of the graft may vary, and the graft copolymer often carries a fair amount of homopolymer.

The "graft-onto" scheme. In this scheme, the polymer backbone carried a randomly distributed reactive functional group X. This reacts with another polymer molecule carrying functional groups Y, located selectively at the chain ends, as follows:

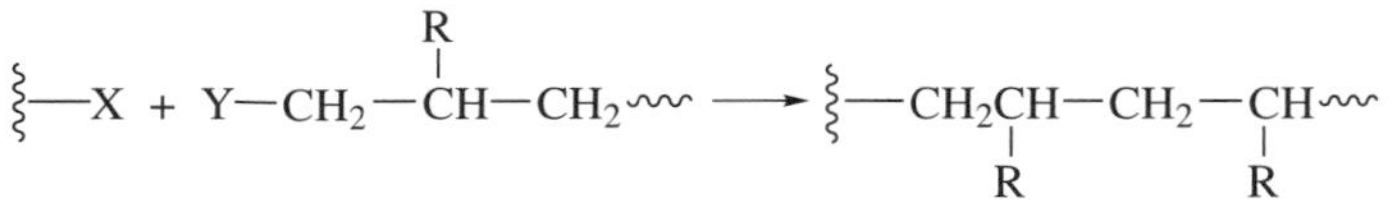

$$ \text{---X} + \text{Y--CH}_2\text{--CH--CH}_2 \longrightarrow \text{---CH}_2\text{CH--CH}_2\text{--CH} \qquad (1.2.13) $$

In this case, grafting does not involve a chain reaction and is best carried out in a common solvent homogeneously. An advantage of this technique is that it allows structural characterization of the graft copolymer formed because the backbone and the pendant graft are both synthesized separately. If the molecular weight of each of these chains and their overall compositions are known, it is possible to determine the number of grafts per chain and the average distance between two successive grafts on the backbone.

The "grafting-through" scheme. In this scheme, polymerization with a macromer is involved. A macromer is a low-molecular-weight polymer chain with unsaturation on at least one end. The formation of macromers has recently been reviewed and the techniques for the maximization of macromer amount discussed therein [4]. A growing polymer chain can react with such an unsaturated site, resulting in the graft copolymer in the following way:

$$\substack{\S}-CH=CH_2 + \sim CH_2-\overset{\overset{\text{o}}{\|}}{C}-\!\!\left[CH_2CH\right]_n \longrightarrow \substack{\S}-CH-\!\!\left[CH_2-CH\right]_m$$

$$\qquad\qquad\qquad\quad \underset{R}{|} \qquad \underset{R}{|} \qquad\qquad -\!\!\left[CH_2-CH\right]_n$$

$$\qquad\qquad\qquad\qquad\qquad\qquad\qquad\qquad\qquad\quad \underset{R}{|}$$

(1.2.14)

This type of grafting can introduce linkages between individual molecules if the growing sites happen to react with an unsaturated site belonging to two or more different backbones. As a result, cross-linked structures are also likely to be formed, and measures must be taken to avoid gel formation.

There are several industrial applications (e.g., paints) that require us to prepare *colloidal dispersions* of a polymer [5]. These dispersions are in a particle size range from 0.01 to 10 μm; otherwise, they are not stable and, over a period of time, they sediment. If the polymer to be dispersed is already available in bulk, one of the means of dispersion is to grind it in a suitable organic fluid. In practice, however, the mechanical energy required to reduce the particle size below 10 μm is very large, and the heat evolved during grinding may, at times, melt the polymer on its surface. The molten surface of these particles may cause agglomeration, and the particles in colloidal suspensions may grow and subsequently precipitate this way, leading to colloidal instability. As a variation of this, it is also possible to suspend the monomer in the organic medium and carry out the polymerization. We will discuss these methods in considerable detail in Chapter 7 ("emulsion and dispersion" polymerization), and we will show that the problem of agglomeration of particles exists even in these techniques.

Polymer colloids are basically of two types: lyophobic and lyophilic. In *lyophilic* colloids, polymer particles interact with the continuous fluid and with other particles in such a way that the forces of interaction between two particles lead to their aggregation and, ultimately, their settling. Such emulsions are unstable in nature. Now, suppose there exists a thermodynamic or steric barrier between two polymer particles, in which case they would not be able to come close to each other and would not be able to agglomerate. Such colloids are *lyophobic* in nature and can be stable for long periods of time. In the technology of polymer colloids, we use special materials that produce these barriers to give the stabilization of the colloid; these materials are called stabilizers. If we wanted to prepare colloids in water instead of an organic solvent, then we could use soap (commonly used for over a century) as a *stabilizer*. The activity of soap is due to its lyophobic and lyophilic ends, which give rise to the necessary barrier for the formation of stable colloids.

In several recent applications, it has been desired to prepare colloids in media other than water. There is a constant need to synthesize new stabilizers for a specific polymer and organic liquid system. Recent works have shown that the block and graft copolymers [in Eqs. (1.2.5) and (1.2.11)] give rise to the needed stability. It is assumed that the A block is compatible with the polymer to be suspended and does not dissolve in the organic medium, whereas the B block dissolves in the organic medium and repulses polymer particles as in Figure 1.1. Because of the compatibility, the section of the chain consisting of A-repeat units gets adsorbed on the polymer particle, whereas the section of the chain having B-repeat units projects outward, thus resisting coalescence.

Example 1.1: Micellar or ampliphilic polymers (having hydrophobic as well as hydrophilic fragments in water) have the property of self-organization. What are these and how are they synthesized?

Solution: Micellar polymers have properties similar to surfactant molecules, and because of their attractive properties, they are used as protective colloids, emulsifiers, wetting agents, lubricants, viscosity modifiers, antifoaming agents, pharmaceutical and cosmetic formulating ingredients, catalysts, and so forth [8].

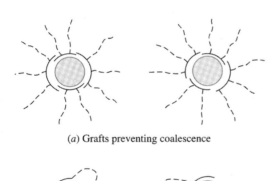

(*a*) Grafts preventing coalescence

(*b*) Block copolymers serving as stabilizers

——— Medium insoluble portion of chain
– – – Medium soluble portion of chain

FIGURE 1.1   Stabilizing effect of graft and block copolymers.

Micellar polymers can have six types of molecular architecture, and in the following, hydrophobic and hydrophilic portions are shown by a chain and a circle, respectively, exactly as it is done for ordinary surfactant (i.e., tail and head portions).

(a) Block copolymer

$$OH-(CH_2CH_2O)_{\overline{m}}(\overset{\overset{\displaystyle CH_3}{|}}{CH}-CH_2-O)_n$$

(b) Star copolymer

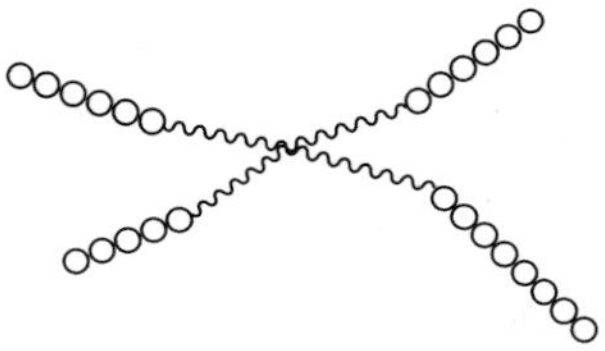

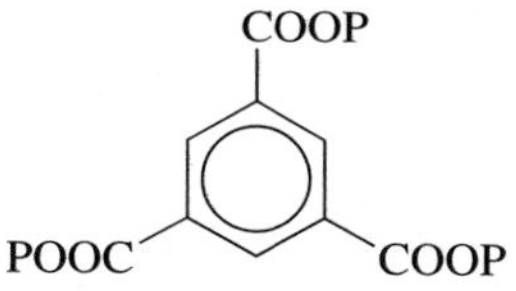

COOP

POOC   COOP

where

$$P = -[(CH_2)_4-O]_m-[CH_2CH_2\overset{\overset{\displaystyle C(CH_3)_3}{|}}{\underset{\underset{\displaystyle X^-}{|}}{N}H^+}]_n$$

(c) Graft polymer

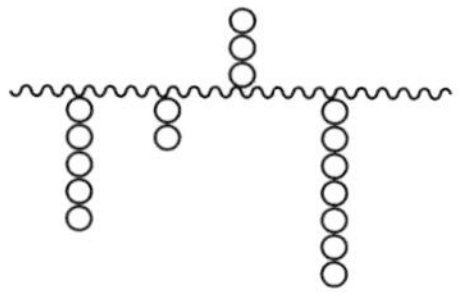

$$-[CH_2CH=CH-CH_2]_n-[\underset{\underset{\displaystyle [NH-CH_2CH_2]_x}{|}}{CH}-CH=CHCH_2]_m-$$

(d)   Dendrimer

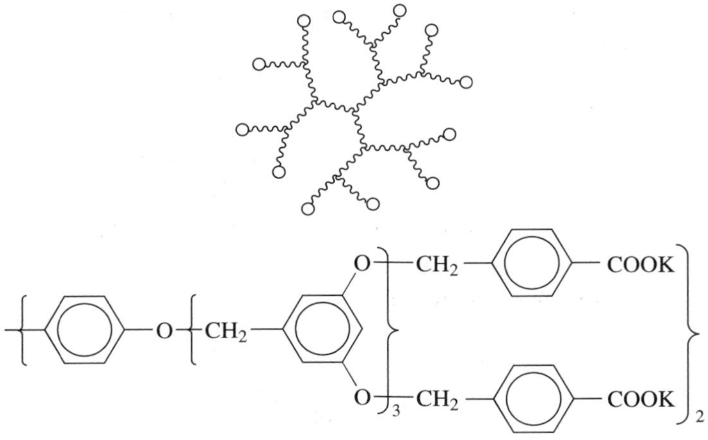

(e)   Segmented block copolymer

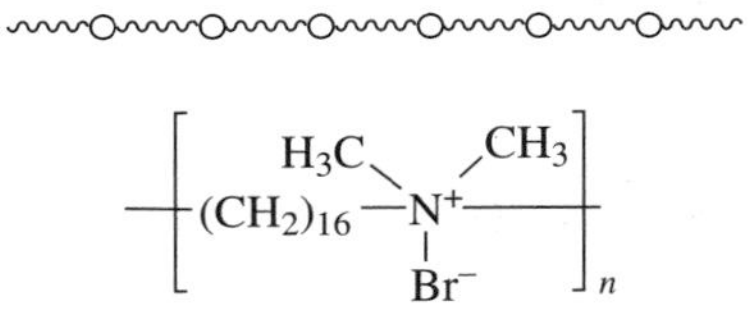

(f)   Polysoap

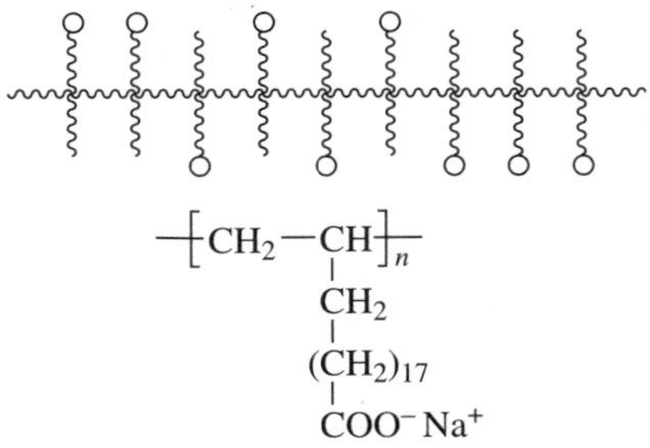

Example 1.2: Describe polymers as dental restorative materials and their requirements.

Solution: The dental restorative polymers must be nontoxic and exhibit long-term stability in the presence of water, enzymes, and various oral fluids. In addition, it should withstand thermal and load cycles and the materials should be easy to work with at the time of application. The first polyacrylolyte material used for dental restoration was zinc polycarboxylate. To form this, one uses zinc oxide powder which is mixed with a solution of polyacrylic acid. The zinc ions cross-link the polyacid chains and the cross-linked chains form the cement.

Another composition used for dental restoration is glass ionomer cement (GIC). The glass used is fluoroaluminosilicate glass, which has a typical composition of 25–25 mol% SiO_2, 14–20 wt% Al_2O_3, 13–35 wt% CaF_2, 4–6 wt% AlF_3, 10–25% $AlPO_4$, and 5–20% Na_3AlF_6. In the reaction with poly-acrylic acid, the latter degrades the glass, causing the release of calcium and aluminum ions which cross-link the polyacid chains. The cement sets around the unreacted glass particles to form a reaction-bonded composite. The fluorine present in the glass disrupts the glass network for better acid degradation.

Completely polymeric material used for dental restoration is a polymer of methyl methacrylate (MMA), bisphenol-A, glycidyl methacrylate (bis GMA), and triethylene glycol dimethacrylate (TEGDMA). The network thus formed has both hydrophilic as well as hydrophobic groups and can react with teeth as well, giving a good adhesion. In order to further improve the adhesion by interpenetration and entanglements into dental surfaces, sometimes additives like 4-META (4-methoxyethyl trimellitic anhydride), phenyl-P (2-methacryloxy ethyl phenyl hydrogen phosphate), or phenyl-P derivatives are added.

Example 1.3: Anticancer compounds used in chemotherapy are low-molecular-weight compounds, and on its ingestion, it is not site-specific to the cancerous tissues leading to considerable toxicity. How can polymer help reduce toxicity? How does this happen? Give a few examples.

Solution: Macromolecules are used as carriers, on whose backbone both the anticancerous compounds as well as the targeting moieties are chemically bound. As a result of this, the drug tends to concentrate near the cancerous tissues. The targeting moieties are invariably complementary to cell surface receptors or antigens, and as a result of this, the carrier macromolecule can recognize (or biorecognize) cancerous tissues. The polymer-mediated drug now has a considerably altered rate of uptake by body cells as well as distribution of the drug within the body.

Some of the synthetic polymers used as drug carriers are HPMA (poly 2-hydroxy propyl methacrylamide), PGA (poly L-glutamic acid), poly(L-lysine), and Block (polyethylene glycol coaspartic acid). Using HPMA, the following drugs have been synthesized [9]:

Drug	Targeting moiety
Abriamycin	Galactosamine
Duanomycin	Anti-Ia^k antibodies
Chlorin e$_6$	anti-Thy 1.2 antibody

By putting the targeting moiety to the polymer, one has created an ability in the polymer to differentiate between different biological cells and recognize tumour cells [10]. This property is sometimes called molecular recognition and this technique can also be used for separating nondesirable components from foods or fluids (particularly biological ones).

The general technique of creating molecular recognition (having antibody-like activity) is called molecular imprinting. Templates are defined as biological macromolecules, micro-organisms, or whole crystals. When functional monomers are brought in contact with the templates, they adhere to it largely because of noncovalent bonding. These could now be cross-linked using a suitable cross-linking agent. If the templates are destroyed, the resulting cross-link polymer could have a mirror-image cavity of the template, functioning exactly like an antibody.

1.3 CHEMICAL CLASSIFICATION OF POLYMERS BASED ON POLYMERIZATION MECHANISMS

In older literature, it was suggested that all polymers could be assigned to one of the two following classes, depending on the reaction mechanism by which they are synthesized.

1.3.1 Addition Polymers

These polymers are formed by sequential addition of one bifunctional or polyfunctional monomer to growing polymer chains (say, P_n) without the elimination of any part of the monomer molecule. With the subscript n representing the chain length, the polymerization can be schematically represented as follows:

$$P_n + M \longrightarrow P_{n+1} \tag{1.3.1}$$

M represents a monomer molecule; this chain growth step is usually very fast.

The classic example of addition polymerization is the preparation of vinyl polymers. Vinyl monomers are unsaturated organic compounds having the following structure:

$$CH_2\!=\!\underset{\underset{R}{|}}{CH} \tag{1.3.2}$$

where R is any of a wide variety of organic groups: a phenyl, a methyl, a halide group, and so forth. For example, the polymerization of vinyl chloride to give poly(vinyl chloride) can be written in the simplified form

$$n\,CH_2\!=\!\underset{\underset{Cl}{|}}{CH} \longrightarrow \left[\!CH_2\!-\!\underset{\underset{Cl}{|}}{CH}\right]_n \tag{1.3.3}$$

Ring-opening reactions, such as the polymerization of ethylene oxide to give poly(ethylene oxide), offer another example of the formation of addition polymers:

$$n\,CH_2\!-\!CH_2 \longrightarrow \left[\!CH_2\!-\!CH_2\!-\!O\!-\!\right]_n \tag{1.3.4}$$

The correct method of naming an addition polymer is to write poly( ), where the name of the monomer goes into the parentheses. If $-R$ in compound (1.3.2) is an aliphatic hydrocarbon, the monomer is an olefin as well as a vinyl compound; these polymers are classified as *polyolefins*. In the case of ethylene and propylene, the parentheses in the names are dispensed with and the polymers are called polyethylene and polypropylene.

1.3.2 Condensation Polymers

These polymers are formed from bifunctional or polyfunctional monomers with the elimination of a small molecular species. This reaction can occur between any two growing polymer molecules and can be represented by

$$P_m + P_n \rightleftharpoons P_{m+n} + W \tag{1.3.5}$$

where P_m and P_n are polymer chains and W is the condensation product.

Polyesterification is a good example of condensation polymerization. In the synthesis of poly(ethylene terephthalate), ethylene glycol reacts with terephthalic acid according to the following scheme:

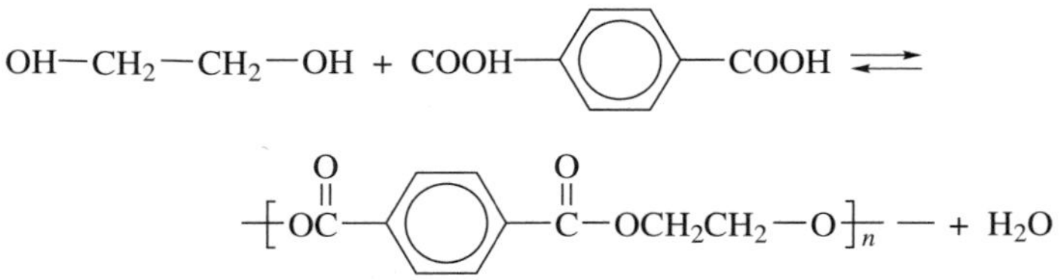

$$OH-CH_2-CH_2-OH + COOH-\langle\bigcirc\rangle-COOH \rightleftharpoons$$

(1.3.6)

$$\left\{OC-\langle\bigcirc\rangle-\overset{O}{\overset{\parallel}{C}}-OCH_2CH_2-O\right\}_n - + H_2O$$

As indicated by the double arrow, polyesterification is a reversible reaction. Polyamides (sometimes called nylons) are an important class of condensation polymers that are formed by reaction between amine and acid groups, as in

$$NH_2-(CH_2)_6-NH_2 + COOH-(CH_2)_4-COOH \rightleftharpoons$$

Hexamethylene   Adipic
diamine   acid

(1.3.7a)

$$\left\{NH-(CH_2)_6-NHCO-(CH_2)_6\right\} + H_2O$$

Nylon 66

$$NH_2-(CH_2)_5-COOH + H_2O \rightleftharpoons \left\{NH-(CH_2)_5-CO\right\}_n$$ (1.3.7b)

ω-Aminocaproic acid   Nylon 6

Both of these polymers are classified as polyamides because the repeat units contain the $-[CO-NH]-$ amide group.

Naming of condensation polymers is done as follows. The polymer obtained from reaction (1.3.6) is called poly(ethylene terephthalate) because the repeat unit is the ester of ethylene glycol and terephthalic acid. Similarly, the polymer in Eq. (1.3.7b) is called poly(ω-aminocaproic acid). The product in Eq. (1.3.7a) is called poly(hexamethylene adipamide), in which the hexamethylene part of the name is associated with the diamine reactant, and the adipamide part is associated with the amide unit in the backbone.

As researchers learned more about polymerization chemistry, it became apparent that the notion of classifying polymers this way was somehow inconsistent. Certain polymer molecules could be prepared by more than one mechanism. For example, polyethylene can be synthesized by either of the two mechanisms:

$$CH_2=CH_2 \longrightarrow \left\{CH_2-CH_2\right\}_n$$ (1.3.8a)

$$mBr\left[CH_2\right]_{10}Br + 2mH_2 \longrightarrow \left\{CH_2-CH_2\right\}_{5m}$$ (1.3.8b)

The latter is neither addition nor condensation polymerization. Likewise, the following reaction, which is a typical addition polymerization, gives the same polyamide as reaction (1.3.7b):

$$\begin{array}{c} CH_2\!-\!(CH_2)_4\!-\!CO \\ \rule{0pt}{0pt}\rule{1.2cm}{0.4pt}NH\rule{1.6cm}{0.4pt} \end{array} \longrightarrow \quad \substack{\displaystyle -\!\!\left[NH\!-\!(CH_2)_5\!-\!CO\right]_n} \tag{1.3.9}$$

ε-Caprolactam   Nylon 6

Similarly, the polymerization of polyurethane does not involve the evolution of a condensation product, even though its kinetics can be described by that of condensation polymerization. Clearly, it is not correct to classify polymers according to the scheme discussed earlier. It is now established that there are two classes of polymerization mechanisms:

1. *Chain-growth polymerization:* an alternative, but more chemically consistent name for addition polymerization.
2. *Step-growth polymerization:* mechanisms that have kinetics of this type exhibited by condensation polymerization but include reactions such as that in (1.3.9), in which no small molecular species are eliminated.

This terminology for discussing polymerization will be used in this textbook.

In chain-growth polymerization, it is found that individual molecules start growing, grow rapidly, and then suddenly stop. At any time, therefore, the reaction mass consists of mainly monomer molecules, nongrowing polymer molecules, and only a small number of rapidly growing polymer molecules. In step-growth polymerization, on the other hand, the monomer molecules react with each other at the beginning to form low-molecular-weight polymer, and the monomer is exhausted very quickly. They initially form low-molecular-weight polymer molecules then continue to react with each other to form continually growing chains. The polymers formed from these distinct mechanisms have entirely different properties due to differences in molecular-weight distribution, which is discussed in the following section.

1.4 MOLECULAR-WEIGHT DISTRIBUTIONS

All commercial polymers have a molecular-weight distribution (MWD). In Chapters 3–7, we will show that this is completely governed by the mechanism of polymerization and reactor design. In Chapter 8, we give some important experimental techniques to determine the molecular-weight distribution and its averages, and in view of the importance of this topic, we give some of the basic concepts here. The chain length n represents the number of repeat units in a given polymer molecule, including units at chain ends and at branch points (even though these units have a somewhat different chemical structure than the rest of the repeat units). For chain molecules with molecular weights high enough to be classified as true polymer molecules, there are at least one order of magnitude more repeat units than units at chain ends and branch points. It is therefore possible to write (with negligible error)

$$M_n = nM_0 \tag{1.4.1}$$

where M_n is the molecular weight of a polymer molecule and M_0 is that of a single repeat unit.

In reality, the average chain length of all polymer molecules in the reaction mass must be equal to some whole number. The product of a given polymerization reaction can be thought of as having a distribution of the degrees of polymerization (DPs), which is given by a histogram, as shown in Figure 1.2. In this representation, W_n^* is the weight of a species of degree of polymerization n such that

$$W_t = \text{Total weight of polymer}$$
$$= \sum_{n=1}^{\infty} W_n^* \tag{1.4.2}$$

By definition, the weight-average molecular weight, M_w, is given by

$$M_w = \frac{\sum_{n=1}^{\infty} W_n^* M_n}{W_t} \tag{1.4.3}$$

where M_n is the molecular weight of a species of chain length equal to n. For sufficiently high molecular weight, M_n is, for all practical purposes, identical to M_n of Eq. (1.4.1). For lower-molecular-weight species, the molecular weights of end units and branch points would have to be considered in determining M_n. Because polymers of high molecular weight are usually of interest, this complexity is normally ignored in the analysis.

Although Eqs. (1.4.1)–(1.4.3) serve as the starting point for this discussion, it is more useful to define a weight distribution of degrees of polymerization W_n by the equation

$$W_n = \frac{W_n^*}{W_t^*} \tag{1.4.4}$$

Alternatively, W_n can be interpreted as the fraction of the mass of the polymer, with the degreee of polimerization (DP) equal to n or a molecular weight of nM_0. The weight-average chain length, μ_w, is now defined by

$$\mu_w = \frac{M_w}{M_0} = \sum_{n=1}^{t} nW_n \tag{1.4.5}$$

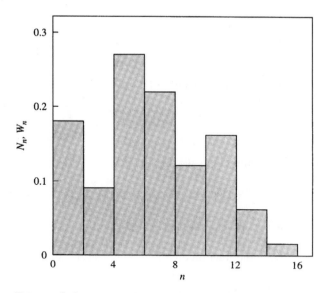

FIGURE 1.2 A typical histogram of the degree of polymerization.

It is thus seen that μ_w is just the first moment of the weight distribution of the degree of polymerization.

There is an alternative but equivalent method of describing distributions of molecular weight. If N_n^* is the total number of moles of a polymer of chain length equal to n in a given sample, one can write

$$N_n^t = \frac{W_n^*}{M_n} \tag{1.4.6}$$

The total number of moles of polymer, N_t, can then be written as

$$N_t = \sum_{n=1}^{\infty} N_n^* \tag{1.4.7}$$

By definition, the number-average molecular weight, μ_n, is given by

$$\mu_n = \sum_{n=1}^{\infty} \frac{M_n N_n^*}{N_t} \tag{1.4.8}$$

It is convenient, however, to define a number distribution of the degree of polymerization (DP) N_n as

$$N_n = \frac{N_n^*}{N_t} \tag{1.4.9}$$

such that

$$\sum_{n=1}^{\infty} N_n = 1 \qquad (1.4.10)$$

Because N_n is also the fraction of the molecules of polymer of DP equal to n or molecular weight of nM_0, Eq. (1.4.8) then becomes

$$M_n = \sum_{n=1}^{\infty} M_n N_n \qquad (1.4.11)$$

which gives the number-average chain length, μ_n, as

$$\mu_n = \frac{M_n}{M_0} = \sum_{n=1}^{\infty} n N_n \qquad (1.4.12)$$

and, as before, we see that μ_n is just the first moment of the distribution function N_n.

The higher moments of the mole fraction distribution N_n can be defined as

$$\lambda_k = \sum_{n=1}^{\infty} n^k N_n \qquad k = 0, 1, 2, \ldots \qquad (1.4.13)$$

where λ_k represents the kth moment. The zeroth moment (λ_0) is, according to Eq. (1.4.10), unity. The first moment (λ_1) is the same as μ_n in Eq. (1.4.12). The second moment (λ_2) is related to μ_w by

$$\mu_w = \frac{\lambda_2}{\lambda_1} \qquad (1.4.14)$$

The polydispersity index Q of the polymer is defined as the ratio of μ_w and μ_n by the following relation:

$$Q = \frac{\mu_w}{\mu_n} = \frac{\lambda_2 \lambda_0}{\lambda_1^2} \qquad (1.4.15)$$

The polydispersity index is a measure of the breadth of mole fraction (or molecular weight) distribution. For a monodisperse polymer, Q is unity; commercial polymers may have a value of Q lying anywhere between 2 and 20.

1.5 CONFIGURATIONS AND CRYSTALLINITY OF POLYMERIC MATERIALS

So far, we have examined the broader aspects of molecular architecture in chain-like molecules, along with the relationship between the polymerization mechanism and the repeat units making up the chain. We have introduced the concept of distribution of molecular weights and molecular-weight averages.

As expected, the architectural features (branching, extent of cross-linking, nature of the copolymer) and the distribution of molecular weight play an important role in determining the physical properties of polymers. In addition, the geometric details of how each repeat unit adds to the growing chain is an important factor in determining the properties of a polymer. These geometric features associated with the placement of successive repeat units into the chain are called the configurational features of the molecules, or, simply, chain configuration. Let us consider the chain polymerization of vinyl monomers as an example. In principle, this reaction can be regarded as the successive addition of repeat units of the type

$$-CH_2-\underset{\underset{R}{|}}{CH}- \qquad (1.5.1)$$

where the double bond in the vinyl compound has been opened during reaction with the previously added repeat unit. There are clearly three ways that two contiguous repeat units can be coupled.

Head-to-tail

$$-(CH_2-\underset{\underset{R}{|}}{CH}-CH_2-\underset{\underset{R}{|}}{CH}-)$$

Head-to-head

$$-(CH_2-\underset{\underset{R}{|}}{CH}-\underset{\underset{R}{|}}{CH}-CH_2-) \qquad (1.5.2)$$

Tail-to-tail

$$-(\underset{\underset{R}{|}}{CH}-CH_2-CH_2-\underset{\underset{R}{|}}{CH}-)$$

The head of the vinyl molecule is defined as the end bearing the organic group R. All three linkages might appear in a single molecule, and, indeed, the distribution of occurrence of the three types of linkage would be one way of characterizing the molecular structure. In the polymerization of vinyl monomers, head-to-tail placement is favored, and this structural feature normally dominates.

A more subtle structural feature of polymer chains, called *stereoregularity*, plays an important factor in determining polymer properties and is explained as follows. In a polymer molecule, there is usually a backbone of carbon atoms linked by covalent bonds. A certain amount of rotation is possible around any of these backbone covalent bonds and, as a result, a polymer molecule can take several shapes. Figure 1.3a shows three possible arrangements of the substituents of any one carbon atom with respect to those of an adjacent one when viewed end-on, such that the two consecutive carbon atoms C_n and C_{n-1} appear one behind the other. The potential energy associated with the rotation of the C_n-C_{n-1} bond is shown in Figure 1.3b and is found to have three angular positions of minimum energy. These three positions are known as the gauche-

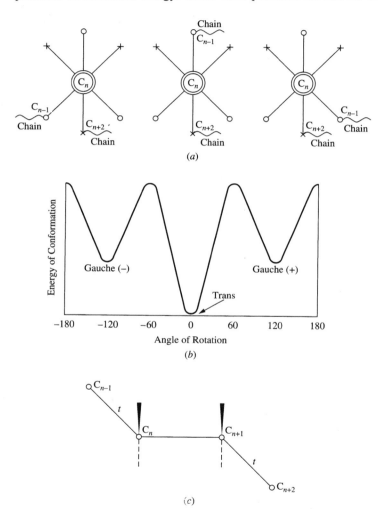

FIGURE 1.3 Different conformations in polymer chains and potential energies associated with them.

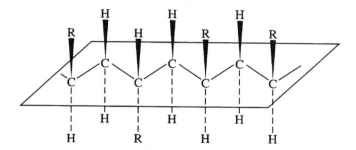

FIGURE 1.4 Spatial arrangement of $[C_2CHR]_n$ when it is in a planar zigzag conformation: actactic when R is randomly distributed, isotactic when R is either above or below the plane, and syndiotactic when R alternates around the plane.

positive (g^+), trans (t), and gauche-negative (g^-) conformations of the bond; the trans state is the most probable one by virtue of having the lowest potential energy.

Substituted polymers, such as polypropylene, constitute a very special situation. Because the polymer is substituted, the conformation of each of the backbone bonds is distinguishable. Each of the C−C backbone bonds can take up any one of the three (g^+, t, and g^-) positions. Because the polymer is a sequence of individual C−C bonds, the entire molecule can be described in terms of individual bond conformations. Among the various conformations that are possible for the entire chain, there is one in which all the backbone atoms are in the trans (t) state. From Figure 1.3c, it can be observed that if bonds C_n-C_n and $C_{n+1}-C_{n+2}$ are in the trans conformation, carbon atoms C_{n-1}, C_n, C_{n+1}, and C_{n+2} all lie in the same plane. By extending this argument, it can be concluded that the entire backbone of the polymer molecule would lie in the same plane, provided all bonds are in the trans conformation. The molecule is then in a planar zigzag form, as shown in Figure 1.4. If all of the R groups now lie on the same side of the zigzag plane, the molecule is said to be isotactic. If the R groups alternate around the plane, the molecule is said to be syndiotactic. If there is no regularity in the placement of the R groups on either side of the plane, the molecule is said to be atactic, or completely lacking in order. A given vinyl polymer is never 100% tactic. Nonetheless, polymers can be synthesized with high levels of stereoregularity, which implies that the molecules have a long block of repeat units with completely tactic placement (isotactic, syndiotactic, etc.), separated by short blocks of repeat units with atactic placement. Indeed, one method of characterizing a polymer is by its extent of stereoregularity, or tacticity.

Further, when a diene is polymerized, it can react in the two following ways by the use of the appropriate catalyst:

1,2 Polymerization

$$n\underset{1}{C}=\underset{2}{C}-\underset{3}{C}=\underset{4}{C} \longrightarrow -[C-\underset{\substack{| \\ C \\ \| \\ C}}{C}-]_n$$

1,4 Polymerization

$$n\underset{1}{C}=\underset{2}{C}-\underset{3}{C}=\underset{4}{C} \longrightarrow -[C-C=C-C-]_n \qquad (1.5.3)$$

The 1,2 polymerization leads to the formation of substituted polymers and gives rise to stereoregularity, as discussed earlier (Fig. 1.5). The 1,4 polymerization, however, yields double bonds on the polymer backbone. Because rotation around a double bond is not possible, polymerization gives rise to an inflexible chain backbone and the g^+, t, and g^- conformations around such a bond cannot occur. Therefore, if a substituted diene [e.g., isoprene $(CH_2=CH-C(CH_3)=CH_2)$] is polymerized, the stereoregularity in molecules arises in the following way. It is known that the double-bond formation occurs through sp hybridization of molecular orbitals, which implies that in Figure 1.4, carbon atoms C_{n-1}, C_n, C_{n+1}, and C_{n+2}, as well as H and R groups, all lie on the same plane. Two configurations are possible, depending on whether H and CH_3 lie on the same side or on opposite sides of the double bond. If they lie on the same side, the polymer has *cis configuration*; if they lie on opposite sides, the polymer has *trans configuration*. Once again, it is not necessary that all double bonds have the same configuration; if a variety of configurations can be found in a polymer molecule, it is said to have mixed configuration.

The necessary condition for chainlike molecules to fit into a crystal lattice is that they demonstrate an exactly repeating molecular structure along the chain. For vinyl polymers, this prerequisite is met only if they have predominantly head-to-tail placement and are highly tactic. When these conditions are satisfied, polymers can, indeed, form highly crystalline domains in the solid state and in concentrated solution. There is even evidence of the formation of microcrystalline

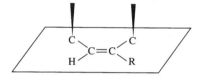

FIGURE 1.5 Spatial arrangement of diene polymers.

regions in moderately dilute solutions of a highly tactic polymer. Formation of highly crystalline domains in a solid polymer has a profound effect on the polymer's mechanical properties. As a consequence, new synthesis routes are constantly being explored to form polymers of desired crystallinity.

A qualitative notion of the nature of crystallinity in polymers can be acquired by considering the crystallization process itself. It is assumed that a polymer in bulk is at a temperature above its melting point, T_m. As the polymer is cooled, collections of highly tactic repeat units that are positioned favorably to move easily into a crystal lattice will do so, forming the nuclei of a multitude of crystalline domains. As the crystalline domains grow, the chain molecules must reorient themselves to fit into the lattice.

Ultimately, these growing domains begin to interfere with their neighbors and compete with them for repeat units to fit into their respective lattices. When this begins to happen, the crystallization process stops, leaving a fraction of the chain segments in amorphous domains. How effectively the growing crystallites acquire new repeat units during the crystallization process depends on their tacticity.

Furthermore, chains of low tacticity form defective crystalline domains. Indeed, after crystallization has ceased, there may be regions of ordered arrangements intermediate between that associated with a perfect crystal and that associated with a completely amorphous polymer. The extent and perfection of crystallization even depends on the rate of cooling of the molten polymer. In fact, there are examples of polymers that can be cooled sufficiently rapidly that essentially no crystallization takes place. On the other hand, annealing just below the melting point, followed by slow cooling, will develop the maximum amount of crystallinity (discussed in greater detail in Chapter 11). Similarly, several polymers that have been cooled far too rapidly for crystallization to take place can be crystallized by mechanical stretching of the samples.

1.6 CONFORMATION OF POLYMER MOLECULES

Once a polymer molecule has been formed, its configuration is fixed. However, it can take on an infinite number of shapes by rotation about the backbone bonds. The final shape that the molecule takes depends on the intramolecular and intermolecular forces, which, in turn, depend on the state of the system. For example, polymer molecules in dilute solution, melt phase, or solid phase would each experience different forces. The conformation of the entire molecule is first considered for semicrystalline solid polymers. Probably the simplest example is the conformation assumed by polyethylene chains in their crystalline lattice (planar zigzag), as illustrated in Fig. 1.4. A polymer molecule cannot be expected to be fully extended, and it actually assumes a chain-folded conformation, as described in detail in Chapter 11. The most common conformation for amorphous bulk polymers and most polymers in solution is the random-flight (or random-coil) conformation, which is discussed in detail later.

In principle, it is possible for a completely stereoregular polymer in a dilute solution to assume a planar zigzag or helical conformation—whichever represents the minimum in energy. The conformation of the latter type is shown by biological polymers such as proteins and synthetic polypeptides. Figure 1.6 shows a section of a typical helix, which has repeat units of the following type.

$$\left.+\text{NH}-\underset{\underset{\text{R}}{|}}{\text{CH}}-\overset{\overset{\text{O}}{\|}}{\text{C}}-\text{C}\right.+$$

The best known example is deoxyribonucleic acid (DNA), which has a weight-average molecular weight of 6–7 million. Even in aqueous solution, it is locked

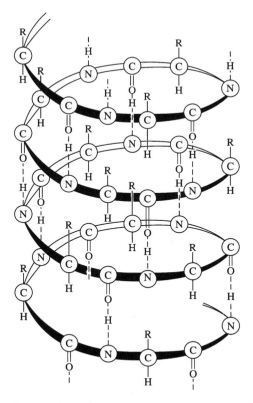

FIGURE 1.6   The helical conformation of a polypeptide polymer chain.

into its helical conformation by intramolecular hydrogen bonds. Rather than behaving as a rigid rod in solution, the helix is disrupted at several points: It could be described as a hinged rod in solution. The helical conformation is destroyed, however, if the solution is made either too acidic or too basic, and the DNA reverts to the random-coil conformation. The transformation takes place rather sharply with changing pH and is known as the helix–coil transition. Sometimes, the energy required for complete helical transformation is not enough. In that case, the chain backbone assumes short blocks of helices, mixed with blocks of random-flight units. The net result is a highly extended conformation with most of the characteristics of the random-flight conformation.

1.7 POLYMERIC SUPPORTS IN ORGANIC SYNTHESIS [11–13]

In conventional organic synthesis, organic compounds (say, A and B) are reacted. Because the reaction seldom proceeds up to 100% conversion, the final reaction mass consists of the desired product (say, C) along with unreacted reactants A and B. The isolation of C is normally done through standard separation techniques such as extraction, precipitation, distillation, sublimation, and various chromatographic methods. These separation techniques require a considerable effort and are time consuming. Significant advancements have been made by binding one of the reactants (A or B) through suitable functional groups to a polymer support that is insoluble in the reaction mass. To this, the other reactant (B or A) is introduced and the synthesis reaction is carried out. The formed chemical C is bound to the polymer, which can be easily separated.

The polymer support used in these reactions should have a reasonably high degree of substitution of reactive sites. In addition, it should be easy to handle and must not undergo mechanical degradation. There are several polymers in use, but the most common one is the styrene–divinyl benzene copolymer.

$$(1.7.1)$$

Because of the tetrafunctionality of divinyl benzene, the polymer shown is a three-dimensional network that would swell instead of dissolving in any solvent. These polymers can be easily functionalized by chloromethylation, hydrogenation, and metalation. For example, in the following scheme, an organotin reagent is incorporated:

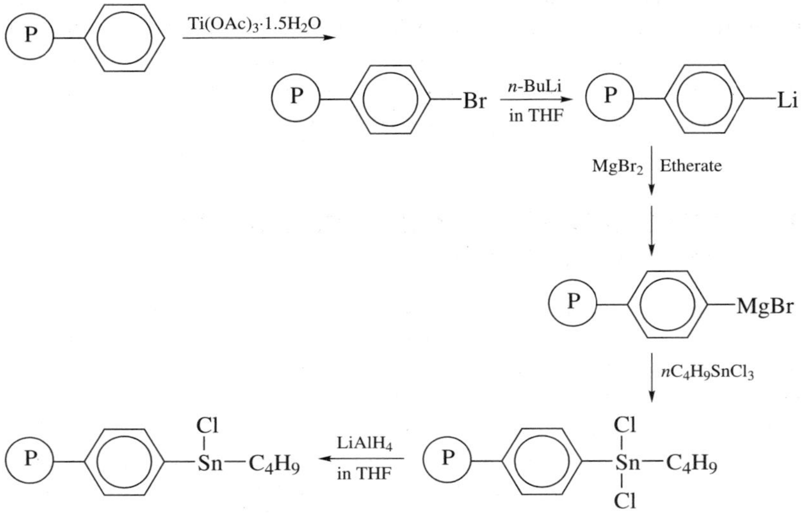

$$(1.7.2)$$

Because the cross-linked polymer molecule in Eq. (1.7.1) has several phenyl rings, the reaction in Eq. (1.7.2) would lead to several organotin groups distributed randomly on the network polymer molecule.

Sometimes, ion-exchanging groups are introduced on to the resins and these are synthesized by first preparing the styrene–divinyl benzene copolymer [as in Eq. (1.7.1)] in the form of beads, and then the chloromethylation is carried out. Chloromethylation is a Friedel–Crafts reaction catalyzed by anhydrous aluminum, zinc, or stannous chloride; the polymer beads must be fully swollen in dry chloromethyl methyl ether before adding the catalyst, $ZnCl_2$. Normally, the resin has very small internal surface area and the reaction depend heavily on the degree of swelling. This is a solid–liquid reaction and the formed product can be shown to be

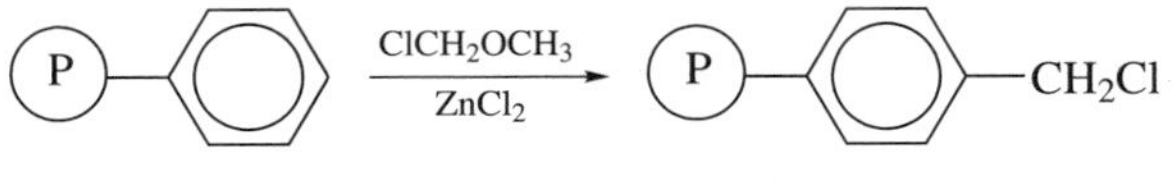

$$(1.7.3)$$

This reaction is fast and can lead to disubstitution and trisubstitution on a given phenyl ring, but monosubstitution has been found to give better results. The chloromethylated resin in Eq. (1.7.3) is quaternized using alkyl amines or ammonia. This reaction is smooth and forms a cross-linked resin having anions groups within the matrix:

$$\text{(P)}-\text{⬡}-\text{CH}_2\text{Cl} + \text{NH}_3 \longrightarrow \text{(P)}-\text{CH}_2\text{NH}_4^+\text{Cl}^- \qquad (1.7.4)$$

which is a commercial anion-exchange resin.

It is also possible to prepare anion-exchange resins by using other polymeric bases. For example beads of cross-linked polyacrylonitrile are prepared by using a suitable cross-linking agent (say, divinyl benzene). The polymer bead can then be represented as $\text{(P)}-\text{CN}$, where the cyanide group is available for chemical reaction, exactly as the phenyl group in Eq. (1.7.3) participated in the quaternization reaction. The cyanide group is first hydrogenated using a Raney nickel catalyst, which is further reacted to an alkyl halide, as follows:

$$\text{(P)}-\text{CN} \xrightarrow{\text{NiH}} \text{(P)}-\text{CH}_2-\text{NH}_2 \xrightarrow{\text{C}_2\text{H}_5\text{Br}} \text{(P)}-\text{CH}_2\underset{\underset{\text{C}_2\text{H}_5}{|}}{\overset{\overset{\text{C}_2\text{H}_5}{|}}{\text{N}^+}}\text{Br}^- \qquad (1.7.5)$$

Instead of introducing active groups into an already cross-linked resin, it is possible to polymerize monomeric bases with unsaturated groups or salts of such bases. For example, we first copolymerize p-dimethyl aminostyrene with divinyl benzene to form a polymer network as in Eq. (1.7.1):

$$\underset{\text{N(CH}_3)_2}{\overset{\text{CH}=\text{CH}_2}{\text{⬡}}} + \underset{\text{CH}=\text{CH}_2}{\overset{\text{CH}=\text{CH}_2}{\text{⬡}}} \longrightarrow \text{(P)}-\overset{\overset{\text{CH}_3}{|}}{\text{N}}-\text{CH}_3 \qquad (1.7.6)$$

The resulting network polymer in the form of beads is reacted with dimethyl sulfonate to give a quaternary group, which is responsible for the ion-exchange ability of the resin:

$$\text{(P)}-\text{N}\overset{\text{CH}_3}{\underset{\text{CH}_3}{\diagdown}} + (\text{CH}_3)_2-\text{SO}_4 \longrightarrow \text{(P)}-\overset{\overset{\text{CH}_3}{|}}{\underset{\underset{\text{CH}_3}{|}}{\text{N}^+}}\text{HSO}_4^- \qquad (1.7.7)$$

Sometimes, we want to prepare a quaternary salt of the vinyl monomer and then copolymerize this with divinyl benzene to form the network polymer resin shown in the following diagram:

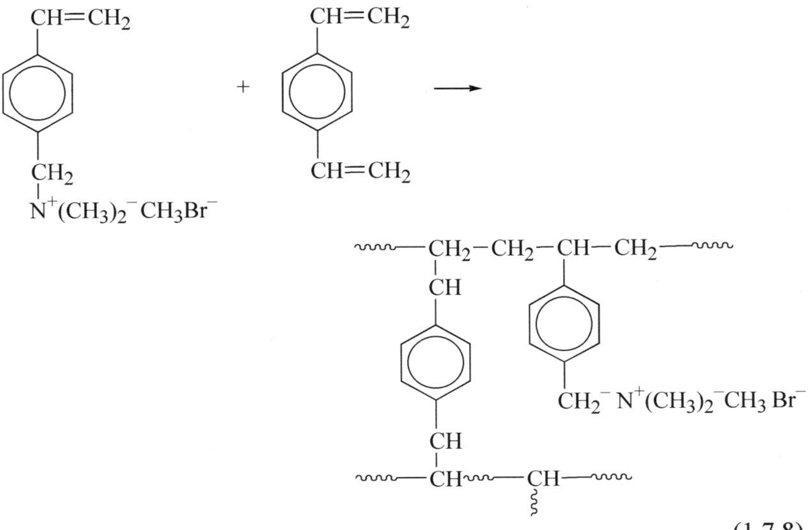

$$(1.7.8)$$

Evidently, this polymer resin has a greater number of sites because the quaternary group is present at every alternate covalent bond on the backbone.

The other support materials that are commonly used are Tenta Gel resins which are obtained by grafting the styrene–divinyl benzene copolymer [of Eq. (1.7.1)] with polyethylene glycol (PEG). Due to the grafts of PEG, the support is polar in nature and it easily swells in water, methanol, acetonitrile, dimethyl formamide, and dichloroethane. Crowns/pins (CP) are another kind of support which consists of radiation grafted polyethylene or polypropylene materials. Polymer formed from monomer polyethylene glycol dimethacrylamide is a network because of the two acrylamide molecules are chemically bound to the two ends of polyethylene glycol and is sometimes abbreviated as PEGA support. This is highly polar, swelling extensively in water, having extremely flexible interior, and suitable for reactions in which it is desired for large macromolecules like enzymes to enter into the matrix of the support.

The synthesis using polymer supports can be of the following two types. In the first one, the catalyst metal is covalently linked to the support and this covalently bound metal serves as a catalyst in a given reaction. In the following example, the supported metal is utilized as a hydrogenation catalyst:

$$\text{(P)}-\text{Cat} + H_2 + H_2C{=}CH_2 \longrightarrow \text{(P)}-\text{Cat} + CH_3{-}CH_3 \qquad (1.7.9)$$

In the second type (called organic synthesis on solid support), one of the reactants (say, X) is first reacted to the support (say, step a) and then the excess reagent X is removed (say, step b). The resultant resin is then reacted to the second reactant (called step c; in this way, X and Y chemically bonding to the support), and after this, the resultant support is reduced (called step d). This reduction process should be such that the product of the surface-reacted X and Y cleave efficiently from the support without affecting the support. Such supports are regenerable and can be utilized in several cycles of chemical reaction between X and Y. For practical reasons, such supports have specific functional groups (called linkers) which are chemically stable during the synthesis of the product X−Y. In addition, the linker group is spaced from the surface of the support and it could be represented by (P)−spacer-linkers, whereas the chemical reaction between X and Y can be written as

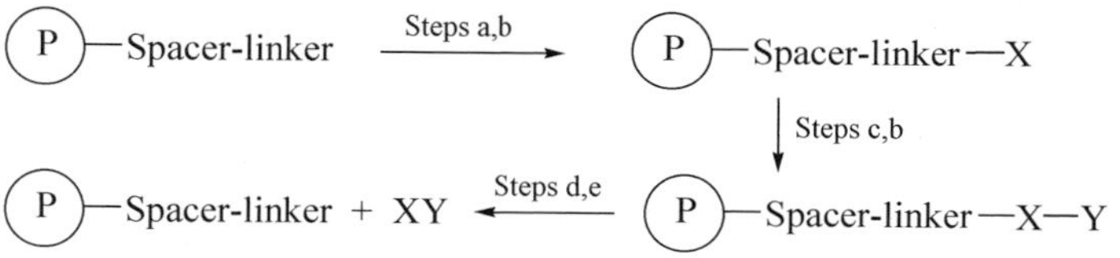

$$\text{(P)}-\text{Spacer-linker} \xrightarrow{\text{Steps a,b}} \text{(P)}-\text{Spacer-linker}-X$$
$$\Big\downarrow \text{Steps c,b} \qquad (1.7.10)$$
$$\text{(P)}-\text{Spacer-linker} + XY \xleftarrow{\text{Steps d,e}} \text{(P)}-\text{Spacer-linker}-X-Y$$

Principally, the purpose of the spacer is to alter the swelling properties of the resin, in this way imparting the resin a better solvent compatibility. For example, in Tenta Gel resin, the graft polyethylene glycol serves as a spacer and makes the styrene–divinyl benzene copolymer swell in presence of water, which otherwise would not do.

The organic synthesis on solid support was first carried out by Merrifield in 1963 for synthesis of Peptide with a well-defined sequence of amino acids. As an example, the support used for the synthesis was the styrene–divinyl benzene copolymer having the following structure:

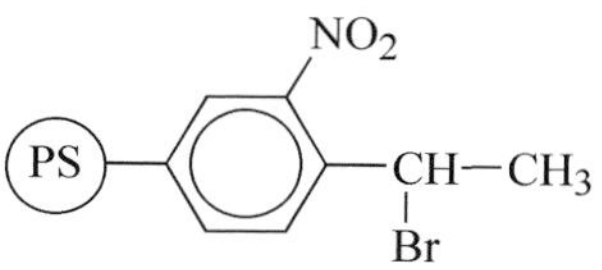

$$\text{(PS)}-\underset{}{\bigcirc}-\underset{\underset{Br}{|}}{CH}-CH_3 \qquad \overset{NO_2}{} \qquad (1.7.11)$$

where $C_6H_3(NO_2)-CH(CH_3)Br$ serves as the integral linker with no spacer. In order to load the resin with the first amino acid $NH_2-CH(R_1)-COOH$, the amino group of the latter is first blocked with the benzyloxycarbonyl group (Cbz) and then reacted to the resin as follows:

$$
\text{(PS)}\!-\!\!\langle\text{benzene ring, NH}_2\rangle\!-\!CH_2Cl \;+\; COOH\!-\!\underset{\underset{R_1}{|}}{CH}\!-\!NH\!-\!Cbz \;\longrightarrow
$$

$$ (1.7.12) $$

$$
\text{(PS)}\!-\!\!\langle\text{benzene ring, NO}_2\rangle\!-\!CH_2O\overset{O}{\overset{\|}{C}}\!-\!\overset{R_1}{\underset{|}{C}}NNH\!-\!Cbz
$$

Resin 1

For resin 1 to react with another amino acid molecule, $NH_2\!-\!CH(R_2)\!-\!COOH$, the Cbz−NH− group of the former must be deprotected (using HBr in glacial acetic acid and then neutralizing) and the amine group of the latter should be protected using Cbz as follows:

$$
\text{Resin 1} \;+\; COOH\!-\!\underset{\underset{R_2}{|}}{CH}\!-\!NH\!-\!Cbz \quad\xrightarrow[\substack{\text{2) Neutralization}\\\text{3) Amino acid coupling}}]{\text{1) 10\%HBr/AcOH}}
$$

$$
\text{(PS)}\!-\!\!\langle\text{benzene ring, NO}_2\rangle\!-\!CH_2O\overset{O}{\overset{\|}{C}}\!-\!\overset{R_1}{\underset{|}{C}}\!-\!NH\overset{O}{\overset{\|}{C}}\!-\!\overset{R_2}{\underset{|}{CH}}\!-\!NH\!-\!Cbz
$$

Resin 2

$$
\text{Resin 2} \;\xrightarrow[\text{2) 2N,NaOH/EtOH}]{\text{1) 10\%HBr/AcOH}}\; \text{(PS)}\!-\!\!\langle\text{benzene ring, NO}_2\rangle\!-\!CH_2Cl \;+\; NH_2\!-\!\overset{R_1}{\underset{|}{CH}}\!-\!\overset{O}{\overset{\|}{C}}NH\!-\!\overset{R_2}{\underset{|}{CH}}\!-\!COOH
$$

$$ (1.7.13) $$

Another very active area of research where polymer supports are utilized is the combinatorial synthesis methods, applied to the synthesis of biologically active compounds. The sources of the latter has always been the nature itself and all natural products are mixture of several compounds. A considerable amount of work is required to identify and isolate the active component which serves as the target molecule. Because this is in small amount, invariable having an extremely complex structure, it cannot be easily synthesized, and therefore, as such, it cannot be adopted for commercial application. In view of this, a new active substance, based on the study of the target molecule, is found by trial and error; this has a comparable biological activity but simpler molecular structure so that it could be manufactured commercially. The identification of the new active substance (having simpler molecular structure) evidently requires extensive organic synthesis followed by purification of the compounds formed and their identification. After these are synthesized, they are then tested for the biological activity and we wish to find that new active substance which has the highest biological activity. Evidently, in order to achieve this extensive organic synthesis, traditional procedures of organic synthesis reaches the limit of time and effort. The speed of synthesis is a new dimension in which many structurally diverse substances are synthesized and are subjected to high characterization and screening throughout.

The goal of the combinatorial approach is to produce many different products with defined structures and bind them chemically with a polymer support through their linkers. The set of supports storing these chemicals is known as a library, totally in analogy with a library of books. Suppose that there is an unknown molecule (assuming that it is available in pure form) whose molecular structure is to be determined. One determines either its high pressure liquid chromatography (HPLC) or its mass spectra. One could compare these with various known compounds from the library as follows. One releases the bound compound by breaking the bond with the linkers of the support and then compares the spectra of the unknown compound with the spectra of this. In this way, one could determine the molecular weight as well as the chemical structure of the unknown compound. This is also not a simple task, but using the following scheme (called combinatorial scheme), this task can be considerably simplified as follows.

Suppose the unknown product is known to be an amide formed by the reaction of an acyl chloride with an amine. Let us also say that there are 10 types of acyl chloride (A_1 to A_{10}) and 10 types of amine (B_1 to B_{10}) and the products formed are represented by

$$(A_1 \text{ to } A_{10})COCl + (B_1 \text{ to } B_{10})NH_2 \rightarrow (A_1 \text{ to } A_{10})CNH(B_1 \text{ to } B_{10})$$

$$(1.7.14)$$

In the combinatorial scheme, there are 100 products and they can be carried out in 10 reaction steps as follows. We prepare a mixture of B_1-NH_2 to $B_{10}-NH_2$ in equal proportion and this mixture is reacted in 10 broths, each containing A_1COCl to $A_{10}COCl$. In this way, we generated 10 mixtures which contain

$$A_1B_1 + A_1B_2 + A_1B_3 + \cdots + A_1B_{10} \qquad (1.7.15a)$$
$$A_2B_1 + A_2B_2 + A_2B_3 + \cdots + A_2B_{10} \qquad (1.7.15b)$$

$$\vdots$$

$$\vdots$$

$$A_{10}B_1 + A_{10}B_2 + A_{10}B_3 + \cdots + A_{10}B_{10} \qquad (1.7.15c)$$

These 10 mixtures are then stored (by chemically binding) on 10 different supports. In this case, the library consists of 10 supports and the compounds released from the linkers is a mixture of 10 amides. However, there is never a confusion, simply because the peak positions of the 10 amides in HPLC experiments are unique and are known a priori. Similarly in its mass spectra, the molecular weights of each components and their fragments are precisely known beforehand. It is thus seen that the unknown compound can be easily identified precisely with 10 HPLC (or mass spectral) experiments alone.

The success of the combinatorial scheme discussed above lies in the fact that the mixture could be easily bound to the polymer support covalently as well as they should easily be cleaved from the linker completely. In addition to this, the attachment points of the linker (or spacer) with the polymer support should be chemically stable during the binding and cleaving of the mixture. In past years, several linkers have been developed allowing many multi step organic synthesis and cleavage efficiently. The conditions of the reactions are found to depend not only upon the linker and spacer but also upon the type of resin, its extent of loading and the nature of compound. In light of this, in the present context, this is a rapidly growing area of research [11,12].

Example 1.4: Discuss different methods of functionalizing the styrene–divinyl benzene polymer.

Solution: A wide variety of vinyl-derivitized monomers are available commercially and some of these are 4-vinylimidazole, vinyl pyridines, and acryloyl morpholine [13]. This can be terpolymerized with styrene and divinyl benzene to obtain polymer gel having the following structures:

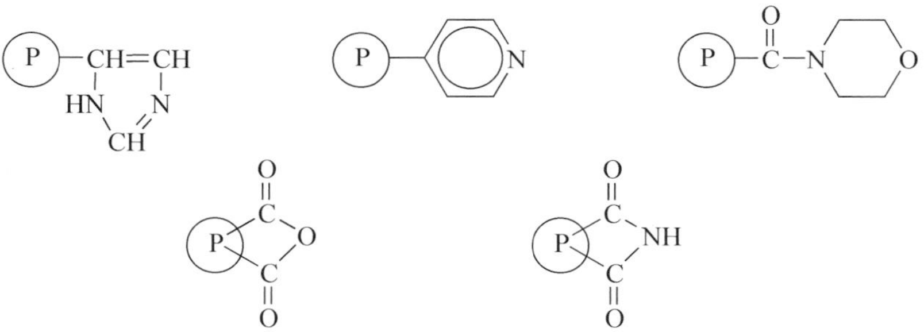

Once the gel is formed, they can be functionalized only through chemical modification, and normally this is done via chloromethylation using chloromethyl methyl ether and lithiation by lithium–bromine exchange as follows:

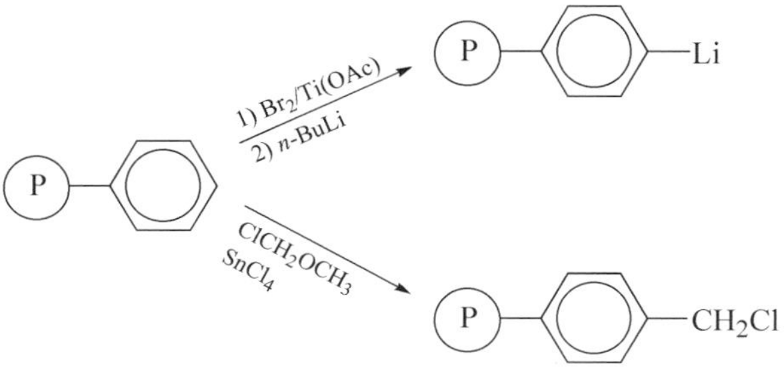

Chloromethylation is a versatile and reliable reaction, even through chloromethyl methyl ether is highly carcinogaric reaction.

Example 1.5: Define photoconductivity and how it achieved in polymers.

Solution: Photoconductivity is defined as a significant increase in conductivity caused by illumination and is attributed to increase in charge carriers (electrons or holes due to it). For polymers to be photoconductive, mobile charge carriers must be generated with light. The resulting charge pair may then separate, and either a positive charge, an electron, or both may migrate in a polarizing electrical field as follows:

$$D + A \xrightarrow{h\nu} [D^+A^0]^* \xrightarrow{\text{electric field}} D^{\circ+} + A^{\circ-}$$

Polymers are normally insulators with negligible conduction and the latter can be achieved by (1) addition of a small molecular dopants or (2) chemical modification of the polymer.

The dopants (e.g., dyes like rose bengal, methyl violet, methylane blue, etc.) re charge-transfer agents. The dyes have the ability to absorb light and sensitize the polymer by the addition of electron transfer. The technique of chemical modification improves the spectral response, giving a higher speed of movement of holes.

Example 1.6: Describe the various steps of photocopying.

Solution: There are four steps:

1. The surface of a metal drum is coated with photoconductive material [selenium or polyvinyl carbazole (PVK)]. This is charged in the dark by spraying ions under corona discharge. This gives a uniform charge on the surface.

2. The image to be copied is projected on the drum, and by this, different areas of the drum are discharged in the light signal. Charge is retained in areas not illuminated, and in this way, an electrical pattern is generated on the drum.

3. The developer consists of two components: carrier and toner. The carrier is metal beads and the toner is a polymer with black dye. On shaking, the toner becomes positively charged, which, on exposure to the drum, accumulates around dark areas. The paper is exposed to a high voltage to make it negatively charged and the toner shifts from the drum surface to the paper.

4. The image transferred to the paper is fixed by heating the paper, and the polymer particles of toner are then sintered.

Example 1.7: After separating gas, gasoline, naphtha, and gas oil from crude petroleum, the residue is depolymerized and then further distilled for these materials. Discuss the depolymerization of long-chain hydrocarbon (called Visbreaking step). Sometimes, a catalytic cracking process is used. Discuss the difference between the two.

Solution: Under depolymerization, a $C-C$ bond breaks homolytically to give two radicals:

$$R_1CH_2CH_2R_2 \rightarrow R_1CH_2^{\cdot} + R_2CH_2^{\cdot}$$

They produce ethylene under β-scission:

$$R_1-CH_2CH_2CH_2^{\cdot} \rightarrow RCH_2^{\cdot} + CH_2{=}CH_2$$

The radical need not necessarily be at the chain end alone. For production of propylene, we need to have the radical at the third carbon and then there is β-scission reaction:

$$R_1CH_2CH_2CH_2CH_3 + {}^{\cdot}CH_3 \rightarrow RCH_2CH^{\cdot}CH_2CH_3 + CH_4$$
$$R_2CH_2CH^{\cdot}CH_2CH_3 \rightarrow RCH_2^{\cdot} + CH_2{=}CH-CH_3$$

Catalytic cracking in the petroleum industry is a very important process in which heavy oils are converted into gasoline and lighter products. This occurs through carboneum ion formation in the presence of zeolite catalyst (Lewis acid, L) as follows:

$$RH + L \rightleftharpoons LH^- + R^{\cdot}$$

In the case of R^+ isomerization occurs easily and a shift of the double bond and a shift of the methyl group are commonly seen. The aromatic formation on zeolite occurs as follows:

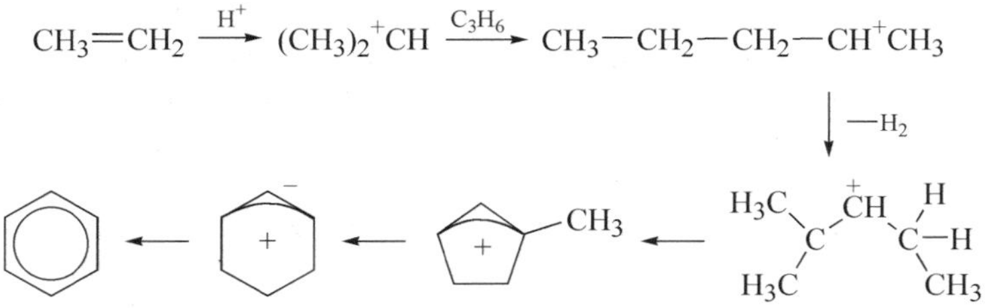

1.8 CONCLUSION

In this chapter, various methods for classifying polymers have been discussed and basic concepts regarding molecular-weight distribution, mechanism of polymerization, molecular conformations, configurations, and crystallinity have been presented. These concepts will be amplified in the rest of the book. It may be mentioned that synthetic polymers are mainly emphasized in this textbook, but the concepts developed here can be easily extended to naturally occurring polymers such as proteins, nucleic acids, cotton, silk, wool, and paper. No single textbook can do justice to so many fascinating areas of research. This is the only justification for the exclusion of natural polymers from this book.

REFERENCES

1. Allen, G., and J. C. Bevington, Comprehensive Polymer Science, Pergamon, New York, 1989, Vols. 1–7.
2. Kumar, A., and S. K. Gupta, Fundamentals of Polymer Science and Engineering, Tata McGraw-Hill, New Delhi, 1978.
3. Tsuruta, T., Contemporary Topics in Polymeric Materials for Biomedical Applications, Adv. Polym. Sci., 126, 1–54, 1996.
4. Curran, D. P., Strategy Level Separation in Organic Synthesis: From Planning to Practice, Angew. Chem. Int. Ed. Eng., 37. 1174–1196, 1998.
5. Barrett, K. E. J., Dispersion in Organic Media, Wiley, New York, 1982.
6. Fischer, M., and F. Vogtle, Dendrimers: From Design to Applications—A Progress Report, Angew. Chem. Int. Ed. Eng., 38, 885–905, 1999.
7. Hazer, B., Synthesis and Characterization of Block Copolymers, in Handbook of Polymer Science and Technology, N. Cheremisnoff (ed.), Marcel Dekker, New York, 1989, Vol. 1.
8. Loschewsky, A., Molecular Concept, Self Organization and Properties of Polysoaps, Adv. Polym. Sci., 124, 1–86, 1995.
9. Putnam, D., and J. K. Kopacek, Polymer Conjugates with Anticancer Activity, Adv. Polym. Sci., 122, 55–124, 1995.
10. Sellergren, B., Imprinted Polymers with Memory for Small Molecules, Protein or Crystals, Angew. Chem. Int. Ed. Eng., 39, 1031–1037, 2000.
11. Laszlo, P., Preparative Chemistry Using Supported Reagents, Academic Press, San Diego, CA, 1987.
12. Guillier, F., D. Orain, and M. Bradley, Linkers and Cleavage Strategies in Solid Phase Organic Synthesis and Combinatorial Chemistry, Chem. Rev. 100, 2091–2157, 2000.
13. Hodge, P., and D. C. Sherrington, Polymer Supported Reactions in Organic Synthesis, Wiley, Chichester, 1980.

PROBLEMS

1.1. The molecular functionality of real systems is usually expressed by a fraction in view of the possible occurrence of side processes involving cyclization, cross-linking, and so forth. When more than one monomer is involved, we talk of average functionality $\bar{f}$ defined as

$$\bar{f} = \frac{\sum f_i N_i}{\sum N_i}$$

where f_i is the functionality of the ith monomer whose N_i moles are present in the reaction mixture. Calculate $\bar{f}$ for a mixture of glycerol–phthalic anhydride in the ratio of $2:3$. For branched and network polymer, $\bar{f}$ must be greater than 2.

1.2. In the polymerization of phthalic anhydride with glycerol, one adds ethylene glycol also. Why? Assuming their concentrations in the ratio of $2:3:1$, find their average functionality. What is the main difference between this polymer and the one in Eq. (1.22)?

1.3. Methyl methacrylate $[CH_2{=}C(CH_3)COOCH_3]$ is randomly copolymerized with maleic anhydride in the first stage, and the resulting copolymer is reacted with polyvinyl alcohol $[CH_2{-}CH{-}(OH)]_n$ in the second stage. Write down all of the chemical reactions involved and the molecular structure of the resulting polymers in both stages.

1.4. 12-Hydroxystearic acid $[CH_3{-}(CH_2)_5{-}CH(OH){-}(CH_2)_{10}{-}COOH]$ is polymerized through the step-growth mechanism (ARB monomer type reaction) in stage 1. In stage 2, this is reacted with glycidyl methacrylate

$$[CH_2{=}C(CH_3)COOCH_2{-}CH{-}CH_2; \ GMA],$$
$$\diagdown \ \diagup$$
$$O$$

which produces the end-capping of the polymer formed at the end of stage 1. In this reaction, the epoxy group of the GMA reacts, keeping the double bond safe. This is copolymerized with methyl methacrylate in stage 3, and in this stage, the double bond of the GMA reacts. Write down the structure of the polymer at the end of each stage.

1.5. A small amount of grafting of polyethylene (PE) changes the properties of the polymer. The polymer in the molten stage is mixed with a suitable monomer (e.g., acrylic acid $CH_2{=}CHCOOH$, fumaric acid $COOHCH_2{=}CHCOOH$, or maleic anhydride $(O){-}CH{=}CH{-}C(O){-}O$ and an initiator [e.g., dicumylperoxide; $C_6H_5C(CH_3)_2OOC(CH_3)_2C_6H_6$]. Explain why the polymer formed would be a mostly grafted one and what its structure would be. Discuss other possible products.

1.6. Polyvinyl alcohol (PVA) is normally atactic and has a high concentration of head-to-head defects. However, PVA in the crystalline state is in the all-trans stage and gives tough fibers. PVA fibers can also absorb water and, in this way, lose strength. To avoid this, one usually treats the polymer with formaldehyde and the latter reacts (the reaction is called ketalization) with the two adjacent hydroxyl groups of the polymer. What changes are produced in the polymer through this treatment and why?

1.7. Polyvinyl amine $[-CH_2-CH(NH_2)-]_n$ can be easily prepared through Hoffmann degradation of polyacrylamide $[-CH_2-CH(CONH_2)-]_n$. Write down the basic reactions using sodium hydrochloride (NaCl) in methanol medium. The modified polymer now is an ammonium chloride salt and has to undergo an ion exchange to obtain the desired polymer.

1.8. Linear polyacetylene is known to be an intrinsically electrically conductive polymer. The configuration could be cis, trans, or mixed. Do you think that the configuration would have any influence on the electrical conductivity?

1.9. Polyphenylene sulfide is an important electronic polymer and can be synthesized using p-dichlorobenzene with Na_2S in n-pyrolidone (Campbell method). Its initiation step is as follows:

Initiation

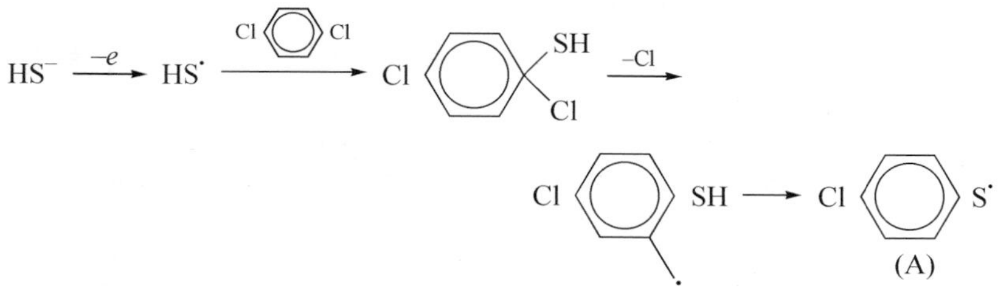

Propagation

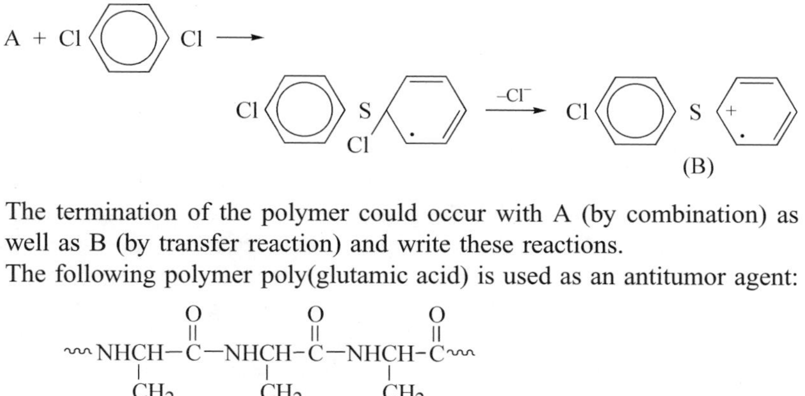

The termination of the polymer could occur with A (by combination) as well as B (by transfer reaction) and write these reactions.

1.10. The following polymer poly(glutamic acid) is used as an antitumor agent:

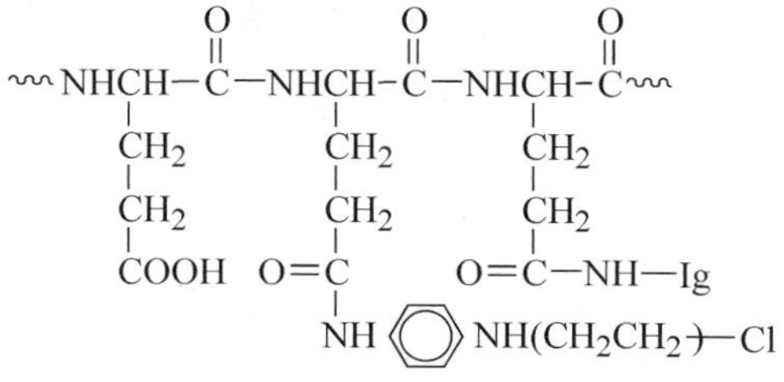

where Ig is a protein immunoglobin. In this polymer molecule, identify various groups serving as the solubilizer, pharmacon, and the homing device.

1.11. Poly(2-vinylpyridine-1-oxide) (PVNO) is a polymer that dissolves in water and serves as a medicinal drug:

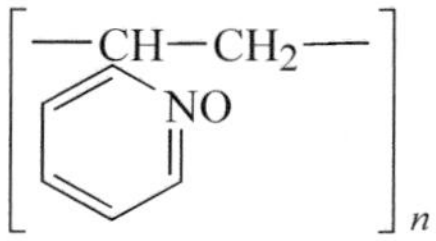

$$\left[\begin{array}{c} -CH-CH_2- \\ \underset{NO}{\bigcirc} \end{array} \right]_n$$

The polymer prevents the cytotoxic action of quartz dust, and its activity is found for a molecular weight above 30,000. Assuming that the monomer is available, how would you form the required polymer?

1.12. Penicillin has the following chemical formula:

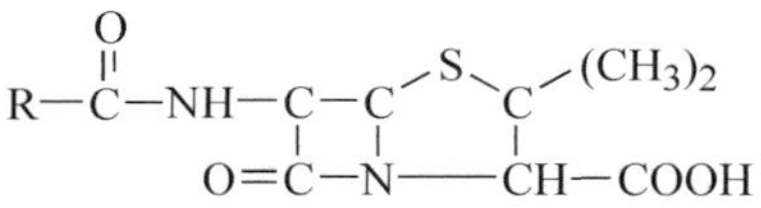

$$\begin{array}{c} O \\ \parallel \\ R-C-NH-C-C \overset{S}{\diagdown} C-(CH_3)_2 \\ \quad\quad\quad | \quad | \quad\quad | \\ \quad\quad O=C-N\!\!-\!\!-\!\!CH-COOH \end{array}$$

On binding it to a polymer, the activity of the drug is found to increase. Suggest a suitable polymer on which this can be bound.

1.13. Heparin is highly acidic dextrorotatory copolymer of glucosamine and glucuronic acid having the following structure:

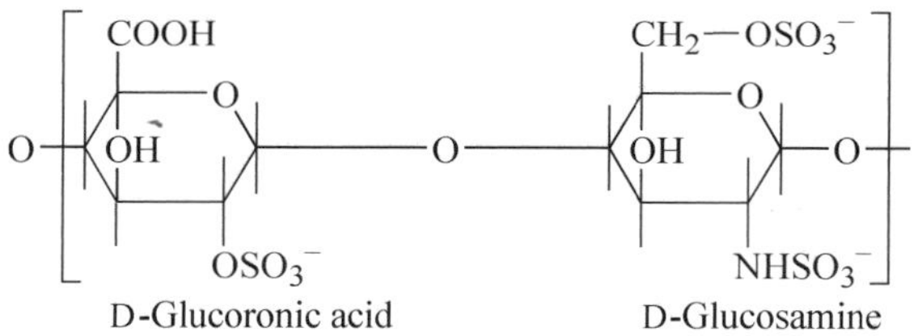

D-Glucoronic acid   D-Glucosamine

It is naturally present in blood and inhibit its clotting. It is desired to use PVC bags for storing blood and this can be done only if its inner surface is passivated with heparin. Suggest a method by which it can be done.

1.14. In diabetes (where there is poor insulin delivery by the pancreas in response to glucose) or Parkinson's disease (where there is poor release of dopamine in response to potassium), it has been recommended to transplant encapsulated animal cells in the human body to supplement the existing deficiency. The limitation of this cell transplant is the immune-mediated rejection, and to overcome this, the cells are microencapsulated with acrylates and methacrylates, which cannot be penetrated by large antibodies but insulin or dopamine can diffuse out easily. Readily available methacrylates $[CH_2{=}C(CH_3){-}C(O)OR]$ are MAA (methacrylic acid with R as H), MMA (methyl methacrylate with R as CH_3), HEMA (2-hydroxyethyl methacrylate with R as $-CH_2CH_2OH$), HPMA (2-hydroxy methyl methacrylate with R as $-CH_2CH(OH){-}CH_3$), and DMAEMA [dimethylaminoethyl methacrylate with R as $-CH_2CH_2N(CH_3)_2$]. Find out which monomer would give acidic, basic, or neutral film.

1.15. When acidic and basic polymeric polyelectrolytes are mixed, they form a complex which precipitates. Based on this, mammalian cells are coated with sodium alginate (natural polymer) solution (in water) and then put in a suitable water-soluble acrylate solution. This forms a hard encapsulation around the mammalian cell; this is impervious to antibodies. Suggest a suitable acrylate (as methacrylate) for the encapsulation.

1.16. One of the synthetic blood plasmas is poly[N-(2-hydroxypropyl)] methacrylamide:

$$\left[CH_2-\underset{\underset{O=C-NH-CH_2-\underset{\underset{OH}{|}}{C}-CH_3}{|}}{\overset{\overset{CH_3}{|}}{C}} \right]_n$$

Suggest some (at least two) plausible ways of making this material.

1.17. In following multifunctional initiators, the azo (at 50°C) and peroxide groups (at 90°C) decompose at separate temperatures and the monomer M_1 is polymerized first:

$$R'O\overset{\overset{O}{||}}{O}C-RN=NR-\overset{\overset{O}{||}}{C}-OOR'$$

Considering that polymer radicals due to monomer M_1 terminates (say, P_{M_1}) by recombination alone and that due to M_2 (say, P_{M_2}), terminates by recombination alone, what kinds of polymer that would be formed. What would be the nature of the polymer if the polymer radicals due to monomer M_2 terminate by disproportionation only.

1.18. Suppose in Problem 1.17 that the polymer radical due to monomer M_1 terminate by disproportionation and that due to M_2 terminates by combination, what is the nature of the final polymer? Suppose P_{M_2} terminates by disproportionation only; what would be the nature of the polymer then?

1.19. Iodine transfer polymerization requires a peroxide ($R-O-O-R$) and an alkyl iodide (R_FI). The primary radicals are generated and the polymer radicals ($R_F = M_n'$) undergo transfer reaction as follows:

$$R-O-O-R \xrightarrow{heat} RO^{\cdot}$$
$$RO^{\cdot} + R_F \longrightarrow Product + R_F^{\cdot}$$
$$R_FM_n' + R_FI \longrightarrow R_FM_nI + R_F^{\cdot}$$

Assuming termination to occur by combination, give the complete mechanism of polymerization.

1.20. Synthesis gas is a mixture of carbon monoxide and hydrogen. In the Fischer–Tropsch process, the formation of alkane, alkene, alcohol of large chain-length occurs, depending on the catalyst used, as follows:

1. Formation of alkane and alkene of larger chain length:

$$(n + 1)H_2 + 2nCO \rightleftharpoons C_nH_{2n+2} + nCO_2$$
$$nH_2 + 2nCO \rightleftharpoons C_nH_{2n} + nCO_2$$

Hydrocarbons up to C_{20} (e.g., gasoline, diesel, or aviation fuel) are produced using a (Co, Fe, Ru) catalyst.

2. Alcohol formation:

$$2nH_2 + nCO \rightleftharpoons C_nH_{2n+1}OH + (n - 1)H_2O$$

Ethanol using a Co catalyst, higher alcohols using a (Fe, Co, Ni) catalyst, and ethylene glycol using a Rh catalyst are produced.

In the Fischer–Tropsch process, carbon monoxide is absorbed on a metal (M) as one of its ligands, as follows:

$$CO + M \longrightarrow M{=}C{=}O$$
$$M{=}C{=}O + M \longrightarrow OM{-}C + M{-}O$$

It is these M−C ligands that give the growth of chains:

$$M{-}C \xrightarrow{H_2} M{=}CH_2 \xrightarrow{H_2} M{-}CH_3$$

Now write down the propagation and termination reactions in the Fischer–Tropsch process.

1.21. Liquid hydrocarbons have been prepared using the ZSM-5 zeolite catalyst. At 600°C, the following initiation reaction occurs:

$$2CH_3OH \underset{CH_3-O-CH_3}{\overset{}{\updownarrow}} \xrightarrow[+H^+]{-H_2O} CH_3CH_2{}^+ \underset{+H}{\overset{-H^+}{\rightleftharpoons}} C_2H_4$$

With the help of your understanding of Example 1.7, explain the formation of C_3H_8, paraffin, paraffin isomers, and aromatics.

2

Effect of Chemical Structure on Polymer Properties

2.1 INTRODUCTION

In the previous chapter, we discussed different ways of classifying polymers and observed that their molecular structure plays a major role in determining their physical properties. Whenever we wish to manufacture an object, we choose the material of construction so that it can meet design requirements. The latter include temperature of operation, material rigidity, toughness, creep behavior, and recovery of deformation. We have already seen in Chapter 1 that a given polymer can range all the way from a viscous liquid (for linear low-molecular-weight chains) to an insoluble hard gel (for network chains), depending on how it was synthesized. Therefore, polymers can be seen to be versatile materials that offer immense scope to polymer scientists and engineers who are on the lookout for new materials with improved properties. In this chapter, we first highlight some of the important properties of polymers and then discuss the many applications.

2.2 EFFECT OF TEMPERATURE ON POLYMERS [1–4]

We have observed earlier that solid polymers tend to form ordered regions, such as spherulites (see Chapter 11 for complete details); these are termed *crystalline polymers*. Polymers that have no crystals at all are called *amorphous*. A real

45

polymer is never completely crystalline, and the extent of crystallization is characterized by the percentage of crystallinity.

A typical amorphous polymer, such as polystyrene or polymethyl methacrylate, can exist in several states, depending on its molecular weight and the temperature. In Figure 2.1, we have shown the interplay of these two variables and compared the resulting behavior with that of a material with moderate crystallinity. An amorphous polymer at low temperatures is a hard glassy material which, when heated, melts into a viscous liquid. However, before melting, it goes through a rubbery state. The temperature at which a hard glassy polymer becomes

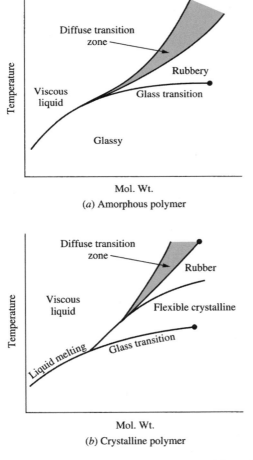

(a) Amorphous polymer

(b) Crystalline polymer

FIGURE 2.1 Influence of molecular weight and temperature on the physical state of polymers.

a rubbery material is called the *glass transition temperature*, T_g (see Chapter 12 for the definition of T_g in terms of changes in thermodynamic and mechanical properties; there exists a sufficiently sharp transition, as seen in Fig. 2.1a). There is a diffuse transition zone between the rubbery and liquid states for crystalline polymers; the temperature at which this occurs is called the *flow temperature*, T_f. As the molecular weight of the polymer increases, we observe from Figure 2.1 that both T_g and T_f increase. Finally, the diffuse transition of the rubber to the liquid state is specific to polymeric systems and is not observed for low-molecular-weight species such as water, ethanol, and so forth, for which we have a sharp melting point between solid and liquid states.

In this section, only the effect of chain structure on T_g is examined—other factors will be discussed in Chapters 10–12. In order to understand the various transitions for polymeric systems, we observe that a molecule can have all or some of the following four categories of motion:

1. Translational motion of the entire molecule
2. Long cooperative wriggling motion of 40–50 C–C bonds of the molecule, permitting flexing and uncoiling
3. Short cooperative motion of five to six C–C bonds of the molecule
4. Vibration of carbon atoms in the polymer molecule

The glass transition temperature, T_g, is the temperature below which the translational as well as long and short cooperative wriggling motions are frozen. In the rubbery state, only the first kind of motion is frozen. The polymers that have their T_g values less than room temperature would be rubbery in nature, such as neoprene, polyisobutylene, or butyl rubbers. The factors that affect the glass transition temperatures are described in the following subsections.

2.2.1 Chain Flexibility

It is generally held that polymer chains having −C−C− or −C−O− bonds are flexible, whereas the presence of a phenyl ring or a double bond has a marked stiffening effect. For comparison, let us consider the basis polymer as polyethylene. It is a high-molecular-weight alkane that is manufactured in several ways; a common way is to polymerize ethylene at high pressure through the radical polymerization technique. The polymer thus formed has short-chain as well as long-chain branches, which have been explained to occur through the "backbiting" transfer mechanism. The short-chain branches (normally butyl) are formed as follows:

$$(2.2.1)$$

and the long-chain branches are formed through the transfer reaction at any random point of the backbone as

$$
\begin{array}{ccc}
\text{\tiny$\sim\!\!\sim$}\overset{\textstyle\cdot}{\text{CH}}\text{\tiny$\sim\!\!\sim$}\text{CH}_2 & \longrightarrow & \text{\tiny$\sim\!\!\sim$}\overset{\textstyle\cdot}{\text{CH}} \\
\mid\ \ \ \ \ \mid & & \mid \\
\text{H}\ \ \ \text{CH}_2 & & \text{CH}_2 \\
\mid & & \mid \\
\overset{\textstyle\cdot}{\text{CH}}_2\!-\!\text{CH}_2 & & \text{CH}_3
\end{array}
\qquad
\begin{array}{c}
\longrightarrow\ \ \text{\tiny$\sim\!\!\sim$}\text{CH}_2\!-\!\text{CH}\text{\tiny$\sim\!\!\sim$}\text{CH}_2\overset{\textstyle\cdot}{\text{CH}}_2 \\
\mid \\
\text{CH}_2 \\
\mid \\
\text{CH}_3
\end{array}
\qquad (2.2.2)
$$

The polymer has a T_g of about $-20°C$ and is a tough material at room temperature. We now compare polyethylene terephthalate with polyethylene. The former has a phenyl group on every repeat unit and, as a result, has stiffer chains (and, hence, higher T_g) compared to polyethylene. 1,4-Polybutadiene has a double bond on the backbone and similarly has a higher T_g.

The flexibility of the polymer chain is dependent on the free space v_f available for rotation. If v is the specific volume of the polymer and v_s is the volume when it is solidly packed, then v_f is nothing but the difference between the two $(v - v_s)$. If the free space v_f is reduced by the presence of large substituents, as in polyethylene terephthalates, the T_g value goes up, as observed earlier.

2.2.2 Interaction Between Polymers

Polymer molecules interact with each other because of secondary bondings due to dipole forces, induction forces, and/or hydrogen bonds. The dipole forces arise when there are polar substituents on the polymer chain, as, for example, in polyvinyl chloride (PVC). Because of the substituent chlorine, the T_g value of PVC is considerably higher than that of polyethylene. Sometimes, forces are also induced due to the ionic nature of substituents (as in polyacrylonitrile, for example). The cyanide substituents of two nearby chains can form ionic bonds as follows:

$$
\begin{array}{l}
\text{\tiny$\sim\!\!\sim$}\text{CH}_2\!-\!\text{CH}\!-\!\text{\tiny$\sim\!\!\sim$} \\
\quad\quad\quad\ \mid \\
\quad\quad\quad\ \overset{+}{\text{C}}\cdots\cdots\overset{-}{\text{N}} \\
\quad\quad\quad\ \mid\mid\quad\quad\mid\mid \\
\quad\quad\quad\ \overset{-}{\text{N}}\quad\quad\overset{+}{\text{C}} \\
\quad\quad\quad\quad\quad\quad\ \mid \\
\quad\text{\tiny$\sim\!\!\sim$}\text{CH}_2\!-\!\text{CH}\!-\!\text{\tiny$\sim\!\!\sim$}
\end{array}
\qquad (2.2.3)
$$

Hydrogen bonding has a similar effect on T_g. There is an amide ($-CONH-$) group in nylon 6, and it contributes to interchain hydrogen- bonding, increasing the glass transition temperature compared to polyethylene. In polytetrafluoroethylene, there are van der Waals interaction forces between fluorine atoms and, as a result, it cannot be melted:

$$
\begin{array}{c}
\text{F—C—F} \cdots \text{F—C—F} \\
\text{F—C—F} \cdots \text{F—C—F}
\end{array}
\tag{2.2.4}
$$

Even though the energy required to overcome a single secondary-force interaction is small, there are so many such secondary forces in the material that it is impossible to melt it without degrading the polymer.

2.2.3 Molecular Weight of Polymers

Polymers of low molecular weight have a greater number of chain ends in a given volume compared to those of high molecular weight. Because chain ends are less restrained, they have a greater mobility at a given temperature. This results in a lower T_g value, as has been amply confirmed experimentally. The molecular-weight dependence of the glass transition temperature has been correlated by

$$
T_g = T_g^\infty - \frac{K}{\mu_n}
\tag{2.2.5}
$$

where T_g^∞ is the T_g value of a fictitious sample of the same polymer of infinite molecular weight and μ_n is the number-average chain length of the material of interest. K is a positive constant that depends on the nature of the material.

2.2.4 Nature of Primary Bondings

The glass transition temperature of copolymers usually lies between the T_g values of the two homopolymers (say, T_{g1} and T_{g2}) and is normally correlated through

$$
\frac{1}{T_g} = \frac{w_1}{T_{g1}} + \frac{(1 - w_1)}{T_{g2}}
\tag{2.2.6}
$$

where w_1 is the weight fraction of one of the monomers present in the copolymer of interest. With block copolymers, sometimes a transition corresponding to each block is observed, which means that, experimentally, the copolymer exhibits two T_g values corresponding to each block. We have already observed that, depending on specific requirements, one synthesizes branch copolymers. At times, the long branches may get entangled with each other, thus further restraining molecular motions. As a result of this, Eq. (2.2.6) is not obeyed and the T_g of the polymer is expected to be higher. If the polymer is cross-linked, the segmental mobility is further restricted, thus giving a higher T_g. On increasing the degree of cross-linking, the glass transition temperature is found to increase.

The discussion up to now has been restricted to amorphous polymers. Figure 2.1b shows the temperature–molecular weight relation for crystalline polymers. It has already been observed that these polymers tend to develop crystalline zones called "spherulites." A crystalline polymer differs from the amorphous one in that the former exists in an additional flexible crystalline state before it begins to behave like a rubbery material. On further heating, it is converted into a viscous liquid at the melting point T_m. This behavior should be contrasted with that of an amorphous polymer, which has a flow temperature T_f and no melting point.

The ability of a polymeric material to crystallize depends on the regularity of its backbone. Recall from Chapter 1 that, depending on how it is polymerized, a polymeric material could have atactic, isotactic, or syndiotactic configurations. In the latter two, the substituents of the olefinic monomer tend to distribute around the backbone of the molecule in a specific way. As a result (and as found in syndiotactic and isotactic polypropylene), the polymer is crystalline and gives a useful thermoplastic that can withstand higher temperatures. Atactic polymers are usually amorphous, such as atactic polypropylene. The only occasion when an atactic material can crystallize is when the attached functional groups are of a size similar to the asymmetric carbon. An example of this case is polyvinyl alcohol, in which the hydroxyl group is small enough to pack in the crystal lattice. Commercially, polyvinyl alcohol (PVA1c) is manufactured through hydrolysis of polyvinyl acetate. The commonly available PVA1c is always sold with the percentage alcohol content (about 80%) specified. The acetate groups are large, and because of these residual groups, the crystallinity of PVA1c is considerably reduced.

It is now well established that anything that reduces the regularity of the backbone reduces the crystallinity. Random copolymerization, introduction of irregular functional groups, and chain branchings all lead to reduction in the crystalline content of the polymer. For example, polyethylene and polypropylene are both crystalline homopolymers, whereas their random copolymer is amorphous rubbery material. In several applications, polyethylene is partially chlorinated, but due to the presence of random chlorine groups, the resultant polymer becomes rubbery in nature. Finally, we have pointed out in Eqs. (2.2.1) and (2.2.2) that the formation of short butyl as well as long random branches occurs in the high-pressure process of polyethylene. It has been confirmed experimentally that short butyl branches occur more frequently and are responsible for considerably reduced crystallinity compared to straight-chain polyethylene manufactured through the use of a Ziegler–Natta catalyst.

2.3 ADDITIVES FOR PLASTICS

After commercial polymers are manufactured in bulk, various additives are incorporated in order to make them suitable for specific end uses. These additives have a profound effect on the final properties, some of which are listed for polyvinyl chloride in Box 2.1. PVC is used in rigid pipings, conveyor belts, vinyl floorings, footballs, domestic insulating tapes, baby pads, and so forth. The required property variation for a given application is achieved by controlling the amount of these additives. Some of these are discussed as follows in the context of design of materials for a specific end use.

Plasticizers are high-boiling-point liquids (and sometimes solids) that, when mixed with polymers, give a softer and more flexible material. Box 2.1 gives dioctyl phthalate as a common plasticizer for PVC. On its addition, the polymer (which is a hard, rigid solid at room temperature) becomes a rubberlike

---

Box 2.1
Various Additives to Polyvinyl Chloride

Commercial polymer	Largely amorphous, slightly branched with monomers joined in head-to-tail sequence.
Lubricant	Prevents sticking of compounds to processing equipment. Calcium or lead stearate forms a thin liquid film between the polymer and equipment. In addition, internal lubricants are used, which lower the melt viscosity to improve the flow of material. These are montan wax, glyceryl monostearate, cetyl palmitate, or aluminum stearate.
Filler	Reduces cost, increases hardness, reduces tackiness, and improves electrical insulation and hot deformation resistance. Materials used are china clay for electrical insulation and, for other works, calcium carbonate, talc, magnesium carbonate, barium sulfate, silicas and silicates, and asbestos.
Miscellaneous additives	Semicompatible rubbery material as impact modifier; antimony oxide for fire retardancy; dioctyl phthalate as plasticizer; quaternary ammonium compounds as antistatic agents; polyethylene glycol as viscosity depressant in PVC paste application; lead sulfate for high heat stability, long-term aging stability, and good insulation characteristics.

material. A plasticizer is supposed to be a "good solvent" for the polymer; in order to show how it works, we present the following physical picture of dissolution. In a solvent without a polymer, every molecule is surrounded by molecules (say, z in number) of its own kind. Each of these z nearest neighbors interacts with the molecule under consideration with an interaction potential E_{11}. A similar potential, E_{22}, describes the energy of interaction between any two nonbonded polymer subunits. As shown in Figure 2.2, the process of dissolution consists of breaking one solvent–solvent bond and one interactive bond between two nonbonded polymer subunits and subsequently forming two polymer–solvent interactive bonds. We define E_{12} as the interaction energy between a polymer subunit and solvent molecule. The dissolution of polymer in a given solvent depends on the magnitudes of E_{11}, E_{22}, and E_{12}. The quantities known as solubility parameters, δ_{11} and δ_{22}, are related to these energies. Their exact relations will be discussed in Chapter 9. It is sufficient for the present discussion to know that these can be experimentally determined; their values are compiled in *Polymer Handbook* [4].

We have already observed that a plasticizer should be regarded as a good solvent for the polymer, which means that the solubility parameter δ_{11} for the former must be close ($=\delta_{22}$) to that for the latter. This principle serves as a guide for selecting a plasticizer for a given polymer. For example, unvulcanized natural rubber having δ_{22} equal to 16.5 dissolves in toluene ($\delta_{11} = 18.2$) but does not dissolve in ethanol ($\delta_{11} = 26$). If a solvent having a very different solubility parameter is mixed with the polymer, it would not mix on the molecular level. Instead, there would be regions of the solvent dispersed in the polymer matrix that would be incompatible with each other.

Fillers are usually solid additives that are incorporated into the polymer to modify its physical (particularly mechanical) properties. The fillers commonly used for PVC are given in Box 2.1. It has been found that particle size of the filler has a great effect on the strength of the polymer: The finer the particles are, the

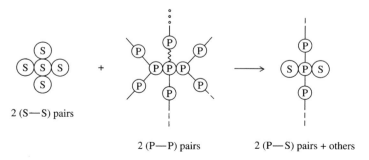

2 (S—S) pairs

2 (P—P) pairs   2 (P—S) pairs + others

FIGURE 2.2   Schematic diagram of the process of polymer dissolution.

higher the hardness and modulus. Another factor that plays a major role in determining the final property of the polymer is the chemical nature of the surface. Mineral fillers such as calcium carbonate and titanium dioxide powder often have polar functional groups (e.g., hydroxyl groups) on the surface. To improve the wetting properties, they are sometimes treated with a chemical called a coupling agent.

Coupling agents are chemicals that are used to treat the surface of fillers. These chemicals normally have two parts: one that combines with the surface chemically and another that is compatible with the polymer. One example is the treatment of calcium carbonate filler with stearic acid. The acid group of the latter reacts with the surface, whereas the aliphatic chain sticks out of the surface and is compatible with the polymer matrix. In the same way, if carbon black is to be used as a filler, it is first mixed with benzoyl peroxide in alcohol at 45°C for at least 50 h and subsequently dried in vacuum at 11°C [5]. This activated carbon has been identified as having $C-OH$ bonds, which can lead to polymerization of vinyl monomers. The polymer thus formed is chemically bound to the filler and would thus promote the compatibilization of the filler with the polymer matrix. Most of the fillers are inorganic in nature, and the surface area per unit volume increases with size reduction. The number of sites where polymer chains can be bound increases, and, consequently, compatibility improves for small particles.

For inorganic fillers, silanes also serve as common coupling agents. Some of these are given in Table 2.1. The mechanism of the reaction consists of two steps; in the first one, the silane ester moiety is hydrolyzed to give

$$(C_2H_5O)_3-Si-(CH_2)_3-NH_2 + 3H_2O$$
$$\longrightarrow (OH)_3-Si-(CH_2)_3-NH_2 + C_2H_5OH \qquad (2.3.1)$$

These subsequently react with various OH groups of the surface, Sur-$(OH)_3$:

$$Sur\begin{array}{l}-OH\\-(OH)\\-OH\end{array} + (OH)_3-Si-(CH_2)_3-NH_2 \longrightarrow$$

$$(2.3.2)$$

$$Sur-O-\underset{\underset{O}{|}}{\overset{\overset{O}{|}}{Si}}-[CH_2]_3-NH_2$$

Silane coupling agents can have one to three of these bonds, and one would ideally like to have all of them reacted. The reaction of OH groups on Si is a competitive one; because of steric factors, not all of them can undergo reaction. The net effect of the reaction in Eq. (2.3.2) is to give chemically bonded silane molecules on the surface of glass or alumina particles. The amine group now

TABLE 2.1 Silane Coupling Agents

Name	Formula
γ-Aminopropyl triethoxy silane	$(C_2H_5O)_3-Si-(CH_2)_3-NH_2$
γ-Chloropropyl triethoxy silane	$(C_2H_5O)_3-Si-(CH_2)_3-Cl$
γ-Cyanopropyl trimethoxy silane	$(CH_3O)_3-Si-(CH_2)_3-CN$
γ-Glycidoxypropyl trimethoxy silane	$(CH_3O)_3-Si-(CH_2)_3-CH_2-CHCH_2$ with epoxide O
γ-Mercaptopropyl trimethoxy silane	$(CH_3-O)_3-Si-(CH_2)_3-SH$
γ-Methacryloxypropyl trimethoxy silane	$(CH_3-O)_3-Si-(CH_2)_3-OC(=O)-C(CH_3)=CH_2$

Some Silanization Procedures

Using γ-aminopropyl triethoxy silane

Glass. One gram of glass beads is added to 5 mL of 10 solution of the coupling agent at pH 5 (adjusted with acetic acid). The reaction is run for 2 h at 80°C. The silanized glass beads are then washed and dried at 120°C in an oven for 2 h.

Alumina

One gram of alumina is added to 5 mL of the coupling agent in toluene. The reaction mixture is refluxed for about 2 h. Alumina is washed with toluene, then with acetone, and finally dried in oven at 120°C for 2 h.

Using γ-mercaptopropyl trimethoxy silane

Glass. One gram of porous glass is added to 5 mL of 10 solution of the coupling agent at pH 5 (adjusted with 6 N HCl). The mixture is heated to reflux for 2 h. The glass beads are washed with pH 5 solutions, followed by water, and ultimately dried in an oven for 2 h at 120°C.

bound to the surface is a reactive one and can easily react with an acid or an aldehyde group situated on a polymer molecule.

Recently, Goddart et al. [6] reported a polyvinyl alcohol–copper(II) initiating system, which can produce branched polymers on surfaces. The initiating system is prepared by dissolving polyvinyl alcohol in water that already contains copper nitrate (or copper chloride). The calcium carbonate filler is dipped into the solution and dried. If this is used for polymerization of an olefin (say, styrene), it would form a polymer that adheres to the particles, ultimately encapsulating them. The mechanical properties of calcium-carbonate-filled polystyrene have been found to depend strongly on filler–matrix compatibility, which is considerably improved by this encapsulation.

Polymers also require protection against the effect of light, heat, and oxygen in the air. In view of this, polymers are mixed with antioxidants and stabilizers in low concentrations (normally less than 1%). If the material does not have these compounds, a polymer molecule M_n of chain length n interacts with light (particularly the ultraviolet portion of the light) to produce polymer radicals P_n, as follows:

$$M_n \xrightarrow{hv} P_n \qquad (2.3.3)$$

The polymer radicals thus produced interact with oxygen to form alkyl peroxy radicals $(P_{n1}-O_2)$ that can abstract hydrogen of the neighboring molecules in various ways, as shown in the mechanism of the auto-oxidation process of Table 2.2. The formation of hydroperoxide in step C of the sequence of reactions is the most important source of initiating radicals. In practice, the following three kinds of antioxidant and stabilizer are used. Peroxide decomposers are materials that form stable products with radicals formed in the auto-oxidation of Table 2.2;

TABLE 2.2 Mechanism of Auto-oxidation and Role of Antioxidants

Initiation	$M_n \xrightarrow{hv} P_n$ $P_n + O_2 \longrightarrow P_n-O_2$ $P_n + O_2 + M_nH \longrightarrow M_nO_2H \ldots M_n$
Propagation	$M_nO_2H \ldots M_n$ → $P_nO + P_m-O_2 + H_2O$ $P_n-O + P_m + O\dot{H}$ $P_n-O_2 + P_m-O + H_2O$
Termination	$P_mO_2 + P_2O_2 \longrightarrow$ inert $P_mO_2 + P_n \longrightarrow M_m-O_2-M_m$ $P_m + P_n \longrightarrow M_m + M_n$ $P_n =$ a polymer radical $[CH_2-CH]_{n-1}CH_2-\dot{CH}$ with R, R
Peroxide decomposers	Mercaptans, sulfonic acids, zinc alkyl thiophosphate, zinc dimethyldithiocarbamate, dilauryl thiodipropionate
Metal deactivators	Various chelating agents that combine with ions of manganese, copper, iron, cobalt, and nickel; e.g., N,N',N,N-tetrasalicylidene tetra (aminomethyl) methane, 1,8-bis(salicylidene amino)-3,6-dithiaoctane
Ultraviolet light adsorbers	Phenyl salicylate, resorcinol monobenzoate, 2-hydroxyl-4-methoxybenzophenone, 2-(2-hydroxyphenyl)-benzotriazole, etc.

chemical names of some of this class are given therein. Practice has also shown that the presence of manganese, copper, iron, cobalt, and nickel ions can also initiate oxidation. As a result, polymers are sometimes provided with metal deactivators. These compounds (sometimes called chelating agents) form a complex with metal ions, thus suppressing auto-oxidation. When the polymer is exposed to ultraviolet rays in an oxygen-containing atmosphere, it generates radicals on the surface.

The ultraviolet absorbers are compounds that react with radicals produced by light exposures. In the absence of these in the polymer, there is discoloration, surface hardening, cracking, and changes in electrical properties.

Once the polymer is manufactured, it must be shaped into finished products. The unit operations carried out in shaping include extruding, kneading, mixing, and calendering, all involving exposure to high temperatures. Polymer degradation may then occur through the following three ways: depolymerization, elimination, and/or cyclization [7,8]. Depolymerization is a reaction in which a chemically inert molecule, M_n, undergoes a random chain homolysis to form two polymer radicals, P_r and P_{n-r}:

$$M_n \longrightarrow P_r + P_{n-r} \qquad (2.3.4)$$

A given polymer radical can then undergo intramolecular as well as intermolecular transfer reactions. In the case of intramolecular reactions, monomer, dimer, trimer, and so forth are formed as follows:

$$P_r \begin{cases} \longrightarrow P_{r-1} + M \\ \longrightarrow P_{r-2} + M_2 \\ \longrightarrow P_{r-3} + M_3 \quad \text{etc.} \end{cases} \qquad (2.3.5)$$

In the case of the latter, however, two macroradicals interact to destroy their radical nature, thus giving polymers of lower molecular weight:

$$P_r + P_m \longrightarrow M_r + M_m \qquad (2.3.6)$$

This process is shown in Box 2.2 to occur predominantly for polyethylene. Elimination in polymer degradation occurs whenever the chemical bonds on substituents are weaker than the $C-C$ backbone bonds. As shown in Box 2.2, for PVC (or for polyvinyl acetate), the chloride bond (or acetate) breaks first and HCl (or acetic acid) is liberated. Normally, the elimination of HCl (or acetic acid) does not lead to a considerable decrease in molecular weight. However, because of the formation of double bonds on the backbone, cross- linking occurs as shown. Intramolecular cyclization in a polymer is known to occur at high temperatures

> **Box 2.2**
> **Thermal Degradation of Some Commercial Polymers**
>
> *Polymethyl methacrylate (PMMA).* The degradation occurs around 290–300°C. After homolysis of polymer chains, the macroradicals depropagate, giving a monomer with 100% yield.
> *Polystyrene.* Between 200°C and 300°C, the molecular weight of the polymer falls, with no evolution of volatile products. This suggests that polymers first undergo homolysis, giving macroradicals, which later undergo disproportionation.
>
>
> Above 300°C, polystyrene gives a monomer (40–60%), toluene (2%), and higher homologs. Polymer chains first undergo random homolytic decomposition.
>
> $$M_n \longrightarrow P_m + P_{n-m}$$
>
> The macroradicals then form monomers, dimers, and so forth, by intramolecular transfer.
>

Polyethylene. Beyond 370°C, polyethylene degrades, forming low-molecular-weight (through intermolecular transfer) and volatile (through intramolecular transfer) products.

Hindered phenols such as 2,6-di-*t*-butyl-4- methylphenol (BHT) are effective melt stabilizers.

Polyacrylonitrile (PAN). On heating PAN at 180–190°C for a long time (65 h) in the absence of air, the color changes to tan. If it is heated under controlled conditions at 1000°C, it forms carbon fibers. The special properties of the latter are attributed to the formation of cyclic rings through the combination of nitrile groups as follows:

Polyvinyl chloride (PVC). At 150°C, the polymer discolors and liberates chlorine. The reaction is autocatalytic and occurs as follows:

The polymer thus formed has several double bonds on the backbone during HCl loss. It can undergo intermolecular cross-linking through a Diels–Alder type reaction as follows:

Some of the melt stabilizers for PVC are lead carbonate and dialkyl carboxylate.

whenever substituents on it can undergo further reactions. The most common example in which cyclization occurs predominantly is found in nitrile polymers, whose cyanide groups are shown in Box 2.2 to condense to form a cyclic structure. The material thus formed is expected to be strong and brittle, a fact which is utilized in manufacturing carbon fiber used in polymer composites.

Finally, there are several applications in packaging (e.g., where it is desirable that a polymeric material easily burn in fire). On the other hand, several other applications, such as building furniture and fitting applications, require that the material have a sufficient degree of fire resistance. Fire retardants are chemicals that are mixed with polymers to give this property; they produce the desired effect by doing any combination of the following:

1. Chemically interfering with the propagation of flame
2. Producing a large volume of inert gases that dilute the air supply
3. Decomposing or reacting endothermally
4. Forming an impervious fire-resistant coating to prevent contact of oxygen with the polymer

Some of the chemicals (such as ammonium polyphosphate, chlorinated *n*-alkanes for polypropylene, and tritolyl phosphate) are used in PVC as fire retardants.

Example 2.1: Describe a suitable oxidation (or etching) method of polyethylene and polypropylene surfaces. Also, suggest the modification of terylene with nucleophilic agents like bases.

Solution: A solution of $K_2Cr_2O_7 : H_2O : H_2SO_4$ in the ratio of $4.4 : 7.1 : 88.5$ by weight at $80°C$ gave carboxylic groups on the surface which can be further functionalized as follows:

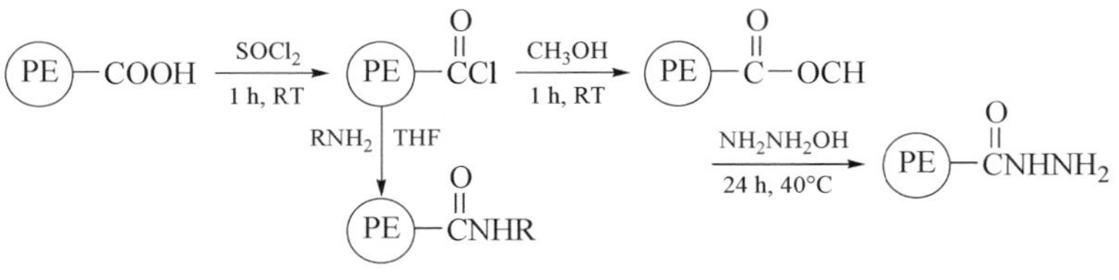

This surface treatment increases the wettability of polyethylene and can also be done by a $KMnO_4$, H_2SO_4 mixture. The hydrazine modified polyethylene can further be reacted with many reagents.

The polyester can be easily reacted on surfaces with 4% caustic soda solution at $100°C$:

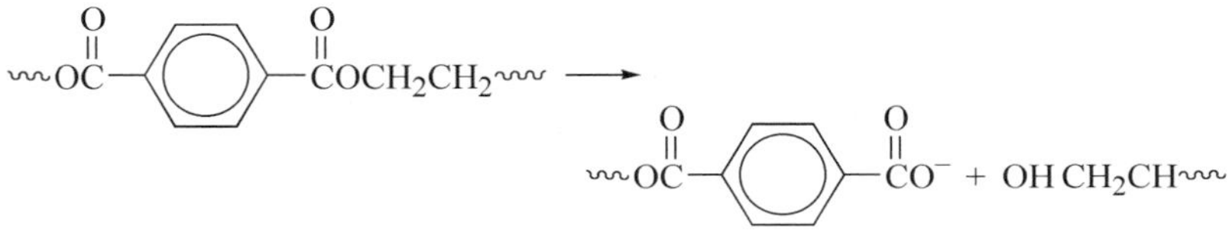

There is 30% loss in weight in 2 h and excessive pitting and roughening of the surface occurs.

Example 2.2: Fiberglass-reinforced composites (FRCOs) are materials having an epoxy resin polymer matrix which embeds glass fabric within it. In order to compatibilize glass fabric, a thin layer of polymer could be chemically bound to it in order to improve fracture toughness. Suggest a suitable method of grafting polymer on glass fabric.

Solution: All commercially available glass fabrics are already silanated using aminopropyl triethoxysilate and can serve as points where initiators can be chemically bound. For this purpose, we can prepare a dichlorosuccinyl peroxide initiator starting from succinic anhydride. The latter is first reacted with hydrogen peroxide at room temperature and then reacted with thionyl chloride as follows:

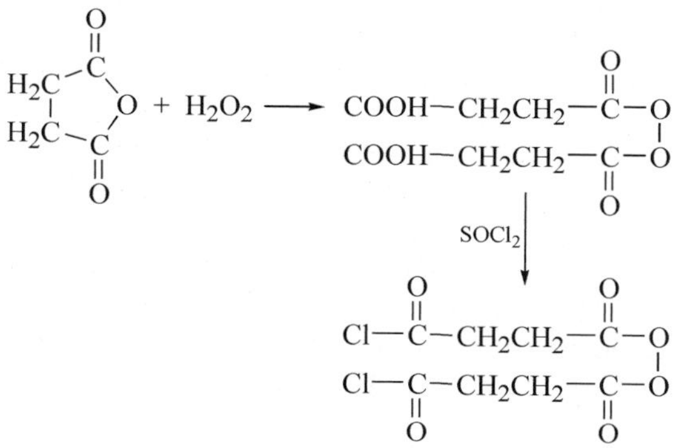

(A)

This initiator can be immobilized on glass fabric and the MMA can be easily polymerized using the modified fabric as follows:

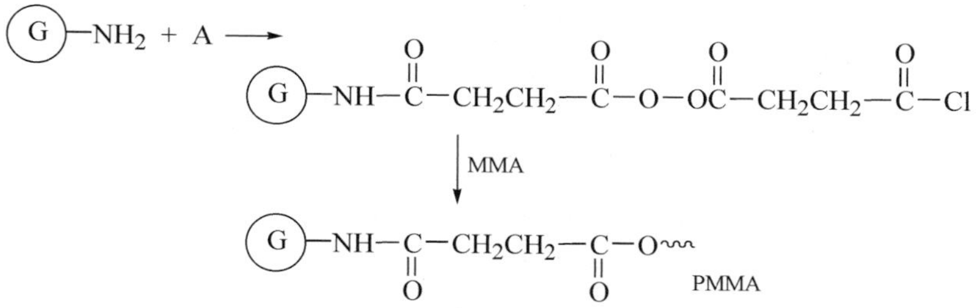

In grafting polymers, we need to covalently bind on suitable initiator on the surface as it has been done in this example.

2.4 RUBBERS

Natural and synthetic rubbers are materials whose glass transition temperatures T_g are lower than the temperature of application. Rubber can be stretched up to 700% and exhibit an increase in modulus with increasing temperature.

2.4.1 Natural Rubber

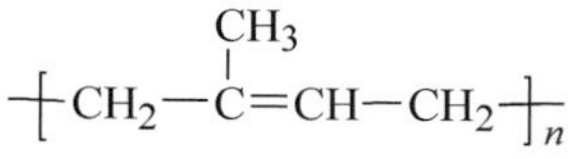

On gouging the bark of *Hevea brasiliensis*, hevea latex is collected, which has close to a 33% dry rubber content. Natural rubber, a long-chain polyisoprene, given by is produced by coagulating this latex (e.g., using acetic acid as the coagulating agent) and is used in adhesives, gloves, contraceptives, latex foam, and medical tubing. Ribbed smoked sheets (RSSs) are obtained by coagulating rubber from the latex, passing it through mill rolls to get sheets and then drying it at 43°C to 60°C in a smokehouse. Crepes are obtained by washing the coagulum to remove color impurity and β-carotene, and then bleaching with xylyl mercaptan. Comminuted rubbers are produced by drying the coagulum and then storing them in bales.

Natural rubber displays the phenomenon of natural tack and therefore serves as an excellent adhesive. Adhesion occurs because the ends of rubber molecules penetrate the adherend surfaces and then crystallize. The polymer has the following chemical structure, having a double bond at every alternate carbon atom:

$$\begin{array}{c} CH_3 \\ | \\ \left[CH_2-C=CH-CH_2 \right]_n \end{array}$$

$$\begin{array}{ccc} CH_3 & H & CH_3 \\ | & | & | \\ \sim\sim C=C-C=C-C=C\sim\sim \\ & | & | & | \\ & H & CH_3 & H \end{array}$$

$$(2.4.1)$$

and it can react with sulfur (in the form of sulfur chloride) to form a polymer network having sulfur bridges as follows:

$$\begin{array}{c} C-S_n-C \\ C-S_x-C \end{array} \quad x, n = 1, 2, \ldots$$

$$(2.4.2)$$

This process is known as *vulcanization*. The polymer thus formed is tough and is used in tire manufacture.

In ordinary vulcanized rubber used in tire industries, the material contains about 2–3% sulfur. If this sulfur content is increased to about 30%, the resultant material is a very hard nonrubbery material known as *ebonite* or "hard rubber." The double bonds of natural rubber can easily undergo addition reaction with hydrochloric acid, forming rubber hydrochloride:

$$\begin{array}{ccc} CH_3 & & CH_3 \\ | & & | \\ \sim\sim CH_2-C=CHCH_2\sim\sim + HCl \longrightarrow \sim\sim CH_2C-CH_2CH_2\sim\sim \\ & & | \\ & & Cl \end{array}$$

$$(2.4.3)$$

Rubber hydrochloride

If natural rubber is treated with a proton donor such as sulfuric acid or stannic chloride, the product is cyclized rubber (empirical formula of $-C_5H_8-$), having the following molecular structure:

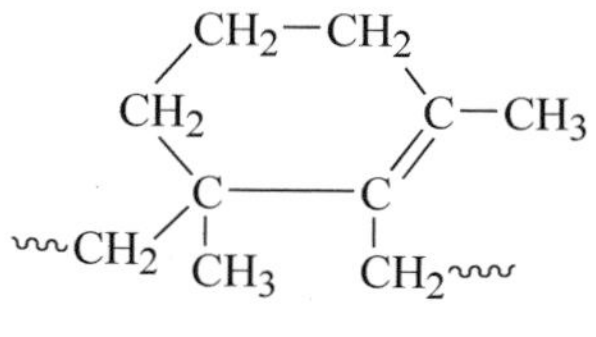

The polymer is inelastic, having high density, and dissolves in hydrocarbon solvents only. Treatment of natural rubber with chlorine gives chlorinated rubber, which has the following structure:

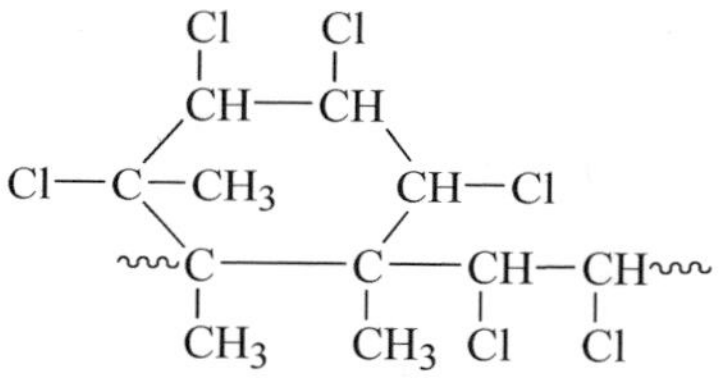

Chlorinated rubber is extensively employed in industry for corrosion-resistant coatings.

There are several other 1,4-polyisoprenes occurring in nature that differ significantly in various properties from those of natural rubbers. One of these is *gutta percha*, which is essentially a nonelastic, hard, and tough material (used for making golf balls). The stereoisomerism in diene polymers has already been discussed in Chapter 1; *gutta percha* has been shown to be mainly *trans*-1,4-polyisoprene. Because of their regular structure, the chains can be packed closely, and this is responsible for the special properties of the polymer.

2.4.2 Polyurethane Rubbers

The starting point in the manufacture of polyurethane rubbers is to prepare a polyester of ethylene glycol with adipic acid. Usually, the former is kept in excess to ensure that the polymer is terminated by hydroxyl groups:

$$OH-CH_2-CH_2-OH + COOH-(CH_2)_4-COOH \longrightarrow$$

Ethylene glycol   Adipic acid

$$OH-\left[CH_2CH_2-O\overset{\overset{\displaystyle O}{\|}}{C}-[CH_2]_4-\overset{\overset{\displaystyle O}{\|}}{C}-O-CH_2CH_2\right]_n-OH \qquad (2.4.4)$$

Polyol

The polyol (denoted OH~P~OH) is now reacted with a suitable diisocyanate. Some of the commercially available isocyanates are tolylene diisocyanate (TDI),

diphenylmethane diisocyanate (MDI),

$$NCO-\underset{\underset{C_6H_5}{|}}{\overset{\overset{C_6H_5}{|}}{C}}-NCO$$

and naphthylene diisocyanate,

When polyol is mixed with a slight excess of a diisocyanate, a prepolymer is formed that has isocyanate groups at the chain ends:

$$OH{\sim}P{\sim}OH \ + \ NCO-\underset{\underset{C_6H_5}{|}}{\overset{\overset{C_6H_5}{|}}{C}}-NCO \ \longrightarrow$$

$$NCO{\sim}\underset{\underset{C_6H_5}{|}}{\overset{\overset{C_6H_5}{|}}{C}}-NHCOO{\sim}P{\sim}OOCNH-\underset{\underset{C_6H_5}{|}}{\overset{\overset{C_6H_5}{|}}{C}}{\sim}NCO \tag{2.4.5}$$

With the use of P to denote the polyester polymer segment, U to denote the urethane $-$CONH linkage, and I to denote the isocyanate $-$NCO linkage, the polymer formed in reaction (2.4.5) can be represented by I$-$PUPUPU$-$I. This is sometimes called a *prepolymer* and can be chain-extended using water, glycol, or amine, which react with it as

$$\sim\!\!\sim I + H_2O \ \longrightarrow \ \sim\!\!\sim UPUP-\overset{\overset{\displaystyle O}{\|}}{C}-PUPUPU\!\sim\!\!\sim \ + CO_2 \tag{2.4.6a}$$

$$\sim\!\!\sim I + OH-R-OH \ \longrightarrow \ \sim\!\!\sim UPUP-R-PUPU\!\sim\!\!\sim \tag{2.4.6b}$$

$$\sim\!\!\sim I + NH_2-R-NH_2 \ \longrightarrow \ \sim\!\!\sim UPUP-R-PUPU\!\sim\!\!\sim \tag{2.4.6c}$$

Experiments have shown that the rubbery nature of the polymer can be attributed to the polyol "soft" segments. It has also been found that increasing the "size" of R contributed by the chain extenders tends to reduce the rubbery nature of the polymer. The urethane rubber is found to have considerably higher tensile strength and tear and abrasion resistance compared to natural rubber. It has found extensive usage in oil seals, shoe soles and heels, forklift truck tires, diaphragms, and a variety of mechanical applications.

2.4.3 Silicone Rubbers

Silicone polymers are prepared through chlorosilanes, and linear polymer is formed when a dichlorosilane undergoes a hydrolysis reaction, as follows:

$$Cl-\underset{\underset{CH_3}{|}}{\overset{\overset{CH_3}{|}}{Si}}Cl \xrightarrow{H_2O} \quad \sim\sim O-\underset{\underset{CH_3}{|}}{\overset{\overset{CH_3}{|}}{Si}}-O\sim\sim \qquad (2.4.7)$$

Silicone rubbers are obtained by first preparing a high-molecular-weight polymer and then cross-linking it. For this, it is important that the monomer not have trichlorosilanes and tetrachlorosilanes even in trace quantity. The polymer thus formed is mixed with a filler (a common one for this class of polymer is fumed silica), without which the resultant polymer has negligible strength. The final curing is normally done by using a suitable peroxide (e.g., benzoyl peroxide, t-butyl perbenzoate, dichlorobenzoyl peroxide), which, on heating, generates radicals (around $70°C$).

$$\underset{\text{Benzoyl peroxide}}{C_6H_5COO-OO-CC_6H_5} \longrightarrow 2C_6H_5-COO\cdot \qquad (2.4.8)$$

The radicals abstract hydrogen from the methyl groups of the polymer. The polymer radical thus generated can react with the methyl group of another molecule, thus generating a network polymer:

$$\underset{\underset{CH_3}{|}}{\overset{\overset{CH_3}{|}}{\sim\sim Si}}-O\sim\sim + C_6H_5-COO\cdot \longrightarrow C_6H_5-COOH + \underset{\underset{\cdot CH_2}{|}}{\overset{\overset{CH_3}{|}}{\sim\sim Si}}-O\sim\sim \quad (2.4.9a)$$

$$\underset{\underset{\cdot CH_2}{|}}{\overset{\overset{CH_3}{|}}{\sim\sim Si}}-O\sim\sim + \underset{\underset{\cdot CH_2}{|}}{\overset{\overset{CH_3}{|}}{\sim\sim Si}}-O\sim\sim \longrightarrow \underset{\underset{\underset{\underset{\underset{\underset{CH_3}{|}}{\sim\sim Si-O\sim\sim}}{|}}{CH_2}}{|}}{\overset{\overset{CH_3}{|}}{\sim\sim Si}}-O\sim\sim \qquad (2.4.9b)$$

Silicone rubbers are unique because of their low- and high-temperature stability (the temperature range for general applications is −55°C to 250°C), retention of elasticity at low temperature, and excellent electrical properties. They are extremely inert and have found several biomedical applications. Nontacky self-adhesive rubbers are made as follows. One first obtains an OH group at chain ends through hydrolysis, for which even the moisture in the atmosphere may be sufficient:

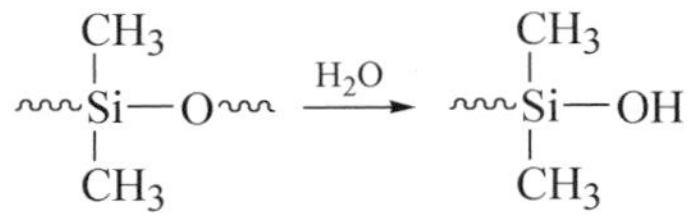

$$
\underset{\substack{|\\ CH_3}}{\overset{\substack{CH_3\\ |}}{\sim\sim Si-O\sim\sim}} \xrightarrow{H_2O} \underset{\substack{|\\ CH_3}}{\overset{\substack{CH_3\\ |}}{\sim\sim Si-OH}}
\tag{2.4.10}
$$

On reacting this product with boric acid, there is an end-capping of the chain, yielding the self-adhesive polymer. On the other hand, "bouncing putty" is obtained when −Si−O−B− bonds are distributed on the backbone of the chain.

2.5 CELLULOSE PLASTICS

Cellulose is the most abundant polymer constituting the cell walls of all plants. Oven-dried cotton consists of lignin and polysaccharides in addition to 90% cellulose. On digesting it under pressure and a temperature of 130–180°C in 5–10% NaOH solution, all impurities are removed. The residual α-cellulose has the following structure:

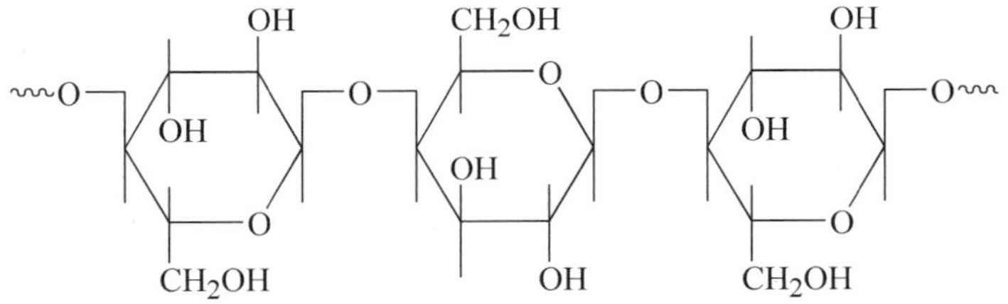

Every glucose ring of cellulose has three −OH functional groups that can further react. For example, cellulose trinitrate, an explosive, is obtained by nitration of all OH groups by nitric acid. Industrial cellulose nitrate is a mixture of cellulose mononitrate and dinitrate and is sold as celluloid sheets after it is plasticized with camphor. Although cellulose does not dissolve in common solvents, celluloid dissolves in chloroform, acetone, amyl acetate, and so forth. As a result, it is used in the lacquer industry. However, the polymer is inflammable and its chemical resistance is poor, and its usage is therefore restricted.

Among other cellulosic polymers, one of the more important ones is cellulose acetate. The purified cellulose (sometimes called *chemical cellulose*) is pretreated with glacial acetic acid, which gives a higher rate of acetate formation and more even substitution. The main acetylation reaction is carried out by acetic anhydride, in which the hydroxyl groups of cellulose (denoted X—OH) react as follows:

$$X-OH + CH_3\overset{O}{\overset{\|}{C}}-O-\overset{O}{\overset{\|}{C}}-CH_3 \longrightarrow X-O\overset{O}{\overset{\|}{C}}-CH_3 + CH_3COOH \qquad (2.5.1)$$

If this reaction is carried out for long times (about 5–6 h), the product is cellulose triacetate. Advantages of this polymer include its water absorptivity, which is found to reduce with the degree of acetylation, the latter imparting higher strength to the polymer. The main usage of the polymer is in the preparation of films and sheets. Films are used for photographic purposes, and sheets are used for glasses and high-quality display boxes.

Cellulose ethers (e.g., ethyl cellulose, hydroxyethyl cellulose, and sodium carboxymethyl cellulose) are important modifications of cellulose. Ethyl cellulose is prepared by reacting alkali cellulose with ethyl chloride under pressure. If the etherification is small and the average number of ethoxy groups per glucose molecule is about unity, the modified polymer is soluble in water. However, as the degree of substitution increases, the polymer dissolves in nonpolar solvents only. Ethyl cellulose is commonly used as a coating on metal parts to protect against corrosion during shipment and storage.

Sodium carboxymethyl cellulose (CMC) is prepared through an intermediate alkali cellulose. The latter is obtained by reacting cellulose $[X-(OH)_3]$ with sodium hydroxide as follows:

$$X-(OH)_3 + 3NaOH \longrightarrow X-(ONa)_3 + 3H_2O \qquad (2.5.2)$$

which is further reacted with sodium salt of chloroacetic acid ($Cl-CH_2COONa$), as follows:

$$X-[ONa]_3 + 3ClCH_2COONa \longrightarrow X-[OCH_2COONa]_3 + NaCl \qquad (2.5.3)$$

Commercial grades of CMC are physiologically inert and usually have a degree of substitution between 0.5 and 0.85. CMC is mainly used in wallpaper adhesives, pharmaceutical and cosmetic agents, viscosity modifiers in emulsions and suspensions, thickener in ice cream industries, and soil- suspending agents in synthetic detergents.

It has already been pointed out that naturally occurring cellulose does not have a solvent and its modification is necessary for it to dissolve in one. In certain applications, it is desired to prepare cellulose films or fibers. This process involves first reacting it to render it soluble, then casting film or spinning fibers, and, finally, regenerating the cellulose. Regenerated cellulose (or rayon) is manufactured by reacting alkali cellulose [or $X-(ONa)_3$] with carbon disulfide to form sodium xanthate:

$$X-[ONa]_3 + CS_2 \longrightarrow R-[O\overset{\overset{\displaystyle S}{\|}}{C}-ONa] \tag{2.5.4}$$

which is soluble in water at a high pH; the resultant solution is called *viscose*. The viscose is pushed through a nozzle into a tank with water solution having 10–15% H_2SO_4 and 10–20% sodium sulfate. The cellulose is immediately regenerated as fiber of foil, which is suitably removed and stored.

2.6 COPOLYMERS AND BLENDS [9–11]

Until now, we have considered homopolymers and their additives. There are several applications in which properties intermediate to two given polymers are required, in which case copolymers and blends are used. Random copolymers are formed when the required monomers are mixed and polymerization is carried out in the usual fashion. The polymer chains thus formed have the monomer molecules randomly distributed on them. Some of the common copolymers and their important properties are given in Box 2.3.

Polymer blends are physical mixtures of two or more polymers and are commercially prepared by mechanical mixing, which is achieved through screw compounders and extruders. In these mixtures, different polymers tend to separate (instead of mixing uniformly) into two or more distinct phases due to incompatibility. One measure taken to improve miscibility is to introduce specific interactive functionalities on polymer pairs. Hydrogen-bondings have been shown to increase miscibility and, as a consequence, improve the strength of the blends. Eisenberg and co-workers have also employed acid–base interaction (as in sulfonated polystyrene with polyethylmethacrylate–Co–4-vinyl pyridine) and ion–dipole interaction (as in polystyrene–Co–lithium methacrylate and polyethylene oxide) to form improved blends.

Commonly, the functional groups introduced into the polymers are carboxylic or sulfonate groups. The following are the two general routes of their synthesis:

1.   Copolymerization of a low level of functionalized monomers with the comonomer
2.   Direct functionalization of an already formed polymer

Because of the special properties imparted to this new material, called an ionomer, it has been the subject of vigorous research in recent years. Ionomers are used as compatibilizing agents in blends and are also extensively employed in permselective membranes, thermoplastic elastomers, packaging films, and viscosifiers. Carboxylic acid groups are introduced through the first synthetic route by

---

Box 2.3
Some Commercial Copolymers

Ethylene–vinyl acetate copolymer (EVA). Vinyl acetate is about 10–15 surface gloss, and melt adhesive properties of EVA.

Ethylene–acrylic acid copolymer. Acrylic acid content varies between 1 and 10 polymer. When treated with sodium methoxide or magnesium acetate, the acid groups form ionic cross-linking bonds at ambient condition, whereas at high temperature these break reversibly. As a result, they behave as thermosetting resins at low temperatures and thermoplastics at high temperatures.

Styrene–butadiene rubber (SBR). It has higher abrasion resistance and better aging behaviour and is commonly reinforced with carbon black. It is widely used as tire rubber.

Nitrile rubber (NBR). In butadiene acrylontrile rubber, the content of the acrylonitrile lies in the 25–50 range for its resistance to hydrocarbon oil and gasoline. It is commonly used as a blend with other polymers (e.g., PVC). Low-molecular weight polymers are used as adhesives.

Styrene–acrylonitrile (SAN) copolymer. Acrylonitrile content is about 20–30 grease, stress racking, and crazing. It has high impact strength and is transparent.

Acrylonitrile–butadiene–styrene (ABS) terpolymer. Acrylonitrile and styrene are grafted on polybutadiene. It is preferred over homopolymers because of impact resistance, dimensional stability, and good heat-distortion resistance. It is an extremely important commercial copolymer and, in several applications, it is blended with other polymers (e.g., PVC or polycarbonates) in order to increase their heat-distortion temperatures. When methyl methacrylate and styrene are grafted on polybutadiene, a methyl methacrylate–butadiene–styrene MBS copolymer is formed.

Vinylidene chloride–vinyl chloride copolymer. Because of its toughness, flexibility, and durability, the copolymer is used for the manufacture of filaments for deck chair fabrics, car upholstery, and doll's hair. Biaxially stretched copolymer films are used for packaging.

---

employing acrylic or methacrylic acids as the comonomer in small quantity. Sulfonate groups are normally introduced by polymer modification; they will be discussed in greater detail later in this chapter.

A special class of ionomers in which the functional groups are situated at chain ends are telechelic ionomers. The technique used for their synthesis depends on the functional groups needed; the literature reports several synthesis routes. The synthesis via radical polymerization can be carried out either by using a large amount of initiator (sometimes called *dead-end polymerization*) or by using a suitable transfer agent (sometimes called *telomerization*). If a carboxylic acid group is needed, a special initiator–3,3-azobis (3-cyanovaleric acid) should be used:

$$
\begin{array}{c}
\overset{\displaystyle CH_3}{\underset{\displaystyle CH_2CH_2-COOH}{NC-C}}-N{=}N-\overset{\displaystyle CH_3}{\underset{\displaystyle CH_2CH_2-COOH}{C}}-NC
\end{array}
$$

For a hydroxyl end group, 4,4-azobis 2(cyanopentanol) could be employed:

$$
OH-CH_2-CO-\overset{\displaystyle CH_3}{\underset{\displaystyle CH_3}{C}}-N-N-\overset{\displaystyle CH_3}{\underset{\displaystyle CH_3}{C}}-CO-CH_2OH
$$

We will show in Chapter 5 that using a large amount of initiator gives polymer chains of smaller length and is therefore undesirable. Instead, radical polymerization in the presence of transfer agents can be performed. The best known transfer agent is carbon tetrachloride, which can abstract an electron from growing polymer radicals, P_n, as follows:

$$P_n + CCl_4 \longrightarrow M_n-Cl + Cl_3-C^{\cdot} \tag{2.6.1}$$

The CCl_3 radical can add on the monomer exactly as P_n, but the neutral molecule M_n-Cl is seen to contain the chloride group at one of its ends. This chloride functional group can subsequently be modified to hydroxy, epoxide, or sulfonate groups, for example, as follows:

$$ Cl\text{\textasciitilde\textasciitilde\textasciitilde} Cl \xrightarrow{\;-HCl\;} \quad \tag{2.6.2}$$

$$ \tag{2.6.3}$$

Synthesis of telechelics through anionic polymerization is equivalently convenient; interested readers should consult more advanced texts [11].

We have already indicated that incompatibility in polymer blends causes distinct regions called *microphases*. The most important factor governing the mechanical properties of blends is the interfacial adhesion between microphases. One of the techniques to improve this adhesion is to bind the separate micro-phases through chemical reaction of functional groups. Figure 2.3 shows a styrene copolymer containing oxazoline groups and an ethylene copolymer with acrylic acid as a comonomer. These polymers are represented as follows:

$$P_1 \mathord{\sim\!\!\!\sim} \overbrace{\underset{O}{\overset{N}{\bigg\langle}}}^{} \quad \text{and} \quad P_2 \mathord{\sim\!\!\!\sim} COOH$$

The following reaction of functional groups occurs at the microphase boundaries:

$$P_1 \mathord{\sim\!\!\!\sim} \overbrace{\underset{O}{\overset{N}{\bigg\langle}}}^{} + P_2 \mathord{\sim\!\!\!\sim} COOH \longrightarrow P_1 \mathord{\sim\!\!\!\sim} CONH-CH_2CH_2-O-CO \mathord{\sim\!\!\!\sim} P_2$$

$$(2.6.4)$$

The two polymers are blended in an extruder and, due to this reaction, there is some sort of freezing of the microphases, thus giving higher strength. Another interesting example that has been reported in the literature is the compatibiliza-tion of polypropylene with nylon 6. The latter is a polyamide that has a carboxylic acid and an amine group at chain ends; in another words, it is a telechelic. We then prepare a copolymer of polypropylene with 3% maleic anhydride. The melt extrusion of these polymers would lead to a blend with frozen matrices, as shown in Figure 2.4.

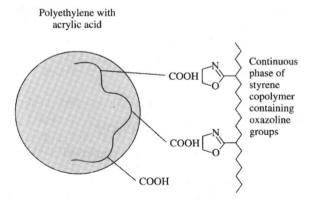

FIGURE 2.3 Polymer compatibilization through chemical reaction of functional groups.

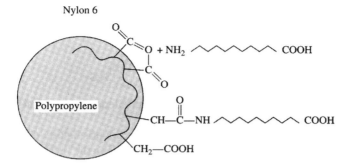

FIGURE 2.4   Use of maleic anhydride to compatibilize polypropylene and nylon 6.

2.7   CROSS-LINKING REACTIONS

We have already discussed the fact that a polymer generated from monomers having a functionality greater than 2 is a network. This is called a *cross-linking* or *curing reaction*. The cured polymer, being a giant molecule, will not dissolve in any solvent. Some of the applications of the polymer that utilize curing are adhesives, paints, fiber-reinforced composites, ion-exchange resins, and polymeric reagents. We will discuss these in the rest of the chapter.

Adhesives are polymers that are initially liquid but solidify with time to give a joint between two surfaces [12,13]. The transformation of fluid to solid can be obtained either by evaporation of solvent from the polymer solution (or dispersion) or by curing a liquid polymer into a network. Table 2.3 lists some common adhesives, which have been classified as nonreactive and reactive systems. In the former, the usual composition is a suitable quick-drying solvent consisting of a polymer, tackifiers, and an antioxidant. Tackifiers are generally low-molecular-weight, nonvolatile materials that increase the tackiness of the adhesive. Some tackifiers commonly used are unmodified pine oils, rosin and its derivatives, and hydrocarbon derivatives of petroleum (petroleum resins). Several polymers have their own natural tack (as in natural rubber), in which case additional tackifiers are not needed.

Before adhesion occurs, wetting of the surface must occur, which implies that the molecules of the adhesives must come close with those of the surface to interact. After the solvent evaporates, a permanent bond sets between the surfaces to be joined. Pressure-sensitive adhesives are special nonreacting ones that do not lose their tackiness even when the solvent evaporates. This is because the polymer used is initially in the liquid stage and it remains so even after drying. The most common adhesive used industrially is polymer dispersion of a copolymer of 2-ethyl hexyl acrylate, vinyl acetate, and acrylic acid in water

Type	Structure	Remarks
	Nonreactive Adhesives	
Hot SBR	Styrene–butadiene copolymer	Its solution in hexane or toluene is used as tile cement and wallpaper adhesive. Its ability to stick on a surface is considerably improved if SBR is a terpolymer with a monomer having carboxylic acid (say, acrylic acid).
Nitrile rubber	Copolymer of butadiene and acrylonitrile (20–40%)	Used with any nonpolar solvent; provides good adhesion with surfaces.
Polyvinyl acetate and its copolymers	Copolymerized with acrylates and maleates to improve T_g, tack, and compatibility	Common household glue (white glues). It resists grease, oil, and hydrocarbon solvents; has poor resistance to weather and water. Copolymerization is done to improve this.
Polyvinyl acetals		Polyvinyl formal (R=H) is used as a structural adhesive in the aircraft industry. Polyvinyl butyral (R–C_3H_7) is used as the interlayer in safety glasses in the automobile industry.
	Pressure-Sensitive Adhesives	
Polyacrylates	Water emulsions of copolymer of 2-ethyl acrylate (352 parts), vinyl acetate (84 parts), and acrylic acid (4 parts)	Pressure-sensitive adhesive used for labels. They have permanent tack, and labels with this glue can be refused.
Silicone rubbers	$$OH-(Si-O-Si-O)_n-H$$ with R groups on each Si. R: methyl or phenyl $M = 500\text{–}600$	The tack is considerably improved by the phenyl group. It can produce adhesion with any surface, including Teflon. Polymer-coated polyester films are used in plating operations and insulations.

(*continued*)

TABLE 2.3 (continued)

Type	Structure	Remarks
Polyvinyl ether	$\sim\sim(CH_2-CH)\sim\sim$ | OR R: methyl, ethyl, or isobutyl	These polymers are frequently used in pressure-sensitive adhesive applications, as in cellophane tapes and skin bandages.
Reactive adhesives		
Two-component polyurethane adhesives	Prepolymer NCO$\sim$NCO with polyol OH$\sim$OH hardener	Used as structural adhesive. Usually the curing is slow and the joint has low modules.
Epoxy adhesives	Diglycidyl ether of bisphenol-A, $CH_2-CH\sim\sim CH-CH_2$ __O__/ __O__/ with triamines R$-(NH_2)_3$	Two parts epoxy resins are mixed before use. It exhibits excellent adhesion to metals, plastics, woods, glass, ceramics, etc. It is unaffected by water, and its major use is in aerospace, automotive, electrical, and electronics industries.
Anaerobic acrylic	Polyethylene glycol Bismethacrylates with a hydroperoxide catalyst $$CH_2=C-C-(CH_2CH_2O)_n-CC=CH_2$$ (with O above each C, Me below each C)	It cures at room temperature through a free-radical mechanism in contact with metal without air. Originally used as sealant but now also used as structural adhesive. Curing is sensitive to substrate.
Cyanoacrylates	Methyl or ethyl cyanoacrylates $$CH_2=C\begin{smallmatrix}COOMe\\ CN\end{smallmatrix}$$	It polymerizes on a surface with a slight amount of moisture. It joins any surface except polyethylene, polypropylene, and Teflon.

prepared through the emulsion polymerization technique. The other polymeric materials that give permanent tack are natural rubber, polyvinyl ethyl, isobutyl ethers, and silicone rubbers, all of which are commercially available. The silicone polymers, in addition, have considerable thermal stability and are known to be used at low as well as high temperatures ($-75°C$ to $250°C$).

Reactive adhesives are those liquid materials that are cured (or cross-linked) into a solid network in situ. For example, epoxy adhesives consist of two components, one of which is a prepolymer formed by the reaction of an excess of epichlorohydrin with bisphenol-A, as follows [14]:

$$CH_2\text{—}CH\text{—}CH_2Cl + OH\text{—}\bigcirc\text{—}\underset{\underset{CH_3}{|}}{\overset{\overset{CH_3}{|}}{C}}\text{—}\bigcirc\text{—}OH \longrightarrow$$

(2.7.1)

$$CH_2\text{—}CHCH_2O\text{—}\bigcirc\text{—}\underset{\underset{CH_3}{|}}{\overset{\overset{CH_3}{|}}{C}}\text{—}\bigcirc\text{—}O\text{—}CH_2CH\text{—}CH_2$$

The diglycidyl ether of bisphenol-A is a liquid that is mixed with a polyether triamine:

$$CH_3CH_2\text{—}\underset{\underset{CH_2-[OCH_2CH(CH_3)]_z-NH_2}{|}}{\overset{\overset{CH_2-[OCH_2CH(CH_3)]_x-NH_2}{|}}{C}}\text{—}CH_2\text{—}[OCH_2CH(CH_3)]_y\text{—}NH_2$$

The curing reaction occurs at room temperature, and it normally takes around 4–5 h to set into a network.

Anaerobic adhesives are single-component adhesives that are normally multifunctional acrylates or methacrylates; for example, polyethylene glycol bismethacrylate:

$$CH_2{=}\underset{\underset{CH_3}{|}}{C}\text{—}\overset{\overset{O}{||}}{C}\text{—}O\text{—}[CH_2CH_2O]_n\text{—}\overset{\overset{O}{||}}{C}\text{—}\overset{\overset{CH_3}{|}}{C}{=}CH_2$$

This adhesive has two double bonds and is therefore tetrafunctional. Its curing reaction is known to be suppressed by oxygen of the air, but it can undergo redox reaction with metals. This property leads to its polymerization through the radical mechanism. As a result, it is used for locking threaded machine parts (e.g., lock-nuts, lock-screws, pipe fittings, and gaskets). Cyanoacrylates (a variant of the acrylates) are also room-temperature adhesives, but they polymerize through anionic mechanism. The initiation of the polymerization occurs through the surface, and the liquid material turns into a solid quite rapidly.

TABLE 2.4   Common Terminology Used in Paints Industry

Common names	Description	Remarks
Lacquer	Consists of a polymer solution with a suitable pigment. The solvent used is organic in nature, having a high vapor pressure.	The chosen polymer should form a tough film on drying and should adhere to the surface. Acrylic polymers are preferred because of their chemical stability.
Oil paints	A suspension in drying oils (e.g., linseed oil). Cross-linking of oil occurs by a reaction involving oxygen.	Sometimes, a catalyst such as cobalt naphthenate is used to accelerate curing.
Varnish	A solution of polymer–either natural or synthetic-in-drying oil. When cured, it gives a tough polymer film.	Ordinary spirit varnish is actually a lacquer in which shellac is dissolved in alcohol.
Enamel	A pigmented oil varnish.	It is similar to nature to oil paint. Sometimes, some soluble polymer is added to give a higher gloss to the dried film.
Latex paint	Obtained by emulsion polymerizing. A suitable monomer in water. The final material is a stable emulsion of polymer particles coalesce, giving a strong film with a gloss.	To give abrasion resistance to the film, sometimes inorganic fillers such as $CaCO_3$ are added. Because of their chemical stability, acrylic emulsions are preferred.

Paints are utilized mainly for covering open surfaces to protect them from corrosion and to impart good finish. They are further classified as lacquers, oil paints, etc.; their differences are highlighted in Table 2.4. The main property requirements for these are fast drying, adhesion to the surface, resistance to corrosion, and mechanical abrasion. Various paints available in industry are based mainly on (1) alkyd and polyester resins, (2) phenolic resins, (3) acrylic resins, and (4) polyurethanes, which we now discuss in brief.

Alkyd resins are polyesters derived from a suitable dibasic acid and a polyfunctional alcohol. Instead of using a dibasic acid, for which the polymerization is limited by equilibrium conversion, anhydrides (e.g., phthalic and maleic anhydrides) are preferred; among alcohols, glycerine and pentaerythritol are employed. Drying oils (e.g., pine oil, linolenic oil, linseed oil, soybean oil, etc.)

are esters of the respective acids with glycerine. For example, linolenic acid is R_1—COOH, where R_1 is

$$CH_3CH_2CH_2=CHCH_2CH=CHCH_2CH=CH(CH_2)_6-CH_3$$

and the linolenic oil is

```
        O
        ‖
CH₂—O—C—R₁
|       O
|       ‖
CH—O—C—R₁
|       O
|       ‖
CH₂—O—C—R₁
```

Evidently, the drying oil has several double bonds, which can give rise to cross-linking. At times, a hydroperoxide catalyst is added to promote curing of the drying oil.

Phenolic resins are obtained by polymerizing phenol with formaldehyde. When polymerized at low pH (i.e., acidic reaction medium), the resultant material is a straight-chain polymer, normally called novolac. However, under basic conditions, a higher-branched polymer called resole is formed. To cure novolac, a cross-linking agent, hexamethylenetetramine, is required, which has the following chemical formula:

```
            N
          /  |  \
 CH₂    CH₂    CH₂
 |        N
 |      /  \
 |   CH₂   CH₂
 |   /        \
 N              N
  \            /
       CH₂
```

During curing, ammonia and water are released. Because low-molecular-weight reaction products are formed, the film thickness must be small ($< 25\,\mu m$); otherwise, the film would develop pinholes or blisters. The curing of resole, on the other hand, does not require any additional curing agent. It is heat cured at about 150°C to 200°C and its network polymer is called resite. Curing at ambient conditions can be done in the presence of hydrochloric acid or phosphoric acid. The film of the polymer is generally stable to mineral acid and most of the organic solvents. It has good electrical insulation properties and is extremely useful for corrosion-resistant coatings.

Acrylic paints are normally prepared through the emulsion polymerization of a suitable acrylic monomer. In this process, the monomer (sparingly soluble in water) is dispersed in water and polymerized through the free-radical mechanism using a water-soluble initiator such as sodium persulfate. The main advantage of emulsion paint is its low viscosity, and after the water evaporates, the polymer particles coalesce to give a tough film on the surface. In several applications, it is desired to produce cross-linked film, for which the polymer must be a thermosetting acrylic resin. This can be done by introducing functional groups onto polymer chains by copolymerizing them with monomers having reactive functional groups. For example, acrylic acid and itaconic acids have carboxylic acid groups, vinyl pyridine has amine groups, monoallyl ethers of polyols have hydroxyl groups, and so forth.

Composites are materials that have two or more distinct constituent phases in order to improve mechanical properties such as strength, stiffness, toughness, and high-temperature performance [15]. Polymer composites are those materials that have a continuous polymer matrix with a reinforcement of glass, carbon, ceramic, hard polymeric polyaramid (commercially known as Kevlar) fibers, hard but brittle materials such as tungsten, chromium, and molybdenum, and so forth. These can be classified into particle-reinforced or fiber-reinforced composites, depending on whether the reinforcing material is in the form of particles or long woven fibers.

In polymer composites, the common reinforcing materials are glass particles or fibers; we will restrict our discussion to glass reinforcements only in this chapter. In our earlier discussion of fillers, we recognized that surface treatment is required in order to improve their compatibility. During the forming of glass, it is treated with γ-amino propyl ethoxy silane, which forms an organic coating to reduce the destructive effect of environmental forces, particularly moisture. We have already discussed that the glass surfaces have several $-OH$ groups that form covalent bonds with the silane compound. The dangling amine functional groups on the glass later react with the polymer matrix, giving greater compatibility with the glass and, hence, higher strength.

The cheapest glass-reinforcement material is E-glass, often used as a roving, or a collection of parallel continuous filaments. Among the polymer matrices, polyester and epoxy resins, which we discuss shortly, are commonly employed. An unsaturated polyester prepolymer is first prepared by reacting maleic acid with diethylene glycol:

$$COOH-CH=CH-COOH + OH-CH_2CH_2-OCH_2CH_2OH \longrightarrow$$

Maleic acid · · · Diethylene glycol

$$OH-\left[CH_2CH_2O-CH_2CH_2O\overset{O}{\overset{\|}{C}}-CH=CH-\overset{O}{\overset{\|}{C}}-O\right]_n-H$$

(2.7.2)

The polyester prepolymer is a solid and, for forming the composite matrix, it is dissolved in styrene, a small amount of multifunctional monomer divinyl benzene, and a free-radical peroxide initiator, benzoyl peroxide. The resultant polymer is a network, and the curing reaction is exothermic in nature. The final properties of the polyester matrix depend considerably on the starting acid glycols, the solvent monomer, and the relative amount of the cross-linking agent divinyl benzene. In this regard, it provides an unending opportunity to the polymer scientists and engineers to be innovative in the selection of composition and nature of reactants.

We have already discussed the chemistry of epoxide resins. The properties of the cured epoxy resin depend on the epoxy prepolymer as well as the curing agent used. Epoxy resin is definitely superior to polyester because it can adhere to a wide variety of fibers and has a higher chemical resistance. Polyimides and phenolic resins have also been used as matrix material. The former has higher service temperature (250–300°C), but during curing, it releases water, which must be removed to preserve its mechanical properties. Many thermoplastic polymers have also been used as matrix material for composites. They are sometimes preferred because they can be melted and shaped by the application of heat and can be recycled; however, they give lower strength compared to thermosetting resins.

Example 2.3: Fiberglass composites are prepared by coating unidirectional fiberglass with epoxy prepolymer and then heating until it forms a hard matrix. Present a simple stress analysis of this under loading in the direction of fibers.

Solution: Let us assume that there is perfect bonding between fiber and matrix with no slippage at the interface:

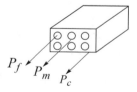

P_f: Force on fiber, cross-sectional area A_f
P_m : Force on matrix, cross-sectional area A_m
P_m : Force on composite, cross-sectional area A_c

Due to continuity, strains in the matrix (ε_m) and fibers (ε_f) must be equal. Therefore, forces shared by the matrix (P_m) and the fiber (P_f) are related to the stresses σ_f (in fibers) and σ_m (in the matrix) through the following relations:

$$P_f = \sigma_f A_f = E_f \varepsilon_f A_f$$
$$P_m = \sigma_m A_m = E_m \varepsilon_m A_m$$

where A_f (of fibers) and A_m (of the matrix) are cross-sectional areas. It has been assumed that both fibers (of modules E_f) and the matrix (of modulus E_m) behave elastically. if the composite as a whole has a cross-sectional area of A_c and a stress σ_c in it, then

$$P_c = \sigma_c A_c = \sigma_f A_f + \sigma_m A_m$$

2.8 ION-EXCHANGE RESINS

Ion-exchange materials are insoluble solid materials that carry exchangeable cations or anions or both [16–18]. Materials having exchangeable cations are cation exchangers, those having exchangeable anions are anion exchangers, and those having both are called amphoteric exchangers. These materials have a porous framework held together by lattice energy, with labile functional groups that can be exchanged. There are naturally available aluminosilicates with ion-exchange properties. Commonly called zeolites, these are relatively soft materials. In recent years, several synthetic zeolites (sometimes called molecular sieves) have been developed that are now available commercially.

Among all exchangers, the most important are organic ion exchangers, which are cross-linked polymeric gels. When the polymer matrix carries ions such as $-SO_3^{1-}$, $-COO^{1-}$, PO_3^{2-}, AsO_3^{2-}, and so forth, it is called a cation exchanger; when it has $-NH_4^{1+}$, $-NH_2^{2+}$, $-N^+-$, $-S^+$, and so forth, it is called an anion exchanger. The organic material most commonly in use is a copolymer gel of styrene and divinyl benzene (DVB), and the general-purpose resin contains about 8–12% of the latter. As the DVB content is reduced, the degree of cross-linking reduces, and at around 0.25% DVB, the polymeric gel swells strongly to give a soft, gelatinous material. As DVB is increased (at about 25%), the polymer swells negligibly and is a mechanically tough material.

The copolymer beads of ion-exchange resins are prepared by the suspension polymerization scheme [16,19]. In this technique, monomers styrene and divinyl benzene are mixed with a suitable initiator such as benzoyl peroxide and suspended in water under constant stirring. This produces small droplets that are prevented from coagulation by dissolving a suspension stabilizer (e.g., gelatin, polyvinyl alcohol, sodium oleate, magnesium silicate) in water. The particle size of the resin depends on several factors—in particular, the choice of the suspension stabilizer. Normally, a bead size of 0.1–0.5 mm is preferred. After the beads are formed, the polymer can be conveniently sulfonated by concentrated sulfuric acid or chlorosulfonic acid. The sulfonation starts from the resin surface, and the reaction front marches inward. It has been shown that this reaction introduces one group per benzene ring, and more than one group per ring only under extreme conditions. Sulfonation is an exothermic process—which means that if the resin particles are not swollen beforehand, they can crack under the stress generated by local heating and swelling caused by the substitution of the groups.

Let us now examine the physical nature of the resin beads formed during suspension copolymerization. Because of stirring and the suspension stabilizer, the organic phase consisting of monomers and initiator breaks into small droplets. Under heat, the initiator decomposes into radicals, which gives rise to polymerization as well as cross-linking in the medium of the monomer. As higher conversion is approached, monomers begin to diminish and the solvation reduces, ultimately vanishing with the monomers. With the reduction of solvation, polymer chains start collapsing, eventually forming a dense glasslike resin. When the cross-link density is small, these glasslike resins can once again swell with the addition of a good solvent. Such materials are called xerogels. For styrene–divinyl benzene, the *xerogel* beads are formed for DVB content less than 0.2%. As the DVB content is increased, the polymer chains, in addition to cross-linking, start getting entangled; if the gel collapses once, it does not swell again to the same level. Good solvents for the styrene–DVB system are toluene and diethyl benzene. If the suspension polymerization is carried out in their presence, the chains do not collapse. This gives high porosity to the beads, and the resultant product is called *macroporous resin*.

Solvents such as dodecane and amyl alcohol are known to mix with styrene and divinyl benzene in all proportions. However, if polymerization is carried out in the presence of these solvents, the polymer chains precipitate because of their limited solubility. Such a system is now subjected to suspension polymerization. The process of bead formation is complicated due to precipitation, and the polymer chains are highly entangled. Each resin particle has large pores filled with the solvent. Unlike macroporous particles, these are opaque and retain their size and shape even when the diluent is removed. These are called *macroreticular resins* and will absorb any solvent filling their voids.

From this discussion, it might appear as if styrene–divinyl benzene copolymer is the only accepted resin material. In fact, a wide range of materials have been used in the literature, among which are the networks formed by phenol and formaldehyde, acrylic or methacrylic acids with divinyl benzene, and cellulose. Ion-exchange cellulose is prepared by reacting chemical cellulose with glycidyl methacrylate using hydrogen peroxide, ferrous sulfate, and a thiourea dioxide system [20]. The grafted cellulose,

$$Cell-CH\overset{\displaystyle O}{\overset{\diagup\diagdown}{\underline{\hspace{1cm}}}}CH_2$$

is reacted with aqueous ammonia, with which amination, cross-linking, and a hydrolysis reaction occur, as follows:

Amination

$$Cell-\underset{\underset{O}{\diagdown \diagup}}{CH}-CH_2 + NH_3 \longrightarrow Cell-\underset{\underset{OH}{|}}{CH}-CH_2-NH_2 \qquad (2.8.1a)$$

Cross-linking

$$Cell-\underset{\underset{O}{\diagdown \diagup}}{CH}-CH_2 + Cell-\underset{\underset{OH}{|}}{CH}-CH_2-NH_2 \longrightarrow$$

$$\qquad\qquad (2.8.1b)$$

$$Cell-\underset{\underset{OH}{|}}{CH}-CH_2-NH-CH_2\underset{\underset{OH}{|}}{CH}-Cell$$

Hydrolysis

$$Cell-\underset{\underset{O}{\diagdown \diagup}}{CH}-CH_2 + H_2O \longrightarrow Cell-\underset{\underset{OH}{|}}{CH}-CH_2-OH \qquad (2.8.1c)$$

In several applications, it is desired to introduce some known functional groups into the ion-exchange resins. Introduction of a halogen group through chloromethyl styrene or acenaphthene, carboxylic acid through acrylic or methacrylic acid, and so forth have been reported in literature [19]. It can be seen that these functional groups could serve as convenient points either for polymer modification or for adding suitable polymer chains.

The classical application of ion-exchange resin has been in the treatment of water for boilers, for which the analysis of the column has now been standardized [18]. It is suggested that a packed bed of these resins first be prepared and the water to be processed pumped through it. Because ion resin particles are small, the resistance to the flow of water through the colunm is high. It would be desirable to add these particles into a vessel containing impure water, whereupon the former would absorb the impurities [21,22]. Because these particles are small, their final separation from water is difficult; to overcome this handling difficulty, the exchangeable groups are sometimes attached to magnetic particles such as iron oxide. These particles are trapped in polyvinyl alcohol cross-linked by dialdehyde (say, gluteraldehyde). These resin beads are mixed with the water to be purified and, after the exchange of ions has occurred, are collected by bringing an external magnet. The bead material is highly porous but has the disadvantage of its exchanged salt clogging the holes, thus giving reduced capacity to exchange. An alternative approach that has been taken is to first prepare a nonporous resin of polyvinyl alcohol cross-linked with a dialdehyde. A redox initiating system is subsequently used to prepare grafts of copolymer of acrylic acid and acrylamide. The resultant material, sometimes known as whisker resin (Fig. 2.5), is known to give excellent results.

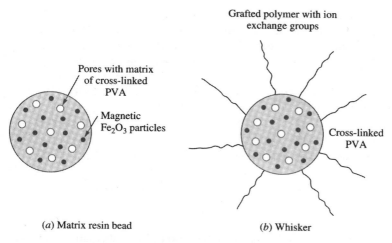

(a) Matrix resin bead   (b) Whisker

FIGURE 2.5 Two possible forms of ion-exchange resins used for water treatment.

We have already observed that cation exchange resins have bound ions like $\text{(P)}-SO_3^-$, $\text{(P)}-COO^-$, $\text{(P)}-SO_3^-$, $\text{(P)}-COO^-$, $\text{(P)}-PO_3^{2-}$, and $\text{(P)}-AsO_2^-$. These are present as salts with sodium counterion. If water has calcium chloride (hard water) as the impurity to be removed, then calcium ion is exchanged as follows:

$$\text{(P)}-SO_3^- Na^+ + CaCl_2(aq) \rightleftharpoons \left(\text{(P)}-SO_3^-\right)_2 Ca + 2NaCl \qquad (2.8.2)$$

It is thus seen that calcium is retained by exchanger resin. The separation, as shown, can be done for any other salt, as long as it reacts with an SO_3^- group and displaces sodium. The specificity of a resin toward a specific metal ion can be improved by altering the exchanging ions.

For the separation of metals, organic reagents that form a complex with them are used, ultimately precipitating from the solution [23–28]. These are called chelating agents. It is well known that the functional groups are responsible for their properties. Some of the chelating functional groups are given in Table 2.5. There are several techniques by which these could be affixed on polymer gel:

1. Polymerization of functional monomers
2. Grafting of second functional monomers on already prepared polymer, followed by second-stage polymerization
3. Immobilization of chelating organic reagents onto polymer
4. Polymerization of a nonfunctional monomer followed by modification

The preparation of chelating resins is still an area of active research, so it cannot be discussed in detail in the limited scope of this chapter. However, let us consider one example to illustrate the technique in which a hydroxamic acid group has been introduced into the polymer matrix. In terpolymerization of styrene, divinyl benzene, and acrylic acid, the final polymer is a network resin with carboxylic acid groups on the chain (represented by [P]−COOH) [16]. This polymer is subjected to the following modifications:

$$\begin{array}{ccc}
\boxed{P}-COOH & \xrightarrow{SOCl_2} & \boxed{P}-COCl & \xrightarrow{NH_2OH} & \boxed{P}-CO-NHOH \\
\downarrow^{CH_2N_2} & & & & \\
& \boxed{P}-COCH_3 & \xrightarrow{NH_2OH} & \boxed{P}-CO-NHOH &
\end{array}$$

(2.8.3)

This resin has been shown to be specific to Fe^{3+} ions. In an alternative technique [29], cross-linked polyacrylamide is prepared by maintaining a solution of acrylamide, N,N'-methylenebisacrylamide with ammonium persulfate at 25°C. A solution of hydroxylamine hydrochloride is added to the gel, and the pH of the reaction mass is raised to 12 by adding sodium hydroxide. The reaction is carried out for 24 h, and ammonia is released as the hydroxamic acid groups are formed on the matrix of the gel. The polyacrylamide gel $P-CO-NH_2$ is modified through the following mechanism:

$$\boxed{P}-\overset{\overset{O}{\|}}{C}-NH_2 + NH_2OH \rightleftharpoons \boxed{P}-\underset{\underset{^+NH_2OH}{|}}{\overset{\overset{O^-}{|}}{C}}-NH_2 \rightleftharpoons$$

$$\boxed{P}-\overset{\overset{O}{\|}}{C}-NHOH + NH_3 \quad (2.8.4a)$$

$$\boxed{P}-\overset{\overset{O}{\|}}{C}-NHOH + \xrightarrow{NaOH/H_2O} \boxed{P}-\overset{\overset{O^-}{|}}{C}-NH-O^-Na^+ \quad (2.8.4b)$$

In another interesting application of chelating ion-exchange resin, uranium from seawater can be recovered [30]. Uranium in seawater is present in a trace concentration of 2.8–3.3 mg/cm^3. A macroreticular acrylonitrile–divinyl benzene resin is prepared by suspension polymerization with toluene as a diluent and benzoyl peroxide as initiator. Within 4 h at 60°C, fine macroreticular beads are produced. A solution of sodium hydroxide in methanol is added to the solution of

TABLE 2.5 Some Chelating Functional Groups

Name	Formula	
β-Diketones	$\begin{array}{c}-C=CH-C-\\ \mid\quad\quad\parallel\\ OH\quad\quad O\end{array}$ enol	$\begin{array}{c}-C-CH_2-C-\\ \parallel\quad\quad\quad\parallel\\ O\quad\quad\quad O\end{array}$ keto
Dithiozone	$\begin{array}{c}\quad\quad\quad N=N-\\ -H-S-C\\ \quad\quad\quad\quad N-NH-\end{array}$ enol	$\begin{array}{c}\quad\quad N=N-\\ S=C\\ \quad\quad\quad NH-NH-\end{array}$ keto
Monoximes	$\begin{array}{c}-CH-C-\\ \mid\quad\quad\parallel\\ OH\quad N-OH\end{array}$	
Dioximes	$\begin{array}{c}-C-C-\\ \parallel\quad\parallel\\ OH-N\quad N-OH\end{array}$	
Nitrosophenol	$\begin{array}{c}-C-C-\\ \mid\quad\parallel\\ OH\quad N-OH\end{array}$	
Nitrosoaryl hydroxylamine	$\begin{array}{c}-N-N\\ \mid\quad\parallel\\ OH\quad O\end{array}$	
Hydroxamic acid	$\begin{array}{c}-C-N-\\ \parallel\quad\mid\\ O\quad OH\end{array}$	
Dithiocarbamates	$\begin{array}{c}\diagdown\quad\quad\quad S\\ N-C\diagup\\ \diagup\quad\quad\quad SH\end{array}$	
Amidoxime	$\begin{array}{c}-C-NH_2\\ \parallel\\ NOH\end{array}$	

hydroxyl amine hydrochloride in methanol. This is reacted with the gel and the resin, forming well-defined pores as follows:

$$
CH_2=\overset{\overset{\displaystyle CN}{\mid}}{CH} \;+\; \text{(styrene)} \;\longrightarrow\; \boxed{P}-CN \qquad (2.8.5a)
$$

Polymer resin

$$
\boxed{P}-CN \xrightarrow{NH_2OH} \boxed{P}-\underset{\underset{\displaystyle NH_2}{\mid}}{C}=N-OH \qquad (2.8.5b)
$$

The easily recognized oxime group shown forms a complex with the uranium salt present in seawater.

Most polymeric surfaces are hydrophobic in nature. In order to improve adhesion (adhesion with other surfaces, adhesion with paints or heparin for biomedical applications), this trait must be modified [31]. The most common method of doing this is by oxidation of the surface, which can be carried out by either corona discharge, flame treatment, plasma polymerization at the surface, grafting reactions, or blending the polymer with reactive surfactants that enrich at polymer interfaces. It has been shown that benzophenone under ultraviolet irradiation can abstract hydrogen from a polymer surface:

$$\text{(2.8.6a)}$$

$$\text{(2.8.6b)}$$

$$\text{(2.8.6c)}$$

The polymer radical generated at the surface can add on any monomer near the surface through the radical mechanism, as shown. Figure 2.6 presents the schematic diagram showing the setup needed for grafting. The chamber is maintained at around 60°C, at which benzophenone gels into the vapor phase

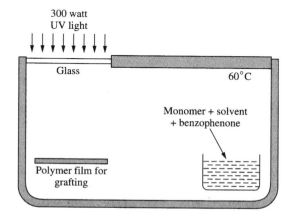

FIGURE 2.6   Grafting of benzophenone on the surface of polyethylene.

and interacts with the polymer surface. By this method, it is possible to obtain a thin layer of the grafted polymer on polyethylene.

Ion-exchange resins have also served as catalysts [32–35]. However, the resin gets completely deactivated at around 200°C, and the safe working temperature is around 125°C. Strongly acidic resins are prepared by sulfonation of polystyrene gels. Strongly basic resins are obtained by the amination of chloromethylated resins by tertiary amines such as trimethyl amine:

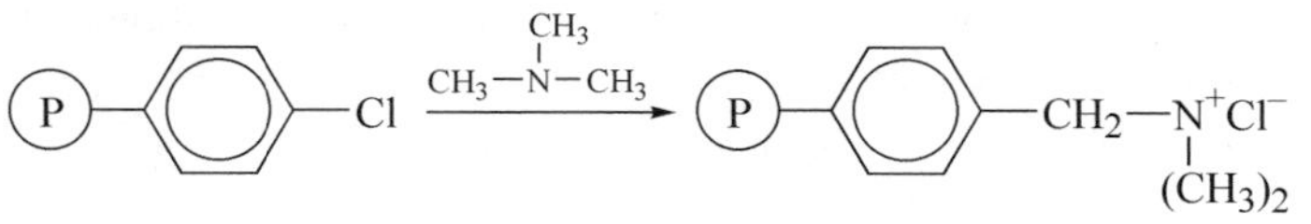

$$(2.8.7)$$

The literature is full of reactions carried out in the presence of polymer catalysts. A full discussion on this matter is beyond the scope of the present discussion. It might suffice here to state that virtually all of those organic reactions that have been carried out in the presence of homogeneous acids or bases are also catalyzed by polymer catalysts.

Example 2.4: Give the mechanism of esterification reaction with certain exchanger catalyst and mathematically model the overall heterogeneous reaction.

Solution: The mechanism of esterification of stearic acid with butanol can be written as

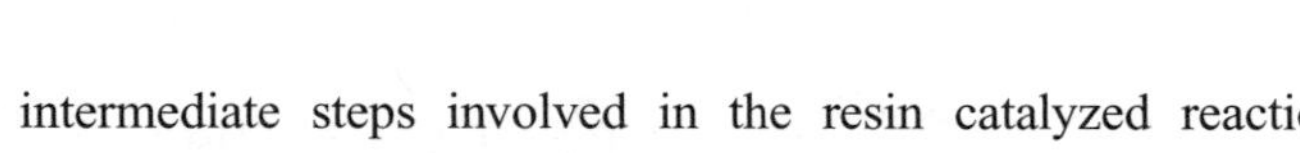

$$C_{17}H_{33}COOH + H^+ \rightleftharpoons C_{17}H_{33}COOH_2^+$$
$$C_{17}H_{33}COOH_2^+ + C_4H_9OH \rightleftharpoons C_{17}H_{33}COOC_4H_9 + H_3^+O$$
$$H_3^+O \rightleftharpoons H_2O + H^+$$

Different intermediate steps involved in the resin catalyzed reaction are as follows:

1. Diffusion of reactants across the liquid film adhering to the surface
2. Diffusion of reactants to the active sites of the resin
3. Adsorption of reactants to the active sites of the resin
4. Chemical reaction at the active sites of the resin
5. Desorption of the products

Let us say that the chemical reaction at the active sites is the rate-determining step, in which case the rate of reaction can be written as

$$x_f = k_s \left(C_{OS} C_{BS} - \frac{C_{WS} C_{ES}}{K_S} \right)$$

where C_{OS}, C_{BS}, C_{WS}, and C_{ES} are concentrations of stearic acid, butanol, water, and the ester at the active sites, respectively. The rest of the above intermediate steps must be at equilibrium and these can be related to bulk concentration C_{Ob}, C_{Bb}, C_{Wb}, and C_{Eb} as follows. Let C_L be the total molar concentration and the C_v of adsorption r_A can he written as

$$v_{A,O} = k_A C_{Ob} C_v - k'_A C_{Ds} = 0$$

where C_v is yet to be determined. Similarly for other components,

$$C_{Os} = K_O C_{Ob} C_v$$
$$C_{Bs} = K_B C_{Bb} C_v$$
$$C_{Wb} = K_w C_{Wb} C_v$$
$$C_{Eb} = K_E C_{Eb} C_v$$
$$C_v = C_L - C_v (K_O C_{Ob} + K_B C_{Bb} + K_W C_{Wb} + K_E C_{Eb})$$

From these equations, one can solve for C_{OS} in terms of bulk concentrations:

$$C_{OS} = \frac{C_L K_O C_{Ob}}{1 + K_O C_{Ob} + K_B C_{Bb} + K_W C_{Wb} + K_E C_{Eb}}$$

Example 2.5: A commercial styrene–divinyl benzene (SFDVB) anion exchanger has an exchange capacity of 1.69 mEq/wet gram having 42% moisture content. Relate this exchange capacity information to average member exchanging groups per repeat unit of the resin.

Solution: Anion-exchange resin is prepared by chloromethylating SFDVB resin using chloromethyl methyl ether (CMME) and then quarternizing it with trimethyl amine (TMA) as follows:

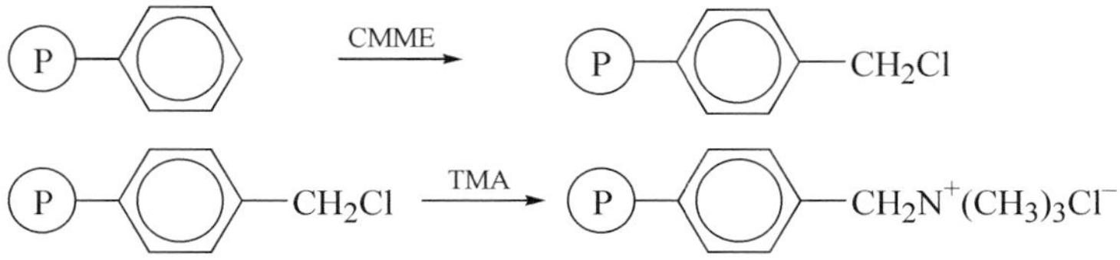

Here, the exchanging group is Cl^-. Let us say that on a given chain there are N_0 (this being a very large value for network) repeat units and all repeat units have one exchanging group. The molecular weight of the repeat unit is 184.5 and the molecular weight of the polymer is $184.5 N_0$. If it is assumed that all repeat units have one exchanging group, its exchange capacity in milliequivalents per dry gram of the resin would be $N_0/184.5 N_0$ or 5.42 mEq/dry g. The exchange capacity of commercial resin is 1.69 mEq/wet g or 2.91 $(= 1.69/0.58)$ mEq/dry g, which suggests that about every second repeat unit should be having one exchanging group.

Example 2.6: Polymer membranes are commonly used in barrier separation. Reverse osmosis (RO) and ultrafiltration (OF) both utilize the pressure gradient, causing separation of solutions (usually water as the solvent). Give a simple analysis of transport salt (species 1) and solvent (species 2) through membrane for both these cases.

Solution: In reverse osmosis, the membrane is nonporous in nature. A molecule is transported across it because a driving force (F) acts on it and the flux is proportional to it:

$$\text{Flux } (J) = [\text{proportionality } (A)][\text{driving force } (X)]$$

If t is the thickness of the membrane, then across it, there may exist a concentration (say ΔC), pressure (say ΔP), and electrical potential (say ΔE) difference. The average driving force (F_{av}), therefore, would be

$$F_{av} = \frac{RT}{l} \frac{\Delta C_i}{C_i} + \frac{Z_i \xi}{l} \Delta E + \frac{v_i}{l} \cdot \Delta p$$

where v_i is the specific volume of the solute. The first term arises because chemically potential $\mu_i = \mu_i^0 + RT \ln C_1$ and $\Delta \ln C_1 = 1/C_i$.

As opposed to this, in ultrafiltration, membranes are porous in nature and the pore diameter varies between 2 nm and 10 μm. The simplest representation of the membrane would be a set of parallel cylindrical pores, and based on Kozeny–Carman relationship, the flux could be written as

$$J = \frac{\varepsilon^3}{K \mu S^2 (1 - \varepsilon^2)^2} \frac{\Delta p}{\Delta x}$$

where ε is the volume fraction of pores, K is a constant, and S is the internal surface area.

2.9 CONCLUSION

In this chapter, we have examined polymers as useful materials, specifically focusing on the effect of the chemical structure on properties. Because of their high mechanical strength and easy moldability, polymers are used as structural materials, replacing metals in several applications. Because a polymer can be dissolved in a suitable solvent, it can be used as a paint. It also forms a network, for which it conveniently serves the purpose of polymer- supported reagents and a catalyst.

REFERENCES

1. Brydson, J. A., Plastics Materials, 4th ed., Butterworth, London, 1982.
2. Rosen, S. L., Fundamental Principles of Polymeric Materials, Wiley–Interscience, New York, 1982.
3. Kumar, A., and S. K. Gupta, Fundamentals of Polymer Science and Engineering, Tata McGraw-Hill, New Delhi, 1978.
4. Brandrup, J., and E. H. Immergut, Polymer Handbook, 3rd ed., Wiley–Interscience, New York, 1989.
5. Tsubokawa, N., F. Fujiki, and Y. Sone, Graft Polymerization of Vinyl Monomers onto Carbon Black by Use of the Redox System Consisting of Ceric Ions and Carbon Black, Carrying Alcoholic Hydroxyl Groups, J. Macromol. Sci. Chem., A-25, 1159–1171, 1988.
6. Goddart, P., J. L. Wertz, J. J. Biebuyck, and J. P. Mercier, Polyvinyl Alcohol–Copper II Complex: Characterization and Practical Applications, Polym. Eng. Sci., 29, 127–133, 1989.
7. Al-Malaika, S., Effects of Antioxidants and Stabilizers, in Comprehensive Polymer Science, G. Allen and J. C. Bevington (eds.), Pergamon, London, 1989, Vol. 6, pp. 539–578.
8. McNeill, I. C., Thermal Degradation, in Comprehensive Polymer Science, G. Allen and J. C. Bevington (eds.), Pergamon, London, 1989, Vol. 6, pp. 451–500.
9. Eisenberg, A., and M. King, Ion-Containing Polymers: Physical Properties and Structure, Academic Press, New York, 1977.
10. Utracki, L. A., and R. A. Weiss (eds.), Multiphase Polymers: Blends and Ionomers, ACS Symposium Series, Vol. 395, American Chemical Society, Washington, 1989.
11. Percec, V., and C. Pugh, Macromonomers, Oligomers, and Telechelic Polymers, in Comprehensive Polymer Science, G. Allen and J. C. Bevington (eds.), Pergamon, London, 1989, Vol. 6, pp. 281–358.
12. Melody, D. P., Advances in Room Temperature Curing Adhesives and Sealants—A Review, Br. Polym. J., 21, 175–179, 1989.
13. Fabris, H. J., Synthetic Polymeric Adhesives, in Comprehensive Polymer Science, G. Allen and J. C. Bevington (eds.), Pergamon, London, 1989, Vol. 7, pp. 131–178.
14. Morgan, R. J., Structure–Property Relations of Epoxies Used as Composite Matrices, Adv. Polym. Sci., 72, 1–44, 1985.
15. Agarwal, B. D., and L. J. Broutman, Analysis and Performance of Fiber Composites, Wiley, New York, 1980.
16. Heifferich, F., Ion Exchange, McGraw-Hill, New York, 1962.
17. Streat, M. (ed.), Ion Exchange for Industry, Ellis Horwood, Chichester, 1988.

18. Solt, G. S., A. W. Nowosielski, and P. Feron, Predicting the Performance of Ion Exchange Columns, Chem. Eng. Res. Des., 66, 524–530, 1988.
19. Balakrishnan, T., and W. T. Ford, Particle Size Control in Suspension Copolymerization of Styrene, Chloromethylstyrene and Divinylbenzene, J. Appl. Polym. Sci., 27, 133–138, 1982.
20. Khalil, M. I., A. Wally, A. Kanouch, and M. H. Abo-Shosha, Preparation of Ion Exchange Celluloses, J. Appl. Polym. Sci., 38, 313–322, 1989.
21. Bolto, B. A., Novel Water Treatment Processes which Utilize Polymers, J. Macromol. Sci. Chem., A14, 107–120, 1980.
22. Hodge, P., B. J. Hunt, and I. H. Shakhier, Preparation of Crosslinked Polymers Using Acenaphthylene and Chemical Modification of These Polymers, Polymer, 26, 1701–1707, 1985.
23. Marcus, Y., and A. S. Kertes, Ion Exchange and Solvent Extraction of Metal Complexes, Wiley–Interscience, London, 1969.
24. Dey, A. K., Separation of Heavy Metals, Pergamon, London, 1961.
25. Streat, M., and D. Naden (eds.), Ion Exchange and Sorption Processes, in Hydrometallurgy, Crit. Rep. Appl. Chem., Wiley, Chichester, 1987, Vol. 19.
26. Hodge, P., and D. C. Sherrington, Polymer Supported Reactions in Organic Synthesis, Wiley, Chichester, 1980.
27. Laszlo, P., Preparative Chemistry Using Supported Reagents, Academic Press, San Diego, CA, 1987.
28. Carraher, C. E., and J. A. Moore (eds.), Modification of Polymers, Plenum, New York, 1983.
29. Domb, A. J., E. G. Cravalho, and R. Langer, Synthesis of Poly(hydroxamic Acid) from Polyacrylamide, J. Polym. Sci. Polym. Chem. 26, 2623–2630, 1988.
30. Egawa, H., M. Makayama, T. Nonaka, and E. Sugihara, Recovery of Uranium from Sea Water: Influence of Crosslinking Reagent on Uranium Adsorption of Macroreticular Chelating Resin Containing Amidoxime Group, J. Appl. Polym. Sci., 33, 1993–2005, 1987.
31. Allmer, K., A. Hult, and B. Ramby, Surface Modification of Polymers: I. Vapour Phase Photografting with Acrylic Acid, J. Polym. Sci. A, 26, 2099–2111, 1988.
32. Nguyen, H. A., and E. Marechal, Synthesis of Reactive Oligomers and Their Use in Block Polycondensation, J. Macromol. Sci. Rev. Macromol. Phys., C28, 187–191, 1988.
33. Montheard, J. P., and M. Chatzopoulos, Chemical Transformations of Chloromethylated Polystyrene, JMS—Rev. Macromol. Chem. Phys., C28, 503–592, 1988.
34. Bradbury, J. H., and M. C. S. Perera, Advances in the Epoxidation of Unsaturated Polymers, Ind. Eng. Chem. Res., 27, 2196–2203, 1988.
35. Alexandratos, S. D., and D. W. Crick, Polymer Supported Reagents: Application to Separation Science, Ind. Eng. Chem. Res., 35, 635–644, 1996.

PROBLEMS

2.1. In Example 2.4, we have evaluated C_v and C_{OS} analytically. In the dual-site mechanism, the surface reaction between adsorbed A and adsorbed B is the rate-determining step. The expression for the rate has been derived to be

$$r = \frac{\bar{k} K_A K_B (C_{Ai} C_{Bi} - C_{Ri} C_{zi}/K)}{(1 + C_{Bi} K_B + C_{Ai} K_A + C_{Ri} K_R + C_{Si} K_S)^2}$$

In a study, oleic acid (109 g) was esterified at 100°C using butanol at three different concentrations (166 g for Expt. 1, 87 g for Expt. 2, and 31 g for Expt. 3). The X-8 cation-exchange resin (4 g) has the exchange capacity of 4.3 mEq/g and an average particle diameter of 0.48 mm. The dynamic analysis has yielded some of these constants as follows:

	Fractional conversion of oleic acid		
Time, min	Expt. 1	Expt. 2	Expt. 3
0.00	0.00	0.00	0.00
60.0	0.1419	0.1254	0.1063
120	0.2517	0.2396	0.2068
180	0.3541	0.3411	0.2989
240	0.4410	0.4108	0.3576
300	0.5149	0.4806	0.4081
360	0.5712	0.5399	0.4590
420	0.6271	0.5863	0.5392
480	0.6901	0.6202	0.5783
600	0.7406	0.6907	0.5787
∞	0.9129	0.8369	0.7212
K_A	8.08	13.76	21.91
K_B	22.58	14.78	0.46
K_R	12.39	12.03	11.49
K_S	12.39	12.03	11.49
$\bar{k}$	?	?	?

Plot the kinetic data and determine the initial slope. From these, evaluate the initial rate r_0 and determine the missing constants of the above model. Show that the model is not consistent and should be rejected.

2.2. The kinetic data of oleic acid esterification in Problem 2.1 is next evaluated against the single-site model in which the adsorption of B is controlling. The rate expression can be easily determined to be

$$r = \frac{k_B C_L (C_{Bi} - C_{Ri} C_{Si}/C_{Ai} K)}{1 + (C_{Ri} C_{Si} K_B / C_A K) + C_{Ri} K_R + C_{Si} K_S}$$

The fitting of the conversing versus time data has yielded the following:

	Expt. 1	Expt. 2	Expt. 3
K_A	24.1	20.16	21.62
K_R	0.2	0.1	0.3
K_S	?	?	?
$k_B C_L$	0.081	0.52	0.046

Find the missing constants using the initial rate information of Problem 2.1. Show whether the model is consistent or not consistent.

2.3. For a surface-reaction-controlling, single-site model, the rate of reaction can be derived as

$$r = \frac{\bar{k} K_A (C_{Bi} C_{Ai} - C_{Ri} C_{Si}/K)}{1 + C_{Ai} K_{Ai} + C_{Si} K_S + C_{Rl} K_R}$$

The fitting of time-conversion data yield the following constants:

	Expt. 1	Expt. 2	Expt. 3
K_A	21.56	21.06	21.73
K_R	9.63	8.74	8.17
K_S	9.63	8.74	8.17
k	?	?	?

Determine the missing rate constant and show that it could serve as a plausible model for the esterification of oleic acid.

2.4. The oxidative coupling of 2,6-dimethyl phenol (DMP) has been studied by Challa:

The catalyst was prepared by first synthesizing a copolymer of styrene and N-vinyl imidazole. The polymeric catalyst was prepared in situ by dissolving copolymer in toluene and adding $CuCl_2$–isopropanol solution. The catalyst activity is attributed to the following complex:

$$\begin{bmatrix} (PS-Im) & OH & (PS-Im) \\ \diagdown & \diagdown & \diagup \\ & Cu & Cu \\ \diagup & \diagdown & \diagup \diagdown \\ (PS-Im) & OH & (PS-Im) \end{bmatrix}^{+2}$$

The above oxidative coupling reaction has been explained by the following Michaelis–Menten-type mechanism.

$$E + DMP \underset{k_{-1}}{\overset{k_1}{\rightleftharpoons}} E \cdot DMP \overset{k_2}{\longrightarrow} E^* + PPO + DPQ$$

$$E^* + O_2 \overset{k_{reox}}{\longrightarrow} E + H_2O$$

Derive an expression for the rate of consumption of DMP.

2.5. The cellulose–polyglycidyl methacrylate $(Cell-\underset{\underset{O}{\diagdown\diagup}}{CH}-CH_2)$ copolymer was prepared by grafting glycidyl methacrylate on cellulose using the hydrogen peroxide–ferrous sulfate thiourea dioxide system as the initiator. The resultant copolymer is reacted with a mixture of ammonia and ethyl amine. Write down all possible reactions, including the one leading to cross-linking. Notice that the reactions are similar to curing of epoxy resin consisting of amination and hydrolysis reaction with water.

2.6. Polymer surface properties control wettability, adhesion, and friction, and, in some cases, electronic properties. Gas-phase chlorination of polyethylene surfaces is done just for this purpose, and the reaction can be followed using x-ray photoelectron spectroscopy (XPS). The XPS technique can identify various chemical species within 10–$70\,\mu m$ of the surface. In the chlorination of polyethylene, the species are $-CH_{2-}$, $-CHC-$, $-CCl_2-$, $-CH-CH-$, and $-CH-CX-$. Observe that the chlorination proceeds through a radical mechanism. The mechanism of polymerization, assuming that all reaction steps are reversible, can be represented by

Initiation

$$Cl_2 \underset{k_2}{\overset{k_1}{\rightleftharpoons}} 2Cl^{\bullet} \tag{1}$$

Propagation

$$-CH_2- + Cl^{\cdot} \underset{k_4}{\overset{k_3}{\rightleftharpoons}} HCl + -\dot{C}H- \tag{2}$$

$$-\dot{C}H- + Cl_2 \underset{k_6}{\overset{k_5}{\rightleftharpoons}} -CHCl- + Cl^{\cdot} \tag{3}$$

$$-CHCl- + Cl^{\cdot} \underset{k_8}{\overset{k_7}{\rightleftharpoons}} -\dot{C}Cl \longrightarrow + HCl \tag{4}$$

$$-\dot{C}Cl- + Cl_2 \underset{k_{10}}{\overset{k_9}{\rightleftharpoons}} -CCl- + Cl^{\cdot} \tag{5}$$

Termination

$$2-CH^{\cdot}- \underset{k_{12}}{\overset{k_{11}}{\rightleftharpoons}} \diagdown CH-CH \diagup \tag{6}$$

$$-CH^{\cdot}- + -\dot{C}Cl- \underset{k_{14}}{\overset{k_{13}}{\rightleftharpoons}} \diagdown CH-\underset{\underset{Cl}{|}}{C} \diagup \tag{7}$$

$$-CH^{\cdot}- + Cl^{\cdot} \underset{k_{16}}{\overset{k_{15}}{\rightleftharpoons}} -CHCl- \tag{8}$$

$$-\dot{C}Cl- + Cl \underset{k_{18}}{\overset{k_{17}}{\rightleftharpoons}} -CCl_2- \tag{9}$$

Assuming a thin surface layer as a batch reactor, write mole balance for each species.

2.7. Let us make the following simplifying assumptions regarding Problem 2.6.

1. All intermediate radical species have small but time-invariant concentrations.

2. Reactions involving $(-\dot{C}H-$ and $\dot{C}l)$ and $(-CHCl-$ and $Cl')$ are irreversible.

3. Neglect termination reaction [reactions (8) and (9)] between $(-CH-$ and $Cl')$ and $(-CCl-$ and $Cl')$.

4. Reaction (CH and Cl_2) is essentially irreversible $(k_s \gg k_6, k_9 \gg k_7)$.

5. The rate of formation of $-CCl-$ controls the $-CCl_2-$ formation.

Assuming that the thin layer of the polythene surface could be described by a batch reactor, find the concentration of $[-CH_2-]$, $[-CHCl-]$, and $[-CCl_2-]$ analytically as a function of time.

2.8. Oxazoline–polystyrene (commercially called OPS) is a copolymer of styrene and vinyl oxazoline:

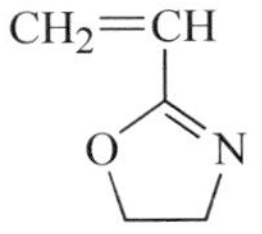

$$CH_2\!=\!CH$$

Solid polyethylene pieces are mixed with lupersol 130 (LPO) and maleic anhydride and reacted at 120°C in 1,2- dichlorobenzene (DCB) solvent. The resultant polymer is then mixed and extruded with OPS. Elaborate what precisely would happen in the extruder. Write down the mechanism of the reaction occurring in DCB. The initiator LPO is a solution of 2,5-di(t-butylperoxy)-2,5-dimethyl-3-hexyne with a half-life of about 12 min at 165°C.

2.9. Melt-mixed blends of polyvinyl chloride and carboxylated nitrile rubber cross-link by themselves. Such blends are found to have good oil resistance, high abrasion resistance, and high modulus with moderate tensile and tear strength. Write down all reactions occurring therein.

2.10. A mixture of methyl methacrylate, N-vinyl pyrrolidone [$CH_2\!=\!CH\!-N\!-\!(CH_2)_3C(O)$] divinyl benzene, ethyl acrylate and benzoyl peroxide has been polymerized between two glass plates. The resultant polymer can incorporate water within its matrix, and because of this property, it is sometimes called a *hydrogel*. In order to incorporate a drug into the hydrogel, the polymer was dipped in a solution of erythromycin estolate. The hydrogel is transparent initially but becomes opaque on incorporation of the drug. If this is now kept in physiological saline water (containing 0.9% NaCl and 0.08% NaHCO$_3$), the drug is leached out and the hydrogel begins to regain its transparency. The release of drug depends on the diffusion of erythromycin through the matrix. Because the diffusion coefficient of the drug depends on the matrix property, we can manipulate the rate of release of drug. Assuming that the entire polymer has uniform drug concentration, determine the rate of release of the drug. Then, solve this problem analytically.

2.11. Assume in Problem 2.10 that a quasi-steady-state exists and the concentration (C_d) profile is time invariant, given by the following diagram where x is the distance measured from the surface of rectangular hydrogel sheet and L is the value of x at the center. Find x as a function of time.

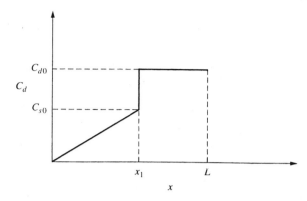

2.12. Polysulfone membranes are commonly used in ultrafiltration and are a copolymer of bisphenol-A and dichlorophenyl sulfone, having the molecular structure $[-OC_6H_4-SO_2-C_6H_4-OO-C_6H_4-C(CH_3)_2-C_6H_4-O-]_n$. In an experiment, five of these membranes having each a dry mass of 0.763 g were nitrated for different times. These were then aminated using hydrazine hydrate and the resultant material had $-NH_2^+Cl^-$ exchanging groups. The following results were reported:

Duration of modification (h)	0.5	1.0	2	3	4
Accurate exchange capacity (mEq/dry g)	0.810	1.420	1.723	1.741	1.681

Determine the average number of NH_2 groups per repeat unit (in fractions) as a function of time of nitration and plot your results.

2.13. Polystyrene pellets have been nitrated using similar procedure and then aminated:

$$\text{(PS)}-C_6H_4-NO_2 \longrightarrow \text{(PS)}-C_6H_4-NH_2$$
$$R_1 \qquad\qquad\qquad R_2$$

However, R_2 and R_1 resins were found to exchange only once and the one-time capacity of these were 1.63 mEq/wet g and moisture content of 40%. Explain why this is so and find the number of NH_2 in the R_2 resin per repeat unit.

In an alternate experiment, the R_2 resin is reacted with epichlorohydrin and the oxirane ring hydrolyzed using NH_3. Write down the chemical reactions and predict their capacity. The resultant resin could be regenerated repeatedly.

2.14. The R_2 resin of Problem 2.13 is reacted with dichloroethane and then quarternized using triethyl amine giving R_5 resin. The resultant resin (R_5) has an exchange capacity of 5 mEq/wet g with 69% moisture. Write down chemical reactions forming the R_5 resin and explain the reason for this sudden jump in exchange property.

2.15. The cross-linked polymethyl methacrylate–ethylene dimethacrylate (PMMA–EGDMA) copolymer resin (represented by $\text{(P)}-CH_2C(CH_3)-COOCH_3$) can be similarly nitrated using NO_x and this transformation can be written as

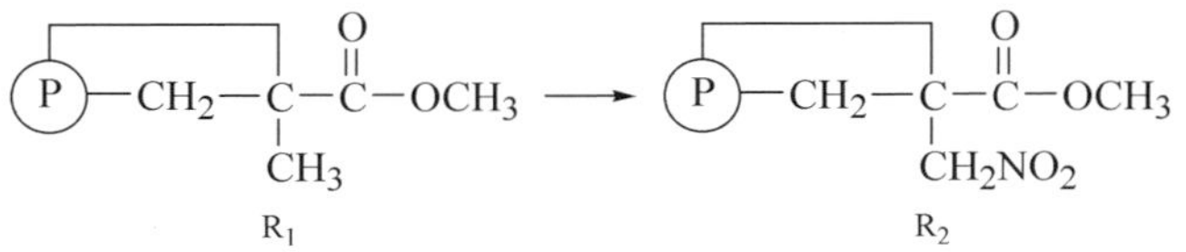

The R_2 resin can similarly be aminated and its exchanging groups are $-NH_3^+Cl^-$. It has an exchange capacity of 4.6 mEq/wet g with 79% moisture. Calculate the extent of nitration of the R_2 resin and suggest why this has become so highly hygroscopic.

2.16. Polymers can be degraded by thermal, oxidative, chemical, radiative, mechanical, and biological agents. In the photo-oxidative degradation of polyethylene (PE), radicals are first formed anywhere on the chain, which combine with oxygen to give a peroxy radical. This peroxy radical is converted to a carbonyl group. On further exposure to light, the following reactions occur:

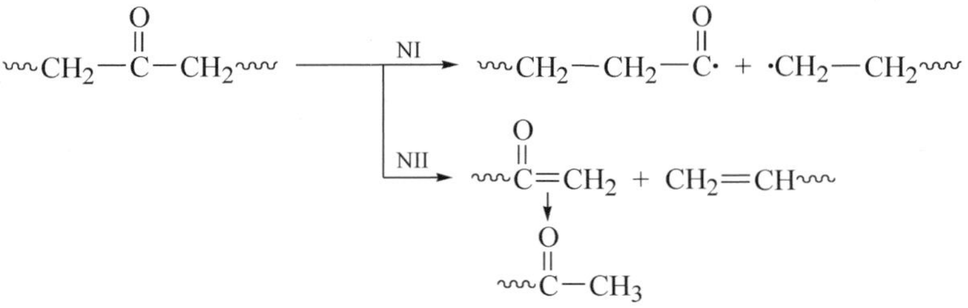

These are called Narish type I and II (NI and NII) degradations. Develop the reaction mechanism. Show how an ester group could be formed. You can see that the photo-oxidation embrittles the polymer and makes the polymer hydrophilic also.

2.17. The micro-organisms that degrade paraffins (straight-chain polymers) are mycobacteria, nocardia, candida, and pseudomonas. However, these do not react with branched polyethylene. In the biodegradation of polyethylene in the presence of ultraviolet (UV) light seems to proceed as in Problem 2.16, yielding carbonyl groups. As soon as this happens, these are attacked by micro-organisms that degrade the shorter segments of PE and form CO_2 and H_2O. Carbonyl groups are converted (unlike in Problem 2.16, where NI and NII reactions occur) to a carboxylic group, which is attacked by CoASH enzyme produced by the bacteria. This gives rise to a β-oxidation, giving a double bond that combines with water, ultimately being converted to another carbonyl group. In addition, the following additional reaction occurs:

$$\sim\sim CH_2-\overset{\overset{\displaystyle O}{||}}{C}-CH_2-\overset{\overset{\displaystyle O}{||}}{C}\sim SCoA + CoASH \longrightarrow$$

$$\sim\sim CH_2-\overset{\overset{\displaystyle O}{||}}{C}\sim SCoA + CH_3-\overset{\overset{\displaystyle O}{||}}{C}\sim SCoA$$

$$\text{citric acid cycle}$$

$$CO_2 + H_2O$$

Write the full mechanism.

2.18. Plastics are reinforced wth fillers to give higher strength and stiffness and reduced thermal expansion. Leading examples of reinforced polyesters are sheet-molding compounds (SMC) and bulk-molding compounds (BMC). Typical SMC consists of filler calcium carbonate (47.5%), chopped glass rovings (29%), fumerate or malleate polyester (13%), maturation agent magnesium oxide, catalyst t-butyl perbenzoate, low-profile additive (PVAc + styrene, 8%), internal mold-release agent zinc stearate (0.8%), and carrier resin (PVAc). Write the formation of maleate polyester (and fumerate polyester) with propylene glycol. Show how branching and lactone formation can occur.

2.19. Explain the need for various ingredients of SMC polyester described in Problem 2.18. The maturation agent participates in the polymerization; some believe it does so as follows:

$$\sim\sim COOH + MgO \longrightarrow \sim\sim COOMgOH$$
$$\sim\sim COOMgOH + \sim\sim COOH \longrightarrow \sim\sim COOMgOOC\sim\sim + H_2O$$

However, some scientists feel that there is only coordination complex formation, as follows:

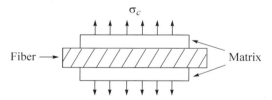

How would you propose to confirm which mechanism represents the true configuration?

2.20. We define v_f and v_m as volume fractions of fibers and the matrix, respectively, and assume that the fibers are laid parallel longitudinally. Rewrite σ_c in Example 2.3 in terms of these. Calculate the fraction of load carried by the fibers in composites of glass fibers and epoxy resin containing 16% fibers. $E_f = 72\ \text{GN/m}^2$ and $E_m = 3.6\ \text{GN/m}^2$.

2.21. Repeat the earlier problem for carbon fibers which has $E_f = 437\ \text{GN/m}^2$. In Example 2.3, we assumed only one kind of fiber material. Suppose there are n materials and a determine relation similar to that in Problem 2.20.

2.22. Consider a transverse loading of unidirectional loading composite as follows:

σ_c

Fiber $\longrightarrow$ Matrix

In this case, the elongation in the composite (δ_c) is the sum of the elongation in the fiber (δ_f) and the matrix (δ_m). Determine the transverse modulus E_c in terms of E_m and E_f.

2.23. Unidirectional composites have longitudinal (α_L) and transverse (α_L) coefficients of thermal coefficients given by

$$\alpha_L = \frac{\alpha_f E_f v_f + \alpha_m E_m v_m}{E_c}$$

$$\alpha_T = (1 + v_f)\alpha_f V_f + (1 + v_m)\alpha_m V_m - \alpha_L v_f v_f + v_m v_m)$$

where α_f and α_m are coefficients of thermal expansion for fiber and matrix, E_c is the elastic modulus of composite in the longitudinal direction, and v_f and v_m are the Poisson ratios of the fibers and the composites, respectively. Plot α_L and α_T as a function of v_f for the following properties:

$$\alpha_f = 0.5 \times 10^{-5}/°C \qquad\qquad \alpha_m = 6.0 \times 10^{-5}/°C$$
$$E_f = 70\,GN/m^2 \qquad\qquad\qquad E_m = 3.5\,GN/m^2$$
$$v_f = 0.20 \qquad\qquad\qquad\qquad v_m = 0.35$$

2.24. The thermal conductivities (longitudinal, k_4 and transverse k_T in W/m °C) composites are determined using the following relations:

$$k_L = V_f k_f + V_m k_m$$
$$k_T = k_m \frac{1 + \xi \eta V_f}{1 - \eta V_f}$$

where

$$\eta = \frac{k_f/k_m - 1}{k_f/k_m + \xi}$$
$$\log \xi = \sqrt{3} \log\left(\frac{a}{b}\right)$$

where k_f and k_m are transfer coefficients for the fiber and matrix, respectively. For $V_f = 0.6$, $k_m = 0.25$ W/m °C, and $k_f = 1.05$ W/m °C (for glass fibers), determine k_L and k_T. What would be their values, if the carbon fibers ($k_f = 12.5$ W/m °C) are used in place of glass fibers.

2.25. The process for reverse osmosis (used to get pure water from sea) can be schematically shown as

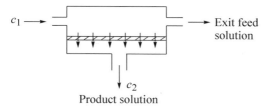

$c_1 \longrightarrow$ $\longrightarrow$ Exit feed solution

c_2

Product solution

Calculate the osmotic pressure (π, in atmospheres) of the NaCl solution with $C_1 = 10$ kg NaCl/m³ solution (density $\rho_1 = 1004$ kg solution/m³) using the following relation:

$$\pi = \frac{nRT}{V_m}$$

where n is kilogram mole of solute, V_m is the volume of pure solvent water (in m³), R is the gas constant (82.057×10^{-3} m³ atm/kg mol K), and T is the temperature (in °K). The density of pure water is given as 997.

2.26. The flux of water, N_w, and solute, N_s (in kg/m^2 S) are given by

$$N_w = A_w(\Delta P - \Delta \pi)$$
$$N_s = A_s(C_1 - C_2)$$

where $\Delta \pi = \pi_1 - \pi_2$, A_w and A_s are the solvent and solute permeability constants, respectively, and for the cellulose acetate membrane, these are 2.039×10^{-4} kg solvent/S m^2 atm and 3.896×10^{-7} m/sec. Calculate these fluxes if C_1 and C_2 are 10 kg NaCl/m^3 and 0.39 kg NaCl/m^3 and the applied pressure (ΔP) is 50 atm.

3

Step-Growth Polymerization

3.1 INTRODUCTION

As described in the previous chapters, the properties of polymeric materials depend considerably on their molecular-weight distribution (MWD). This in turn is completely determined by the mechanism of polymerization. There are various mechanisms by which polymer chains grow; this chapter focuses on one of them: step-growth polymerization.

Monomer molecules consisting of at least two functional groups can undergo step-growth polymerization. In order to keep mathematics tractable, this chapter will focus on polymerization of bifunctional monomers. The two reacting functional groups can either be on the same monomer molecule, as in amino caproic acid, $NH_2-(CH_2)_5-COOH$, or on two separate molecules, as in the reaction between ethylene glycol, $OH-(CH_2)_2-OH$, and adipic acid, $COOH-(CH_2)_4-COOH$. If they are located on the same monomer molecule, represented schematically as ARB, the concentrations of the functional groups remain equimolar throughout the course of the reaction, which can be schematically represented as follows [1–7]:

$$n(A-R-B) \rightarrow A-R-B[A-R-B]_{n-2}-A-R-B \qquad (3.1.1)$$

Here, R represents an alkyl or aryl group to which the two functional groups A and B are attached. In case the functional groups are located on two different monomers, $A-R-A$ and $B-R'-B$, an analysis similar to the one for ARB

103

polymerization can be conducted. As pointed out in Chapter 1, the overall reaction represented by Eq. (3.1.1) consists of several elementary reactions, which can be represented as follows:

$$P_m + P_n \underset{k'_{m+n}}{\overset{k_{m,n}}{\rightleftarrows}} P_{m+n} + W, \quad m, n = 1, 2, \ldots \tag{3.1.2}$$

where P_n represents $A-R-B-(A-R-B)_{n-2}-A-R-B$, P_m represents the $A-R-B-(A-R-B)_{m-2} A-R-B$ molecule, and W represents the condensation product. The forward and reverse rate constants $k_{m,n}$ and k'_{m+n} are, in general, chain-length dependent, as discussed in the following paragraphs.

If two small molecular species, A and B, react as

$$A + B \rightleftarrows \text{Products} \tag{3.1.3}$$

it is evident that the reaction will proceed only after a molecule of A diffuses close to a molecule of B from the bulk. Thus, the overall reaction between A and B consists of two consecutive steps: (1) the diffusion of molecules from the bulk of the mixture to within close proximity of each other and (2) the chemical interaction leading to product formation. This is represented schematically as

$$A + B \underset{}{\overset{\text{Diffusion}}{\rightleftarrows}} [A\ B] \underset{\text{reaction}}{\overset{\text{Chemical}}{\rightleftarrows}} AB(\text{Product}) \tag{3.1.4}$$

However, polymer molecules are very long and generally exist in a highly coiled state in the reaction mass with the functional groups situated at the chain ends. Therefore, in addition to the "bulk" molecular diffusion of P_m and P_n, the chain ends must diffuse close to each other (called segmental motion) before the chemical reaction can occur. This can be represented schematically as

$$P_m + P_n \underset{\text{diffusion}}{\overset{\text{Bulk}}{\rightleftarrows}} [P_m\ P_n] \underset{\text{diffusion}}{\overset{\text{Segmental}}{\rightleftarrows}} [P_m : P_n] \underset{\text{reaction}}{\overset{\text{Chemical}}{\rightleftarrows}} P_{m+n} \tag{3.1.5}$$

Because the bulk and segmental diffusion steps depend on the chain lengths of the two polymer molecules involved, the overall rate constants in Eq. (3.1.2) are, in general, a function of m and n. The exact nature of this dependence can be deduced from the following experiments.

3.2 ESTERIFICATION OF HOMOLOGOUS SERIES AND THE EQUAL REACTIVITY HYPOTHESIS [1,4,5]

The following esterification reactions of monobasic and dibasic acids of homologous series illustrate the effect of molecular size on the rate constants:

$$H-(CH_2)_n-COOH + C_2H_5OH \xrightarrow{\text{HCl}} H-(CH_2)_n-COOC_2H_5 + H_2O$$

$$(3.2.1a)$$

$$COOH-(CH_2)_n-COOH + C_2H_5OH \xrightarrow{\text{HCl}}$$
$$COOC_2H_5(CH_2)_n-COOH + H_2O$$
$$+ COOC_2H_5-(CH_2)_n-COOC_2H_5$$
$$+ H_2O + C_2H_5OH + HCl$$

$$(3.2.1b)$$

These reactions have been carried out in excess of ethanol with HCl catalyst, and the rates of reaction have been measured for various values of the chain length n. The reaction rate constants are evaluated using the following rate expression:

$$r_e = \frac{d[COOH]}{dt} = k_A[-COOH][H^+]$$

$$(3.2.2)$$

where r_e is the rate of esterification and [ ] represents molar concentrations. The concentration of ethanol does not enter into Eq. (3.2.2) because it is present in the reaction mass in large excess.

In Eq. (3.2.2) [−COOH] represents the total concentration of the carboxylic acid groups in the reaction mass at any time, whether present in the form of a monobasic of dibasic acid; this is usually determined by titration. [H$^+$] is the concentration of protons liberated by the hydrochloric acid. Use of the rate equation in the form shown in Eq. (3.2.2), together with experiments on monobasic and dibasic acids having different n, makes it possible to isolate the effect of the size of the molecule on k_A.

The rate constants for various values of n are tabulated in Table 3.1. Two important conclusions can be drawn from the experimental results:

1. The reactivity of larger molecules does not depend on the size of the molecule for $n > 8$. [2,7]
2. For larger molecules, the rate constant is independent of whether there are one, two, or more carboxylic acid groups per molecule.

Similar conclusions have also been obtained on the saponification of esters and etherification reactions [4,5]. If, in the chemical reaction step of Eq. (3.1.5),

TABLE 3.1　Rate Constants for the Esterification of Monobasic

Chain length (n)	$k_A \times 10^4$ (250°C)[a] (monobasic acid)	$k_A \times 10^4$ (250°C)[a] (dibasic acid)
1	22.1	—
2	15.3	6.0
3	7.5	8.7
4	7.45	8.5
5	7.42	7.8
6	—	7.3
8	7.5	—
9	7.47	—
Higher	7.6	—

[a]In liters per mole (of functional group) second.

the reactivity of a −COOH group with an −OH group is assumed to be independent of n, these observations imply that the rate of diffusion of large molecules is not affected by the value of n. However, we know intuitively that the larger the molecule, the slower is its rate of diffusion. Consequently, it is expected that, as n increases, the diffusional rate should decrease, implying that k_A must decrease with increasing n, a conclusion in apparent contradiction with the observed behavior.

As shown in Eq. (3.1.5), there are two types of diffusional mechanisms associated with the reaction of polymer molecules. Although the rate of bulk diffusion of two molecules decreases with n, the rate of the other step, called segmental diffusion, is independent of n. The independence of n is due to the fact that there is some flexibility of rotation around any covalent bond in a polymer molecule (see Chapter 1), and there is restricted motion of a small sequence of bonds near the ends, which constitutes segmental diffusion. This brings the functional groups of two neighboring molecules near each other, regardless of the chain length of the entire molecule. Thus, with increasing n, two polymer molecules diffuse slowly toward each other by bulk diffusion but stay together for a longer time (the two effects canceling out), during which, segmental diffusion may bring the functional groups together for possible reaction.

Based on the experimental results of Table 3.1, we can postulate a simple kinetic model for the study of step-growth polymerization in which all of the rate constants are assumed to be independent of chain length. This is referred to as the equal reactivity hypothesis. The following section shows that this assumption leads to a considerable simplification of the mathematical analysis. However, there are several systems in which this hypothesis does not hold accurately, and the analysis presented here must be accordingly modified [2,8–14].

3.3 KINETICS OF A–R–B POLYMERIZATION USING EQUAL REACTIVITY HYPOTHESIS [2]

A chemical reaction can occur only when the reacting functional groups collide with sufficient force that the activation energy for the reaction is available. The rate of reaction, r, can thus be written as proportional to the product of the collision frequency, ω_{mn}, between P_m and P_n and the probability of reaction, Z_{mn} (which accounts for the fraction of successful collisions), as follows:

$$R = \alpha \omega_{mn} Z_{mn} \tag{3.3.1}$$

where α is a constant of proportionality. According to the equal reactivity hypothesis, Z_{mn} is independent of m and n and is, say, equal to Z. If the functional groups of the two molecules P_m and P_n can react in s distinct ways, the probability of a reaction between P_m and P_n is given by sZ. The collision frequency ω_{mn} between two dissimilar molecules P_m and P_n in the forward step is proportional to $[P_m][P_n]$, whereas that for P_m and P_m is proportional to $-[P_m]^2/2$ (the factor of one-half has been used to avoid counting collisions twice). Thus, if k_p is the rate constant associated with the reaction between functional groups, then under the equal reactivity hypothesis, $k_{m,n}$, the rate constant associated with molecules P_m and P_n in the forward step, is given by

$$k_{m,n} = \begin{cases} \dfrac{r}{[P_m][P_n]} = sk_p, & m \neq n;\, m, n = 1, 2, \ldots \\[2ex] \dfrac{r}{[P_m]^2} = \dfrac{sk_p}{2}, & m = n;\, n = 1, 2, \ldots \end{cases} \tag{3.3.2}$$

For linear chains with functional groups A and B located at the chain ends, there are two distinct ways in which polymer chains can react, as shown in Figure 3.1. This fact implies the following for such cases:

$$k_{m,n} = \begin{cases} 2k_p & m \neq n;\, m, n = 1, 2 \ldots & (3.3.3a) \\ k_p & m = n;\, n = 1, 2, \ldots & (3.3.3b) \end{cases}$$

The various (distinct) elementary reactions in the forward step can now be written as follows:

$$P_m + P_n \xrightarrow{2k_p} P_{m+n} + W, \quad m \neq n;\, n = 1, 2, 3, \ldots \tag{3.3.4a}$$

$$P_m + P_n \xrightarrow{k_p} P_{2m} + W, \quad m = n;\, n = 1, 2, \ldots \tag{3.3.4b}$$

The reverse step in Eq. (3.1.2) involves a reaction between polymer molecule P_n and condensation product W; there is a bond scission in this process. It may be observed that P_n has $n - 1$ equivalent chemical bonds where the reaction can occur with equal likelihood. It is thus seen that if k'_p is the reactivity of a bond

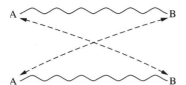

FIGURE 3.1   The two distinct ways in which two linear bifunctional chains can react.

with W, the reactivity of an oligomer P_n is $(n-1)k_p'$. The mole balance equations for various molecular species in a constant-density batch reactor can now be easily written. Species P_1 is depleted in the forward step when it reacts with any other molecule in the reaction mass. However, P_n $(n \geq 2)$ is formed in the forward step when a molecule $P_r(r < n)$ reacts with P_{n-r} and is depleted by reaction with any other molecule. In the reverse step P_n is depleted when any of its chemical bonds are reacted and it is formed whenever a P_q $(q > n)$ reacts at a specified bond position. For example, if we are focusing our attention on the formation of P_4, a molecule having chain length greater than 4, say, P_6, would lead to the formation of P_4 if W reacts at the second or fourth position of P_6. The mole balance relations are therefore given by the following:

$$\frac{d[P_1]}{dt} = -2k_p[P_1]\{[P_1] + [P_2] + \cdots\} + 2k_p'[W]\{[P_2] + [P_3] + \cdots\} \quad (3.3.5a)$$

$$\frac{d[P_n]}{dt} = k_p \sum_{r=1}^{n-1}[P_r][P_{n-r}] - 2k_p[P_n]\{[P_1] + [P_2] + \cdots\}$$

$$- k_p'[W](n-1)[P_n] + 2k_p'[W]\{[P_{n+1}] + [P_{n+2}] + \cdots\},$$

$$n = 2, 3, 4, \ldots \quad (3.3.5b)$$

There is no factor of two in the first term of Eq. (3.3.5b) because of the symmetry, as shown through an example of the formation of P_6. This occurs at a rate given by $(2k_p[P_1][P_5] + 2k_p[P_2][P_4] + 2k_p[P_3][P_3])$. The factor of the first two terms arises because $k_{m,n}$ is $2k_p$, whereas the factor of the last term, $2k_p[P_3]/2$, arises because of the fact that two molecules of P_3 are consumed simultaneously when P_3 reacts with P_3. The first term in Eq. (3.3.5b) for this is $k_p \sum_{r=1}^{5}[P_r][P_{n-r}]$, as shown.

If the concentration of all the reactive molecules in the batch reactor is defined as

$$\lambda_0 = \sum_{n=1}^{\infty} [P_n] \quad (3.3.6)$$

one can sum up the equations in Eq. (3.3.5) for all n to give the following:

$$\frac{d\lambda_0}{dt} = \frac{d[P_1]}{dt} + \frac{d[P_2]}{dt} + \frac{d[P_3]}{dt} + \cdots$$

$$= -2k_p\lambda_0^2 + k_p\lambda_0^2 - k_p'[W] \sum_{n=2}^{\infty}(n-1)[P_n] + 2k_p'[W]\sum_{n=1}^{\infty}\sum_{i=n+1}^{\infty}[P_i]$$

(3.3.7)

It is recognized that

$$\sum_{n=1}^{\infty}\sum_{i=n+1}^{\infty}[P_i] = [P_2] + [P_3] + [P_4] + \cdots + [P_3] + [P_4] + \cdots + [P_4] + \cdots$$

$$= [P_2] + 2[P_3] + 3[P_4] + \cdots$$

$$= \sum_{n=1}^{\infty}(n-1)[P_n]$$

(3.3.8)

Therefore, Eq. (3.3.7) can be written as

$$\frac{d\lambda_0}{dt} = -k_p\lambda_0^2 + k_p'[W]\sum_{n=1}^{\infty}(n-1)[P_n]$$

(3.3.9)

It may be observed that $\sum_{n=1}^{\infty}(n-1)[P_n]$ represents the total number of reacted bonds in the reaction mass. It is thus seen that the infinite set of elementary reactions in step-growth polymerization in Eq. (3.1.2) can be represented kinetically by the following equivalent and simplified equation:

$$-A + -B \underset{\longleftarrow}{\overset{k_p}{\rightleftharpoons}} -AB- + W$$

(3.3.10)

where $-AB-$ represents a reacted bond. The representation of an infinite series of elementary reactions by only one elementary reaction [Eq. (3.3.10)] involving functional groups is a direct consequence of the equal reactivity hypothesis. This leads to a considerable simplification of the mathematical analysis of polymerization reactors.

Example 3.1: Consider the ARB step-growth polymerization in which monomer P_1 reacts with P_n (for any n) with a different rate constant, as follows:

$$P_1 + P_n \overset{k_1}{\rightleftharpoons} P_{n+1} + W, \quad n = 1, 2, 3 \quad \text{(a)}$$

$$P_m + P_n \overset{k_p}{\rightleftharpoons} P_{n+m} + W, \quad m, n = 2, 3, \quad \text{(b)}$$

Derive the mole balance relations for the MWD of the polymer in a batch reactor.

Solution: s in Eq. (3.3.2) is 2 because the polymer chains are linear. Let us first consider the reaction of P_1. In the reactions of P_1 with P_1, similar molecules are involved, and the reactivity would be $2k_1/2$. However, for the reaction of P_1 with any other molecule, the reactivity would be $2k_1$. Therefore,

$$k_{1n} = \begin{cases} k_1 & \text{for } n = 1 \\ 2k_1 & \text{for } n = 1, 2, 3, \ldots \end{cases}$$

The other reactivities remain the same as in Eq. (3.3.3):

$$k_{mn} = \begin{cases} k_p & \text{for } m = n \\ 2k_p & \text{for } m \neq n; \, m, n = 2, 3, \ldots \end{cases}$$

The mole balance of species P_1 is made by observing that two molecules of P_1 are depleted whenever there is a reaction of P_1 with itself in the forward step, whereas only one molecule of P_1 disappears in a reaction with any other molecule. Similarly, in the reverse step, whenever W reacts at the chain ends, P_1 is formed:

$$
\begin{aligned}
\frac{d[P_1]}{dt} &= -(\text{Forward reaction of } P_1 \text{ with } P_1) \\
&\quad - (\text{Forward reaction of } P_1 \text{ with } P_2, P_3 \text{ etc.}) \\
&\quad + (\text{reverse reaction of W at chain ends to give } P_1) \\
&= -2k_1[P_1][P_1] - 2k_1[P_1]\{[P_2] + [P_3] + \cdots\} \\
&\quad + 2k'_p[W]\{[P_2] + [P_3] + \cdots\} \\
&= -2k_1[P_1]\lambda_0 + 2k'_p[W] \sum_{n=2}^{\infty} [P_n]
\end{aligned}
$$

The rate of formation of P_2 is $k_1[P_1]/2$, and P_2 is depleted whenever it reacts with any molecule in the forward step or its bond reacts with W in the reverse step:

$$
\begin{aligned}
\frac{d[P_2]}{dt} &= (\text{Forward reaction forming } P_2) - (\text{Forward reaction of } P_2 \text{ with } P_1) \\
&\quad - (\text{Forward reaction of } P_2 \text{ with } P_2 \text{ with } P_2, P_3, \text{ etc.}) \\
&\quad - (\text{Reverse reaction of bonds of } P_2 \text{ with W}) \\
&\quad + (\text{reverse reaction of W with } P_3, P_4, \text{ etc. to give } P_2) \\
&= k_1[P_1]^2 - 2k_1[P_1][P_2] - 2k_p[P_2]\{[P_2] + [P_3] + \cdots\} \\
&\quad - k'_p[W][P_2] + 2k'_p[W]\{[P_3] + [P_4] + \cdots\} \\
&= k_1[P_1]^2 - 2(k_1 - k_p)[P_2][P_1] - 2k_p[P_2]\lambda_0 - k'_p[W][P_2] \\
&\quad + 2k'_p[W] \sum_{i=3}^{\infty} [P_i]
\end{aligned}
$$

Similarly, the mole balance relaxation for species P_n is given by the following:

$$\frac{d[P_n]}{dt} = (\text{Forward reaction of } P_1, P_2, \text{ etc. with } P_{n-1}, P_{n-2}, \text{ etc.})$$
$$- (\text{Forward reaction of } P_n \text{ with } P_1)$$
$$- (\text{Forward reaction of } P_n \text{ with } P_2, P_3, \text{ etc.})$$
$$- (\text{Reverse reaction of } W \text{ with } n-1 \text{ bonds of } P_n)$$
$$+ (\text{Reverse reaction of } W \text{ with } P_{n+1}, P_{n+2}, \text{ etc. to give } P_n)$$

$$= +2k_1[P_1][P_{n-1}] + k_p \sum_{r=2}^{n-2} [P_r][P_{n-r}] - 2k_1[P_n][P_1]$$
$$- 2k_p[P_n]\{[P_2] + [P_3] + \cdots\} - k'_p[W](n-1)[P_n]$$
$$+ 2k'_p[W]\{[P_{n+1}] + [P_{n+2}] + \cdots\}$$
$$= 2(k_p - k_1)[P_n][P_1] - 2k_p[P_n]\lambda_0 + k_p \sum_{r=1}^{n-1} [P_r][P_{n-r}]$$
$$+ 2(k_1 - k_p)[P_{n-1}][P_1] - k'_p(n-1)[W][P_n] + 2k'_p[W] \sum_{r=n+1}^{\infty} [P_r]$$

The zeroth moment of the MWD can be easily found as follows:

$$\frac{d\lambda_0}{dt} = \frac{d[P_1]}{dt} + \frac{d[P_2]}{dt} + \cdots$$
$$= (k_1 - k_p)[P_1]^2 - k_p\lambda_0^2 - 2(k_1 - k_p)[P_1]\lambda_0 + k'_p[W] \sum_{n=2}^{\infty} (n-1)[P_n]$$

3.4 AVERAGE MOLECULAR WEIGHT IN STEP-GROWTH POLYMERIZATION OF ARB MONOMERS

Having modeled the rate of step-growth polymerization of ARB monomers, we can easily derive an expression for the average molecular weight of the polymer so formed. It is assumed that one starts with pure ARB monomer and that there are N_0 molecules present initially. After polymerization for time t, there would be fewer, say, N molecules, left in the reaction mass. This number N includes both unreacted monomer molecules, P_1, as well as dimers, trimers, tetramers, and so forth. In the computation of the average molecular weight for the system at time t, we could either consider only the dimers, trimers, and all other homologs to constitute molecules of the polymer, or, alternatively, include monomer molecules as well. Naturally, the results using the second approach would be lower than that obtained from the first one. In the following analysis, the monomer is included in the computation of the average molecular weight. This is not a drawback because, for practically important situations, the concentration of P_1 is usually negligible.

It may be observed that during polymerization the total number of repeat units at any time remains unchanged and is equal to the initial number of monomer molecules, N_0. These repeat units, however, are now disturbed over N polymer molecules at time t, so the average number of repeat units per molecule is equal to N_0/N. This is defined as the number-average chain length, μ_n (sometimes called the degree of polymerization), and is given by

$$\mu_n = \frac{N_0}{N} = \frac{[A]_0}{[A]} = \frac{[B]_0}{[B]} \tag{3.4.1}$$

where $[A]_0$, $[B]_0$, and $[A]$ and $[B]$ are the concentrations of the functional groups A and B at times $t = 0$ and $t = t$, respectively. It is convenient to work in terms of the (fractional) conversion of functional group A (or B), defined as

$$p \equiv \frac{[A]_0 - [A]}{[A]_0} = \frac{N_0 - N}{N_0} \tag{3.4.2}$$

which gives

$$\mu_n = \frac{1}{1 - p} \tag{3.4.3}$$

Integration of Eq. (3.3.9) can be carried out by observing that for every chemical bond formed, one molecule of condensation product, W, is formed. If $[P_1]_0$ and $[W]_0$ moles of monomer and condensation product are initially present in a batch reactor and W does not leave the reaction mass, then stoichiometry of polymerization gives

$$[W] + \lambda_0 = [W]_0 + [P_1]_0 \tag{3.4.4}$$

where $[W]$ and λ_0 are the concentrations of condensation product and polymer at any instant of time. We substitute $[W]$ from this equation into Eq. (3.3.9) to obtain

$$\frac{d\lambda_0}{dt} = -k_p\lambda_0^2 + k_p'\{[W]_0 + [P]_0 - \lambda_o\}\left\{\sum_{n=1}^{\infty} n[P_n] - \sum_{n=1}^{\infty} [P_n]\right\} \tag{3.4.5}$$

We further observe that $\sum_{n=1}^{\infty} n\,[P_n]$ is the first moment of the MWD and is equal to the total number of repeat units, which means that the first moment, λ_{10}, is time invariant. Therefore, Eq. (3.4.5) becomes

$$\frac{d\lambda_0}{dt} = -k_p\lambda_0^2 + k_p'\{[W]_0 + [P_1]_0 - \lambda_0\}(\lambda_{10} - \lambda_0) \tag{3.4.6}$$

which can be integrated as follows:

$$\int \frac{d\lambda_0}{-k_p\gamma_0^2 + k_p'\{[W]_0 + [P_1]_0 - \lambda_0\}(\lambda_{10} - \lambda_0)} = \int dt \tag{3.4.7}$$

where the denominator is a quadratic expression that can be easily factorized and then written in partial fractions. It can then be integrated to give

$$q = q_0 e^{-\delta t} \tag{3.4.8}$$

where

$$m_0 = k_p'([W]_0 + [P_1]_0)[P_1]_0 \tag{3.4.9a}$$

$$m_1 = k_p'([W]_0 + 2[P_1]_0) \tag{3.4.9b}$$

$$m_2 = (k_p - k_p') \tag{3.4.9c}$$

$$\delta = (m_1^2 + 4m_0 m_2)^{1/2} \tag{3.4.9d}$$

$$q = \frac{2m_2\lambda_0 + m_1 - \delta}{2m_2[P_1]_0 + m_1 + \delta} \tag{3.4.9e}$$

$$q_0 = \frac{2m_2[P_1]_0 + m_1 - \delta}{2m_2[P_1]_0 + m_1 + \delta} \tag{3.4.9f}$$

The number-average molecular weight can be easily obtained by multiplying μ_n by the molecular weight of ARB (because the molecular weight of W is usually small).

Example 3.2: Suppose N_{A0} moles of AR_1A monomer are reacted with N_{B0} moles of BR_2B monomer to form the polymer. Derive an expression for the average molecular weight of the polymer formed.

Solution: We first observe that there are $2N_{A0}$ moles of A functional groups and $2N_{B0}$ moles of B functional groups present at time $t = 0$. Whenever a functional group A (or B) reacts, the total number of molecules in the reaction mass decreases by 1. Let us, for the moment, assume that N_{B0} is greater than N_{A0}.

In order to determine the molecular weight, we needed to determine the total number of molecules at time t when the conversion of A functional groups is p_A:

$$p_A = \frac{2N_{A0} - 2N_A}{2N_{A0}}$$

The total number of moles of unreacted A functional groups at time t is equal to

$$2N_{A0}(1 - p_A)$$

The total number of moles of unreacted B functional groups at time t is equal to

$$(2N_{B0} - 2N_{A0}p_A)$$

At any time t molecules of (A A), (A B), and (B B) types are present, and all of these are equally likely to occur. If we know the total number of moles of unreacted A and B functional groups, the total number of moles of polymer is simply half of this. In other words, the total number of moles of polymer, N, at time t is equal to $\frac{1}{2}\{2N_{A0}(1 - p_A) + 2N_{B0} - 2N_{A0}p_A\}$. Similarly, the total number of moles of polymer initially, N_0, is equal to $N_{A0} + N_{B0}$:

$$\mu_n = \frac{N_0}{N} = \frac{N_{A0} + N_{B0}}{N_{A0} + N_{B0} - 2p_A N_{A0}} = \frac{1 + r}{1 + r - 2rp_A}$$

where $r = N_{A0}/N_{B0}$.

Observe that even when 100% conversion of A functional groups (i.e., $p_A = 1$) is achieved, the average chain length μ_n has a limiting value of $(1 + r)/(1 - r)$ instead of ∞, as predicted by Eq. (3.4.3). It is thus seen that an equimolar ratio (i.e., $r = 1$) is desirable for the formation of polymer of high molecular weight.

Example 3.3: The polyester PET, commonly used in the manufacture of synthetic fibers, is prepared through polymerization of bis-hydroxyethyl terephthalate (BHET). During polymerization, several side reactions occur, but if these are ignored, PET formation can be modeled by ARB kinetics as discussed. Experiments have shown that

$$k_p = 4.0 \times 10^4 \ \exp(-15 \times 10^3/1.98T) \ \text{L/mol min}$$

and the equilibrium constant K_p is

$$K_p = \frac{k_p'}{k_p} = 0.5 \ \text{(independent of temperature)}$$

For the initial monomer concentration, $[P_1]_0 = 4.58$ g mol/L, find the conversion, the average chain length, and the polydispersity index Q after 10 min of polymerization at 280°C and 200°C.

Solution: At 280°C

$$k_p = 4.0 \times 10^4 \exp\left(\frac{15,000}{1.98(273 + 280)}\right) = 4.49 \times 10^{-2} \text{L/mol min}$$

$$k_p' = 2.25 \times 10^{-2} \ \text{L/mol min}$$

Because $t = 0$, BHET does not have any W,

$$[W]_0 = 0, \qquad [P]_0 = [P_1]_0 = 4.58 \text{ g mol/L}$$
$$m_0 = k_p'[P]_0^2, \qquad m_1 = 2k_p'[P_1]_0$$
$$m_2 = k_p'\left(\frac{1}{K_p} - 1\right)$$

$$\delta = \left[4k_p'^2[P_1]_0^2 + 4k_p'^2\left[[P_1]_0^2\left(\frac{1}{K_p} - 1\right)\right]\right]^{1/2} = \frac{2k_p'[P_1]_0}{k_p^{1/2}} = 0.291$$

$$q_0 = \frac{2k_p'[P_1]_0(1/K_p - 1) + 2k_p'[P_1]_0 - 2k_p'[P_1]_0/K_p^{1/2}}{2k_p'[P_1]_0(1/K_p - 1) + 2k_p'[P_1]_0 - 2k_p'[P_1]_0/K_p^{1/2}}$$

$$= \frac{1/K_p - 1/K_p^{1/2}}{1/K_p + 1/K_p^{1/2}} = \frac{1 - K_p^{1/2}}{1 + K_p^{1/2}} = \frac{1 - 0.71}{1 + 0.71} = 0.172$$

$$q = \frac{2k_p'[P_1]_0(1/K_p - 1) + 2k_p'[P_1]_0 - 2k_p'[P_1]_0/K_p^{1/2}}{2k_p'[P_1]_0(1/K_p - 1) + 2k_p'[P_1]_0 - 2k_p'[P_1]_0/K_p^{1/2}}$$

$$= \frac{\lambda_0/[P_1]_0(1/K_p - 1) + (1 - 1/K_p^{1/2})}{1/K_p - 1/K_p^{1/2}}$$

$$= \frac{(2 - 1)\lambda_0/[P_1]_0 + (1 - 1.41)}{2 - 1.41}$$

$$= \frac{\lambda_0/[P_1]_0 - 0.41}{0.59}$$

After 10 min,

$$\frac{\lambda_0/[P_1]_0 - 0.41}{0.59} = 0.172 r^{-2.91}$$

$$\therefore \frac{\lambda_0}{[P_1]_0} = 0.41 + 0.172(0.59)(0.0545) = 0.416$$

Conversion $= 1 - \dfrac{\lambda^0}{[P_1]_0} = 0.584$

$$\mu_n = 2.40, \qquad Q = 1 + p = 1.584$$

At 200°C,

$$k_p = 4.43 \times 10^{-3}, \qquad k_p' = 2.22 \times 10^{-3}$$

$$\delta = 2(2.22 \times 10^{-3})\left(\frac{4.58}{0.71}\right) = 0.0287$$

$$\frac{\lambda_0/[P_1]_0 - 0.41}{0.59} = 0.172 \times e^{-0.287} = 0.7505$$

$$\frac{\lambda_0}{[P_1]_0} = 0.41 + 0.59(0.7505) = 0.853$$

Conversion $= 0.147$

$$\mu_n = \frac{1}{0.853} = 1.172, \qquad Q = 1 + 0.147 = 1.147$$

3.5 EQUILIBRIUM STEP-GROWTH POLYMERIZATION [15–18]

As in chemical reactions of small molecules, condensation polymerizations also have an equilibrium. In fact, in several cases (e.g., polyethylene terephthalate polymerization), equilibrium is attained at very low values of μ_n, and high vacuum must be applied to drive the reaction in the forward direction to get polymer of high enough molecular weight to be of commercial interest.

When the reaction mass attains equilibrium, the rates of formation of all polymeric species in Eq. (3.3.5) are all zero. In other words,

$$\frac{d[P_1]_e}{dt} = 0 = -2k_p[P_1]_e\lambda_{0e} + 2k_p'[W]_e \sum_{i=2}^{\infty} [P_i]_e$$

$$\frac{d[P_n]_e}{dt} = 0 = -2k_p[P_n]_e\lambda_{0e} + 2k_p' \sum_{r=1}^{n-1} [P_r]_e[P_{n-r}]_e$$

$$- k_p'[W]_e(n-1)[P_n]_e + 2k_p'[W]_e \sum_{i-n+1}^{\infty} [P_i]_e, \quad n = 2, 3, \dots$$

$$(3.5.1)$$

We want to find the molecular-weight distribution satisfying Eq. (3.5.1). Let us assume that it is given by

$$[P_n]_e = xy^{n-1} \tag{3.5.2}$$

where x and y are some parameters that are not dependent on chain length n. We have already observed that the first moment λ_1 is time invariant and is the same as the initial value λ_{10}. Therefore, Eq. (3.5.2) must satisfy the following relation:

$$\lambda_{10} = \sum n[P_n] = x(1 + 2y + 3y^2 + \cdots) = \frac{x}{(1-y)^2} \tag{3.5.3}$$

At equilibrium, the total moles of polymer, λ_{0e}, can be obtained by equating $(d\lambda_{0e}/dt)$ equal to zero in Eq. (3.3.9):

$$-k_p \lambda_{0e}^2 + k'_p[W]_e \sum_{n=1}^{\infty} (n-1)[P_n] = -k_p \lambda_{0e}^2 + k'_p[W]_e (\lambda_{10} - \lambda_{0e}) = 0 \tag{3.5.4}$$

This is a quadratic equation and can be easily solved. In addition, we can also find λ_{0e} from the assumed form of the MWD in Eq. (3.5.2) as follows:

$$\lambda_{0e} = x(1 + y + y^2 + \cdots) = \frac{x}{1-y} \tag{3.5.5}$$

Between Eqs. (3.5.3) and (3.5.5), we get

$$\lambda_{0e} = \lambda_{10}(1-y) \tag{3.5.6}$$

and the MWD in Eq. (3.5.2) is given by

$$[P_n]_e = \lambda_{10} \left(\frac{\lambda_{0e}}{\lambda_{10}} \right)^2 \left(1 - \frac{\lambda_{0e}}{\lambda_{10}} \right)^{n-1} \tag{3.5.7}$$

Now, we show that Eq. (3.5.1) is identically satisfied by Eq. (3.5.2) as follows:

$$\sum_{r=1}^{n-1}[P_r][P_{n-r}]_e = \sum_{r=1}^{n-1} x^2 y^{r-1} y^{n-r-1}$$
$$= x^2 y^{n-2}(n-1) \tag{3.5.8}$$

$$\sum_{i=n+1}^{\infty}[P_i]_e = \sum_{n=1}^{\infty}[P_i]_e - \sum_{n=1}^{n}[P_i]_e - \lambda_{0e} - x \sum_{i=1}^{n} y^{i-1}$$
$$= \frac{1}{1-y} - \frac{x(1-y^n)}{1-y} = \frac{xy^n}{1-y} \tag{3.5.9}$$

On substituting these in Eq. (3.5.1), we have the following:

$$- 2k_p[P_n]_e\lambda_0 + k_p \sum_{r=1}^{n-1} [P_r]_e[P_{n-r}]_e - k_p'[W]_e(n-1)[P_n]_e$$

$$+ 2k_p'[W]_e \sum_{i=n+1}^{\infty} [P_i]_e$$

$$= -2k_p \frac{x^2}{1-y}y^{n-1} + k_px^2(n-1)y^{n-2}$$

$$+ 2k_p'[W]\frac{xy^n}{1-y} - k_p'[W](n-1)y^{n-1}$$

$$= 2(-k_px + k_p'W_ey)xy^{n-1} + (n-1)xy^{n-2}\{-k_px + k_p'W_ey\}$$

$$(3.5.10)$$

However, from Eq. (3.5.4), we have

$$-k_p \frac{x^2}{(1-y)^2} + k_p'\left(\frac{x}{(1-y)^2} - \frac{x}{1-y}\right) = 0 \qquad (3.5.11)$$

or

$$-k_px + k_p'[W]_ey = 0$$

which means that Eq. (3.5.10) is identically satisfied by the assumed equilibrium MWD in Eq. (3.5.2).

3.6 MOLECULAR-WEIGHT DISTRIBUTION IN STEP-GROWTH POLYMERIZATION

Let us consider step-growth polymerization in a batch reactor having feed of the following composition. At $t = 0$,

$$[P_n] = [P_n]_0, \quad n = 1, 2, \ldots \qquad (3.6.1)$$

One of the ways to solve such problems is to utilize the technique of the generating function, which is described in Appendix 3.1. With the help of the generating function $G(s, t)$, the mole balance relations in Eq. (3.3.5) are combined into one partial differential equation that has a numerical solution only. However, if the feed to the batch reactor is a pure monomer or has a distribution given by Eq. (3.5.7), it is possible to obtain an analytical solution.

Let us assume that the feed to the batch reactor is a pure monomer, which means that the initial condition is given at $t = 0$ by the following:

$$[P_1] = [P_1]_0 \qquad (3.6.2a)$$
$$[P_n]_0 = 0, \quad n = 2, 3, \ldots \qquad (3.6.2b)$$

We guess the form of the MWD as

$$[P_n] = [P_1]_0(1-p)^2 p^{n-1} \tag{3.6.3}$$

where p is the conversion of the functional groups defined in Eq. (3.4.2). We further observe [steps of the derivation are identical to those in Eqs. (3.5.5) and (3.5.6)] the following:

$$\lambda_{10} = [P_1]_0 \tag{3.6.4a}$$
$$\lambda_0 = [P_1]_0(1-p) \tag{3.6.4b}$$

On substituting these in Eq. (3.3.5), we find

$$\frac{d\lambda_0}{dt} = -k_p \lambda_0^2 + k_p'[W](\lambda_{10} - \lambda_0) \tag{3.6.5}$$

which is the same as Eq. (3.3.9), indicating that the assumed form of the MWD is correct.

The molecular-weight distribution of ARB polymerization was originally derived by Flory using statistical arguments and is presented here for its historical significance [17]. A polymer molecule of chain length n has $n-1$ reacted A (or B) groups and one unreacted A (or B). Therefore, the probability of obtaining a sequence of $n-1$ reacted and one unreacted A group in a polymer molecule of size n would be $p^{n-1}(1-p)$ and the number of molecules of size n, N_n, would be given by the product of this probability and the total number of molecules present in the reaction mass at that time; that is,

$$N_n = Np^{n-1}(1-p) \tag{3.6.6}$$

Using Eq. (3.4.3), we obtain

$$\frac{N_n}{N_0} = (1-p)^2 p^{n-1} \tag{3.6.7}$$

which is identical to $[P_n]/[P_1]_0$, given by Eq. (3.6.3).

If M_1 is the molecular weight of the monomeric repeat unit, the weight fraction, W_n, of a molecule of size n would be given by

$$W_n = \frac{(nM_1)N_n}{M_1 N_0} = \frac{nN_n}{N_0} \tag{3.6.8}$$

and using Eq. (3.6.7),

$$W_n = n(1-p)^2 p^{n-1} \tag{3.6.9}$$

The theoretical number and weight fraction distributions have been plotted in Figures 3.2 and 3.3, respectively, for several values of the conversion p. This is sometimes called *Flory's distribution*. It is observed from Figure 3.3 that, as time progresses, the conversion p increases and the molecular-weight distribution not

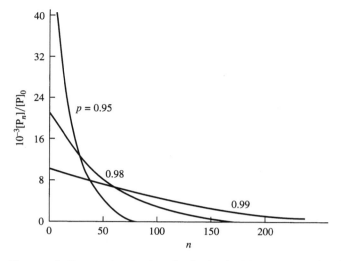

FIGURE 3.2 Number fraction distribution in ARB step-growth polymerization in batch reactors using pure monomer feed. [Reprinted from P. J. Flory, Chem. Rev., 39, vol. 137 (1946) with permission of American Chemical Society.]

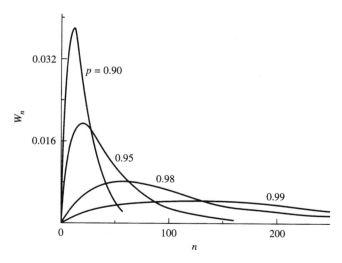

FIGURE 3.3 Weight fraction distribution in ARB step-growth polymerization in batch reactors using pure monomer feed. [Reprinted from P. J. Flory, Chem. Rev., 39, vol. 137 (1946) with permission of American Chemical Society.]

only shifts to higher and higher molecular weight but also broadens out. Figures 3.2 and 3.3 dictate that even though the *concentrations* of the low-molecular-weight homologs P_1, P_2, and so forth are always the highest, their weight fractions decrease significantly as p changes from 0.95 to 0.99.

It may be re-emphasized that in deriving Eq. (3.6.3) it is essential to assume that the feed to the batch reactor is a pure monomer. If higher homologs are present in the feed, as would be encountered in any intermediate reactor in a sequence of batch reactors, the molecular-weight distribution would be different and the polydispersity index (PDI) of the polymer formed would not necessarily be restricted to the limiting value of 2, as shown in Appendix 3.1. As a matter of fact, one of the practical methods of achieving a PDI of more than 2 is to partially recycle a portion of the product stream, as shown in Figure 3.4 [19–22]. Polymerization is carried out in a tubular reactor, and a fraction F of the product is mixed with the monomer feed. The mole balance equations for tubular reactors under suitable variable transformations become identical to those for batch reactors [23]; these must be solved simultaneously, along with mole balance equations for the mixer. The solution of polymerization with mixing is involved and has, therefore, been omitted in this book.

Example 3.4: Prove the following summations:

$$S_1 = \sum_{n=2}^{\infty} n^k \sum_{n=1}^{n-1} P_m P_n = \sum_{n=1}^{\infty} \sum_{n=1}^{\infty} (m+n) P_n P_n \tag{1}$$

$$S_2 = \sum_{n=1}^{\infty} n^k \sum_{m=n+1}^{\infty} P_m = \tag{2}$$

$$S_2 = \sum_{n=2}^{\infty} \sum_{J=1}^{n-1} \sum_{M=N-J+1}^{\infty} P_J P_M = \tag{3}$$

Solution: The way of proving these identities is to expand the left-hand side term-by-term and rearranging. For example,

$$S_1 = \sum_{n=2}^{\infty} n^k \sum_{n=1}^{n-1} P_m P_n = 2^k (P_2 P_1) + 3^k (P_3 P_1 + P_2 P_2 + P_1 P_3) + 4^k (\cdots) \cdots$$

$$= P_1 (2^k P_2 + 3^k P_3 + \cdots) + P_2 (3^k P_3 + \cdots) + \cdots$$

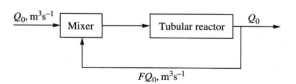

$Q_0, m^3 s^{-1}$ → Mixer → Tubular reactor → Q_0

$FQ_0, m^3 s^{-1}$

FIGURE 3.4 Use of recycle and a mixture M to obtain PDIs of values more than 2.

The general term of this series is $(m + n)^k P_m P_n$ and, therefore, the sum S_1 is given by

$$S_1 = \sum_{n=2}^{\infty} \sum_{n=1}^{\infty} (m + n)^k P_m P_n$$

Suppose we are interested in the special case of $k = 2$, then

$$S_1 = \sum_{n=2}^{\infty} \sum_{n=1}^{\infty} (m + n)^2 P_m P_n$$

$$= \sum_{n=2}^{\infty} \sum_{n=1}^{\infty} (m^2 + n^2 + 2mn) P_m P_n$$

$$= \sum_{n=1}^{\infty} (m^2 P_m \lambda_0 + 2n\lambda_1 P_m + Q P_m \gamma_0^2) = 2\lambda_0 \lambda_2 + 2\lambda_1^2$$

In the second summation, we have

$$S_2 = \sum_{n=1}^{\infty} n^k \sum_{m=n+1}^{\infty} P_m = 1^k (P_2 + P_3 + P_4 + \cdots) + 2^k$$

$$\times (P_3 + P_4 + P_5 + \cdots) + 3^k (P_4 + P_5 + \cdots)$$

$$= [P_2](1^k) + [P_3](1^k + 2^k) + \cdots$$

$$= \sum_{n=2}^{\infty} \left(\sum_{m=1}^{n-1} m^k \right) [P_n]$$

Let us say that $k = 2$, in which case

$$\sum_{m=1}^{n-1} n^k = \frac{(n-1)n(2n-1)}{6} = \frac{(2h^3 - 3n^2 + n)}{6}$$

Therefore

$$S_2 = \sum_{n=2}^{\infty} \frac{2n^3 - 3n^2 + n}{6} [P_n] = \frac{2\lambda_3 - 3\lambda_2 + \lambda_1}{6}$$

Summation S_3 is the kind which appears in redistribution reaction

$$S_3 = \sum_{n=2}^{\infty} n^k \sum_{J=1}^{n-1} \sum_{M=N-J+1}^{\infty} [P_J][P_M] = 2^k[P_1]([P] + [P_3] + [P_4] + \cdots)$$
$$+ 3^k P_1(P_3 + P_4 + \cdots) + 3^k[P_2](P_2 + P_3 + P_4+) + \cdots$$
$$= [P_1]\{[P_2] + [P_3](2^k + 3^k) + P_4(2^k + 3^k + 4^k) + \cdots\} + \cdots$$
$$+ P_n\{P_2(n+1)^k) + P_3\{(n+1)^2 + (n+2)^2\} + \cdots\}$$
$$= \sum_{n=1}^{\infty} P_n \sum_{j=2}^{\infty} \left(\sum_{i=1}^{j-1}(n+i)kP_j\right)$$

For example, for $k = 2$, S_3 is given by

$$S_3 = \sum_{n=1}^{\infty} P_n \sum_{j=2}^{\infty} \left(\sum_{i=1}^{j-1}(n+i)^2\right)P_j$$
$$= \sum_{n=1}^{\infty} P_n \sum_{j=1}^{\infty} \left(\sum_{i=1}^{j-1}(n^2 + 2ni + i^2)\right)P_j$$
$$= \sum_{n=1}^{\infty} P_n \sum_{j=1}^{\infty} \left(n^2(j-1) + 2n\frac{(j-1)j}{2} + \frac{(j-1)j(2j-1)}{6}\right)P_j$$
$$= \sum_{n=1}^{\infty} P_n\left\{n^2(\lambda_1 - \lambda_0) + 2n\frac{\lambda_2 - \lambda_1}{2} + \frac{(2\lambda_3 - 3\lambda_2 + 1)}{6}\right\}$$
$$= (\lambda_1 - \lambda_0)\lambda_2 + (\lambda_2 - \lambda_1)\lambda_1 + \frac{\lambda_0}{6}(2\lambda_3 - 3\lambda_2 + 1)$$

Example 3.5: Consider the polymerization of AA + BC monomers where both B and C react with A at different rates. Determine the number-average molecular weight of the polymer.

Solution: Let us say that at time $t = 0$, the concentrations of A, B, and C are $[A]_0$, $[B]_0$, and $[C]_0$, respectively, such that the total number of molecules, N_0, per unit volume is

$$N_0 = \frac{[A]_0}{2} + [B]_0$$

and

$$[B]_0 = [C]_0$$

However, as time progresses, these concentrations become different due to different reactivities, and the number average molecular weight cannot be obtained from pure kinetic analysis of functional groups. The analysis presented is strictly true for batch reactors where probabilities can be equated to conversions.

Let us define probabilities p_A, p_B, and p_C for finding reacted A, B, and C functional groups, respectively, at time t. These can be taken as equal conversions as follows:

$$p_A = \frac{[A]_0 - [A]}{[A]_0}$$

$$p_B = \frac{[B]_0 - [B]}{[B]_0}$$

$$p_C = \frac{[C]_0 - [C]}{[C]_0}$$

Because A can react with B and C functional groups, by stoichiometry, one has

$$p_A = \frac{[B]_0}{[A]_0}(p_B + p_C)$$

and in terms of N_0, one has

$$[A]_0 = \frac{2N_0}{1 + 2p_A/(p_B + p_C)}$$

$$[B]_0 = N_0\left\{1 - \frac{1}{2p_A/(p_B + p_C)}\right\}$$

In order to find total number of molecules at time t, N_t, it is observed that the total number of molecules is reduced by 1 whenever an A functional group reacts (either with B or C). This gives

$$N_t = N_0\left\{1 - \frac{2p_A}{1 + 2p_A/(p_B + p_C)}\right\}$$

In order to find the total weight, W_t, we define M_{AA} and M_{BC} as the molecular weights of the repeat units formed through monomers AA and BC, respectively. W_t is given by

$$W_t = \text{(Total number of molecules and of AA monomer)}\, M_{AA}$$
$$+ \{\text{Total number of molecules of BC monomer } ([A]_0/2)M_{AA}$$
$$+ [B]_0 M_{BC}\}$$
$$= \frac{N_0}{2p_A/(p_B + p_C)\left\{M_{AA} + \dfrac{2p_A}{p_B + p_C}M_{BC}\right\}}$$

The number-average molecular weight μ_A is

$$\mu_A = \frac{W_t}{N_t} = \frac{(p_B + p_C)M_{AA} + 2p_A m_{BC}}{2p_A + p_B = p_C - 2p_A(p_B + p_C)}$$

3.7 EXPERIMENTAL RESULTS

In previous sections, the set of infinite elementary reactions occurring in step-growth polymerization was shown to reduce kinetically to a single reaction involving functional groups. It was essential to assume the equal reactivity hypothesis in order to achieve this simplification. Thus, tests against various experimental results are needed to confirm the equal reactivity hypothesis. In this section, however, experimental results on several important commercial systems are presented, and we find that the simple model presented earlier needs to be substantially extended to explain them. In fact, in several situations the equal reactivity hypothesis itself is inapplicable.

In order to confirm the equal reactivity hypothesis, Flory originally studied the polymerization of adipic acid with decamethylene glycol in the absence of a strong acid [4,5]. The course of the polymerization was followed by titrating the carboxylic end group. Flory assumed that the carboxylic groups act as a catalylst, and he represented the polymerisations as

$$-COOH + -OH \xrightarrow{-COOH} \overset{\overset{\displaystyle O}{\displaystyle \|}}{-C}-O- + H_2O \tag{3.7.1}$$

with the rate of reactions given by

$$-\frac{d[COOH]}{dt} = k_p[-COOH]^2[-OH] \tag{3.7.2}$$

In this experiment, conditions were maintained such that reaction (3.7.1) remained irreversible during the entire period of study. If the hydroxyl and the carboxylic acid groups are present in an equimolar ratio, Eq. (3.7.2) can be integrated to give

$$\frac{1}{[-COOH]^2} = 2k_p t + const_1 \tag{3.7.3}$$

If the initial concentration of the carboxylic acid group is $[-COOH]_0$, then its concentration at any time can be expressed as a function of the conversion, p, as follows:

$$[-COOH] = (1 - p)[-COOH]_0 \tag{3.7.4}$$

On substituting this into Eq. (3.7.3), we obtain the following:

$$\frac{1}{(1-p)^2} = 2k_p t[-COOH]_0^2 + \text{const}_2 \tag{3.7.5}$$

A plot of $1(1-p)^2$ versus time should be linear; in Figure 3.5, it is found to be so after $1/(1-p)^2$ values of about 25.7. This means that this kinetic representation is valid only after 80% conversion. If the reaction is catalysed by a strong acid

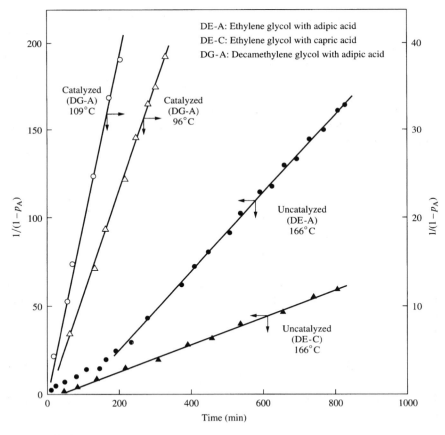

FIGURE 3.5 Catalyzed and uncatalyzed polymerization of ethylene glycol–adipic acid (DE-A) and ethylene glycol–caproic acid (DE-C).

(e.g., toluene sulfonic acid), the reaction rate represented by Eq. (3.7.2) then has to be modified to

$$\frac{d[-COOH]^2}{dt} = k_p^*[H^+][-COOH][-OH] \tag{3.7.6}$$

where $[H^+]$ is the concentration of the acid. Because the concentration of hydrogen ions remains constant during polyesterification, $[H^+]$ can be absorbed with k_p^*, and Eq. (3.7.6) can be easily integrated. If the concentrations of functional groups $-COOH$ and $-OH$ are again equal, the following integrated form is obtained:

$$\frac{1}{[-COOH]} = 2k_pt + \text{const}_1 \tag{3.7.7}$$

where k_p is equal to $k_p^*[H^+]$ and is a constant. Equation (3.7.7) can be rewritten as follows:

$$\mu_n = \frac{1}{(1-p)} = k_p[-COOH]_0 t + \text{const}_2 \tag{3.7.8}$$

The acid-catalyzed polyesterification of ethylene glycol and adipic acid has been studied by Flory. The data are plotted in Figure 3.5. This figure also reveals that Eq. (3.7.8) holds after about 80% conversion. More extensive experimental data [24–27] do not, however, confirm Flory's conclusions. It has been argued that only a limited amount of adipic acid dissociates in ethylene glycol (0.390 moles per mole ethylene glycol), and only this acid contributes to the catalysis in the polymerization *without* strong acid. Hence, instead of Eq. (3.7.2), it has been proposed that

$$-\frac{d[COOH]}{dt} = k_p[COOH][OH]^2 \tag{3.7.9}$$

Similarly, for acid-catalyzed polymerization, instead of Eq. (3.7.6),

$$-\frac{d[COOH]}{dt} = k''[COOH]^2 \tag{3.7.10}$$

has been proposed. If adipic acid and ethylene glycol are fed at a molar ratio of $1:r$, that is,

$$\frac{[OH]_0}{[COOH]_0} = r \tag{3.7.11}$$

then Eqs. (3.7.9) and (3.7.10) can easily be integrated after using appropriate stoichiometric relations between $[-COOH]$ and $[-OH]$. The fit of the experimental results for both uncatalyzed and catalysed polymerisations to the theory is shown in Figures 3.5 and 3.6, where straight-line plots are predicted theoretically and p is the conversion of $[-COOH]$. It is unfortunate that Eq. (3.7.10) shows k'' to depend not only on temperature but also on r. A semiempirical method has been suggested to account for this.

Commercially, the most common polyester in use is polyethylene terephthalate (PET), which is prepared from dimethyl terephthalate (DMT),

$$CH_3OC-\bigcirc-COCH_3$$

in the following three stages [28–30]:

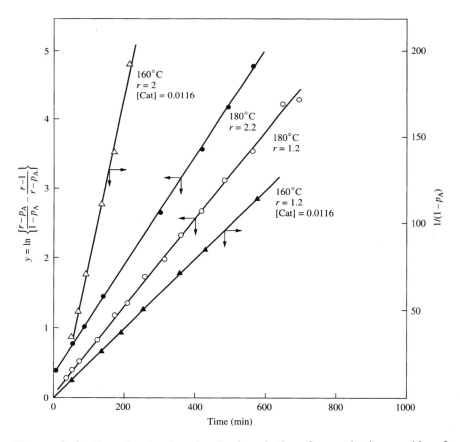

FIGURE 3.6 Uncatalyzed and catalysed polymerization of nonequimolar quantities of adipic acid and ethylene glycol.

1. Transesterification of DMT with ethylene glycol to produce bis-hydroxyethyl terephthalate (BHET):

$$OHCH_2CH_2O\overset{\overset{\displaystyle O}{\|}}{C}-\!\!\!\left\langle\bigcirc\right\rangle\!\!\!-\overset{\overset{\displaystyle O}{\|}}{C}OCH_2CH_2OH$$

at about 200°C and 1 atm pressure with continuous removal of methanol.
2. Polycondensation of BHET at 260–290°C at a vacuum of about 1 mm Hg, to remove the ethylene glycol produced.
3. Polymerization in the final stage, where special wiped film reactors are used. Temperatures around 350°C and a vacuum of about 1 mm Hg are maintained.

At the usual temperatures of the first and second stages of polymerization, there are several side reactions that determine the final properties of the polyester. Therefore, they cannot be ignored and must be accounted for in any realistic analysis of reactors. The various reactions occurring in the transesterification stage are summarized in Table 3.2. The only difference in the mechanism between this stage and the polycondensation stage lies in the fact that, in the latter, reactions (1) and (2) do not occur. In order to simulate the formation of PET in commercial reactors, we need to rewrite the rate of polymerization first, which is done as follows. A careful study of Table 3.2 reveals that the reaction mass consists of linear polyester molecules having different chain lengths and functional groups at chain ends. As a first approximation, we assume that a given functional group reacts with a rate constant that is independent of the chain length of the polymer molecule on which it is situated. As a result [e.g., in Eq. (3) of Table 3.2], the rate of formation of Z would be $k_3[COOH]^2 - k_3[Z][G]/K$. The analysis of the reactor can be performed only numerically in view of the set of nonlinear differential equations for the balance of functional groups.

Polyamides are formed by the polymerization of a diamine and a dicarboxylic acid and are commonly known as nylons. Among the various nylons, nylon 66 and nylon 6 are commercially important polymers. It is shown here that in these cases the application of the equal reactivity hypothesis is no more than an approximation. Nylon 66 is prepared in two stages. In the first stage, the monomers hexamethylene diamine $[NH_2-(CH_2)_6-NH_2]$ and adipic acid $[COOH-(CH_2)-COOH)]$ are reacted to form hexamethylene adipamide (sometimes called nylon 66 salt). It is known that the amino and carboxylic groups ionise in the molten state and the ionised species do not participate in the step-growth polymerization [7,31–33]. In other words, the polymerization can be

TABLE 3.2 Various Reactions of Functional Groups in the Transesterification Stage of PET Formation from DMT

Main reactions

---

1. Ester interchange

$$E_m + G \underset{k_1/K_1}{\overset{k_1}{\rightleftharpoons}} E_g + M$$

3. Polycondensation

$$2E_g \underset{k_3/K_3}{\overset{k_3}{\rightleftharpoons}} Z + G$$

2. Transesterification

$$E_m + E_g \underset{k_2/K_2}{\overset{k_2}{\rightleftharpoons}} Z + M$$

Important side reactions

---

4. Acetaldehyde formation

$$E_g \overset{k_4}{\longrightarrow} E_c + A$$

6. Water formation

$$E_c + G \underset{k_2/K_4}{\overset{k_7}{\rightleftharpoons}} E_g + W$$

$$E_C + E_g \underset{k_8/K_5}{\overset{k_8}{\rightleftharpoons}} Z + M$$

5. Diethylene glycol formation

(a) $E_g + G \overset{k_5}{\longrightarrow} E_c + D$

(b) $E_g \overset{k_6}{\longrightarrow} E_c + E_d$

7. Vinyl group formation

(a) $Z \overset{k_9}{\longrightarrow} E_c + E_v$

(b) $E_g + E_v \overset{k_3}{\longrightarrow} Z + A$

Symbols

$A = CH_3CHO$

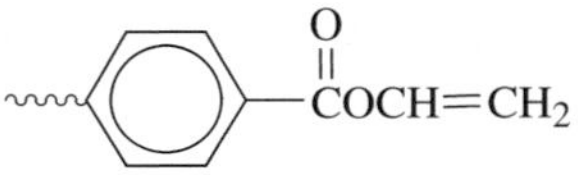

$G = OH-CH_2CH_2-OH$

$D = OHCH_2CH_2OCH_2OH$

$E_m = $ $-COCH_3$ (with O double bond)

$E_g = $ $-COCH_2CH_2OH$

$E_c = $ $-COH$

$E_v = $ $-COCH=CH_2$

$E_d = $ $-COCH_2CH_2OCH_2CH_2OH$

$M = CH_3OH$

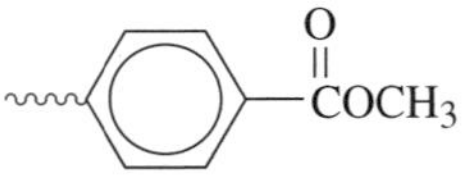

$W = H_2O$

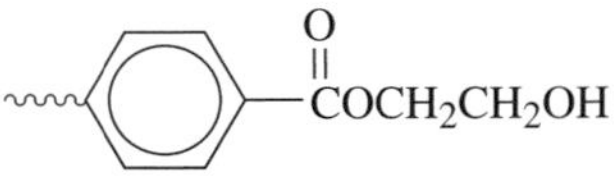

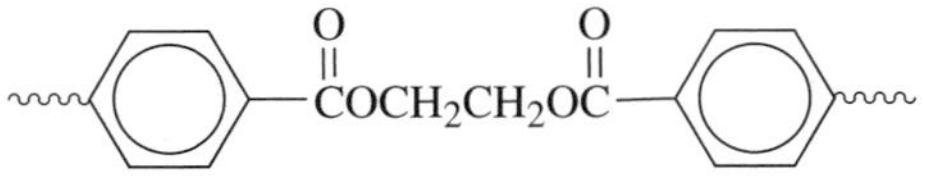

$Z = $ $-COCH_2CH_2OC-$

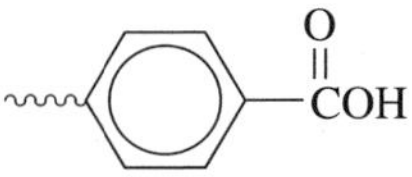

Source: Symbols were reprinted from Ref. 2 with the permission of Plenum Publishing Corporation.

represented as follows:

$$
\begin{array}{cc}
-COO- & -NH_3^+ \\
\updownarrow\uparrow & \updownarrow\downarrow \\
-COOH + -NH_2 & \underset{k_p'}{\overset{k_p}{\rightleftharpoons}} \quad -CONH- + H_2O
\end{array}
\tag{3.7.12}
$$

Two equilibrium relations can be written after some manipulations:

$$
K_i = \frac{[-NH_3^+][-COO^-]}{\dfrac{[-NH_2][-COO^-]}{[-NH_2][-COOH]}}
\tag{3.7.13a}
$$

$$
K_a = \frac{k_p [-CONH-][H_2O]}{k_p' [-NH_2][-COOH]}
\tag{3.7.13b}
$$

The ionisation constant K_i depends upon the dielectric constant of the reaction mass (and so, on the water concentration) because it involves ionised species. The dielectric constant of the reaction mass is increased by the addition of water. As a result, the concentration of the ionised species increases (due to shielding of charges) and that of the product $[-NH_2][-COOH]$ reduces, thus lowering the reaction rate in the forward direction. If one ignores the presence of the ionised species $[-NH_3^+]$ and $[-COO^-]$ and correlates the rate of reaction without correcting for the decrease in the concentrations of $-NH_2$ and $-COOH$ groups, the effect is the lowering of the apparent rate constant k.

The ionisation constant K_i cannot be measured experimentally since it is not possible to measure concentrations of ionised species in the reaction mass. Because of difficulties in experimentally distinguishing these species from the molecular ones, the ionisation reactions in Eq. (3.7.12) are ignored and K is calculated by considering the equilibrium of the amino–carboxylic reaction only. This K would be some combination of K_i and K_a; it is defined by the following equation,

$$
K = \frac{[-COOH]_{eq}[-NH_2]_{eq}}{[-H_2O][-CONH-]_{eq}}
$$

where $[-COOH]_{eq}$, $[-NH_2]_{eq}$, and $[-CONH-]_{eq}$ are the measured concentrations of the species in the reaction mass at equilibrium.

Finally, the effect of pH on the yield of the polymer can be deduced as follows. A high pH (low $[H^+]$) would suppress the formation of $[-COOH]$, with $[-NH_2]$ being limited, and a low pH (high $[H^+]$) would suppress the concentration of $-NH_2$, with $[-COOH]$ being limited, as seen from Eq. (3.7.12). Therefore, there is an optimal pH when the product $[-COOH][-NH_2]$ is maximum and the polymer yield measured by $[-CONH-]_{eq}$ in the above equation is maximum.

The monomer for nylon 6 is ε-caprolactam (C_1) and its polymerization is much more complicated than the simple polyamidation reaction. Water is the catalyst by which the caprolactam ring is opened to give aminocaproic acid, P_1. Aminocaproic acid can undergo addition as well as condensation reaction, as follows [33,34]:

Ring opening

$$C_1 + H_2O \underset{k_1'}{\overset{k_1}{\rightleftharpoons}} P_1 \qquad\qquad (3.7.14)$$

Polycondensation

$$P_m + P_n \underset{k_2'}{\overset{k_2}{\rightleftharpoons}} P_{m+n} + H_2O \qquad\qquad (3.7.15)$$

Polyaddition

$$C_1 + P_n \underset{k_3'}{\overset{k_3}{\rightleftharpoons}} P_{n+1} \qquad\qquad (3.7.16)$$

Industrially, caprolactam (C_1) is almost always polymerised with a monofunctional acid (denoted A_1) so as to control the molecular weight of the final polymer. These acid molecules can react with various polymer molecules P_m to give higher oligomers (denoted A_n), which are "capped" at one end. In addition to these reactions, P_n has been known to undergo cyclization reactions, and higher cyclic oligomers so formed causes problems in the spinning of the nylon. The complete polymerization mechanism is given in Table 3.3. Once again, as in the case of PET formation, it is observed that many more reactions must be incorporated in the kinetic scheme. The only way to analyze the reactor is to solve for the MWD and calculate the various moments from these. It may be added that, for nylon 6 polymerization, various rate constants in Table 3.3 are catalysed by the acid end-group concentrations and are usually expressed as follows:

$$k_i = k_{i0} \exp\left(-\frac{E_t^0}{RT}\right) + k_{ic} \exp\left(-\frac{E_i^c}{RT}\right)[-COOH] \qquad\qquad (3.7.17)$$

The values of k_{i0}, k_{ic}, E_i^0, and E_1^c used in various simulation studies of nylon 6 have been reviewed and can be obtained from the literature [33,34].

Polyurethanes are polymers with characteristic linkage $-NH-C\,O-O-$ and are formed by the step-growth polymerization of a diol and a diisocyanate as follows:

$$\underset{\text{Diisocyanate}}{OCN-R-NCO} + \underset{\text{Diol}}{OH-R'-OH} \;\rightleftharpoons \qquad\qquad (3.7.18)$$

$$\begin{array}{cc} & \overset{O}{\overset{\|}{}} \qquad\qquad \overset{O}{\overset{\|}{}} \\ \left[\!\!\left[O-C-NH-R-NHCOR' \right]\!\!\right]_n \end{array}$$

TABLE 3.3 The Kinetic Scheme for Nylon 6 Polymerization

1. Ring opening

$$C_1 + W \underset{k_1'}{\overset{k_1}{\rightleftarrows}} P_1$$

2. Polycondensation

$$P_n + P_m \xrightarrow{\quad 2k_2\ (n \neq m)\ \text{or}\ k_2\ (n=m) \quad} P_{n+m} + W$$

$$P_1 + P_{n+m-1} \overset{2k_2'}{\longleftarrow}$$

$$P_2 + P_{n+m-2} \overset{2k_2'}{\longleftarrow}$$

$$\vdots$$

$$2P_{(n+m)/2} \overset{k_2'\ (n+m\ \text{even})}{\longleftarrow}$$

$$2P_{(n+m-1)/2} + P_{(n+m-1)/2} \overset{2k_2'\ (n+m\ \text{odd})}{\longleftarrow}$$

3. Polyaddition

$$P_n + C_1 \underset{k_3'}{\overset{k_3}{\rightleftarrows}} P_{n+1}, \quad n = 1, 2, \ldots$$

4. Reaction with monofunctional acid

$$P_n + A_m \xrightarrow{\quad k_2 \quad} A_{n+M} + W$$

$$P_1 + A_{n+m-1} \overset{k_2'}{\longleftarrow}$$

$$P_2 + A_{n+m-2} \overset{k_2'}{\longleftarrow}$$

$$\vdots$$

$$P_{n+m-1} + A_1 \overset{k_2'}{\longleftarrow}$$

Forward reaction: $n, m = 1, 2, \ldots$
Reverse reaction: $n + m = 2, 3, 4, \ldots$

5. Ring opening of cyclic polymer

$$C_n + W \underset{k_4'}{\overset{k_4}{\rightleftarrows}} P_n, \quad n = 2, 3, \ldots$$

6. Polyaddition of C_r

$$P_n + C_r \underset{k_5'}{\overset{k_5}{\rightleftarrows}} P_{n+2}; \quad n = 1, 2, \ldots$$

7. Cyclization

$$P_n \underset{k_6'}{\overset{k_6}{\rightleftarrows}} P_{n-r} + C_r$$

TABLE 3.3 *(continued)*

Symbols

$$W = \text{Water} \qquad P_n = H\!-\!\!\left[NH\!-\!(CH_2)_5\!-\!\overset{\overset{\displaystyle O}{\|}}{C}\right]_n\!\!-\!OH$$

$$A_n = X\!-\!\!\left[\overset{\overset{\displaystyle O}{\|}}{C}\!-\!NH\!-\!(CH_2)_5\right]_{n-1}\!\!-\!COOH \qquad \text{(X is any unreactive group)}$$

$$C_n = H\!-\!N\!-\!(CH_2)_5\!-\!\!\left[\overset{\overset{\displaystyle O}{\|}}{C}\!-\!NH\!-\!(CH_2)_5\right]_{n-1}\!\!-\!\overset{\overset{\displaystyle O}{\|}}{C} \qquad \text{(C}_1 \text{ is the momomer)}$$

Source: Symbols are reprinted from Ref. 2 with the permission of Plenum Publishing Corporation.

with no condensation product formed. Industrially, the diisocyanates used are either 2,4-toluene diisocyanate (TDI) or 4,4′-diphenyl methane diisocyanate (MDI). The diols are polyester diols formed by the polymerization of adipic acid (or phthalic anhydride) in the presence of an excess of ethylene glycol. The formation of polyurethanes cannot be represented by the simple scheme of ARB polymerization because (1) the reaction of a given isocyanate group (i.e., −NCO group) does not follow second-order kinetics and (2) the ability of a −NCO group to react depends to a large extent on the linkage of the other isocyanate group.

In aromatic diisocyanates, such as toluene dissocyanate (TDI), one isocyanate group can modify the activity of the other, and the activity of both groups can depend on the other substituents of the aromatic ring. For a mixture of 2,4 and 2,6 isomers of TDI (industrially, it is difficult to separate the two), 12 reactions with primary and secondary 1OH groups of the polyols have been identified. The rate constants for these reactions have been measured experimentally and are summarized in Table 3.4. Significant differences can be observed in the reactivity of the two −NCO groups; the equal reactivity hypothesis is definitely not followed. There have been several fundamental studies to model the unequal reactivity of functional groups in urethane formation. It has been shown that such reactivity has considerable influence on the polymer formed.

The usual ingredients for forming urethane polymers are a silicone surfactant (sometimes called releasing agent), a flame retardant, a polyol, a diisocyanate, and a suitable catalyst. Polyols used have alcoholic functional groups and are rarely small molecules like ethylene glycol. The formation of the polymer is usually fast, even without catalysts, but recent applications such as reaction injection molding (RIM) require very fast reactions for which a catalyst

TABLE 3.4 Rate Constants for Various Reactions of 2,4 and 2,6 Toluene Diisocyanates with Polyols

Reactions		$10^4 k_t$ L equivalent^{-1} s^{-1}	
Nature of OH	Location of NCO	25°C	60°C
1	Primary hydroxyl	Monomeric para	0.613	4.17
2		Monomeric ortho	0.230	1.67
3		Polymeric para	0.161	1.10
4		Polymeric ortho	0.0605	0.439
5	Secondary hydroxyl	Monomeric para	0.204	1.67
6		Monomeric ortho	0.0273	0.333
7		Polymeric para	0.0538	0.439
8		Polymeric ortho	0.00717	0.0877
9	H of −NHCO− bond	Monomeric para	0.00307	0.0208
10		Monomeric ortho	0.00409	0.00417
11		Polymeric para	0.000807	0.00548
12		Polymeric ortho	0.000108	0.00110

Source: Data from Refs. 35–37.

must be used. Examples of the latter are tertiaryamines such as triethylene diamine and triethyl amine. In producing flexible foams, a blowing agent such as water is used; on reaction with water, the isocyanate group produces carbon dioxide as follows:

$$\text{ⁿⁿ NCO} + H_2O \rightarrow \text{ⁿⁿ NH}_2 + CO_2 \qquad (3.7.19)$$

The carbon dioxide thus liberated initially leaves the reaction mass, but with the progress of polymerization, the viscosity increases and the gas is trapped, giving a cellular structure. Finally, the urethane formed is not necessarily linear, but branches are generated through allophenate and biuret linkages:

```
ⁿⁿ NCONHⁿⁿⁿ
     |                        ⁿⁿ NHCONCONHⁿⁿ
     C=O                                |
     |                                            (3.7.20)
   ⁿⁿ NH
Allophenate linkage          Biuret linkage
```

The kinetics of the step-growth polymerization of formaldehyde with phenol, urea, and melamine are even more complex [27]. Commercially available formaldehyde is sold as a 37% solution in water. The following represents the equilibrium between formaldehyde and water:

$$\underset{\text{H}}{\overset{\text{O}}{\overset{\|}{\text{C}}}}\text{H} + \text{H}_2\text{O} \underset{k_1'}{\overset{k_1}{\rightleftharpoons}} \text{OH}-\text{CH}_2-\text{OH} \qquad (3.7.21)$$
$$\text{Methylene glycol}$$

and the equilibrium constant is such that, under normal operating conditions, formaldehyde is present almost entirely as methylene glycol. Therefore, when the formaldehyde solution undergoes step-growth polymerization, it exhibits a functionality of 2. Phenol (OH$-$) reacts with formaldehyde at its two ortho and one para positions, which means that it has a functionality of 3. Similarly, urea

$$\underset{\text{NH}_2-\text{C}-\text{NH}_2}{\overset{\text{O}}{\overset{\|}{}}}$$

has four reactive hydrogens and, hence, is tetrafunctional, and melamine, with three NH_2 groups, is hexafunctional.

We have already observed that the polymerization of trifunctional (and higher-functionality) monomers leads to branched polymers, which ultimately form network molecules. The main commercial interest in the polymers of phenol and melamine has been in producing molded objects that exhibit high chemical and environmental resistance. These are network polymers and are formed in two stages. In the first step, a prepolymer is prepared that is, in the second step, cross-linked to the desired shape in a mold in the presence of a suitable cross-linking agent. The urea formaldehyde polymer has found extensive use in plywood industries; in its first stage, a syrupy prepolymer is prepared that is cross-linked between the laminates of the plywood.

Commercially, two grades of prepolymers (novolacs and resoles) are made through the polymerization of phenol and formaldehyde. Novolacs are linear polymer chains with little branching and are formed when the pH of the reaction mass is low (2 to 3). Resole prepolymers are manufactured at high pH (9 to 11) and are highly branched. The characteristics of the prepolymer formation are complex; some of these are given in Table 3.5. The important feature of the polymerization, as can be seen from the table, is the different reactivities of the sites.

To model the prepolymer formation, the usual approach taken is to work in terms of functional groups. These are defined as entities, the use of which preserves the characteristics of the reaction steps leading to the formation of the

TABLE 3.5 Polymerization Characteristics of Phenol, Urea, and Melamine with Formaldehyde

Phenol	1. Trifunctional. 2. Ortho and para positions have different reactivities. 3. In acidic medium, novolacs (essentially linear polymers) are formed. In basic medium, resoles (essentially branched polymers) are formed.
Urea	1. Tetrafunctional, but only three of its sites participate in polymerization. 2. Polymerization could be acid or base catalysed. The equilibrium constant is independent of pH. 3 All sites have different reactivities.
Melamine	1. Hexafunctional. 2. Polymerization is reversible. The primary H's react at different rates compared to secondary ones.

polymer. For example, let us consider the formation of the urea formaldehyde prepolymer. The chemical formula of urea is

$$NH_2-\underset{\underset{\displaystyle \|}{\displaystyle O}}{C}-NH_2$$

and experiments have shown that the hydrogens of the two amine groups combine with the hydroxyl groups of the methylene glycol. Even though urea has four reactive hydrogens, it has been shown that after three of its hydrogens have reacted, the fourth one remains inert, as mentioned in Table 3.5. The four possible functional groups A, B, C, and D are shown in Figure 3.7 and are obtained by assuming bonds at the various reactive sites of the urea molecule. Subsequently, we observe that methylene glycol molecule has two $-OH$ groups and reacts in two steps. When it reacts for the first time, it gives rise to $-CH_2OH$ groups that can react further to give a $-CH_2-$ methylene bridge. In defining species A to D, no distinction has been made as to whether the linkages at the reacted sites are a $-CH_2-$ bond or $-CH_2OH$ groups.

Species A to D can be used to represent any polymer molecule. For example,

$$CH_2OH-NH-CO-NH-CH_2-NH-CO-\underset{\underset{\displaystyle |}{\displaystyle CH_2OH-N-CO-NH-CH_2-NH-CO-NH_2}}{N}-CH_2-\underset{\underset{\displaystyle |}{\displaystyle CH_2OH}}{N}-CO-NH_2$$

$$(3.7.22)$$

can be represented by

$$
\begin{array}{c}
\text{C}-\text{D}-\text{C} \\
\quad | \\
\text{D}-\text{A}
\end{array}
\tag{3.7.23}
$$

Instead of attempting to find the concentration of different isomers in a polymer, we make an effort here to determine the conversion of urea and formaldehyde in the reaction mass as a function of time. When polymerization is carried out for some time, starting with a feed consisting of urea and formaldehyde, polymers of various lengths and structures are formed. One plausible description of the progress of reaction might be to follow the concentration of species A to D in the reaction mass. The overall polymerization represented by the reaction of functional groups can be written in terms of the following rate constants:

k_1 = Rate constant for the reaction of primary hydrogen of urea with the OH groups

k_2 = Rate constant for the reaction of secondary or tertiary hydrogen of urea with the OH group

k_3 = Rate constants for the reverse reaction occurring between a reacted $-\text{CH}_2-$ bond (denoted Z) and a water molecule

It is now possible to write the polymerization of urea with formaldehyde as follows. The forward reactions can be easily written in terms of A to D (the formation of tetrasubstituted urea does not occur) as follows:

$$
\text{U} + \text{F} \xrightarrow{8k_1} \text{A} + \text{CH}_2\text{OH} + \text{H}_2\text{O}
\tag{3.7.24a}
$$

$$
\text{U} + \text{CH}_2\text{OH} \xrightarrow{4k_1} \text{A} + \text{H}_2\text{O} + \text{Z}
\tag{3.7.24b}
$$

$$
\text{A} + \text{F} \xrightarrow{2k_2} \text{B} + \text{CH}_2\text{OH} + \text{H}_2\text{O}
\tag{3.7.24c}
$$

$$
\text{A} + \text{CH}_2\text{OH} \xrightarrow{k_2} \text{B} + \text{H}_2\text{O} + \text{Z}
\tag{3.7.24d}
$$

$$
\text{A} + \text{F} \xrightarrow{4k_1} \text{C} + \text{CH}_2\text{OH} + \text{H}_2\text{O}
\tag{3.7.24e}
$$

$$
\text{A} + \text{CH}_2\text{OH} \xrightarrow{2k_1} \text{C} + \text{H}_2\text{O} + \text{Z}
\tag{3.7.24f}
$$

$$
\text{C} + \text{F} \xrightarrow{4k_2} \text{D} + \text{CH}_2\text{OH} + \text{H}_2\text{O}
\tag{3.7.24g}
$$

$$
\text{C} + \text{CH}_2\text{OH} \xrightarrow{4k_2} \text{D} + \text{H}_2\text{O} + \text{Z}
\tag{3.7.24h}
$$

$$
\text{B} + \text{F} \xrightarrow{2k_2} \text{D} + \text{CH}_2\text{OH} + \text{H}_2\text{O}
\tag{3.7.24i}
$$

$$
\text{B} + \text{CH}_2\text{OH} \xrightarrow{2k_1} \text{D} + \text{H}_2\text{O} + \text{Z}
\tag{3.7.24j}
$$

In writing these reactions, it has been assumed that the overall reactivity of a given reaction is completely governed by the site involved. Therefore, when urea consisting of four hydrogens reacts with formaldehyde (or methylene glycol) having two $-OH$ groups, it forms species A, as in Eq. (3.7.22a), with the overall reactivity 2 $(4k_1)$, or $8k_1$. The reactivity of other steps can be decided similarly. Finally, because species D does not have any reactive site left, it is assumed no longer to react.

Because in species A to D, the $-CH_2-$ bond and the CH_2OH linkage have not been distinguished, it is not possible to write the mechanism of the reverse reaction exactly. In view of this problem, two extreme possibilities have been proposed. In the first one (model I), the various linkages of species A to D have all been assumed to be mainly CH_2OH groups. In the second one (model II), the linkage of species A to D have all been assumed to be mainly reacted $-CH_2-$ bonds. It is assumed that model I is a better representation of the situation in the initial phases of polymerization, whereas model II gives a better description in the final stages of polymerization. The mechanism of polymerization for these two models is as follows.

Reverse reaction for model I. All linkages are assumed to be reacted $-CH_2OH$ groups.

$$A + H_2O \xrightarrow{k_3} U + F-(CH_2OH) \tag{3.7.25a}$$

$$B + H_2O \xrightarrow{2k_3} A + F-(CH_2OH) \tag{3.7.25b}$$

$$C + H_2O \xrightarrow{2k_3} A + F-(CH_2OH) \tag{3.7.25c}$$

$$D + H_2O \xrightarrow{k_3} B + F-(CH_2OH) \tag{3.7.25d}$$

$$D + H_2O \xrightarrow{2k_3} C + F-(CH_2OH) \tag{3.7.25e}$$

In Eq. (3.7.25a), when species A reacts, U and F both are formed and a CH_2OH group simultaneously disappears.

Reverse reaction for model II. All linkages are assumed to be $-CH_2-$.

$$A + H_2O \xrightarrow{k_3} U + (CH_2OH)-Z \tag{3.7.26a}$$

$$B + H_2O \xrightarrow{2k_3} A + (CH_2OH) - Z \tag{3.7.26b}$$

$$C + H_2O \xrightarrow{2k_3} A + F + (CH_2OH)-Z \tag{3.7.26c}$$

$$D + H_2O \xrightarrow{k_3} B + F + (CH_2OH)-Z \tag{3.7.26d}$$

$$D + H_2O \xrightarrow{2k_3} C + F + (CH_2OH)-Z \tag{3.7.26e}$$

In Eq. (3.7.26a), when species A reacts, U and a CH_2OH group is formed and Z simultaneously disappears.

3.8 CONCLUSION

In this chapter, we have presented the kinetics of reversible step-growth poly-merization based on the equal reactivity hypothesis. We have found that the polymerization consists of infinite elementary reactions that collapse into a single one involving reaction between functional groups. This kinetic model has been tested extensively against experimental data. It is found that in most of the systems involving step-growth polymerization, there are either side reactions or the equal reactivity hypothesis does not hold well. This chapter presents the details of chemistry for some industrially important systems; motivated readers are referred to advanced texts for mathematical simulations.

APPENDIX 3.1: THE SOLUTION OF MWD THROUGH THE GENERATING FUNCTION TECHNIQUE IN STEP-GROWTH POLYMERIZATION

The generating function, $G(s, t)$, is defined as

$$G(s, t) = \sum_{n=1}^{\infty} s^n [P_n] \tag{A3.1.1}$$

where s is an arbitrary parameter whose value lies between 0 and 1. On multiplying Eq. (3.3.5a) by s and (3.3.5b) by s^n and adding the equations for all values of n, we find that

$$\frac{\partial G(s, t)}{\partial t} = k_p \sum_n s^n \sum_{i=1}^{n-1} [P_i][P_{n-i}] - 2k_p \lambda_0 \sum_{n=1}^{\infty} s^n [P_n]$$

$$- k_p'[W] \sum_{n=1}^{\infty} (n-1)s^n [P_n] + 2k_p'[W] \sum_{n=1}^{\infty} s^n \sum_{i=n+1}^{\infty} [P_i] \tag{A3.1.2}$$

The following can be derived from Eq. (A.3.1.1):

$$s\frac{\partial G(s, t)}{\partial t} = s \sum_{N=1}^{\infty} ns^{n-1}[P_n] = \sum_{n=1}^{\infty} ns^n[P_n] \tag{A3.1.3}$$

$$\sum_{n=1}^{\infty} s^n \sum_{i=n+1}^{\infty} [P_i] = s([P_2] + [P_3] + [P_4] + \cdots) + s^2([P_3] + [P_4] + \cdots)$$

$$= \sum_{n=2}^{\infty} \left(\sum_{i=1}^{n-1} s^i \right)[P_n] = \sum_{n=2}^{\infty} \frac{(1 - s^n)}{(1 - s)}[P_n]$$

$$= \frac{s\lambda_0 - G}{1 - s} \tag{A3.14}$$

On substituting Eqs. (A3.1.3) and (A3.1.4) into Eq. (2.5.3), we obtain a partial differential equation governing the time variation of G:

$$\frac{\partial G}{\partial t} = k_p G^2 - 2k_p \lambda_0 G + k'_p[W]\frac{\lambda_0 - G}{1 - s} - k'[W]\left\{\frac{\partial G}{\partial s} - G\right\} \tag{A3.1.5}$$

For arbitrary feed to batch or semibatch reactors, Eq. (A.3.1.5) has not yet been solved analytically. If the step-growth polymerization is irreversible, k'_p in Eq. (A3.1.1) is zero and the time variation of the moment-generating function, $G(s, t)$, is as follows [2,5]:

$$\frac{\partial G}{\partial t} = k_p G^2 - 2k_p G\lambda_0 \tag{A3.1.6}$$

This has been solved in the literature by defining

$$y = \frac{G(s, t)}{\lambda_0} \tag{3.1.7}$$

and observing that, for irreversible polymerization,

$$\frac{d\lambda_0}{dt} = -k_p \lambda_0^2 \tag{3.1.8a}$$

$$\frac{\partial y}{\partial t} = k_p \frac{G^2}{\lambda_0} - k_p G \tag{A3.1.8b}$$

From these, one obtains

$$\frac{\partial y}{\partial \lambda_0} = \frac{y(1 - y)}{\lambda_0} \tag{A3.1.9}$$

For any arbitrary feed having a moment-generating function $g_0(s)$ at $t = 0$, Eq. (A.3.1.9) can easily be integrated to yield

$$G(s, t) = \frac{g_0(s)(\lambda_0/g_0(1))}{1 - [1 - \lambda_0/g_0(1)]g_0(s)/g_0(1)} \tag{A3.1.10}$$

where $g_0(s)$ is the value of G for the feed and $g_0(1)$ is defined as

$$g_0(1) = \lim_{s \to 1} g_0(s) \tag{A3.1.11}$$

If the feed is a pure monomer at concentration $[P_1]_0$, then

$$g_0(s) = [P_1]_0 s \tag{A3.1.12a}$$

$$g_0(1) = [P_1]_0 \tag{A3.1.12b}$$

Equation (A3.1.10) gives

$$G(s, t) = \frac{s\{\lambda_0/[P_1]_0\}^2 a}{1 - \{1 - \lambda_0/[P]_0\}s}$$

$$= [P_1]_0 \sum_{n=1}^{\infty} \left\{\frac{[\lambda_0]}{[P_1]_0}\right\}^2 \left\{1 - \frac{[\lambda_0]}{[P_1]_0}\right\}^{n-1} s^n \qquad \text{(A3.1.13)}$$

On comparison of this result with Eq. (A3.1.1), the molecular-weight distribution (MWD) is obtained as follows:

$$[P_n] = [P_1]_0(1 - p)^2 p^{n-1} \qquad \text{(A3.1.14a)}$$

where

$$p = 1 - \frac{\lambda_0}{[P_1]_0} \qquad \text{(A3.1.14b)}$$

Here, p is the same as the conversion of functional groups defined in Eq. (3.4.2).

It is interesting to observe from these equations that the MWD of the polymer formed in batch reactors using a pure monomer feed is a function of only one variable (viz. the conversion, p of functional groups). Thus, an engineer has only one design variable in his control and cannot choose p as well as the product MWD independently.

The number-average and weight-average chain lengths, μ_n and μ_w, respectively, can be found if $G(s, t)$ is known. We observe that

$$\lim_{s \to 1} G(s, t) = \lim_{s \to 1} \sum_{n=1}^{\infty} s^n[P_n] = \sum_{n=1}^{\infty}[P_n] = \lambda_0 \qquad \text{(A3.1.15)}$$

$$\lim_{s \to 1} \frac{\partial G}{\partial s} = \sum_{n=1}^{\infty} ns^{n-1}[P_n]|_{s=1} = \lambda_1 \qquad \text{(A3.1.16)}$$

$$\lim_{s \to 1} s^2 \frac{\partial^2 G}{\partial s^2}\bigg|_{s=1} = \sum_{n=1}^{\infty} n(n-1)s^{n-2}[P_n]_s|_{s=1} = \lambda_2 - \lambda_1 \qquad \text{(A3.1.17)}$$

The second moment, λ_2, is therefore given by the following:

$$\lambda_2 = \sum_{n=1}^{\infty} n^2[P_n] = \frac{\partial^2 G}{\partial s^2}\bigg|_{s=1} + \frac{\partial G}{\partial s}\bigg|_{s=1} \qquad \text{(A3.1.18)}$$

Using the expression for $G(s, t)$ given in Eq. (A3.1.13), the number-average and weight-average chain lengths, μ_n and μ_w, respectively, can be obtained as follows:

$$\mu_n = \frac{\lambda_1}{\lambda_0} = \frac{[P]_{01}}{[P]} = \frac{1}{1 - p} \qquad \text{(A3.1.19a)}$$

$$\mu_w = \frac{\lambda_2}{\lambda_1} = \lim_{s \to 1} \frac{(\partial^2 G/\partial s^2) + (\partial G/\partial s)}{\partial G/\partial s} = 1 + \frac{2p}{1 + p} \qquad \text{(A3.1.19b)}$$

The polydispersity index (PDI), Q can be derived as

$$Q \equiv \frac{\mu_w}{\mu_n} = 1 + p \qquad (A3.1.20)$$

which implies that at 100% conversion (i.e., $p = 1$), the polydispersity index for batch reactors with pure monomer feed attains a maximum value of 2.

REFERENCES

1. Odian, G., Principles of Polymerization, 2nd ed., Wiley, New York, 1981.
2. Gupta, S. K., and A. Kumar, Reaction Engineering of Step Growth Polymerization, Plenum, New York, 1987.
3. Biesenberger, J. A., and D. H. Sebastian, Principles of Polymerization Engineering, Wiley, New York, 1983.
4. Kumar, A., and S. K. Gupta, Fundamentals of Polymer Science and Engineering, Tata McGraw-Hill, New Delhi, 1978.
5. Flory, P. J., Principles of Polymer Chemistry, Cornell University Press, Ithaca, NY, 1953.
6. Throne, J. L., Plastics Process Engineering, Marcel Dekker, New York, 1979.
7. Solomon, D. H. (ed.), Step Growth Polymerization, Marcel Dekker, New York, 1978.
8. Kuchanov, S. I., M. L. Keshtov, P. G. Halatur, V. A. Vasnev, S. V. Vingradova, and V. V. Korshak, On the Principle of Equal Reactivity in Solution Polycondensation, Macromol. Chem., **184**, 105–111, 1983.
9. Ignatov, V. N., V. A. Vasnev, S. V. Vinogradova, V. V. Korshak, and H. M. Tseitlin, Influence of the Far Order Effect on the Reactivity Functional End Groups of Macromolecules in Nonequilibrium Polycondensation, Macromol. Chem., **189**, 975–983, 1988.
10. Hodkin, J. H., Reactivity Changes During Polyimide Formation, J. Polym. Sci., Polym. Chem. Ed., **14**, 409–431, 1976.
11. Kronstadt, M., P. L. Dubin, and J. A. Tyburczy, Molecular Weight Distribution of a Novel Condensation Polymerization Comparison with Theory, Macromolecule, **11**, 37–40, 1978.
12. Tirrell, M., R. Galvan, and R. L. Laurence, Polymerization Reactors, in Chemical Reaction and Reactor Engineering, J. J. Carberry and A. Varma (eds.), Marcel Dekker, New York, 1986.
13. Ray, W. H., On the Mathematical Modeling of Polymerization Reactors, J. Macromol. Sci. Rev., **C8**, 1, 1972.
14. Park, O. O., MWD and Moments for Condensation Polymerization with Variant Reaction Rate Constant Depending on Chain Lengths, Macromolecule, **21**, 732–735, 1988.
15. Kumar, A., Computation of MWD of Reversible Step Growth Polymerization in Batch Reactors, J. Appl. Polym. Sci., **34**, 571–585, 1987.
16. Khanna, A., Simulation, Modelling and Experimental Validation of Reversible Polymerization, Ph.D. thesis, Department of Chemical Engineering, Indian Institute of Technology, Kanpur, India, 1988.

17. Flory, P. J., Fundamental Principles of Condensation Polymerization, Chem. Rev., 39, 137, 1946.
18. Sawada, H., Thermodynamics of Polymerization, Marcel Dekker, New York, 1976.
19. Kilkson, H., Generalization of Various Polycondensation Problem, Ind. Eng. Chem. Fundam., 7, 354–362, 1968.
20. Ray, W. H., and R. L. Laurence, Polymerization Reaction Engineering, in Chemical Reactor Theory, N. R. Amundson and L. Lapidus (eds.), Prentice-Hall, Englewood Cliffs, NJ, 1977.
21. Gupta, S. K., and A. Kumar, Simulation of Step Growth Polymerization, Chem. Eng. Commun., **20**, 1–52, 1983.
22. Kumar, A., MWD of Reversible ARB Polymerization in HCSTRs with Monomers Exhibiting Unequal Reactivity, Macromolecule, **20**, 220–226.
23. Levenspiel, O., Chemical Reaction Engineering, 2nd ed., Wiley, New York, 1975.
24. Chen, S., and J. C. Hsio, Kinetics of Polyesterificaion: I. Dibasic Acid and Glycol Systems, J. Polym. Sci. Polym. Chem., 19, 3136, 1981.
25. Lin, C. C., and K. H. Hsieh, The Kinetics of Polyesterification: I. Dibasic Acid and Glycol Systems, J. Appl. Polym. Sci., **21**, 2711, 1977.
26. Lin, C. C., and P. C. Yu, The Kinetics of Polyesterification: II. Succinic Acid and Ethylene Glycol, J. Polym. Sci. Polym. Chem., **16**, 1005, 1979.
27. Lin, C. C., and P. C. Yu, The Kinetics of Polyesterification: III. Mathematical Model for Quantitative Prediction of Rate Constants, J. Appl. Polym. Sci., **22**, 1797, 1978.
28. Ravindranath, K., and R. A. Mashelkar, Polythylene Terephthalate—Chemistry, Thermodynamics, and Transport Properties, Chem. Eng. Sci., **41**, 2197–2214, 1986.
29. Bensnoin, J. M., and K. Y. Choi, Identification and Characterization of Reaction Byproducts in the Polymerization of Polyethylene Terephthalate, JMS—Rev. Macromol. Chem. Phys., **C29**, 55–81, 1989.
30. Yamada, T., and Y. Imamura, A Mathematical Model for Computer Simulation of a Direct Continuous Esterification Process Between Terephthalic Acid and Ethylene Glycol, Polym. Eng. Sci., **28**, 385–392, 1988.
31. Katz, M., Preparations of Linear Saturated Polyesters, in Polymerization Processes, C. E. Schildknecht and I. Skeist (eds.), Wiley, New York, 1977.
32. Kumar, A., S. Kuruville, A. R. Raman, and S. K. Gupta, Simulation of Reversible Nylon-66 Polymerization, Polymer, **22**, 387–390, 1981.
33. Tai, K., Y. Arai, H. Teranishi, and T. Tagawa, The Kinetics of Hydrolytic Polymerization of n-Caprolactam: IV. Theoretical Aspects of the Molecular Weight Distribution, J. Appl. Polym. Sci., **25**, 1769–1782, 1980.
34. Kumar, A., and S. K. Gupta, Simulation and Design of Nylon-6 Reactors, JMS—Rev. Macromol. Chem. Phys., **C26**, 183–247, 1986.
35. Saunders, T. H., and K. C. Frisch, Polyurethanes, Chemistry, and Technology, Interscience, New York, 1962.
36. Martin, R. A., K. L. Roy, and R. H. Peterson, Computer Simulation of Tolylene Diisocyanate–Polyol Reaction, Ind. Eng. Chem., Prod. R&D, 6, 218, 1967.
37. Mall, S., Modelling of Crosslinked Polyurethane Systems, M. Tech. thesis, Department of Chemical Engineering, Indian Institute of Technology, Kanpur, India, 1982.
38. Pannone, M. C., and C. W. Macosco, Kinetics of Isocyanate Amine Reactions, J. Appl. Polym. Sci., **34**, 2409, 1987.

39. Kumar, A., and A. Sood, Modelling of Polymerization of Urea and Formaldehyde Using Functional Group Approach, J. Appl. Polym. Sci., **40**, 1473–1486, 1989.
40. Kumar, A., and V. Katiyar, Modelling and Experimental Investigation of Melamine Formaldehyde Polymerization, Macromolecule, **23**, 3729–3736, 1989.

PROBLEMS

3.1. In Eq. (3.3.5), assuming the equal reactivity hypothesis, we derived the MWD relations. Using these, derive the following generation relations for λ_1 to λ_4. You would require summations derived in Example 3.4.

$$\frac{d\lambda_1}{dt} = 0$$

$$\frac{d\lambda_2}{dt} = 2k_p\lambda_1^2 + \frac{k_p'W}{3}(\lambda_1 - \lambda_3)$$

$$\frac{d\lambda_3}{dt} = 6k_p\lambda_1\lambda_2 + \frac{1}{2}k_p'W(\lambda_2 - \lambda_4)$$

$$\frac{d\lambda_4}{dt} = k_p(8\lambda_1\lambda_3 + 6\lambda_2^2) - \frac{k_p'}{15}(9\lambda_5 - 10\lambda_3 + \lambda_1)$$

3.2. Determine expressions for generation of λ_1 to λ_4 for the kinetic model of Example 3.1. Also, for the kinetic model

$$k_{p11} = k_{11}$$
$$k_{pmn} = 2k_p \quad m \neq n, m \neq n = 1, 2$$
$$k_{pmm} = k_p, \quad m = 2, 3, \ldots$$

Determine if the following generation relation for $P_n(n \geq 3)$ is valid

$$\frac{dP_n}{dt} = k_p \sum_{m=1}^{n-1} P_m P_{n-m} - 2k_p P_n \lambda_0 + 2k'W \sum_{i=n+1}^{\infty} P_i - k_p'(n-1)WP_n$$

Write the mole balances for P_1 and P_2 and determine the following generation relation for λ_0:

$$\frac{d\lambda_0}{dt} = -R_p(k-1)P_1^2 - k_p\lambda_0^2 - k_p\lambda_0^2 + k_p'W(\lambda_1 - \lambda_0)$$

3.3. For A–R–B polymerization, we showed that complex polymerization can be reduced to the reaction of functional groups. It can be shown that under the equal reactivity hypothesis, even for polymerisations where several side reactions are involved, this nature of step-growth polymerization is preserved. For example, let us analyze the polymerization of dimethyl terephthalate with ethylene glycol, which polymerizes according to the mechanism given in Table 3.2.

Derive the mole balance relation for each functional group for semibatch reactors from which acetaldehyde, ethylene glycol, methanol, and water are continuously flashing.

3.4. Polyesters undergo a redistribution reaction that involves an interaction of reacted bonds of one molecule with the OH functional group of the other. This can be schematically represented as

$$P_x + P_y \rightarrow P_{x-r} + P_{y+r}$$

where x can take on any value except $x = x - y$, when reactants and the products become identical. We observe that P_n is formed in two distinct ways: (1) by a process of elimination in which a reacted bond of P_x having chain length greater than n undergoes a redistribution reaction such that $x - r = n$ and (2) by a process of combination in which a given molecule P_y having chain length less than x combines with a part of chain of the other molecule such that $y + r = n$. If only the redistribution reaction is to occur, the mole balance of species will be given by the following:

$$\frac{d[P_1]}{dt} = k_r \left\{ -4[P_1] \sum_{m=2}^{\infty} (m-1)[P_m] + 4 \sum_{j=1}^{\infty} [P_j] \sum_{m=2}^{\infty} [P_m] \right\}$$

$$\frac{d[P_n]}{dt} = k_r \left\{ -4[P_1] \sum_{m=2}^{\infty} (m-1)[P_m] - 4[P_n](n-1) \sum_{m=1}^{\infty} [P_m] \right.$$

$$\left. + 4 \sum_{m=1}^{\infty} [P_m] \sum_{j=n+1}^{\infty} [P_j] + 4 \sum_{j=1}^{n-1} \sum_{m=n-j+1}^{\infty} [P_j][P_m] \right\}$$

Show that the generation of λ_0 and λ_1 due to this reaction is zero, but that it affects the second moment λ_2 considerably and its generation rate is given by

$$\frac{1}{k_r} \frac{d\lambda_2}{dt} = 4\lambda_0(\lambda_3 - \lambda_2) + 4\lambda_1(\lambda_2 - \lambda_1) + \frac{4}{3}\lambda_0(2\lambda_3 - 2\lambda_2 + \lambda_1)$$

3.5. Sometimes, polyesterificaion of a monomer (P_1) is carried out in the presence of monofunctional compounds such as cetyl alcohol, ethyl benzoate, or ethyl terephthalate. Let us denote these by MF_1. As the polymerization is carried out, the monomer molecules grow in size to give P_n. Large-chain monofunctional polymers MF_n are formed when P_n and MF_i interact. The overall polymerization can be expressed as follows:

$$P_i + P_j \rightarrow P_{i+j} + W$$
$$P_i + MF_j \rightarrow MF_{i+j} + W$$

Show that the mole balance relation can be derived as

$$\frac{d\mathrm{P}_n}{dt} = -4k_p\mathrm{P}_n \sum_{m=1}^{\infty} \mathrm{P}_m + k_p \sum_{r=1}^{n-1} \mathrm{P}_r\mathrm{P}_{n-r} - 2k_m'\mathrm{W}(n-1)\mathrm{P}_n$$

$$4k_p'\mathrm{W} \sum_{i=n+1}^{\infty} \mathrm{P}_i - 2k_m\mathrm{P}_n \sum_{m=1}^{\infty} \mathrm{MF}_m + 2k_p\mathrm{W} \sum_{n+1} \mathrm{MF}_i$$

$$\frac{d\mathrm{MF}_n}{dt} = -2k_m\mathrm{MF}_n \sum_{i=1}^{\infty} \mathrm{P}_i + 2k_m \sum_{m=1}^{n-1} \mathrm{MF}_m\mathrm{P}_{n-m} - 2k_m'\mathrm{W}(n-1)\mathrm{MF}_n$$

$$+ 2k_m'\mathrm{W} \sum_{m=n+1}^{\infty} \mathrm{MF}_m, \quad n \geq 2$$

List your assumptions concerning the reactions of MF_n and P_n. Write the mole balance relations for P_1 and MF_1 also.

3.6. We expect that the number of monofunctional molecules in Problem 3.5 does not change with time. This would mean that the zeroth moment of monofunction molecules, λ_{MF_0}, must be constant. Show this, and then derive the following zeroth, first, and second moment-generation relations from the MWD of Problem 3.5:

$$\frac{d\lambda_{0\mathrm{MF}}}{dt} = 0$$

$$\frac{d\lambda_{1\mathrm{MF}}}{dt} = 2k_m\lambda_1\lambda_{0\mathrm{MF}} - k_p'\mathrm{W}(\lambda_{2\mathrm{MF}} - \lambda_{1\mathrm{MF}})$$

$$\frac{d\lambda_{2\mathrm{MF}}}{dt} = -2k_m\lambda_{0\mathrm{P}}\lambda_{2\mathrm{MF}} + 2k_m(\lambda_{2\mathrm{P}}\lambda_{0\mathrm{MF}} + \lambda_{0\mathrm{P}}\lambda_{2\mathrm{MF}} + 2\lambda_{1\mathrm{P}}\lambda_{1\mathrm{MF}})$$

$$- 2k_m'\mathrm{W}(\lambda_{3\mathrm{MF}} - \lambda_{2\mathrm{MF}}) + \frac{1}{3}k_m'\mathrm{W}(2\lambda_{3\mathrm{MF}} - 3\lambda_{2\mathrm{MF}} + \lambda_{1\mathrm{MF}})$$

$$\frac{d\lambda_{0\mathrm{P}}}{dt} = 2k_p\lambda_{0\mathrm{P}}^2 + 2k_p'\mathrm{W}(\lambda_{1\mathrm{P}} - \lambda_{0\mathrm{P}})$$

$$- 2k_m\lambda_{0\mathrm{MF}}\lambda_{0\mathrm{P}} + 2k_m'\mathrm{W}(\lambda_{1\mathrm{MF}} - \lambda_{0\mathrm{MF}})$$

$$\frac{d\lambda_{1\mathrm{P}}}{dt} = -\frac{d\lambda_{1\mathrm{MF}}}{dt}$$

$$\frac{d\lambda_{2\mathrm{P}}}{dt} = 4k_p\lambda_{1\mathrm{P}}^2 - 2k_p'\mathrm{W}(\lambda_{3\mathrm{P}} - \lambda_{2\mathrm{P}}) - 2k_m\lambda_{2\mathrm{P}}\lambda_{0\mathrm{MF}}$$

$$+ \frac{2}{3}k_p'\mathrm{W}(2\lambda_{3\mathrm{P}} - 3\lambda_{2\mathrm{P}} + \lambda_{1\mathrm{P}})$$

$$+ \frac{1}{3}k_m'\mathrm{W}(2\lambda_{3\mathrm{MF}} - 3\lambda_{2\mathrm{MF}} + \lambda_{1\mathrm{MF}})$$

3.7. Table 3.3 gives the general mechanism of nylon 6 polymerization. Let us assume that reaction with monofunctional polymer and cyclization (i.e., steps 5, 6, and 7) do not occur. Derive the following mole balance relation

P_n species for batch reactors, assuming that there is no flashing of condensation products.

$$\frac{dP_n}{dt} = -2k_2 P_n \sum_{m=1}^{\infty} P_m + k_2 \sum_{m=1}^{n-1} P_m P_{n-m} - k_2' W(n-1)P_n$$

$$+ 2k_2' W \sum_{m=1}^{\infty} P_{n+m} - k_3 P_n C_1 + k_3' P_{n-1} C_1 + k_3' P_{n+1}$$

$$- k_3 P_n n \geq 3$$

Write the mole balance for C_1, P_1, and P_2.

3.7. From the MWD relations derived in Problem 3.7, derive the following moment-generation relations for nylon 6 polymerization in batch reactors:

$$\frac{d\lambda_0}{dt} = -k_2 \lambda_0^2 + k_2' W(\lambda_1 - \lambda_0) + k_1 C_1 W - k_1' P_1$$

$$\frac{d\lambda_1}{dt} = k_1 W C_1 - k_1' P_1 + k_3 \lambda_0 C_1 - k_3'(\lambda_0 - P_1)$$

$$\frac{d\lambda_1}{dt} = k_1 W C_1 - k_1' P_1 + 2k_2 \lambda_1^2 + \frac{k_2'}{3}(\lambda_1 - \lambda_3) + k_3 C_1(\lambda_0 + 2\lambda_1)$$

$$+ k_3'(\lambda_0 - 2\lambda_1 + P_1)$$

Note that we need to know P_1, P_2, and λ_3 in order to solve λ_0, λ_1, and λ_2. However, we also note that the relation for P_2 involves P_3, the one for λ_3 involves λ_4, and so on. This is known as the moment closure problem; it is present in all reversible polymerization analyses. For nylon 6 polymerization, we assume the following relations:

$$[P_3] = [P_2] = [P_1]$$

$$\lambda_3 = \frac{\lambda_2(2\lambda_2\lambda_0 - \lambda_1^2)}{\lambda_1 \lambda_0}$$

This equation is known as the Schultz–Zimm relation.

3.9. To recognize asymmetric functional groups in diisocyanates, we use the notation $A_1 A_2$. If the diol is denoted by B_2, the reaction with hydroxyl groups (denoted by B) can be written as follows:

$$A_1A_2 + B \xrightarrow{k_1} *A_2A_1B$$

$$A_1A_2 + B \xrightarrow{k_2} *A_1A_2B$$

$$A_2A_1{}^* + B \xrightarrow{k_2^*} A_2A_1B$$

$$A_1A_2{}^* + B \xrightarrow{k_2^*} A_1A_2B$$

where A_1^* and A_2^* denote the polymer chain ends to distinguish monomeric functional groups. Carry out the mole balance for each functional group for batch reactors.

3.10. In order to produce flexible urethane foam, a blowing agent (e.g., water) is added to the reaction mass, which reacts with isocyanate functional group, giving CO_2 as follows:

$$NCO + H_2O \rightarrow NH_2 + CO_2$$

The CO_2 first escapes the reaction mass, but, with progress of the reaction, the viscosity increases and gas is trapped, giving cellular structure to the polymer. The amine group produced earlier can react with NCO,

$$NCO + NH_2 \rightarrow NHCONH$$

Water is represented, by CD and the reaction of CD can be written as follows:

$$A_2A_1 + CDR_1 \xrightarrow{R_1k_1} A_2A_1CD$$

$$A_1A_2 + CDR_1k_2 \xrightarrow{R_1k_2^*} A_1A_2CD$$

$$A_1^* + CDR_1 \xrightarrow{R_1k_1^*} A_1CD$$

$$A_2^*CDR_1 \xrightarrow{R_1k_2^*} A_2CD$$

$$A_2A_1 + D \xrightarrow{R_2k_1k_1} A_2D$$

$$A_1A_2 + D \xrightarrow{R_2R_1k_1} A_1D$$

$$A_1^* + D \xrightarrow{R_2R_1k_1^*} A_1D$$

$$A_2^* + D_1 \xrightarrow{R_2R_1k_2^*} A_1D$$

The formation of allophanate (M) and biuret (G) linkages can be schematically represented by the following:

$$A_2A_1 + E \xrightarrow{R_3 k_1} M + {}^*A_2$$

$$A_1A_2 + E \xrightarrow{R_3 k_2} M + {}^*A_1$$

$$A_1^* + E \xrightarrow{R_3 k_1^*} M$$

$$A_2^* + E \xrightarrow{R_3 k_2^*} M$$

$$A_2A_1 + F \xrightarrow{R_4 k_1} {}^*A_2 + G$$

$$A_1A_2 + F \xrightarrow{R_4 k_2} {}^*A_1 + G$$

$$A_1 + F \xrightarrow{R_4 k_1^*} G$$

$$A_2^* + F[R] \xrightarrow{R_4 k_2^*} G$$

Write down the mole balance for each species.

3.11. Write down the mole balance for functional groups in urea formaldehyde polymerization for models I and II given in Eqs. (3.7.24), (3.7.25), and (3.7.26).

3.12. Melamine has the formula

and is therefore hexafunctional. Proceeding in the same way as for urea formaldehyde, we can define 10 functional groups as in Problem 3.11. What are these functional groups.

3.13. For the following unequal reactivities, write down the kinetic mechanism of polymerization of melamine.

k_1 = Reaction between a CH_2OH group and a primary hydrogen.

k_2 = Reaction between a CH_2OH group and a secondary amide group.

k_4 = Reaction between two CH_2OH groups giving a methylene linkage (denoted by z) and a free-formaldehyde molecule. (Note that this does not change the nature of functional groups.)

k_4' = Reaction involving a bond, Z, and a methylene glycol molecule.

k_5 = Reverse reaction involving a bond and a water molecule. Decide the rate constants for each reaction step in terms of k_1 to k_5.

Propose models I and II as we did for the urea formaldehyde system.

3.14. Write down the mole balance for each species in Problem 3.13.

3.15. The epoxy prepolymer is prepared by step-growth polymerization of epichlorohydrin with a diol ($OH-R-OH$), which is normally bisphenol-A. After some time of polymerization, the reaction mass consists of the following:

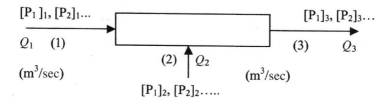

Write the mechanism that explains the formation of these molecules. Model the reaction kinetics after considering the kind of functional groups involved in the reaction.

3.16. For the mechanism developed in Problem 3.16, derive the mole balance for the MWD and functional groups in batch reactors.

3.17. Consider a mixture in which two streams enter and the product is assumed to be a homogeneous mixture of the two:

$$[P_1]_1, [P_2]_1...$$
$$Q_1 \quad (1)$$
$$(m^3/sec)$$

$$(2) \uparrow Q_2$$
$$[P_1]_2, [P_2]_2.....$$

$$[P_1]_3, [P_2]_3...$$
$$(3) \quad Q_3$$
$$(m^3/sec)$$

Develop a relation between the moment-generating functions $G(s, t)$ at points 1, 2, and 3.

3.18. Consider a batch reactor for carrying out irreversible step-growth polymerization described in Appendix 3.1. Let its feed consist of $[P_1]_0 + [P_2]_0$ instead of the pure monomer in Eq. (A3.1.12a). Find its MWD after time t.

3.19. Consider the multifunctional polymerization of RA_f monomers in batch reactors. The mole balance relations for various species are given by the following:

$$\frac{d[P_1]}{dt} = k\frac{f}{2}[P_1]\sum_{m=1}^{\infty}(mf+2-2m)[P_m]$$

$$\frac{d[P_n]}{dt} = \frac{k}{2}\sum_{m=1}^{n-1}(mf+2-2m)\left\{\frac{(n-m)f+2-2(n-m)}{2}\right\}[P_m][P_{n-m}]$$

$$-k(nf+2-2m)[P_n]\sum_{m=1}^{\infty}\frac{mf+2-2m}{2}[P_m],\quad n=2,3,\ldots$$

Derive generation relations for λ_0, λ_1, and λ_2.

3.20. The analytical solution of the MWD in Problem 3.20 has also been derived as

$$\frac{[P_n]}{[P_1]} = \frac{\{n(f-1)\}!f}{n!\{n\{f-2\}+2\}!}p_A^{n-1}(1-p_A)^{n(f-2)+2}$$

when the pure monomer feed has been used and p_A is the conversion of A groups given by

$$p_A = 1 - \frac{[A]}{[A]_0}$$

Derive expressions for λ_0, λ_1, and λ_2.

4

Reaction Engineering of Step-Growth Polymerization

4.1   INTRODUCTION [1,2]

Vessels in which polymerization is carried out are called reactors; they are classified according to flow conditions existing in them. Batch reactors (schematically shown in Fig. 4.1a) are those in which materials are charged initially and polymerization is carried out to the desired time, which in turn, depends on the properties of the material required. We have already observed that polymerization is limited in batch reactors by equilibrium conversion. Because we wish to form polymers of high molecular weights, we can overcome this limitation by applying high vacuum to the reaction mass. On application of low pressures, the reaction mass begins to boil and the condensation product is driven out of the reactor, as shown in Figure 4.1b. Such batch reactors are called *semibatch reactors*.

Batch and semibatch reactors are ideal when the production rate of the polymer needed is small. In larger-capacity plants, continuous reactors are preferred. In these, the raw materials are pumped in continuously while the products are removed at the other end. One example of these is a tubular reactor (shown in Fig. 4.1c). It is like an ordinary tube into which material is pumped at one end. Polymerization occurs in the tubular reactor, and the product stream consists of the polymer along with the unreacted monomer. Sometimes, a stirred vessel (shown in Fig. 4.1d) is employed instead of a tubular reactor. The advantage of such a reactor is that the concentration and temperature variations

153

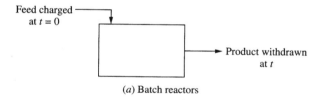

(*a*) Batch reactors

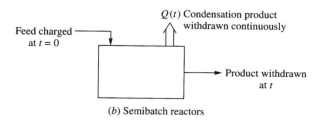

(*b*) Semibatch reactors

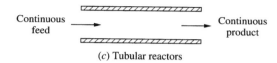

(*c*) Tubular reactors

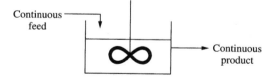

(*d*) Homogeneous continuous-flow stirred-tank reactors

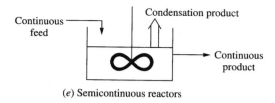

(*e*) Semicontinuous reactors

FIGURE 4.1   Some ideal reactors.

within it are removed due to vigorous stirring, making it possible to control reactor conditions more easily.

It may be mentioned that batch, semibatch, tubular, and stirred-tank reactors serve as mere idealizations of actual reactors. Consider, for example, the industrial V.K. tube (Vereinfacht Kontinuierliches Rohr) reactor, which is used for nylon 6 polymerization. Its schematic diagram is given in Figure 4.2a, in which ε-caprolactam monomer is mixed with water (serving as the ring opener) and introduced as feed. In the top region, the temperature is about 220–270°C and the reaction mass is vigorously boiling. The rising vapors produce intense agitation of the reaction mass and ultimately condense in the reflux exchanger. A small amount of ε-caprolactam also evaporates in this section of the reactor and

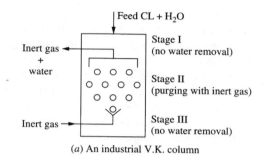

(a) An industrial V.K. column

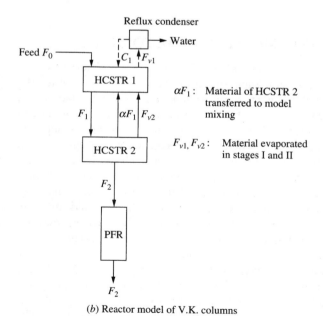

(b) Reactor model of V.K. columns

FIGURE 4.2 Schematic diagram of industrial nylon 6 reactors and reactor model. (Reprinted from Ref. 1 with the permission of Plenum Publishing Corporation.)

is recycled to the reactor, as shown. As the material moves downward, the reactor pressure increases due to gravity and the boiling of the reaction mass stops. In the second stage, most of the ε-caprolactam is reacted, and, in order to push the polymerization to high conversions, it is desired to remove the condensation product (water) from the reaction mass. To facilitate this, the reaction mass is purged with a suitable inert gas (say, nitrogen). In the third stage, the viscosity of the reaction mass is very high and water cannot be removed by purging anymore. Sufficient residence time is provided so as to achieve the desired molecular weight of the polymer. Figure 4.1 shows simple reactors and Figure 4.2b models complex reactors (e.g., V.K. tubes for nylon 6) in terms of a combination of these. Due to intense agitation existing in the first two stages, the entire V.K. column has been viewed as a train of two homogeneous continuous-flow stirred-tank reactors (HCSTRs) followed by a plug flow reactor.

From the example of the V.K. tube for nylon 6, we observe that simple reactors (Fig. 4.1) are building blocks of more complex ones. This chapter focuses on analyzing simple reactors carrying step-growth polymerization. Chapter 3 has already considered polymerization in the batch reactor. We first study the performance of semibatch reactors and examine the effect of flashing of the condensation product on it.

4.2 ANALYSIS OF SEMIBATCH REACTORS [1,3]

We have already observed in earlier chapters that engineering materials should have a large average chain length. Suppose it is desired to have μ_n equal to 100, which would imply a 99.9% conversion of functional groups. Step-growth polymerization is limited by its equilibrium conversion, and there is a need to push the reaction in the forward direction. This is done in industry by applying high vacuum to the reaction mass, whereupon the reaction mass begins to boil under the applied low pressure. We know that polymer chains have very low vapor pressures and, under normal conditions of operation, they do not vaporize; however, the monomer can. This clearly means that in the presence of flashing, the concentration of any given species changes not only by polymerization but also by change in volume V of the reaction mass. We show the schematic diagram of the semibatch reactor in Figure 4.3, and in the analysis presented here, we consider the change in V as an explicit variable. We assume that under the existing reactor conditions, the condensation product W and the monomer P_1 can flash out of the reactor. In all semibatch reactors, the monomer in the vapor phase is condensed in a suitable separator and recycled because of its high cost. It is assumed that the reactor is ε operating isothermally, at total pressure P_T. The volume of the liquid phase of the reactor, V, changes with time as flashing of W and P_1 occurs. We account for this time dependence as follows. We define p_n as the *total moles* of species ($P_n(n = 1, 2)$ and w as total moles of W in the liquid phase. The mole balance relations of these, on the dotted control volume shown in Figure 4.3, are given by

$$\frac{dp_1}{dt} = -\frac{k_p \lambda_0 p_1 - k'_p w(\lambda_0 - p_1)}{V} \tag{4.2.1a}$$

$$dp_n = [-2k_p p_n \lambda_0 + k_p \sum_{r=1}^{n-1} p_r p_{n-r}$$

$$- k'_p w(n-1)p_n + 2k'_p w \sum_{n+1}^{\infty} p_r]V^{-1}, \quad n \geq 2 \tag{4.2.1b}$$

$$\frac{dw}{dt} = \frac{k_p \lambda_0^2 - k'_p w(\lambda_1 - \lambda_0)}{V} - Q_w \tag{4.2.1c}$$

where k_p and k'_p are the forward and reverse rate constants, respectively and λ_0 and λ_1 the zeroth and first moments, which are defined as follows:

$$\lambda_0 = \sum p_n \tag{4.2.2a}$$
$$\lambda_1 = \sum np_n \tag{4.2.2b}$$

The zeroth moment λ_0 gives the total moles of polymer at any time, whereas λ_1 gives the total count of repeat units, which can be shown to be time invariant.

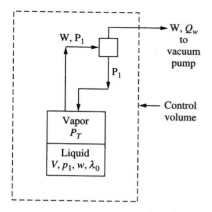

FIGURE 4.3 Schematic diagram of semibatch reactor with monomer and condensation product evaporating.

Equations (4.2.1a) and (4.2.1b) are suitably added to determine the generation relation of the zeroth moment λ_0 and first moment λ_1 as

$$\frac{d\lambda_0}{dt} = -\frac{k_p\lambda_0^2 - k_p'w(\lambda_1 - \lambda_0)}{V} \tag{4.2.3a}$$

$$\frac{d\lambda_1}{dt} = 0 \tag{4.2.3b}$$

Equation (4.2.3b) implies that the first moment λ_1 is time invariant and its value can be obtained from the feed conditions.

In order to solve for the molecular-weight distribution (MWD) of the polymer, as given by Eqs. (4.2.1) and (4.2.2), we must know the volume, V, of the liquid phase of the reactor and the rate of vaporization, Q_w. The rate of change of volume V is given by

$$\frac{dV}{dt} = -v_w Q_w \tag{4.2.4}$$

where v_w is the molar volume of the condensation product W.

In this development, there are seven unknowns [p_1, $p_n(n \geq 2)$, W, λ_0, λ_1, V, and Q_w], but we have only six ordinary differential equations [(4.2.1a)–(4.2.1c), (4.2.3a), (4.2.3b), and (4.2.4)] connecting them. Thus, one more equation is required. This is found by using the appropriate vapor–liquid equilibrium condition. Herein, to keep the mathematics simple, we assume the simplest relation given by Raoult's law.

4.2.1 Vapor–Liquid Equilibrium Governed by Raoult's Law

We assume that all the oligomers, p_n, $n \geq\, = 2$ are nonvolatile and that the condensation product W and the monomer P_1 can vaporize. If P_w^0 and P_{p1}^0 are the vapor pressures and x_w and x_{p1} are the mole fractions of W and P_1 respectively, then the partial pressures are given by Raoult's law as follows:

$$P_w = P_w^0 x_w \tag{4.2.5a}$$

$$P_{p_1} = P_{p_1}^0 x_{p_1} \tag{4.2.5b}$$

where

$$x_w = \frac{w}{\lambda_0 + w} \tag{4.2.6a}$$

$$x_{p_1} = \frac{p_1}{\lambda_0 + w} \tag{4.2.6b}$$

The total pressure P_T is then the sum of partial pressures; that is,

$$P_T = \frac{(P_0^0 P_1 + P_w^0 w)}{\lambda_0 + w} \tag{4.2.7}$$

4.2.2   Volume of Reaction Mass

The previous chapter shows that the MWD of the polymer obtained from batch reactors is given by Flory's distribution. Now, let us show that, in the presence of flashing, the MWD is still given by a similar relation. Let us assume that the feed to the semibatch reactor is pure monomer; that is, at $t = 0$,

$$p_1 = p_{10}, \tag{4.2.8a}$$
$$p_n = 0 \quad \text{for } n \geq 2 \tag{4.2.8b}$$

then

$$\lambda_1 = \lambda_{10} = p_{10} \tag{4.2.9a}$$
$$\lambda_{00} = p_{10} \tag{4.2.9b}$$

We propose that the MWD of the polymer is of the form

$$p_n = x(t)y(t)^{n-1} \tag{4.2.10}$$

where $x(t)$ and $y(t)$ are independent of the chain length n. On direct substitution of Eq. (4.2.10) into Eqs. (4.2.1a) and (4.2.1b), it is seen that the result satisfies the mole balance relation, no matter what the concentration of W. It is thus seen that the form of MWD remains unaffected by flashing. The $x(t)$ and $y(t)$ terms in Eq. (4.2.11), however, are now independent because of the invariance of

$$\sum P_n = \frac{x}{x-y} = \lambda_0 \tag{4.2.11a}$$
$$\sum nP_n = x(1-y)^2 = \lambda_1 \tag{4.2.11b}$$

These give

$$y(t) = 1 - \frac{\lambda_0}{\lambda_{10}} \tag{4.3.12a}$$
$$x(t) = \frac{\lambda_0^2}{\lambda_{10}} \tag{4.2.12b}$$

The addition of Eqs. (4.2.1c) and (4.2.3a) gives

$$\frac{d(w + \lambda_0)}{dt} = -Q_w \tag{4.2.13}$$

which, on substitution into Eq. (4.2.4), yields the following on integration:

$$V - V_0 = -v_w[(w_0 + \lambda_{10}) - (w + \lambda_0)] \tag{4.2.14}$$

Herein, w_0 is the moles of condensation product in the liquid phase having total volume V_0 at time $t = 0$.

4.2.3 Performance of the Semibatch Reactors

We rewrite the vapor–liquid equilibrium in Eq. (4.2.7) as follows:

$$w = \frac{P_T^0 \lambda_0 - P_{p1}^0 p_1}{P_w^0 - P_T} \tag{4.2.15}$$

However, Eq. (4.2.10) gives p_1 as $x(t)$, or λ_0^2/p_{10} [see Eq. (4.2.12)], which on substituting into Eq. (4.2.15), gives

$$W\left[\left(\frac{P_T}{P_w^0 - P_T}\right)\lambda_0\right] - \frac{[P_{p_1}^0/(P_w^0 - P_T)]\lambda_0^2}{p_{10}} \overset{\Delta}{=} a_1 \lambda_0 - a_2 \lambda_0^2 \tag{4.2.16a}$$

where

$$a_1 = \frac{P_{p_1}^0}{(P_w^0 - P_T)p_{10}} \tag{4.2.16b}$$

$$a_2 = \frac{P_{p_1}^0}{(P_w^0 - P_T)p_{10}} \tag{4.2.16c}$$

Between Eqs. (4.2.14) and (4.2.16), it is thus possible to explicitly relate V to λ_0:

$$V = b_0 + b_1 \lambda_0 - b_2 \lambda_0^2 \tag{4.2.17a}$$

where

$$b_0 = V_0 - v_w(w_0 + p_{10}) \tag{4.2.17b}$$
$$b_1 = v_w(a_1 + 1) \tag{4.2.17c}$$
$$b_2 = v_w a_2 \tag{4.2.17d}$$

We can now substitute Eq. (4.2.16a) for w and Eq. (4.2.17a) for V into Eq. (4.2.3a) to obtain the following:

$$(b_0 + b_1 \lambda_0 - b_2 \lambda_0^2)\frac{d\lambda_0}{dt} = k_p \lambda_0^2 + k_p'(\lambda_{00} - \lambda_0)(a_1 \lambda_0 - a_2 \lambda_0^2) \tag{4.2.18}$$

This can be integrated with the initial condition that λ_0 at $t = 0$ is the same as p_{10} and the final result can be derived as

$$A_1 \ln\left(\frac{\lambda_0}{\lambda_{00}}\right) + A_2 \ln\left(\frac{\lambda_0 - d_1}{\lambda_{00} - d_1}\right) - A_3 \ln\left(\frac{\lambda_0 - d_2}{\lambda_{00} - d_2}\right) = t \tag{4.2.19}$$

where

$$d_1 = \frac{g_2 + (g_2^2 - 4g_1g_3)^{0.5}}{2g_1} \tag{4.2.20a}$$

$$d_2 = \frac{g_2 - (g_2^2 - 4g_1g_3)^{0.5}}{2g_1} \tag{4.2.20b}$$

$$A_1 = \frac{b_0}{d_1 d_2 g_1} \tag{4.2.20c}$$

$$A_2 = \frac{b_0 + b_1 d_1 - b_2 d_1^2}{d_1 g_1 (d_1 - d_2)} \tag{4.2.20d}$$

$$A_3 = \frac{b_0 + b_1 d_2 - b_2 d_2^2}{g_1 d_2 (d_1 - d_2)} \tag{4.2.20e}$$

$$g_1 = k_p a^2 \tag{4.2.20f}$$

$$g_2 = k_p' + \lambda_{10} k_p' a_2 + k_p' a_1 \tag{4.2.20g}$$

$$g_3 = k' p a_1 \lambda_{10} \tag{4.2.20h}$$

When monomer P_1 has very low volatility and only water flashes,

$$a_2 = b_2 = 0 \tag{4.2.21a}$$

then

$$w = \frac{P_T \lambda_0}{P_w^0 - P_T} \equiv a_1 \lambda_0 \tag{4.2.21b}$$

$$V_0 = b_0 + b_1 \lambda_0 \tag{4.2.21c}$$

Equation (4.2.18) then becomes

$$(b_0 + b_1 \lambda_0)\frac{d\lambda_0}{dt} = -k_p \lambda_0^2 + k_p'(\lambda_{00} - \lambda_0)a_1 \lambda_0 \tag{4.2.22}$$

which can be integrated to

$$A_4 \ln\left(\frac{\lambda_0}{\lambda_{00}}\right) - \left(\frac{A_5}{g_2}\right) \ln[(g_2\lambda_0 - g_3)/(g_2\lambda_{00} - g_3)] = t \tag{4.2.23}$$

where

$$A_4 = \frac{b_0}{g_3} \tag{4.2.24a}$$

$$A_5 = \frac{g_2 b_0 + b_1 g_3}{g_3} \tag{4.2.24b}$$

Let us consider that some moles of monomer (say, p_{10}) are mixed with some moles (say, w_0) of condensation product before the mixture is charged to the reactor. As long as the constraint of vapor–liquid equilibrium [given in Eq. (4.2.7)] is not satisfied, there is no flashing of W and P_1, and the system behaves like a closed reactor. During polymerization, w increases and λ_0 decreases, and there is a time when the condensation product begins to evaporate. This time can be determined as follows. We observe that there is no flashing for closed reactors,

$$Q_w = 0 \tag{4.2.25a}$$

and Eqs. (4.2.13) and (4.2.14) reduce to

$$w + \lambda_0 = p_{10} + w_0 \tag{4.2.25b}$$

$$V = V_0 \tag{4.2.25c}$$

Substituting these into Eq. (4.2.3a) gives

$$V_0 \frac{d\lambda_0}{dt} = -k_p \lambda_0^2 + k_p'(w_0 + p_{10} - \lambda_0)(p_{10} - \lambda_0) \tag{4.2.26}$$

which can be easily integrated to give λ_0 from

$$\frac{q}{q_0} = \exp\left(-\frac{\delta t}{V_0}\right) \tag{4.2.27}$$

where

$$\delta = (m_1^2 + 4m_0 m_2)^{1/2} \tag{4.2.28a}$$

$$q = \frac{2m_2 \lambda_0 + m_1 - \delta}{2m_2 \lambda_0 + m_1 + \delta} \tag{4.2.28b}$$

$$q_0 = \frac{2m_2 p_{10} + m_1 - \delta}{2m_2 p_{10} + m_1 + \delta} \tag{4.2.28c}$$

$$m_0 = k_p'(w_0 + p_{10})p_{10} \tag{4.2.28d}$$

$$m_1 = k_p'(w_0 + 2p_{10}) \tag{4.2.28e}$$

$$m_2 = (k_p - k_p') \tag{4.2.28f}$$

Two situations are possible, relating to whether the monomer is flashing or not. When only W is evaporating, Eq. (4.2.21) holds for thermodynamic equilibrium and the intersection point is given by

$$\lambda_0^{c_1} = \frac{w_0 + \lambda_{00}}{1 + a_1} \tag{4.2.29}$$

where superscript c_1 stands for this evaporation condition (called case 1). This is now substituted into either Eq. (4.2.27) or Eq. (4.2.24). When P_1 as well as W evaporates (case 2), Eq. (4.2.25) is used to eliminate w from Eq. (4.2.16a), and $\lambda_0^{c_2}$ for this situation is determined from

$$\lambda_0^{c_2} = \frac{[(a_1 + 1) - 4a_2(w_0 + \lambda_{00})]^{1/2}}{2a_2} \tag{4.2.30}$$

This is, once again, substituted into Eq. (4.2.19) to get the time of transition.

Example 4.1: ARB polymerization is being carried out in a semibatch reactor. The feed is assumed to consist of 10 mol of monomer ($w_0 = 0$). At the reactor temperature, let us assume that the reactor pressure is 5 mm Hg and the vapor pressure of the condensation product is 38.483 mm Hg.

1. Determine the time of flashing.
2. Determine the values of λ_0 and w in the reactor at equilibrium and the moles of w flashed.
3. Calculate time taken to reach 101% of the equilibrium λ_0. Assume $k_p = 1$ and $k_p'(\beta) = 0.1$.

Solution: In normal conditions, the number of moles of monomer evaporating is usually small. If the evaporation of P_1 is small, $a_2 \sim 0$ and λ_0 at the transition is calculated from Eq. (4.2.25). Once this is known, the time when the flashing starts can be calculated from either Eq. (4.2.8) or (4.2.20).

The equilibrium in the presence of flashing is reached when $d\lambda_0/dt = 0$ or

$$-k_p \lambda_0^2 + k_p' w(\lambda_{10} - \lambda_0) = 0$$

where w is governed by Eq. (4.2.16a). On eliminating w here, we get

$$-k_p \lambda_{0eq}^2 + k_p' a_1 \lambda_{0eq}(\lambda_{10} - \lambda_0) = 0$$

which gives λ_{0eq} as

$$\lambda_{0eq} = \frac{a k_p' \lambda_{10}}{k_p' + a k_p'}$$

Let us assume that the units of k_p and k_p' are liters moles per hour. Calculations reveal that flashing starts at 0.015 h and λ_0 at this transition point is 8.7 mol/L. The equilibrium values of λ_0 and w are determined to be 0.1471 and 0.022 mol/L and 101% of this λ_0 is 0.1486 mol/L. In order to reach this value, the time needed is 16.62 h and the condensation product flashed is 8.5 mol.

Example 4.2: A mixture of monomer AR_1B is polymerized by the step-growth mechanism with a monofunction compound AR_2B in a batch reactor. The reaction mass consists of two molecules:

$$P_n : A[BA]_{n-1}B$$

and

$$P_{nx} = A[BA]_{n-1}X$$

Determine the MWD of the polymer formed in a batch reactor.

Solution: In reactor applications, recycling is common (see Problem 4.5) and monofunctional compounds are added to control the molecular weight of the formed polymer. The overall polymerization is assumed to be irreversible and can be written as

$$\begin{aligned}
&P_m + P_n \xrightarrow{2k_p} P_{m+n} \\
&P_m + P_n \xrightarrow{k_p} P_{zm} \\
&P_m + P_{nx} \xrightarrow{k_p} P_{(m+n)x}
\end{aligned}$$

The MWD relations for constant reactor volume can be written as

$$\begin{aligned}
\frac{d[P_1]}{dt} &= -k_p[P_1]\lambda_{PO} - k_p[P_1]\lambda_{x0} \\
\frac{d[P_{1x}]}{dt} &= -k_p[P_{1x}]\lambda_{P0} \\
\frac{d[P_n]}{dt} &= k_p \sum_{r=1}^{n-1} [P_r][P_{n-r}] - k_p[P_n]\{2\lambda_{P0} + \lambda_{x0}\} \\
\frac{d[P_{nx}]}{dt} &= k_p \sum_{r=1}^{n-1} [P_{rx}][P_{n-r}] - k_p[P_{nx}]\lambda_{P0}
\end{aligned}$$

where

$$\lambda_{P0} = \sum_{n=1}^{\infty} [P_n]$$

and

$$\lambda_{X0} = \sum_{n=1}^{\infty} [P_{nx}]$$

Let us define $G = \sum s^n[P_n]$ and $G_x = \sum s^n[P_{nx}]$, and with the help of the MWD relations, their time variation can be written as

$$\frac{dG}{d\theta} = -(2\lambda_{P0} + \lambda_{x0})G + G^2$$

$$\frac{dG_x}{d\theta} = -\lambda_{P0}G_x + G + G_x$$

where

$$\theta = k_p t.$$

These equations for $s \to 1$ also yield

$$\frac{d\lambda_{P0}}{d\theta} = -\lambda_{P0}(\lambda_{P0} + \lambda_{x0})$$

and

$$\frac{d\lambda_{x0}}{d\theta} = 0$$

that is, the monofunctional monomers grow in molecular weight but do not increase (or decrease) in total number of moles.

In line with the procedure given in Eq. (A3.1.7), for the solution of these equations, we define

$$y = \frac{G}{\lambda_{P0}}$$

and

$$y_x = \frac{G_x}{\lambda_{x0}}$$

In terms of these,

$$\frac{\partial y}{d\theta} = \frac{1}{\lambda_{P0}}\frac{\partial G}{\partial \theta} - \frac{G}{\lambda_{P0}^2}\frac{d\lambda_{P0}}{d\theta} = \lambda_{P0}y(1 - y)$$

and

$$\frac{\partial y_x}{d\theta} = \lambda_{P0}(y - 1)\lambda_{x0}$$

These can be integrated to give

$$\frac{y}{y - 1} = \frac{y_0}{y_0 - 1}\frac{\lambda_0 + \lambda_{0x}}{\lambda_{00} + \lambda_{0x}} \triangleq \frac{y_0}{y_0 - 1}(1 - p_A)$$

or

$$y(y_0 - 1) = y_0(y - 1)(1 - p_A)$$

or

$$y\{y_0 - 1 - y_0(1 - p_A)\} = y_0(1 - p_A)$$

where

$$1 - p_A = \frac{\lambda_0 + \lambda_{0x}}{\lambda_{00} - \lambda_{0x}}$$

or

$$y = \frac{y_0(1 - p_A)}{1 - y_0 p_A}$$

Similarly,

$$\frac{dy_x}{dy} = \frac{y_x}{y}$$

$$\ln\left(\frac{y_x}{y_{x0}}\right) = \ln\left(\frac{y}{y_0}\right) = \ln\left(\frac{1 - p_A}{1 - y_0 p_A}\right)$$

or

$$y_x = \frac{y_{x0}(1 - p_A)}{1 - p_A y_0}$$

4.3 MWD OF ARB POLYMERIZATION IN HOMOGENEOUS CONTINUOUS-FLOW STIRRED-TANK REACTORS [4]

Chapter 3 derived the MWD of the polymer in batch reactors; Section 4.2 has shown that the flashing of condensation product does not affect the distribution. We have already observed that an HCSTR is a continuous reactor that is employed when large throughputs are required.

The HCSTR shown in Figure 4.1d or 4.1e is assumed to be operated isothermally and under steady-state conditions. For a general feed, there could be higher homologs in addition to the monomer and the product stream consists of various homologs. In an HCSTR, the concentration of various species in the exit stream is equal to the concentration inside the reactor because of its well-mixed condition. As observed earlier, polymerization is, in general, reversible, and depending on the reactor condition existing, the condensation product can flash. In the following, we assume the polymerization to be reversible. The mole balance equations for various oligomers can be derived using the following general relation:

$$V\frac{d[\text{species}]}{dt} = (\text{species in}) - (\text{species out}) + V_{r_B} = 0 \tag{4.3.1}$$

where V is the volume of the reactor and r_B is the rate of formation of the species by chemical reaction. Using this equation, it is possible to derive the following mole balance relations for all oligomers—assummg steady state and no change in density:

$$\frac{[P_1] - [P_1]_0}{\theta} = -2k_p[P_n]\lambda_0 + k_p[W]\sum_{i=2}^{\infty}[P_i] \tag{4.3.2a}$$

$$\frac{[P_n] - [P_n]_0}{\theta} = -2k_p[P_n]\lambda_0 + k_p[W]\sum_{r=1}^{r=1}[P_r][P_{n-r}] - k_p'(n-1)[W][P_n]$$

$$+ 2k_p'[W]\sum_{i=n+1}^{\infty}[P_1], \quad n = 2, 3, 4 \tag{4.3.2b}$$

$$\frac{[W] - [W]_0}{\theta} = -\frac{F_{w_l}}{V} + k_p\lambda_0^2 - k_p'[W](\lambda_{10} - \lambda_0) \tag{4.3.2c}$$

Here, F_{wl} is the rate of flashing of the condensation product from the HCSTR. The parameter θ is known as the reactor residence time and is equal to V/F, where F is the volumetric flow rate of the feed. We have already shown that, for batch reactors, the first moment, λ_1, of the MWD is time invariant. It can similarly be proved using Eqs. (4.3.2a) and (4.3.2b) that the same is true for HCSTRs. This fact means that

$$\lambda_{1,\text{feed}} = \lambda_{1,\text{product}} = \lambda_{10} \quad (\text{say}) \tag{4.3.3}$$

If we add Eqs. (4.3.24a) and (4.3.2b) for all n, we get

$$\frac{1}{\theta}\left(\sum_{n=1}^{\infty}[P_n] - \sum_{n=1}^{\infty}[P_n]_0\right) = -2k_p\lambda_0\sum_{n=1}^{\infty}[P_n] + k_p\sum_{n=2}^{\infty}\sum_{r=2}^{n-1}[P_r][P_{n-r}]$$

$$= -k_p'[W]\sum_{n=2}^{\infty}(n-1)[P_n]$$

$$+ 2k_p'[W]\sum_{n=2}^{\infty}\sum_{i=n+1}^{\infty}[P_i]$$

or

$$\frac{\lambda_0 - \lambda_{00}}{\theta} = -k_p\lambda_0^2 + k_p'[W](\lambda_{10} - \lambda_0) \tag{4.3.4}$$

It can be seen that Eqs. (4.3.2) representing the MWD are nonlinear, coupled algebraic equations, which means that they must be solved simultaneously by trial and error. This problem has been solved in the literature—it has been found that their numerical solution is usually slow and cumbersome. These can, however, be decoupled as follows:

$$\sum_{i=n+1}^{\infty} [\mathrm{P}_i] = \lambda_0 - [\mathrm{P}_n] - \sum_{i=1}^{n-1} [\mathrm{P}_i] \quad \text{for } n = 1, 2, 3 \dots . \tag{4.3.5}$$

Substituting this result into Eqs. (4.3.2) yields the following:

$$[\mathrm{P}_1]\left\{ \frac{1}{\theta} + 2k_p\lambda_0 + 2k'_p[\mathrm{W}] \right\} = \frac{[\mathrm{P}_1]_0}{\theta} + 2k'_p[\mathrm{W}]\lambda_0 \tag{4.3.6a}$$

$$[\mathrm{P}_n]\left\{ \frac{1}{\theta} + 2k_p\lambda_0 + k'_p[\mathrm{W}](n-1) \right\} = \frac{[\mathrm{P}_1]_0}{\theta} + 2k'_p[\mathrm{W}]\lambda_0 + k_p \sum_{r=1}^{n-1} [\mathrm{P}_r][\mathrm{P}_{n-r}]$$

$$- 2k'_p[\mathrm{W}] \sum_{i=1}^{n-1} [\mathrm{P}_i] \quad \text{for } n \geq 2$$

$$\tag{4.3.6b}$$

These equations are now in the decoupled form because $[\mathrm{P}_n]$ can be precisely calculated if $[\mathrm{P}_1], [\mathrm{P}_2], \dots, [\mathrm{P}_{n-1}]$ are known. A sequential computation of the MWD starting from $[\mathrm{P}_1]$ is possible now, provided the concentration of the condensation product, $[\mathrm{W}]$, and λ_0 within the reactor are known. These are determined as follows.

Let us first assume that the condensation product, W, is not flashing from the reactor. This means that in Eq. (4.3.2c), therefore, F_{w1} is zero. On adding this with Eq. (4.3.4), we obtain

$$\lambda_{00} + [\mathrm{W}]_0 = \lambda_0 + [\mathrm{W}] \tag{4.3.7}$$

This result is the same as Eq. (4.2.25b) and can be directly derived from the stoichiometry of polymerization. If $[\mathrm{W}]$ is eliminated between Eqs. (4.3.4) and (4.3.7), we obtain a quadratic expression in λ_0:

$$\lambda_0 = \frac{-e_2 + \sqrt{e_2^2 + 4e_2e_3}}{2e_1} \tag{4.3.8}$$

where

$$e_1 = k_p - k'_p \tag{4.3.9a}$$

$$e_2 = -\left[\frac{1}{\theta} + k'_p(\lambda_{10} + \lambda_{00} + w_0) \right] \tag{4.3.9b}$$

$$e_3 = \frac{\lambda_{00}}{\theta} + k'_p(\lambda_{00} + w_0) \tag{4.3.9c}$$

However, if the condensation product is flashing, there will be a vapor–liquid equilibrium within the reactor. For simplicity, it is assumed that the reaction mass is binary mixture consisting of polymer and condensation product. Their mole fractions, x_p and x_w, are given by

$$x_p = \frac{\lambda_0}{\lambda_0 + [W]} \tag{4.3.10a}$$

$$x_w = 1 - x_p \tag{4.3.10b}$$

If P_T is the reactor pressure and P_r^* is the vapor pressure of W, then the Raoult's law (assuming the polymer cannot be in the vapor phase),

$$A_{eq} = \frac{P_t}{P_r^*} = \frac{[W]}{\lambda_0 + [W]} \tag{4.3.11}$$

From this, [W] can be solved in terms of λ_0:

$$[W] = \frac{A_{eq}}{1 - A_{eq}} \lambda_0 \overset{\Delta}{=} A_x \lambda_0 \tag{4.3.12}$$

In this relation A_{eq} greater than or equal to unity implies that the vapor pressure of W is less than the applied pressure, P_T. This means that the condensation product would not flash from the reactor, and λ_0 is given y Eq. (4.3.8). Otherwise, Eq. (4.3.12) is used to eliminate [W] in Eq. (4.3.4), and λ_0 is solved. The moles of the condensation product, F_{w1}, can be calculated from Eq. (4.3.2) as

$$\frac{F_{w1}}{V} = \frac{[W] - [W]_0}{\theta} + k_p \lambda_0^2 - k_p'[W](\lambda_{10} - \lambda_0) \tag{4.3.13}$$

Consider the following computational scheme to find the MWD of the polymer formed in HCSTRs. First, we find out whether the condensation product is evaporating. If it is, [W] and λ_0 in the product stream are determined and F_{w_1} is calculated using Eq. (4.3.14). However, if F_{w_1} is zero or negative, there is no flashing of the condensation product, and we evaluate λ_0 and [W] using Eqs. (4.3.8) and (4.3.7). Once these are known, the MWD is determined through Eq. (4.3.6) by sequential computations.

4.4 ADVANCED STAGE OF POLYMERIZATION [5–11]

In several cases (e.g., in the manufacture of polyethylene terephthalate), the equilibrium constants of the reactions are such that one must remove the volatile condensation products by application of a vacuum in order to obtain a polymer having long chain lengths. Because the desired degree of polymerization is about 100 for PET, the conversion of over 99% of the functional group must be attained. Under such conditions, the viscosity of the reaction mass is very high and the diffusivities of the volatile condensation product are very low. Special wiped-film reactors operating under high vacuum are then required in order to increase the surface area and reduce the resistance to diffusion. The analysis presented in this section can be contrasted with that in Section 4.2, wherein the mass transfer resistance was assumed to be negligible.

One design of a wiped-film reactor is shown schematically in Figure 4.4, in which the reaction mass in the molten state flows downstream. The reactor is partially full and a high vacuum is applied inside. Inside the reactor, there is a rotating blade (not shown) that continually spreads the molten liquid as a thin film on the reactor wall and, after a certain exposure time, another set of blades scrape it off and mix it with the bulk of the liquid. It is expected that most of the condensation product, W, is removed from the film because it is thin and its area is large. If it is assumed that the material in the bulk is close to equilibrium conditions, the W removed from the film would perturb this equilibrium and, on mixing, the reaction in the bulk would be pushed in the forward direction. This physical picture of the wiped-film reactor suggests that the polymerization in the bulk is different from that occurring in the film. It is necessary that appropriate balance equations for the bulk and the film be written and solved. The solution is usually obtained numerically.

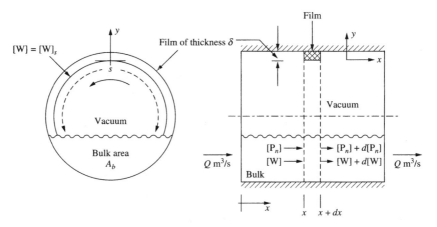

FIGURE 4.4 Schematic diagram of a wiped-film reactor. (Reprinted from Ref. 1 with the permission of Plenum Publishing Corporation.)

Depending on the relative amount of material in the film compared with that in the bulk, two models are possible for the wiped-film reactor. In one model, it is assumed that the entire reaction mass is applied as a thin film and there is no separate bulk phase present in Figure 4.4. This would mean that after some time of exposure, the entire film is well mixed instantaneously and applied once again with the help of the rotating blades. In the second model, it is assumed that the relative amount of material in the film is negligible compared with that in the bulk. Because the removal of condensation product in the film occurs by the mechanism of diffusion in a stationary film, the governing transport equations in the film are partial differential equations (see Appendix 4.1). On the other hand, in the bulk, where there is chemical reaction along with axial transport, the transport equations are ordinary differential equations. The performance of an actual wiped-film reactor lies between these two limiting models. Fortunately, the two limiting models give results that are not significantly different; thus, only one of them (the latter), which is more realistic, is described.

A differential element of the reactor is considered as shown in Figure 4.4. Mole balance equations on the condensation product W and the polymer molecules P_n are written as follows. The moles of W entering this element per unit time are $Q[W]$ and those leaving are $Q([W] + d[W])$. Meanwhile, $r_w A_b \, dx$ mol/sec are produced by polymerization (it is assumed that polymerization occurs primarily in the bulk) and $n_w a_s \, dx$ mol/sec are removed by evaporation from this differential element (through the film). Thus,

$$-Q \, d[W] = r_w A_b \, dx - n_w a_s \, dx = 0 \tag{4.4.1}$$

In Eq. (4.4.1), r_w is the molar rate of production of the condensation product in the bulk, n_w is the time-average removal rate of W from the film at position x, and a_s is the film surface area per unit reactor length. The actual mechanism of mass transfer in the film is extremely complex. Small bubbles are nucleated near the drum surface within the film. As shown in Figure 4.5, there is a diffusion of W into these, and the bubbles grow in size. For simple ARB reversible polymerization, it is observed that every functional group reacted produces a molecule of W. Hence, r_w can be written with the help of Eq. (2.4.4) as

$$r_w = k_p \lambda_0^2 - \frac{[W](\lambda_{10} - \lambda_0)}{K} \tag{4.4.2}$$

where

$$K = \frac{k_p}{k_p'} \tag{4.4.3}$$

λ_{10} is the feed concentration of total $-A$ or $-B$ groups (i.e., both reacted and unreacted), so the term $(\lambda_{10} - \lambda_0)$ is the concentration of the reacted $-AB-$

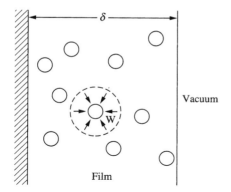

FIGURE 4.5 Diffusion of condensation product, W, toward bubbles moving in the film.

groups. Similarly, it is possible to derive the mole balance equation for P_n in the bulk as

$$-Q\,d[P_n] + r_{P_n} A_b\,dx = 0, \quad n = 1, 2, 3 \tag{4.4.4}$$

where r_{P_n} is the rate of formation of P_n in the bulk, given by

$$r_{P_n} = -2k_p\lambda_0[P_n] + k_p \sum_{r=1}^{n-1} [P_r][P_{n-r}] \tag{4.4.5}$$

$$- k_p'(n-1)[W][P_n] + 2k_p'[W] \sum_{i=n+1}^{\infty} [P_i]$$

Equations (4.4.1) and (4.4.4) can be solved numerically using the Runge–Kutta technique when n_w is known. In order to do so, the mass transfer problem in the film is first solved. This is discussed in Appendix 4.1, where an analytical solution is developed using a similarity transformation. From these results, it is possible to prepare a computer program that gives n_w.

The numerical solution of wiped-film reactors has been obtained by several researchers. It is well recognized that the most important parameter affecting the reactor performance is the film surface area, a_s, in Eq. (4.4.1). Qualitatively, a large a_s would give a higher rate of removal of the condensation product, which would, in turn, push the polymerization in the forward direction. Results of μ_n versus a_s at the reactor outlet are given in Figure 4.6, which shows the increasing trend. However, this increase in μ_n with a_s does not occur for all values of a_s because, beyond a critical value, the rate of mass transfer of W is no longer limiting, and the μ_n versus a_s curve begins to flatten out where the polymer formation is once again overall reaction controlled.

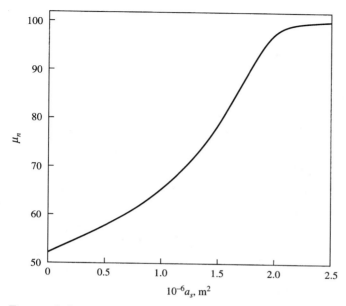

FIGURE 4.6 Average chain length μ_n at the end of the reactor versus surface area, a_s in the film.

Example 4.3: Write material balance equations in terms of Eq. (4.1.6) and solve for λ_{00} at the interface.

Solution: The concentrations of $[W]$ and λ_0 can be written as

$$[W] = [W]_0 = w_\eta([W]_e - [W]_0)$$

and

$$\lambda_{0\eta} = \lambda_{00}^* + \lambda_{0\eta}(\lambda_{0e} - \lambda_{00}^*)$$

where the subscript e denotes the equilibrium values. Then

$$\frac{\partial W}{\partial \theta} = ([W]_e - [W]_0)\frac{\partial w}{\partial \eta}\left(\frac{-y}{\delta^2}\right)\frac{d\delta}{d\theta}$$

$$= ([W]_e - [W])\frac{\eta}{\delta}\frac{\partial w_\eta}{\partial \eta}\frac{d\delta}{d\theta}$$

174

Chapter 4

Similarly

$$\frac{\partial \omega}{\partial y} = ([W]_e - [W]_0) \frac{1}{\delta} \frac{\partial w_\eta}{\partial \eta}$$

and

$$\frac{\partial^2 w}{\partial y^2} = ([W]_e - [W]_0) \frac{1}{\delta^2} \frac{\partial^2 w_\eta}{dy^2}$$

$$\frac{\partial \lambda_0}{\partial \theta} = -(\lambda_{0e} - \lambda_{00})\left(\frac{\eta}{\delta}\right)\frac{\partial \lambda_{0\eta}}{\partial \eta}\frac{d\delta}{d\theta} + (1 - \lambda_{0\eta})\frac{d\lambda_{00}}{d\theta} = -R_w$$

Note that the value of λ_{00} of the interface cannot be time independent if w_0 at the point is assumed to be fixed. Because the variation of λ_0 [i.e., Eq. A4.1.1b) is governed by a partial differential equation which does not have any derivative with respect to y, the time variation of λ_{00}^* at the interface is given by

$$\frac{d\lambda_{00}^*}{dt} = -R_w|_{y=y_0} = -k_p\lambda_{00}^{*2} + k_p'(\lambda_{10} - \lambda_{00}^*)[W]_0$$

If $[W]_0 = 0$, the above differential equation can be easily integrated as

$$\lambda_{00}^* = \frac{\lambda_{0e}}{1 + \theta\lambda_{0e}}$$

The balance relation for the condensation product becomes

$$-([W_e] - [W]_0)\left(\frac{\eta}{\delta}\right)\frac{dw_\eta}{d\eta}\frac{d\delta}{d\theta} = \left(\frac{D}{\delta^2}\right)([W_e] - [W]_0)\frac{d^2w_\eta}{d\eta^2} + r_w$$

4.5 CONCLUSION

This chapter has discussed the analysis of reactors for step-growth polymerization assuming the equal reactivity hypothesis to be valid. Polymerization involves an infinite set of elementary reactions; under the assumption of this hypothesis, the polymerization can be equivalently represented by the reaction of functional groups. The analysis of a batch (or tubular) reactor shows that the polymer formed in the reactor cannot have a polydispersity index (PDI) greater than 2. However, the PDI can be increased beyond this value if the polymer is recycled or if an HCSTR is used for polymerization. A comparison of the kinetic model with experimental data shows that the deviation between the two exists because of (1) several side reactions that must be accounted for, (2) chain-length-dependent reactivity, (3) unequal reactivity of various functional groups, or (4) complications caused by mass transfer effects.

In the final stages of polymerization, in fact, mass transfer of the condensation product must always be considered. Commercially, the reaction is then carried out in special wiped-film reactors. The final transport equations can be numerically solved, and among the various reaction parameters, the film surface area a_s is found to be the most important. The average chain length, μ_n, at the end of the reactor is found to increase with increasing a_s up to some critical value beyond which the overall polymerization becomes reaction controlled.

APPENDIX 4.1 SIMILARITY SOLUTION OF STEP-GROWTH POLYMERIZATION IN FILMS WITH FINITE MASS TRANSFER [12]

In wiped-film reactors, thin films are generated in order to facilitate mass transfer of the condensation product. Because the diameter of the drum in which the film is generated is usually large, we can ignore its curvature, treating it approximately as a flat film.

It may be observed that the flat films in commercial reactor operation are normally heterogeneous in nature. This heterogeneity arises because of the way the condensation product W travels from inside the film to the interface. Bubbles of W are nucleated at the metallic wall of the reactor, and these slowly travel toward the interface. W from the adjoining area diffuses to these, as shown in Figure 4.5. The diffusion of W into a single bubble has been the subject of several studies, but it is difficult to apply this concept to wiped-film reactors. This is because we must know the size and the number density of these bubbles, which is not easily amenable to either experimental measurements or theoretical calculations. In view of this difficulty, it would be erroneous to estimate the surface S per unit area in Eq. (4.4.1) as the wall area as shown in Figure 4.3. In fact, the surface per unit area has been treated as an adjustable parameter, assuming that the film is homogeneous.

For a given level of vacuum applied, the interfacial concentration of W is given by vapor–liquid equilibrium relationships. To keep the mathematics simple, here we assume that $[W]_s$ is governed by Raoult's law. It is further assumed that polymer molecules cannot volatilize and, if the stationary film is treated as a flat plate, the film equation at the axial position x will be given by

$$\frac{\partial [W]}{\partial t} = D^* \frac{\partial^2 [W]_0}{\partial y^2} + k_p \lambda_2^* - k_p' - (\lambda_{10}^* - \lambda_0^*)W \qquad (A4.1.1a)$$

$$\frac{\partial \lambda_0^*}{\partial t} = -k_0 \lambda_0^{*2} + k_p'(\lambda_{10}^* - \lambda_0^*)W \qquad (A4.1.1b)$$

In Eq. (A4.1.1), D^* represents the diffusivity of the condensation product and λ_0^* and λ_1^* are the zeroth and first moments, defined as

$$\lambda_0^* = \sum_{n=1}^{\infty} [P_n], \qquad \lambda_1^* = \sum_{n=1}^{\infty} n[P_n] \qquad (A4.1.2)$$

Because the wall is impervious to w and λ_0^* and there is a vapor–liquid equilibrium existing at the interface, it is possible to write the boundary conditions for Eq. (A4.1.1) as follows:

$$\left.\frac{\partial[W]}{\partial y}\right|_{y=y_0} = 0, \qquad \left.\frac{[W]}{[W] + \lambda_0^*}\right|_{y=0} = \frac{P_T}{P_w^0} \qquad (A4.1.3)$$

Becaue the entire film is initially at equilibrium, the initial conditions at $t = 0$ and at all positions y are given by

$$[W] = [W]_e, \qquad \lambda_0^* = \lambda_{0e}^* \qquad (A4.1.4)$$

Note that λ_1^* represents the concentration of repeat units in the reaction mass and is time and space invariant.

There is no analytical solution to Eq. (A4.1.1); it can only be solved numerically. In these computations, it has been found that, for short times, the film can always be divided into interfacial and bulk regions, as shown in Figure 4.7. The bulk region is the region where the diffusion of W has not had any effect. The thickness of the interfacial region, δ, can be numerically determined by observing that the concentrations of W and polymer P have flat spatial profiles. In

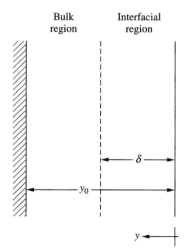

FIGURE 4.7 Schematic diagram shown the interfacial and bulk regions within the films. (Reprinted from Ref. 11 with the permission of VCH Verlagsgesellschaft mbH.)

general, the thickness of the interfacial region for W, δ_w, and that for polymer P, $\delta_{\lambda 0}$, are expected to be different. However, computations have shown that they are equal, as follows:

$$\delta_w = \delta_{\lambda 0} = \delta \quad \text{(say)} \tag{A4.1.5}$$

On application of vacuum in the gas phase, the interface concentrations of W and P, $[W]_0$ and λ_{00}^*, respectively, both change with time. However, due to thermodynamic equilibrium, they are constrained to satisfy Eq. (A4.1.3). We now define the following dimensionless variables:

$$w_\eta = \frac{[W] - [W]_0}{[W]_e - [W]_0}$$

$$\lambda_{0\eta} = (\lambda_0^* - \lambda_{00}^*)(\lambda_{0e}^* - \lambda_{00}^*) \quad \text{and } \eta = \frac{y}{\delta} \tag{A4.1.6}$$

Results for w as a function of η are shown in Figure 4.8; these have been computed numerically according to Eq. (A4.1.6). Similar results are obtained for $\lambda_{0\eta}$. We find that the results are time invariant. For long times, the interfacial

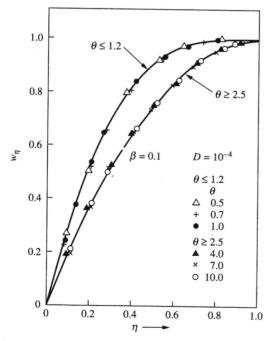

FIGURE 4.8 Similarity profile of condensation product, W, for different dimensionless time θ. (Reprinted from Ref. 11 with the permission of VCH Verlagsgesellschaft.)

region grows up to the wall, and after that, it becomes equal to the film thickness y_0.

In most applications, the film thickness y_0 is large and the situation of the growing interfacial region is more realistic. Based on the previous observations, let us now develop an analytical expression for δ as a function of time and then obtain an expression for the rate of removal of the condensation product.

A.4.1.1 Polynomial Approximation for w_η and $\lambda_{0\eta}$

The time invariance of w_η and λ_η suggest a solution scheme for which the following polynomial approximations are proposed:

$$w_\eta = a_{11}\eta + a_{12}\eta^2 + a_{13}\eta^3$$
$$\lambda_{0\eta} = a_{21}\eta + a_{22}\eta^2 + a_{23}\eta^3 + a_{24}\eta^4 \tag{A4.1.7}$$

These are chosen such that all boundary conditions on w_η and $\lambda_{0\eta}$ are rigorously satisfied. At $\eta = 0$,

$$w_\eta = \lambda_{0\eta} = \frac{\partial \lambda_0}{\partial \eta} = 0 \tag{A4.1.8}$$

At $\eta = 1$,

$$w_\eta = \lambda_{0\eta} = 1, \qquad \frac{\partial \lambda_{w\eta}}{\partial \eta} = \frac{\partial \lambda_{0\eta}}{\partial \eta} = 0$$

On satisfaction of these boundary conditions, we have the following:

$$w_\eta = 3\eta - 3\eta^2 + \eta^3 \quad \text{and} \quad \lambda_{0\eta} = 6\eta^2 - 8\eta^2 + 3\eta^4 \tag{A4.1.9}$$

A.4.1.2 Governing Ordinary Differential Equations for w_η and $\lambda_{0\eta}$

The time invariance of w_η and λ_η is exploited in rewriting the invariant, as shown in Eq. (A4.1.1). When this is done, we find that ordinary differential equations govern w_η and $\lambda_{0\eta}$. These involve time variation of interface concentrations of condensation product ($= [W]_0$) and polymer ($-\lambda_{0e}^*$).

To avoid giving excessive mathematical detail here, we refer interested readers to the literature [12]. To give an outline, we define the following dimensionless variables:

$$\theta = k_p \lambda_{10}^* t, \qquad \beta = \frac{k_p'}{k_p}, \qquad D = \frac{D^*}{k_p \lambda_1^*}$$

$$w_0 = \frac{[W]_0}{\lambda_1^*}, \qquad w_\theta = \frac{[W]_0}{\lambda_1^*}, \qquad \lambda_0 = \frac{\lambda_0^*}{\lambda_1^*}, \qquad \lambda_{0e} = \frac{\lambda_{0e}^*}{\lambda_1^*}$$

TABLE A4.1 Various Constants and Relations Governing Film Thickness

---

Constants arising from averaging of profiles:

$$\begin{aligned} \alpha_0 &= -3 & \alpha_1 &= 0.75 & \alpha_2 &= 0.6 \\ \alpha_3 &= -0.4536 & \alpha_4 &= 2.4857 & \alpha_5 &= 0.25 \end{aligned} \tag{A1}$$

Time variation of zeroth moment, λ_{00}, at the interface:

$$a_w = P_T/(P_w^0 - P_T)$$
$$\beta = k_p'/k_p$$
$$a_i = (1 + \beta a_w)\lambda_{00}^0$$
$$b_i = \beta a_w \tag{A2}$$
$$c_i = (a_i - b_i)$$
$$\lambda_{00} = \frac{b_i \lambda_{00}^0}{a_i - c_i \exp(-b_i \theta)}$$

Film thickness as a function of λ_{00}:

$$a = (1 - 2\alpha_2 + \alpha_4) + a_w \beta(1 - \alpha_1 - \alpha_2 + \alpha_5)$$
$$b = \alpha\beta[\lambda_{0e}(\alpha_2 - \alpha_3) - (1 - \alpha_1)] + \beta w_e(\alpha_1 - \alpha_2) + 2\lambda_{0e}(\alpha_2 - \alpha_4) \tag{A3}$$
$$c = (\alpha_3 \lambda_{0e} - \alpha_1)\beta w_e + \alpha_4 \lambda_{0e}^2$$

$$a_{n1} = 2[a + (1 + b_i)(1 - \alpha_1)a_w]/\alpha_5$$
$$b_{n1} = 2[b - b_i(1 - \alpha_1)a_w]/\alpha_5$$
$$c_{n1} = 2c/\alpha_5$$
$$a_{d1} = a_w(1 + \beta a_w) \tag{A4}$$
$$b_{d1} = -(1 + \beta a_w)w_e + \beta a_w^2$$
$$c_{d1} = \beta a_w w_e$$

$$f_1(\lambda_{00}) = \frac{a_{n1}\lambda_{00}^2 + b_{n1}\lambda_{00} + c_{n1}}{\lambda_{00}\{a_{d1}\lambda_{00}^2 + b_{d1}\lambda_{00} + c_{d1}\}}$$

$$g_1(\lambda_{00}) = \frac{[(a_w\lambda_{00}) - w_e](2D\alpha_0/\alpha_5)}{\lambda_{00}(a_{d1}\lambda_{00}^2)} + b_{d1}\lambda_{00} + c_{d1}\frac{d\delta^2}{d\lambda_{00}} + f_1(\lambda_{00})\delta^2 \tag{A5}$$

TABLE A4.1 *(continued)*

Solution of film thickness, δ, in terms of λ_{00}:

$$u = \lambda_{00} - \lambda_{00}^0$$

$$\delta^2 = \sum_{i=1}^{n_c} r_i u^i \qquad \text{where } n_c \to \infty$$

$$a_{n2} = a_{n1}; \qquad b_{n2} - b_{n1} + 2a_{n1}\lambda_{00}^0; \qquad c_{n2} = a_{n1}\lambda_{00}^{02} + b_{n1}\lambda_{00}^0 + c_{n1}$$

$$a_{d2} = a_{d1}; \qquad b_{d2} = b_{d1} + 3a_{d1}\lambda_{00}^0; \qquad c_{d2} = 3a_{d1}\lambda_{00}^0 + 2b_{d1}\lambda_{00}^0 + c_{d1}$$

$$d_{d2} = a_{d1}\lambda_{00}^{03} + b_{d1}\lambda_{00}^{02} + c_{d1}\lambda_{00}^2; \qquad d_{n2} = \frac{2D\alpha_0 a_w}{\alpha_5}; \qquad e_{n2} = \frac{2D\alpha_0(a_w\lambda_{00}^0 - w_e)}{\alpha_5}$$

$$r_1 = e_{n2}/d_{d2}$$

$$r_2 = (d_{n2} - c_{n2}r_1 - c_{d2}r_1)/2d_{d2}$$

$$r_3 = -[(b_{n2}r_1 + c_{n2}r_1) + (b_{d2}r_1 + 2c_{d2}r_2)]/3d_{d2}$$

$$r_{n+1} = -\{[a_{n2} + (n-2)a_{d2}]r_{i-2} + [b_{n2} + (n-1)b_{d2}]r_{i-1}$$
$$+ [c_{n2} + nc_{d2}]r_i\}/[(n+1)d_{d2}], \qquad i \geq 3$$

Rate of evaporation of W at the interface:

$$n_w = D\frac{dw}{dy}\bigg|_{y=0} = 3D\frac{w_e - w_0}{\delta} \tag{A6}$$

The resultant ordinary differential equations involve interfacial concentrations λ_{00} and w_0, which are known precisely. When the assumed profiles of Eq. (A4.1.7) are substituted in these equations, we obtain the film thickness δ as a function of λ_{00} and various parameters governing it. These all are given in Table A4.1.

Analytical solution of film thickness. The relation governing film thickness $(= \delta)$ involves δ^2 and is a nonlinear first-order ordinary differential equation. The following series solution can easily be developed:

$$\delta^2 = \sum_{i=1}^{\infty} r_i(\lambda_{00} - \lambda_{00}^0)^i$$

The coefficients r_i can easily be obtained by substituting into Eq. (A3) of Table A4.1; these are given in Eq. (A6). Once δ is determined, the rate of removal of the condensation product, n_w, can be evaluated from

$$n_w = 3D(w_e - w_0)/\delta$$

On request the authors can provide a computer program that can be copied onto any personal computer. The program is efficient and always gives a convergent solution.

REFERENCES

1. Gupta, S. K., and A. Kumar, Reaction Engineering of Step Growth Polymerization, Plenum, New York, 1988.
2. Tirrel, M., R. Galvan, and R. L. Laurence, Polymerization Reactors, in Chemical Reaction and Reactor Engineering, J. J. Carberry and A. Verma (eds.), Marcel Dekker, New York, 1986.
3. Kumar, A., MWD in Reversible ARB Polymerization in HCSTRs with Monomers Exhibiting Unequal Reactivity, Macromolecules, 20, 220–226, 1987.
4. Secor, R. M., The Kinetics of Condensation Polymerization, AIChE J. 15(6), 861–865, 1969.
5. Hoftyzer, P. J., and D. W. van Krevelen, The Rate of Conversion in Polycondensation Processes as Determined by Combined Mass Transfer and Chemical Reaction, in Proceedings of Fourth European Symposium on Chemical Reaction Engineering, Brussels, 1968.
6. Shah, Y. T., and M. M. Sharma, Desorption with or Without Chemical Reaction, Trans. Inst. Chem. Eng., 54, 1–41, 1976.
7. Reinisch, G., H. Gajeswaki, and K. Zacharias, Extension of the Reaction Diffusion Model of Melt Polycondensation, Acta Polym. 31, 732–733, 1980.
8. Amon, M., and C. D. Denson, A Study of the Dynamics of Foam Growth: Analysis of the Growth of Closely Spaced Spherical Bubbles, Polym. Eng. Sci., 24, 1026–1034, 1984.
9. Amon, M. W., and C. D. Denson, Simplified Analysis of the Performance of Wiped Film Polycondensation Reactors, Ind. Eng. Chem. Fundam., 19, 415–420, 1980.
10. Kumar, A., S. Madan, N. G. Shah, and S. K. Gupta: Solution of Final Stages of Polyethylene Terephthalate Reactors Using Orthogonal Collocation Technique, Polym. Eng. Sci., 24, 194–204, 1984.
11. Khanna, A., and A. Kumar, Solution of Step Growth Polymerization with Finite Mass in Films with Vapour Liquid Equilibrium at the Interface in Polymer Reaction Engineering, in Polymer Reaction Engineering, K. H. Reichert and W. Griseler (eds.), VCH, Berlin, 1989.

PROBLEMS

4.1. Analyze step-growth polymerization in a tubular reactor and develop relations for the MWD. Carry out a suitable transformation to show that the MWD has the same form as for batch reactors.

4.2. Suppose the vapor–liquid equilibrium in Section 4.2 is governed by the following Flory–Huggins equations:

$$P_w = P_w^0 a_\phi \phi_w$$
$$P_{p_1} = P_{p_1}^0 a_\phi \phi_{p_1}$$

where θ_w and θ_{p1} are the volume fractions given by

$$\phi_\omega = \frac{wv_w}{V}$$

$$\phi_{p_1} = \frac{p_1 v_{p_1}}{V}$$

and a_ϕ is the activity coefficient, which for a high-molecular-weight polymer is

$$a_\phi = \exp(1 + \chi)$$

The term χ is a constant and is normally known. Proceed as in the text and develop the complete solution using the following steps:

(a) Develop an expression for V; similar to Eqs. (4.2.14) and (4.2.17).
(b) Find w similar to Eq. (4.3.15).
(c) Substitute this to get $(d\lambda_0/dt)$ similar to Eq. (4.2.18).

4.3. In Problem 4.2, assume that P_1 does not flash. Now, solve the differential equation governing λ_0 [similar to Eq. (4.2.18)].

4.4 In Problem 4.3, find the transition time [as in Eq. (4.2.29)] from the closed reactor operation to the semibatch reactor operation. Subsequently, develop a similar program on your personal computer.

4.5. Assume irreversible step-growth polymerization in a tubular reactor with recycle as follows:

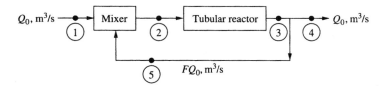

where F is the fraction recycled. Let us say that the feed consists of monomer AR_1B and monofunctional monomer AR_2X in the ratio $1:r$. We define conversion p_B of the B functional group as follows:

$$p_B \overset{\Delta}{=} (1 - f) = 1 - \frac{\lambda_{P_{04}}}{\lambda_{P_{01}}}$$

Find the total flow rate (m^3/sec), moles of B per second, moles of AR_2X per second, and concentrations λ_{0P} and λ_{0P_x} at the five points in the figure.

4.6. Carry out the time-dependent mole balance of an irreversible step-growth polymerization in HCSTRs. Derive the following MWD relations:

$$\frac{dC_1}{dt} = 1 - C_1 - C_1 \lambda_0 \tau^*$$

$$\frac{dC_n}{dt} = -C_n + \frac{1}{2}\tau^* \sum C_r C_{n-r} - C_n \lambda_0 \tau^*$$

where

$$C_n = \frac{[P_n]}{[P_1]_0}, \quad n = 1, 2$$

$$\tau = \frac{t}{V/Q} = \frac{\tau}{\theta}$$

$$\tau^* = 2k_p [P_1]_0$$

4.7. Under unsteady-state operation of an HCSTR in Problem 4.6, the following two common initial conditions arise:

IC1: At $t = 0$,

$$C_n = 0, \quad n = 1, 2$$

IC2: At $t = 0$,

$$C_1 = 1, \quad C_n = 0, \quad n = 1, 2$$

Derive the following general relation for $\lambda_0 (= \sum_{I=1}^{\infty} C_i)$ and show that the following is its solution:

$$\frac{d\lambda_0}{dt} = 1 - \lambda_0 - \frac{1}{2}\tau^{*2}\lambda_0^2$$

$$\lambda_0 = \frac{\theta' - 1}{\tau^*}\frac{1 - e^{-\theta\tau}}{1 - \delta e^{-\theta\tau}} \quad \text{for IC1}$$

$$\lambda_0 = \frac{\theta' - 1}{\tau^*}\frac{1 - (\beta/\delta)e^{-\theta'\tau}}{1 - \delta e^{-\theta'\tau}} \quad \text{for IC2}$$

where

$$\theta' = (1 + 2\tau^*)^{1/2}$$

$$\delta = \frac{1 - \theta'}{1 + \theta'}$$

$$\beta = \frac{1 - \theta' + \tau^*}{1 + \theta' + \tau^*}$$

4.8. Suppose the HCSTR of Problem 4.7 is operating at steady state. Show that the entire MWD could be obtained successively. Through successive elimination, the following solution has been derived in the literature:

$$C_n = g_n \frac{(\tau^*/2)^{n-1}}{(\theta')^{2n-1}}, \quad n = 1, 2, \ldots$$

where

$$g_1 = 1$$

$$g_n = \sum_{r=1}^{n-1} g_r g_{n-r}, \quad n = 2, 3$$

Find g_n up to $n = 11$. Note that g_n is independent of conditions existing in the reactor.

4.9. For Problem 4.8, derive expressions for λ_0, λ_1, and λ_2. Solve these for the steady state and determine the following chain length (μ_n) and weight-average (μ_n) molecular weights and the polydispersity index Q.

$$\mu_{ns} = \frac{\tau^*}{\sqrt{4\tau^{*2} + 1} - 1}, \qquad \mu_{ws} = 1 + \tau^*,$$

$$Q = (\sqrt{4\tau^{*2} + 1} - 1)(1 + \tau^*)$$

Note that Q can take on any value in HCSTRs during irreversible polymerization. It is not limited to a value of 2, as found for batch rreactors.

4.10. Design a computer program implementing Eqs. (4.3.5)–(4.3.13) for reversible step-growth polymerization in HCSTRs. Using the program, evaluate the entire MWD for given k_p, k_p', and θ. From this MWD, evaluate λ_0 and λ_2 and show that the polydispersity index Q for reversible polymerization does not increase forever, as predicted by Problem 4.9.

4.11. Consider the one-dimensional diffusion of condensation product through a film described in Appendix 4.1. The MWD of the polymer would be given by the following:

$$\frac{\partial[W]}{\partial t} = D^* \frac{\partial^2[W]}{\partial y^2} + k_p \lambda_0^{*2} - k_p'(\lambda_1^* - \lambda_0^*)[W]$$

$$\frac{\partial[P_1]}{\partial t} = -2k_p[P_1]\lambda_0^* + 2k_p'[W] \sum_{i=2}^{\infty}[P_i]$$

$$\frac{\partial[P_n]}{\partial t} = -2k_p[P_n]\lambda_0^* + k_p \sum_{i=1}^{n-1}[P_r][P_{n-r}] - k_p'(n-1)[W][P_n]$$

$$+ 2k_p'[W] \sum_{i=1}^{\infty}[P_i], \quad n \geq 2$$

Rewrite these partial differential equations using the following dimensionless variables:

$$\lambda_0 = \frac{\lambda_0^*}{\lambda_1^*}, \qquad \theta = k_p \lambda_1^* t$$

$$W = \frac{[W]}{\lambda_1^*} \qquad \beta = \frac{k_p'}{k_p}$$

$$P_n = \frac{[P_n]}{\lambda_1^*}, \qquad D = \frac{D^*}{k_p \lambda_1^*}$$

Also derive expressions for λ_0 in Eq. (A4.1.1b) starting from these MWD relations. Plausible initial and boundary conditions on the film are as follows:

1. Bulk at equilibrium with reaction and diffusion at the interface
2. Reaction and diffusion in the entire film

Write these initial and boundary conditions mathematically.

4.12. Suppose that we propose that the solution of the MWD in the film is

$$P_n(\theta, y) = X(\theta, y)Y(\theta, y)^{n-1}$$

Show that the MWD relations are satisfied by equations of Pb4.11. Also suggest the simplest form of $X(\theta, y)$ and $Y(\theta, y)$. Also demonstrate that this form is consistent with the equilibrium of polymerization. Try the form given in Eq. (3.5.2) first.

4.13. The rate of evaporation, N_w, of condensation product at $y = 0$ in the film of Figure 4.7 is given by

$$\bar{N}_w = \int_{t=0}^{t_f} -D \frac{\partial W}{\partial y} \big|_{y=0} \, dt$$

Also note that the film at $y = y_0$ is impervious [i.e., $(\partial w/\partial y)_{y=y_0} = 0$]. If we add a governing relation for λ_0 and w in Problem 4.11 and integrate with respect to y, we can determine N_w as

$$N_w = \frac{1}{t_f} \int_0^{t_f} D_w \int_{y=0}^{y_0} \frac{\partial}{\partial t}(w + \lambda_0) \, dy \, dt$$

$$= \frac{1}{t_f} \int_{y=0}^{y_0} [(W + \lambda_0)|_{t=0} - (W + \lambda_0)_{t_f}] \, dy$$

Change the order of integrate first and then complete the above derivation.

4.14. Let us assume that the film is initially at equilibrium and that at $y = 0$ (where vacuum has been applied), the concentration of condensation product is fixed at w_0. We define the following variables:

$$w_\eta = \frac{w - w_0}{w_e - w_0}$$

$$\lambda_{0\eta} = \frac{\lambda_0 - \lambda_{00}}{\lambda_{0e} - \lambda_{00}}$$

$$\eta = \frac{y}{\delta(\theta)}$$

where $\delta(\theta)$ is the thickness of the interfacial region (as shown in Fig. 4.7), which is time dependent. Rewrite the balance relations for w and λ_0 in Problem 4.13.

4.15. If we fix the concentration of the condensation product at w_0 in Problem 4.14, λ_{00} cannot be kept time invariant. At the interface, λ_{00} and w_0 satisfy the following differential equation:

$$\frac{d\lambda_{00}}{d\theta} = -\lambda_{00}^2 + \beta w_0(1 - \lambda_{00})$$

Find the reason for this and integrate this equation.

4.16. An ARB monomer is distributed uniformly in the form of droplets in a medium of high viscosity (such as an agar–agar solution). These droplets would therefore be almost immobile. Assuming the droplets to be of uniform size, obtain the rate of polymerization. If the polymerization is carried out to complete conversion, the resultant polymer would be in the form of beads.

4.17. In Example 3.2, the μ_n for 100% conversion is as follows:

$$\mu_n = \frac{1 + r}{1 - r} = \frac{N_{A0} + N_{B0}}{N_{A0} - N_{B0}} \quad \text{at } p = 1$$

This equation will be valid if the two monomers, A and B, are completely mixed and reacted to 100% conversion. Now, consider the flow reactor,

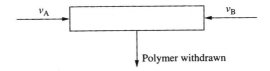

where v_A and v_B are the flow rates of the monomers (mol/sec) into the reactor. Show that the molecular weight of the resultant polymer for 100% conversion is given by

$$\mu_n = \frac{v_A + v_B}{v_A - v_B}$$

4.18. In Problem 4.17, if the reactor size is such that it is not possible to obtain 100% conversion, find the μ_n.

4.19. Consider the condensation polymerization of two monomers A and B that do not mix. In such cases, the monomers diffuse to the interface and polymerize there. Find the molecular weight of the polymer in terms of the diffusivities D_A and D_B if the interfacial reaction rate is very rapid.

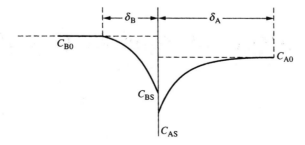

5

Chain-Growth Polymerization

5.1 INTRODUCTION

In step-growth polymerization, reactive functional groups are situated on each of the molecules, and growth of polymer chains occurs by the reaction between these functional groups. Because each molecule has at least one functional group, the reaction can occur between *any* two molecules. In chain-growth polymerization, on the other hand, the monomer polymerizes in the presence of compounds called *initiators*. The initiator continually generates growth centers in the reaction mass, which add on monomer molecules rapidly. It is this *sequential* addition of monomer molecules to growing centers that differentiates chain growth from step-growth polymerization.

Growth centers can either be ionic (cationic or anionic), free radical, or coordinational in nature—depending on the kind of initiator system used. Based on the nature of the growth centers, chain-growth polymerization is further classified as follows [1–3]:

1. Radical polymerization
2. Cationic polymerization
3. Anionic polymerization
4. Coordination or stereoregular polymerization

Initiators for radical polymerization generate free radicals in the reaction mass. For example, in a solution of styrene and benzoyl peroxide, the latter

188

dissociates on heating to benzoyloxy radicals, which combine with the styrene monomer to give growth centers as follows:

$$C_6H_5-\overset{\overset{O}{\|}}{C}-O-O-\overset{\overset{O}{\|}}{C}-C_6H_5 \xrightarrow{60-100°C} 2C_6H_5-\overset{\overset{O}{\|}}{C}-O-\dot{O} \qquad (5.1.1a)$$

$$C_6H_5-\overset{\overset{O}{\|}}{C}-\overset{\circ}{O} + CH_2{=}CH \longrightarrow C_6H_5-\overset{\overset{O}{\|}}{C}-O-CH_2-\overset{\circ}{C}H \qquad (5.1.1b)$$

$$C_6H_5-\overset{\overset{O}{\|}}{C}-O-CH_2-\dot{C}H + CH_2{=}CH \longrightarrow$$

$$(5.1.1c)$$

Growing chain

$$C_6H_5-\overset{\overset{O}{\|}}{C}-O-[CH_2-CH]_{\overline{n-1}}CH_2\dot{C}H$$

It is clear from Eqs. (5.1.1) that there are two types of radicals in the reaction mass:

1. Primary radicals which are generated by the initiator molecules directly, for example,

$$C_6H_5-\overset{\overset{O}{\|}}{C}-O\dot{}$$

2. Growing chain radicals; for example,

$$C_6H_5-\overset{\overset{O}{\|}}{C}-O-[CH_2-CH]_{\overline{n-1}}CH_2\dot{C}H$$

These are generated by the reaction between the primary radicals and the monomer molecules. Growing-chain radicals continue to add monomer molecules sequentially; this reaction is known as *propagation*. Reaction between a primary radical and a polymer radical or between two polymeric radicals would make polymer radicals unreactive by destroying their radical nature. Such *reactions* are called *termination* reactions. Thus, there are five kinds of species in the reaction mass at any time: initiator molecules, monomer molecules, primary radicals, growing-chain radicals, and terminated polymer molecules.

Cationic polymerization occurs in a similar manner, except for the fact that the initiator system produces cations instead of free radicals. Any catalyst system in cationic polymerization normally requires a cocatalyst. For example, protonic acid initiators (or catalysts) such as sulfuric acid, perchloric acid, and trifluoroacetic acid require a cocatalyst (e.g., acetyl perchlorate or water). Together, the two generate cations in the reaction mass. The reaction of boron trifluoride with water as the cocatalyst and styrene as the monomer is an example:

$$BF_3 + H_2O \rightleftharpoons (BF_3OH)^- H^+ \tag{5.1.2a}$$

$$CH_2{=}CH + (BF_3OH)^- H^+ \longrightarrow CH_3{-}CH^+{-}(BF_3OH)^- \tag{5.1.2b}$$

$$CH_3{-}CH^+{-}(BF_3OH)^- + CH_2{=}CH \longrightarrow$$

$$CH_3{-}CH{-}[CH_2{-}CH]_{n-1}{-}CH_2{-}CH^+{-}(BF_3OH)^-$$

$$\tag{5.1.2c}$$

The growth of the polymer chain occurs in such a way that the counterion (sometimes called a gegen ion) is always in the proximity of the growth center.

Anionic polymerization is caused by compounds that give rise to anions in the reaction mass. The compounds normally employed to initiate anionic polymerization are Lewis bases (e.g., primary amines or phosphenes), alkali metals (in the form of suspensions in hydrocarbons), or some organometallic compounds (e.g., butyl lithium). Sodium metal in the presence of naphthalene polymerizes styrene according to the following scheme:

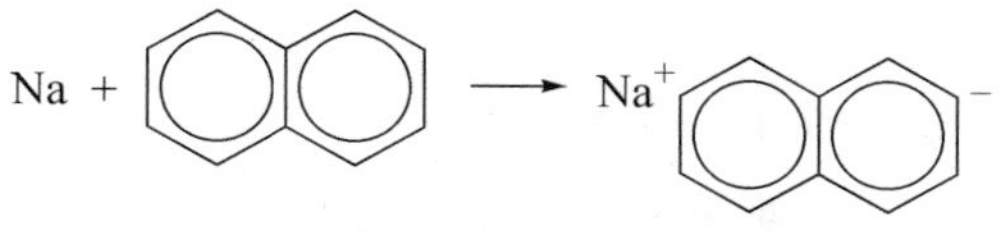

$$Na + \text{(naphthalene)} \longrightarrow Na^+ \text{(naphthalene)}^- \qquad (5.1.3a)$$

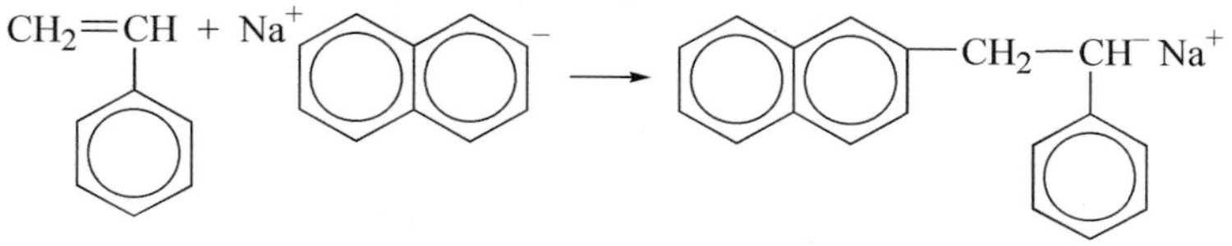

$$CH_2{=}CH(\text{phenyl}) + Na^+ \text{(naphthalene)}^- \longrightarrow \text{(naphthalene)}{-}CH_2{-}CH^-\,Na^+ (\text{phenyl}) \qquad (5.1.3b)$$

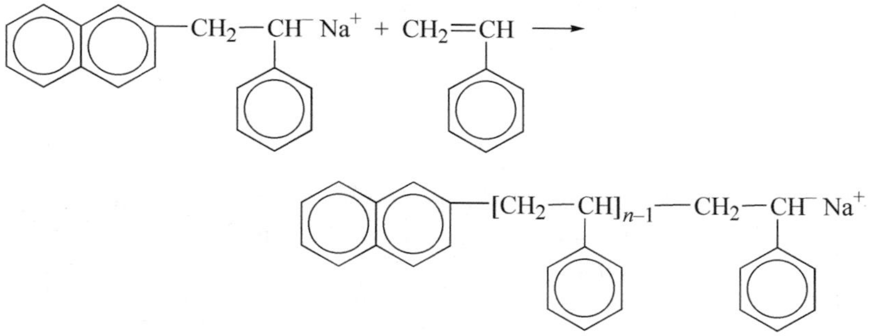

$$\text{(naphthalene)}{-}CH_2{-}CH^-\,Na^+(\text{phenyl}) + CH_2{=}CH(\text{phenyl}) \longrightarrow$$

$$\text{(naphthalene)}{-}[CH_2{-}CH]_{n-1}{-}CH_2{-}CH^-\,Na^+ (\text{phenyl})(\text{phenyl}) \qquad (5.1.3c)$$

As in cationic polymerization, there is a gegen ion in anionic polymerization, and the nature of the *gegen ion* affects the growth of the polymer chains significantly.

Coordination or stereoregular polymerization is carried out in the presence of special catalyst–cocatalyst systems, called Ziegler–Natta catalysts. The catalyst system normally consists of halides of transition elements of groups IV to VIII and alkyls or aryls of elements of groups I to IV. For example, a mixture of $TiCl_3$ and $AlEt_3$ constitutes the Ziegler–Natta catalyst system for the polymerization of propylene.

In all of the four classes of chain-reaction polymerization, the distinguishing feature is the existence of the propagation step between the polymeric growing center and the monomer molecule. This chapter discusses in detail the kinetics of these different polymerizations and the differences between the four modes of chain growth polymerization.

5.2 RADICAL POLYMERIZATION

In order to model radical polymerization kinetically, the various reactions—initiation, propagation, and termination—must be understood.

5.2.1 Initiation

By convention, the initiation step consists of two elementary reactions:

1. Primary radical generation, as in the production of

$$C_6H_5 - \overset{\overset{\displaystyle O}{\|}}{C} - \overset{\displaystyle \cdot}{O}$$

in Eqs. (5.1.1)

2. Combination of these primary radicals with a single monomer molecule, as in the formation of

$$C_6H_5COOCH_2 - \overset{\displaystyle \cdot}{C}H$$

The molecules of the initiator can generate radicals by a homolytic decomposition of covalent bonds on absorption of energy, which can be in the form of heat, light, or high-energy radiation, depending on the nature of the initiator employed. Commercially, heat-sensitive initiators (e.g., azo or peroxide compounds) are employed. Radicals can also be generated between a pair of compounds, called *redox initiators*, one of which contains an unpaired electron. During the initiation, the unpaired electron is transferred to the other compound (called the *acceptor*) and the latter undergoes bond dissociation. An example of the redox initiator is a ferrous salt with hydrogen peroxide:

$$Fe^{++} + H_2O_2 \xrightarrow{\text{Low temperature}} OH^- + Fe^{+++} + {}^{\cdot}OH \tag{5.2.1}$$

This section focuses on heat-sensitive initiators, primarily because of their overwhelming usage in industry. The homolytic decomposition of initiator molecules can be represented schematically as follows:

$$I_2 \xrightarrow{k_1} 2I \tag{5.2.2}$$

where I_2 is the initiator molecule [benzoyl peroxide in Eq. (5.1.1a)] and I is the primary radical [e.g.],

$$C_6H_5 - \overset{\overset{\displaystyle O}{\|}}{C} - \overset{\displaystyle \cdot}{O}$$

in Eq. (5.1.1a)]. The rate of production of primary radicals, r_i', according to Eq. (5.2.2), is

$$r_1' = 2k_1[I_2] \qquad (5.2.3)$$

where $[I_2]$ is the concentration of the initiator in the system at any time. The primary radicals, I, combine with a monomer molecule, M, according to the schematic reaction

$$I + M \xrightarrow{k_1} P_1 \qquad (5.2.4)$$

where P_1 is the polymer chain radical having one monomeric unit [e.g.,

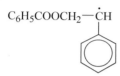

$$C_6H_5COOCH_2 \!-\! \overset{\textstyle\cdot}{C}H$$

in Eq. (5.1.1b)] and k_1 is the rate constant of this reaction. The rate of production, r_1, of the polymer radicals, P_1, can be written as

$$r_1 = k_1[I][M] \qquad (5.2.5)$$

where [I] and [M] are the concentrations of the primary radical and the monomer in the reaction mass, respectively.

Equations (5.2.2) and (5.2.4) imply that all the radicals generated by the homolytic decomposition of initiator molecules, I_2, are used in generating the polymer chain radicals P_1, and no primary radicals are wasted by any other reaction. This is not true in practice, however, and an initiator efficiency is defined to take care of the wastage of the primary radicals.

The initiator efficiency, f, is the fraction of the total primary radicals produced by reaction (5.2.2) that are used to generate polymer radicals by reaction (5.2.4). Thus, the rate of decomposition of initiator radicals is given by

$$r_i = -2fk_1[I_2] \qquad (5.2.6)$$

Table 5.1 gives data on fk_1 for the two important initiators, benzoyl peroxide and azobisdibutyronitrile, in various reaction media. If pure styrene is polymerized with benzoyl peroxide, the value of fk_1 for styrene as the reaction medium must be used to analyze the polymerization. However, if a solvent is also added to the monomer (which is sometimes done for better temperature control), say, toluene in styrene, it is necessary that the fk_1 corresponding to this reaction medium be determined experimentally.

The effect of the reaction medium on the initiator efficiency as shown in Table 5.1 has been explained in terms of the "cage theory." After energy is supplied to the initiator molecules, cleavage of a covalent bond occurs, as shown

TABLE 5.1 Typical Rate Constants in Radical Polymerizations

	Initiation rate constants		
Initiator	Reaction medium	Temp. (°C)	fk_1 (sec^{-1})
Benzoyl peroxide	Benzene	70.0	1.18×10^{-5}
	Toluene	70.3	1.10×10^{-5}
	Styrene^a	61.0	2.58×10^{-6}
	Polystyrene	64.6	1.47×10^{-6}
		56.4	3.8×10^{-7}
	Polyvinyl chloride	64.6	6.3×10^{-7}
Azobisdibutyronitrile	Benzene	69.5	3.78×10^{-5}
	Toluene	70.0	4.0×10^{-5}
	Styrene^b	50.0	2.79×10^{-6}
2-Ethyl hexylperoxy dicarbonate^a (used for polyvinyl chloride formation)		50.0	4.049×10^{-5}

	Initial rate constants k_p and k_t		
Monomer	Temp. (°C)	k_p (L/mol sec)	$k_t \times 10^{-6}$ (L/mol sec)
Acrylic acid	25	13.0	0.018
(n-butyl ester)	35	14.5	0.018
Methacrylic acid	30	369.0	10.2
(n-butyl ester)			
Styrene^b	60	176.0	72.0
	30	55.0	50.5
Vinyl acetate	25	1012.0	58.8
Vinyl chloride^a	50	1717.9	1477.0
Vinylidene	35	36.8	1.80
chloride	25	8.6	0.175

in Eq. (5.2.2). According to this theory, the two dissociated fragments are surrounded by the reaction mass, which forms a sort of cage around them. The two fragments stay inside the cage for a finite amount of time, during which they can recombine to give back the initiator molecule. Those fragments that do not recombine diffuse, and the separated fragments are called *primary radicals*. Various reactions can now occur: The primary radicals from different cages can either recombine to give an initiator molecule or react with monomer molecules to give P_1. If the monomer molecule is very reactive, it can also react with a

TABLE **5.1** (*continued*)

	Transfer rate constants	
	$(k_{tr}S/k_p) \times 10^4$	
Transfer agent	60°C	100°C
Cyclohexane	0.024	0.16
Benzene	0.018	0.184
Toluene	0.125	0.65
Ethylbenzene	0.67	1.62
Iso-propylbenzene	0.82	2.00
Vinyl chloride[a] (in polymerization of vinyl chloride)	14.19	34.59

[a] Calculated from
   $k_{d1} = 1.5 \times 10^{15} \exp(-14554/T)$
   $k_p = 5 \times 10^7 \exp(-3320/T)$
   $k_t = 1.3 \times 10^{12} \exp(-2190/T)$
   $k_{trM1}/k_p = 5.78 \exp(-2768/T)$
   PVC prepared by suspension polymerization
[b] More comprehensive rate constants valid in the entire domain of polymerization of styrene and methylmethacrylate are given in Tables 6.2 and 6.3. These are needed for detailed simulation of reactors. Notice that they are conversion dependent.
Source: Ref. 4.

fragment inside a cage. The cage effect can therefore be represented schematically as follows:

$$\boxed{I_2} \rightleftarrows \boxed{I:I} \qquad \text{Cage formation recombination} \qquad (5.2.7a)$$

$$\boxed{I:I} \rightleftarrows I + I \qquad \text{Diffusion out of cage} \qquad (5.2.7b)$$

$$I + M \longrightarrow P_1 \qquad \text{Formation of primary radicals with} \\ \text{monomer} \qquad (5.2.7c)$$

$$\boxed{I:I} + M \longrightarrow P_1 + I \qquad \text{Reaction with cage} \qquad (5.2.7d)$$

The characteristics of the reaction medium dictate how long the dissociated fragments will stay inside the cage: the medium affects the first and second reactions of Eq. (5.2.7) most significantly. It is therefore expected that, if all other conditions are equal, a more viscous reaction mass will lead to a lower initiator efficiency. This can be observed in Table 5.1 by comparing the values of fk_1 of benzoyl peroxide in styrene and the more viscous polystyrene.

5.2.2 The Propagation Reaction

Polymer chain radicals having a single monomer unit, P_1, are generated by the initiation reaction as previously discussed. The propagation reaction is defined as the addition of monomer molecules to the growing polymer radicals. The reaction mass contains polymer radicals of all possible sizes; in general, a polymer radical is denoted by P_n, indicating that there are n monomeric units joined together by covalent bonds in the chain radical. The propagation reaction can be written schematically as follows:

$$P_n + M \xrightarrow{k_{pn}} P_{n+1}, \quad n = 1, 2, \ldots \quad (5.2.8)$$

where k_{pn} is the rate constant for the reaction between P_n and a monomer molecule. In general, the constant depends on the size of the chain radical. It is not difficult to foresee the increasing mathematical complexity resulting from the multiplicity of the rate constants. As a good first approximation, the principle of equal reactivity is assumed to be valid, even in the case of polymer radicals, which means that

$$k_{p1} = k_{p2} = k_{p3} = k_p \quad (5.2.9)$$

and Eq. (5.2.8) reduces to

$$P_n + M \xrightarrow{k_p} P_{n+1}, \quad n = 1, 2, 3, \ldots \quad (5.2.10)$$

We learned in Chapter 3 that the principle of "equal reactivity" holds well for molecules having reactive functional groups. Even though the nature of the growth centers is different in addition polymerization, segmental diffusion is expected to play a similar role here, justifying the use of equal reactivity for these cases also. The results derived using Eq. (5.2.10) explain experimental data very well, further justifying its use.

5.2.3 Termination of Polymer Radicals

The termination reaction is the one in which polymer chain radicals are destroyed. This can occur only when a polymer radical reacts with another polymer radical or with a primary radical. The former is called mutual termination and the latter is called primary termination. These reactions can be written as follows:

$$P_m + P_n \xrightarrow{k_t} M_{n+m} \quad (5.2.11a)$$

$$P_m + I \xrightarrow{k_{t,1}} M_m, \quad m, n = 1, 2, 3, \ldots \quad (5.2.11b)$$

The term M_{m+n} signifies a dead polymer chain; that is, it cannot undergo any further propagation reaction. In the case of mutual termination, the inactive polymer chains can be formed either by combination or by disproportionation. In combination termination, two chain radicals simply combine to give an inactive chain, whereas in disproportionation, one chain radical gives up the electron to the other and both the chains thus become inactive. These two types of termination can be symbolically written as

$$P_m + P_n \xrightarrow{k_{tc}} M_{m+n} \quad \text{(combination)} \tag{5.2.12a}$$

$$P_m + P_n \xrightarrow{k_{td}} M_m + M_n \quad \text{(disproportionation)} \tag{5.2.12b}$$

where M_m has the saturated chain end and represents the inactive polymer chain to which the electron has been transferred; M_n represents the inactive chain with an unsaturated chain end; k_{tc} is the termination rate constant for the combination step, and k_{td} is the rate constant for the disproportionation step. Once again, the principle of equal reactivity is assumed to be valid in writing Eq. (5.2.12).

Transfer agents (denoted S) are chemicals that can react with polymer radicals, as a result of which S acquires the radical character and can add on the monomer exactly as P_1. The polymer radical thereby becomes a dead chain. The transfer reaction can be represented by

$$P_n S \xrightarrow{k_{tr}S} M_n + P_1, \quad n \geq 2$$

Similar reactions are found to occur quite commonly in radical polymerization with monomer as well as initiator. These are written as follows:

$$P_n + M \xrightarrow{k_{tr}M} M_n + P_1$$

$$P_n + I_2 \xrightarrow{k_{tr}M} M_n + P_1$$

5.3 KINETIC MODEL OF RADICAL POLYMERIZATION [5,6]

If the transfer reaction to the initiator and monomer is neglected, the overall mechanism of polymerization can be expressed as follows:

Initiation

$$I_2 \xrightarrow{k_1} 2I \tag{5.3.1a}$$

$$I + M \xrightarrow{k_1} P_1 \tag{5.3.1b}$$

Propagation

$$P_n + M \xrightarrow{k_{tr}S} P_{n+1}, \quad n = 1, 2, \ldots \tag{5.3.1c}$$

Termination

Chain transfer:

$$P_n + S \xrightarrow{k_{trS}} M_n + P_1 \tag{5.3.1d}$$

Combination:

$$P_n + P_m \xrightarrow{k_{tc}} M_{m+n} \tag{5.3.1e}$$

Disproportionation:

$$P_n + P_m \xrightarrow{k_{td}} M_n + M_m \tag{5.3.1f}$$

The mole balance equations for batch reactors are written for the species I_2, I, P_1, P_2, ..., P_n, as follows:

$$\frac{d[I_2]}{dt} = -k_1[I_2] \tag{5.3.2}$$

$$\frac{d[I]}{dt} = 2fk_1[I_2] - k_1[I][M] \tag{5.3.3}$$

$$\frac{d[P_1]}{dt} = k_1[I][M] - k_p[P_1][M] + k_{trS}(\lambda_{P0} - [P_1])[S]$$
$$- (k_{tc} + k_{td})\lambda_{P0}[P_1] \tag{5.3.4}$$

$$\frac{d[P_n]}{dt} = k_p[M]\{[P_{n-1}] - [P_n]\} - k_{trS}[S][P_n]$$
$$- (k_{tc} + k_{td})\lambda_{P0}[P_n], \quad n \geq 2 \tag{5.3.5}$$

where λ_{P0} is the total concentration of growing polymer radicals $(= \sum_{n=1}^{\infty} [P_n])$ and f is the initiator efficiency. Similar balance equations can be written for the monomer and the dead polymer:

$$\frac{d[M]}{dt} = -k_1[I][M] - k_p[M]\lambda_{P0} \tag{5.3.6}$$

$$\frac{d[M_n]}{dt} = k_{trS}[S][P_n] + k_{td}[P_n]\lambda_{P0} \frac{1}{2} k_{tc} \sum_{m=1}^{n-1} [P_m][P_{n-m}] \tag{5.3.7}$$

Assuming that the quasi-steady-state approximation (QSSA) is valid, the concentration of the intermediate species I can be found from Eq. (5.3.3):

$$[I] = \frac{2fk_1[I_2]}{k_1[M]} \tag{5.3.8}$$

Under QSSA, Eqs. (5.3.4) and (5.3.5) are summed for different values of n to obtain

$$\frac{d[\lambda_{P0}]}{dt} = \frac{d}{dt} \sum_{n=1}^{\infty} [P_n] = k_1[I][M] - 2(k_{tc} + k_{td})\lambda_{P0}^2 = 0 \tag{5.3.9}$$

In radical polymerization, the slowest reaction is the dissociation of initiator molecules. As soon as a primary radical is produced, it is consumed by the reactions of Eq. (5.3.1). Thus, the concentration of I is expected to be much less than that of P; that is,

$$[M] \gg \lambda_{P0} \gg [I] \tag{5.3.10}$$

To determine the rate of monomer consumption r_p for radical polymerization, observe that the monomer is consumed by the second and third reactions of Eq. (5.3.1). Therefore,

$$r_p = -\{k_1[I][M] + k_p[M]\lambda_{P0}\} \tag{5.3.11}$$

From Eq. (5.3.10), this can be approximated as

$$r_p \cong -k_p[M]\lambda_{P0} \tag{5.3.12}$$

To find λ_{P0}, consider the following equation, where [I] in Eq. (5.3.9) is eliminated using Eq. (5.3.8):

$$\lambda_{P0} = \left\{ \frac{fk_1[I_2]}{k_t} \right\}^{1/2} \tag{5.3.13}$$

The rate of propagation is thus given by

$$r_p = k_p \left\{ \frac{fk_1[I_2]}{k_t} \right\}^{1/2} [M] \tag{5.3.14}$$

where

$$k_t = k_{tc} + k_{td} \tag{5.3.15}$$

5.4 AVERAGE MOLECULAR WEIGHT IN RADICAL POLYMERIZATION

The average molecular weight in radical polymerization can be found from the kinetic model, Eq. (5.3.1), as follows. The kinetic chain length, v, is defined as the average number of monomer molecules reacting with a polymer chain radical during the latter's entire lifetime. This will be the ratio between the rate of consumption of the monomer (i.e., r_p) defined in Eq. (5.3.12) and the rate of generation of polymer radicals,

$$v = \frac{r_p}{r_i} \tag{5.4.1}$$

where r_i is the rate of initiation given by

$$r_i = k_1[I][M]$$

From the QSSA, the rate of initiation (r_i) should be equal to the rate of termination ($r_t = k_t\lambda_{P0}^2$). If transfer reactions are neglected,

$$v = \frac{r_p}{r_t} = \frac{k_p[M]\lambda_{P0}}{2k_t\lambda_{P0}^2} = \frac{k_p[M]}{2k_t\lambda_{P0}} = \frac{k_p[M]}{2k_t\lambda_{P0}} \tag{5.4.2}$$

Eliminating λ_{P0} with the help of Eq. (5.3.14),

$$v = \frac{k_p}{2(fk_I)^{1/2}} \frac{[M]}{2k_t\lambda_{P0}k_t^{1/2}} \tag{5.4.3}$$

Equation (5.4.3) shows that the kinetic chain length decreases with increasing initiator concentration. This result is expected, because an increase in $[I_2]$ would lead to more chains being produced. The quantity that is really of interest is the average chain length, μ_n, of the inactive polymers. Average chain length is directly related to ε; the former gives the average number of monomer molecules per *dead polymer* chain, whereas ε gives the average number of monomer molecules per growing polymer radical. To be able to find the exact relationships between the two, the mechanism of termination must be carefully analyzed. If the termination of polymer chain radicals occurs only by combination, each of the dead chains consists of 2ε monomer molecules. If termination occurs only by disproportionation, each of the inactive polymer molecules consists of ε monomer molecules; that is,

$$\mu_n = \begin{cases} 2v & \text{when termination is by combination only} \\ v & \text{when termination is by disproportionation only} \\ \alpha v & \text{when there is mixed termination, } 1 \leq \alpha \leq 2 \end{cases} \tag{5.4.4}$$

As the initiator concentration is increased, the rate of polymerization increases [Eq. (5.3.14)], but ε (and therefore μ_n) decreases [Eq. (5.4.3)]. Therefore, control of the initiator concentration is one way of influencing the molecular weight of the polymer.

Another method of controlling the molecular weight of the polymer is by use of a transfer agent. In the transfer reaction, the total number of chain radicals in the reaction mass is not affected by Eq. (5.3.14). It therefore follows that the presence of a transfer agent does not affect r_p. However, the kinetic chain length,

ε, does change drastically depending on the value of $k_{tr}S$ and [S]. Equation (5.4.2) can easily be modified to account for the presence of transfer agents:

$$v = \frac{r_p}{2k_t[P]^2 + k_{trS}[P][S]}$$ (5.4.5)

or, taking the reciprocal,

$$\frac{1}{v} = \frac{k_{trS}[S]}{k_p[M]} + \frac{2(k_t fk_1)^{1/2}}{k_p} \frac{[I_2]^{1/2}}{[M]}$$ (5.4.6)

Equation (5.4.6) predicts a decrease in μ_n with increasing concentration of the transfer agent.

5.5 VERIFICATION OF THE KINETIC MODEL AND THE GEL EFFECT IN RADICAL POLYMERIZATION

To verify the kinetic model of radical polymerization presented in the last section, the following assumptions must be confirmed:

1. r_p should be independent of time for a given [M] and $[I_2]$ to verify the steady-state approximation.
2. r_p should be first order with respect to monomer concentration.
3. r_p should be proportional to $[I_2]^{1/2}$ if the decomposition of the initiator is first order.

The validity of the steady-state approximation has been shown to be extremely good after about 1–3 min from the start of the reaction [1,3]. The polymerization of methyl methacrylate (MMA) has been carried out to high conversions, and the plot of the percent polymerization versus time is displayed in Figure 5.1 [2]. In Figure 5.2, the corresponding average chain length of the found polymer is shown as a function of time [7–10]. The behavior observed in these figures is found to be typical of vinyl monomers undergoing radical polymerization. On integrating Eq. (5.3.14), we obtain for constant $[I_2]$

$$-\ln(1-x) = k_p \left(\frac{fk_1}{k_t}\right)^{1/2} [I_2]^{1/2} t$$ (5.5.1)

where the conversion x is defined as

$$x = \frac{[M]_0 - [M]}{[M]_0}$$ (5.5.2)

According to this equation, the plot relating the monomer conversion and time should be exponential in nature and independent of $[M]_0$. Because $[M]_0$ depends

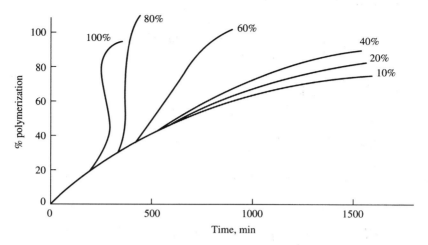

FIGURE 5.1 Polymerization of methyl methacrylate at 50°C with benzoyl peroxide initiator at various monomer concentrations (benzene is the diluent). (From Schultz and Harbart, Makromol. Chem., 1, 106 (1947) with permission from Huthig & Wepf Publishers, Zug, Switzerland.)

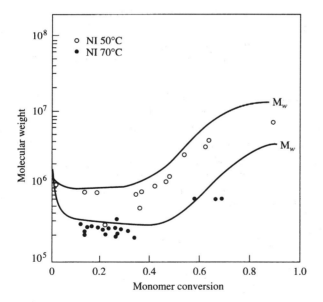

FIGURE 5.2 Experimental results on average molecular weight (measured using intrinsic viscosity) versus monomer conversion for the near-isothermal cases of Figure 5.9. $[I]_0 = 25.8$ mol/m^3. Solid curves are model predictions. (Data from Refs. 7–10.)

on the concentration of the solvent in the reaction mass, Eq. (5.5.2) implies that the plot of conversion versus time should be independent of the solvent concentration. In Figure 5.1, this is found to be so only in the early stages of polymerization. The rate of consumption of the monomer, r_{p0}, for conversion close to zero ($< 10\%$) has been plotted in Figures 5.3 and 5.4 for several cases, and it is found to be consistent with Eq. (5.3.14).

The proportionality of μ_n to the kinetic chain length ε has also been tested by various researchers of radical polymerization [2]. Equation (5.4.2) can be combined with Eq. (5.4.4) to give

$$\mu_n = \alpha v = \alpha \frac{r_p}{r_i} = \alpha \frac{k_p^2 [M]^2}{2 k_t r_{p0}} \tag{5.5.3}$$

According to this equation, μ_n^{-1} (at low conversion) should be proportional to r_{p0} for constant monomer concentration. Figure 5.5 shows that this proportionality [2,3] is followed extremely well for benzoyl peroxide and azobisdibutyronitrile (AZDN) initiators with the methyl methacrylate monomer. The agreement is very poor, however, for other systems, because a transfer reaction occurs between the

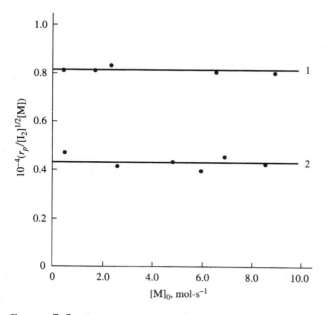

FIGURE 5.3 Dependence of initial rates of polymerization on monomer concentration: (1) MMA in benzene with benzoyl peroxide (BP) initiator at 50°C; (2) styrene in benzene with benzoyl peroxide initiator at 60°C. (From Ref. 2.)

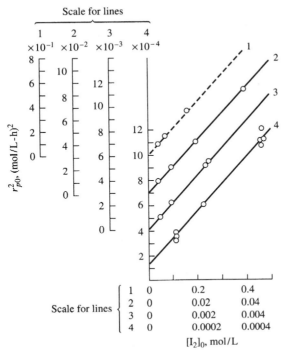

FIGURE 5.4 Log-log plot of r_{p0}^2 versus $[I_2]_0$ for constant $[M]$ for styrene with benzoyl peroxide initiator at 60°C. (Compiled from F. R. Mayo, R. A. Greg, and M. S. Matheson, J. Am. Chem. Soc., 73, 1691 (1951).)

initiator and the polymer radicals. This process leads to a larger number of inactive polymer chains than predicted by Eq. (5.5.3).

The decomposition of initiators invariably releases gaseous products; for example, benzoyl peroxide liberates carbon dioxide, whereas AZDN liberates nitrogen. In the polymerization of various monomers with benzoyl peroxide, the initial rate of monomer consumption is found to be affected by shear rate [11,12]. Recent experiments have shown that r_{p0} for acrylonitrile increases by as much as 400% in the presence of shear, as seen in Figure 5.6. This phenomenon has been attributed to the mass transfer resistance to the removal of carbon dioxide from the reaction mass [13,14]. It may be recognized that the usual geometry of industrial reactors is either tubular or a stirred-tank type, wherein the shear rate varies from point to point. This can profoundly affect the reaction rate; such fundamental information is clearly essential to a rational design of polymerization reactors.

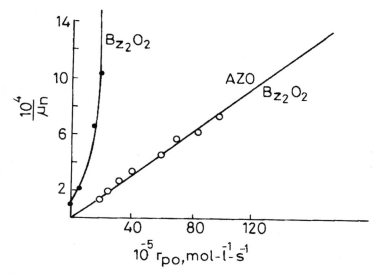

FIGURE 5.5 Average chain length $(1/\mu_{n0})$ versus r_{p0} at 60°C for methyl methacrylate (shown by ○) and styrene (shown by ●) for initiators azobisdibutyronitrile (AZO) and benzoyl peroxide (Bz_2O_2). Rates are varied by changing initiator concentration. Data for styrene are calculated from $10^4/\mu_{no} = 0.6 + 12.05 \times 10^4 r_{p0} + 4.64 \times 10^8 r_{p0}^2$ given in Ref. 2.

The considerable increase in the rate of polymerization in Figure 5.1 and the average chain length μ_n in Figure 5.2 is a phenomenon common to all monomers undergoing radical polymerization. It is called the autoacceleration or gel effect and has been the subject of several studies [15–27]. The gel effect has been attributed to the fall in values of the rate constants k_p and k_t (as shown for methyl methacrylate in Fig. 5.7) in the entire range of polymerization.

During the course of free-radical polymerization from bulk monomer to complete or limiting conversion, the movement of polymer radicals toward each other goes through several regimes of changes. To demonstrate, consider first a solution consisting of dissolved, *nonreacting* polymer molecules. When the solution is very dilute, the polymer molecules exist in a highly coiled state and behave like hydrodynamic spheres. In this regime, polymer molecules can undergo translational motion easily and the overall diffusion is completely governed by polymer–solvent interactions. As the polymer concentration is increased (beyond a critical concentration c^*), the translational motion of a molecule begins to be affected by the presence of other molecules. This effect, absent earlier, constitutes the second regime. On increasing the concentration of the polymer still further (say, beyond c^{**}), in addition to the intermolecular

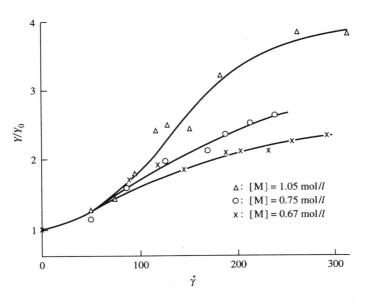

FIGURE 5.6   Effect of shear rate $\dot{\gamma}$ on the rate of solution polymerization of acrylonitrile. $Y = r_p/[M][I_2]$, and Y_0 is the value of Y in the absence of shear. (From Ref. 10.)

interactions in translational motion, polymer chains begin to impose topological constraints upon the motion of surrounding molecules due to their long-chain nature. In other words, polymer molecules become entangled; de Gennes has modeled the motion of polymer chains in this regime through a "tube" defined by the points of entanglement. A polymer molecule can move through this tube only by a snakelike wriggling motion along its length; this mode of motion is sometimes called *reptation*. Finally, at very high concentrations (say, beyond c^{***}), polymer chains begin to exert direct friction upon each other. The values of c^*, c^{**}, and c^{***} have been found to depend on the molecular weight of the polymer. These various regimes are shown schematically in Figure 5.8. As can be seen, for extremely low molecular weights of the polymer, there may not be any entanglement at all.

Demonstrating the correspondence between the polymer solvent system just described and free-radical polymerization, research has shown that gelation starts at the polymer concentration of c^{**}. In fact, it has been shown that k_t changes continuously as the polymerization progresses, first increasing slightly but subsequently reducing drastically at higher conversions [28].

One of the first models (based on this physical picture) was proposed by Cardenas and O'Driscoll, in which two populations of radicals are assumed to exist in the reaction mass [18]. The first are these that are physically entangled

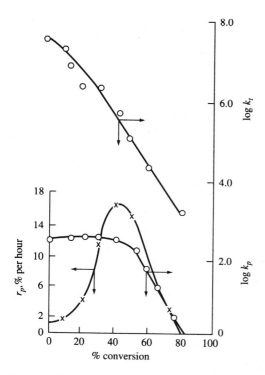

FIGURE 5.7 Bulk polymerization of MMA at 22.5°C with AZDN. The rate of initiation is 8.36×10^{-9} mol/L sec. (From Ref. 23, with the permission of ACS, Washington)

(denoted P_{ne}) and therefore have a lower termination rate constant, k_{te}, than that (k_t) of the second (denoted P_n), which are unentangled. Whenever a polymer radical grows in chain length beyond a critical value n_c, it is assumed that it becomes entangled and its termination rate constant falls from k_t to k_{te}. If it is assumed that the propagation rate constant, k_p, is not affected at all, the overall mechanism of radical polymerization can then be represented by the following:

Initiation

$$I_2 \xrightarrow{2k_1} 2I \qquad (5.5.4a)$$

$$I + M \xrightarrow{k_1} P_1 \qquad (5.5.4b)$$

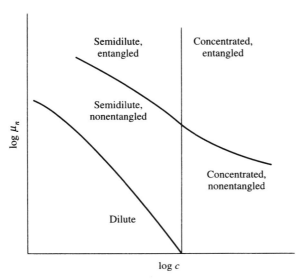

FIGURE 5.8 Molecular-weight-concentration diagram illustrating the dynamic behavior of a polymer–solvent system.

Propagation

$$P_n + M \xrightarrow{k_p} P_{n+1} \tag{5.5.5a}$$

$$P_{ne} + M \xrightarrow{k_p} P_{(n+1)e} \tag{5.5.5b}$$

Termination

$$P_m + P_n \xrightarrow{k_{te}} M_m + M_n \tag{5.5.6a}$$

$$P_m + P_{ne} \xrightarrow{k_{te}} M_m + M_n \tag{5.5.6b}$$

$$P_{me} + P_{ne} \xrightarrow{k_{te}} M_m + M_n \tag{5.5.6c}$$

where the reaction between the entangled and the unentangled radicals is assumed to occur with rate constant k_{tc} lying between k_t and k_{te}. It is possible to derive expressions for r_p and average molecular weight; their comparison with experimental data has been shown to give an excellent fit.

More recently Gupta et al. have used the kinetic scheme shown in Eqs. (5.5.4)–(5.5.6) and modeled the effect of diffusional limitations on f, k_p, and k_t [6–10]. Figures 5.2 and 5.9 show some experimental results on monomer conversion and average molecular weight (as measured using intrinsic viscosities)

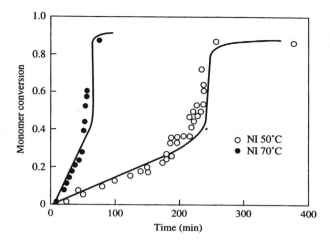

FIGURE 5.9 Experimental monomer conversion histories for the bulk polymerization of methyl methacrylate using AIBN, for two near-isothermal (NI) temperature histories. $[I]_0 = 25.8 \text{ mol/m}^3$. Solid curves are model predictions. (Data from Refs. 6–10.)

under near-isothermal conditions in a 1-L batch reactor. The agreement between the theoretical predictions and experimental data is excellent. Similar agreement between predictions and experimental results has been observed for MMA polymerization with intermediate addition of a solution of initiator (AIBN) in monomer.

Example 5.1: Retarders are molecules which can react with polymer radicals P_n as well as monomer M and slows down the overall rate as follows:

$$P_n + Z \xrightarrow{k_Z} Z_R$$

$$Z_R + M \xrightarrow{k_{Zp}} P_A$$

and

$$Z_R + Z_R \xrightarrow{k_{Zt}} \text{Nonradical species}$$

where Z_R is a reacted radical and has lower reactivity and because $[Z] \gg [Z_R] \gg \lambda_{P0}$, we neglect the reaction between Z_R and P_n. Determine r_p and μ_n in the presence of a retarder.

Solution: Let us say the rate of initiation, $r_1 = (f k_1[I_2])$ is constant and the mole balances of λ_{P0} and Z_R are (assuming QSSA)

$$\frac{d\lambda_{P0}}{dt} = R_i = k_Z \lambda_{P0}[Z] + k_{Zp}[M][Z_R] = 0$$

and

$$\frac{dZ_R}{dt} = k_Z \lambda_{P0}[Z] - k_{Zp}[Z_R][H] - 2k_{Zt}[Z_R]^2 = 0$$

These give λ_{P0} as

$$\lambda_{P0} = \frac{\{R_i + k_{Zp}(R_i/2k_{tZ})^{1/2}[M]\}}{k_Z[Z]}$$

$$[Z_r] = \left(\frac{R_i}{2k_{tZ}}\right)^{1/2}$$

The rate of polymerization, r_p

$$R_p = k_p[M]\lambda_{P0}$$

and the kinetic length, v, is given by

$$v = \frac{k_p[M]\lambda_{P0} + k_{Zp}[Z_R][M]}{k_Z[Z]\lambda_{P0} + k_{Zt}[Z_R]^2}$$

5.6 EQUILIBRIUM OF RADICAL POLYMERIZATION [29]

As for step-growth polymerization, the presentation of the kinetics of radical polymerization must be followed by a description of its equilibrium. The Gibbs free energy, G, for any system at temperature T is defined as $H - TS$, where H and S are the enthalpy and entropy of the system, respectively. The change in Gibbs free energy, ΔG_p, for the formation of a polymer

$$nM \rightarrow M_n \tag{5.6.1}$$

per monomeric unit, can be written as

$$\Delta G_p = \frac{1}{n} G_{polymer} - G_{monomer}$$

$$= \left\{\frac{1}{n}H_{polymer} - H_{monomer}\right\} - T\left\{\frac{1}{n}S_{polymer} - S_{monomer}\right\}$$

$$\equiv \Delta H_p - T\Delta S_p \tag{5.6.2}$$

where ΔH_p and ΔS_p are the enthalpy and entropy of polymerization per monomer unit, respectively. There are four possibilities in Eq. (5.6.2):

1. ΔH_p and ΔS_p are both negative.
2. ΔH_p is negative and ΔS_p is positive.
3. ΔH_p is positive and ΔS_p is negative.
4. ΔH_p and ΔS_p are both positive.

From thermodynamics, we know that a process occurs spontaneously only when ΔG_p is negative and, at equilibrium, ΔG_p is zero. In case 1, ΔG_p would be negative below a certain temperature and positive above it. This implies that the reaction would occur only below this temperature, which is called the ceiling temperature. In case 2, ΔG_p is always negative and, therefore, the polymerization occurs at all temperatures. In case 3, ΔG_p is always positive and therefore the reaction does not go in the forward direction. In case 4, the reaction would occur only when the temperature of the reaction is above a certain value, called the floor temperature.

Almost all radical polymerizations are exothermic in nature. Polymerization is the process of joining monomer molecules by covalent bonds, which might be compared to the threading of beads into a necklace. The final state is more ordered and, consequently, has a lower entropy. Thus, ΔS_p is always negative. The reaction of small molecules differs from polymerization reactions in that the ΔS of the former is invariably negligibly small. However, ΔS is normally a large negative quantity for polymerization and it cannot be neglected. Therefore, most of the monomers undergoing radical polymerization correspond to case 1 and have a ceiling temperature T_c. At this temperature, the monomer and the polymer are in equilibrium:

$$\Delta G_p = 0 \quad \text{at } T_c \tag{5.6.3}$$

From Eq. (5.6.2), it follows that

$$\Delta H_p = T_c \Delta S_p \tag{5.6.4}$$

or

$$T_c = \frac{\Delta H_p}{\Delta S_p} \tag{5.6.5}$$

It may be pointed out that, in general, ΔS_p is a function of the monomer concentration in the system, so the ceiling temperature (or the equilibrium temperature) also depends on the monomer concentration. This dependence is written as

$$\Delta S_p = \Delta S_p^\circ + R \ln[M] \tag{5.6.6}$$

where ΔS_p° is the entropy change when the polymerization is carried out at the standard state. The standard state of a liquid monomer is defined as that at which the monomer concentration is 1 M at the temperature and pressure of polymerization. The standard state for other phases is as conventionally defined in classical thermodynamics. Equation (5.6.5) is now rewritten as

$$T_c = \frac{\Delta H_p^\circ}{R \ \ln[M] + \Delta S_p^\circ} \qquad (5.6.7)$$

where ΔH_p is the same as ΔH_p from its definition. If $[M]_e$ is the concentration of the monomer at equilibrium, then

$$[M]_e = \frac{\Delta H_p^\circ}{RT_c} - \frac{\Delta S_p^\circ}{R} \qquad (5.6.8)$$

The data on T_c for several polymerizations are given in the literature for $[M]_e = 1M$. For example, ΔH (in kcal/mol), ΔS (cal/mol K), and T_c (K at $[M]_e = 1M$) for styrene–polystyrene are -16.7, -25, and 670; for ethylene–polyethylene, they are -25.5, -41.5, and 615; and for α-methyl styrene, they are -8.4, -27.5, and 550, respectively [4].

Equation (5.6.8) represents a very important result in that it has an extra, non-negligible term, $\Delta S_p/R$, which is not present in the corresponding reaction of small molecules. From this equation, we can find the equilibrium monomer concentration at the temperature at which the polymerization is being carried out. It turns out that the equilibrium concentration of monomer is very low at normal temperatures of polymerization that are far below T_c. For example, for styrene at 60°C, $[M]_e$ is obtained using values of ΔH_p and ΔS_p found in Ref. 4 (for liquid styrene and solid amorphous polystyrene) as

$$[M]_e = \exp\left(-\frac{16,700}{1.987 \times 333} + \frac{25}{1.987}\right) = 3.7 \times 10^{-6} \ \text{mol/L} \qquad (5.6.9)$$

It is thus seen that at 60°C, the equilibrium conversion of styrene is close to 100%. In the practical range (25–100°C) of temperatures used, similar computations show that $[M]_e$ is close to 0% for most other systems. However, experimental data of Figures 5.7 and 5.9 show that the terminal monomer conversion is close to 90%, which is far less than values predicted by Eq. (5.6.8). Thus, it is usually not necessary to incorporate reverse reactions in the kinetic mechanism for chain-reaction polymerization.

It has already been observed that there is considerable change in physical properties of the reaction mass as the liquid (or gaseous) monomer is polymerized to the solid polymer. If the temperature of polymerization, T, is greater than the glass transition temperature of the solid polymer (T_g), the terminal conversion is the same as that given by Eq. (5.6.8) [22]. If the temperature of polymerization is less than the glass transition temperature of the solid polymer, the terminal conversion of the monomer is governed by physical factors. When a solvent is mixed with an amorphous polymer, the glass transition temperature of this mixture is known to decrease. It has been demonstrated experimentally that polymerization stops at the moment when the glass transition temperature of the reaction mass is equal to the polymerization temperature. The reaction stops because, at this temperature, molecular motions stop in the matrix of the reaction mass.

Example 5.2: Suppose that there is free-radical equilibrium polymerization with termination by disproportionation alone.

$$I_2 \underset{k_d'}{\overset{k_d}{\rightleftharpoons}} 2I; \qquad K_d = \frac{k_d}{k_d'}$$

$$I + M \underset{k_1'}{\overset{k_1}{\rightleftharpoons}} P_1; \qquad K_1 = \frac{k_1}{k_1'}$$

$$P_n + M \underset{k_p'}{\overset{k_p}{\rightleftharpoons}} P_{n+1}; \qquad n \geq 1, \ K_p = \frac{k_p}{k_p'}$$

$$P_n + P_m \underset{k_{td}'}{\overset{k_{td}}{\rightleftharpoons}} M_n + M_m; \qquad n, \ m \geq 1, \ K_{td} = \frac{k_{td}}{k_{td}'}$$

Establish the molecular-weight distribution (MWD) of P_n and M_n under equilibrium and determine the first three moments of these MWDs.

Solution: From the equilibrium of termination step, one has

$$M_n \lambda_{M0} - k_{td} P_n \lambda_{P0} = 0 \tag{1}$$

The mole balance relations for polymers and radicals under equilibrium are

$$\frac{dI}{dt} = 2\phi k_1[I_2] - k_1[I][M] + k_1[P_1] = 0$$

$$\frac{d[P_1]}{dt} = k_1[I][M] - k_1'[P_1] - k_p[M][P_1] + k_p'[P_2]$$
$$\qquad - k_{td}[P_1]\lambda_{P0} + k_d'[M_1]\lambda_{M0} = 0$$

$$\frac{d[P_n]}{dt} = k_p[M][P_{n=1}] - k_p'[P_n] - k_p[M][P_n] + k_p'[P_{n+1}]$$
$$\qquad - k_{td}[P_n]\lambda_{P0} + k_{td}'[P_{n+1}] = 0, \quad n \geq 2$$

With the help of Eq. (1), one has

$$[P_2] - k_p[P_1] + \frac{k_1}{k'_p}[I][M] - \frac{k'_1}{k'_p}[P_1] = 0 \tag{2}$$

$$[P_{n+1}] - k_p M[P_n] - [P_n] + k_p M[P_{n-1}] = 0 \tag{3}$$

Eq. (3) is an index equation satisfied by

$$[P_n] = (K_p[M])[P_{n-1}] = (K_p[M])^{n-1}[P_1]$$

which is the MWD of polymer radicals. The first three moments are easily obtained by directly summing the geometric senes as

$$\lambda_{P0} = \frac{[P_1]}{1 - K_p[M]}$$

$$\lambda_{P1} = \frac{[P_1]}{(1 - K_p[M]^2}$$

$$\lambda_{P2} = \frac{[P_1]\{1 + K_p[M]\}}{(1 - K_p[M])^3}$$

Equation (1) gives

$$\lambda_{M0} = K_{td}\lambda_{P0}$$
$$\lambda_{M1} = K_{td}\lambda_{P0}\lambda_{P1}$$

and

$$\lambda_{M2} = K_{td}\lambda_{P0}\cdot\lambda_{P2}$$

From these, one can determine the member and weight average molecular weights

$$\mu_n = \frac{\lambda_{M1}}{\lambda_{M0}} = \frac{1}{1 - K_p[M]}$$

and

$$\mu_w = \frac{\lambda_{m2}}{\lambda_{M0}} = \frac{1 + K_p[M]}{1 - K_p[M]}$$

5.7 TEMPERATURE EFFECTS IN RADICAL POLYMERIZATION

In the initial stages of polymerization, the temperature dependence of the rate constants in Eq. (5.3.14) can be expressed through the Arrhenius law:

$$k_I = k_{I0} e^{-E_I/RT} \tag{5.7.1a}$$

$$k_p = k_{p0} e^{-E_p/RT} \tag{5.7.1b}$$

$$k_t = k_{t0} e^{-E_t/RT} \tag{5.7.1c}$$

This representation is completely parallel to the temperature dependence of rate constants for reactions of small molecules. E_I, E_p, and E_t are, therefore, the activation energies for initiation, propagation, and termination reactions, respectively. The values are tabulated extensively in the *Polymer Handbook* [4]. The temperature dependence of r_p and μ_n can be easily found by substituting Eq. (5.7.1) in Eqs. (5.3.14) and (5.4.4) to get

$$r_p = \frac{k_{p0} k_{I0}^{1/2}}{2 k_{t0}^{1/2}} (f[I_2])^{1/2} [M] \exp\left(-\frac{E_p - 0.5E_t + 0.5E_I}{RT}\right) \tag{5.7.2}$$

$$\mu_n = \alpha \frac{k_{p0}}{2 k_{I0}^{1/2} k_t^{1/2}} \exp\left(\frac{E_p - 0.5E_t - 0.5E_I}{RT}\right) \tag{5.7.3}$$

The activation energies are such that the overall polymerization for thermally dissociating initiators is exothermic (i.e., $E_p - 0.5E_t + 0.5E_I$) is normally positive, so the rate increases with temperature. On the other hand, $E_p - 0.5E_t - 0.5E_I$ is usually negative for such cases and μ_n decreases with increasing temperature.

After the gel point sets in, the temperature dependence of k_p and k_t is as follows [14]:

$$\frac{1}{k_t} = \frac{1}{k_{t0}} + \theta_t \frac{\lambda_{m0}}{\exp\{2.303\phi_m/[A(T) + B\phi_m]\}} \tag{5.7.4a}$$

$$\frac{1}{k_p} = \frac{1}{k_{p0}} + \theta_p \frac{\lambda_{m0}}{\exp\{2.303\phi_m/[A(T) + B\phi_m]\}} \tag{5.7.4b}$$

where

$$\phi_m = \frac{1-x}{1+\varepsilon_x} \tag{5.7.5a}$$

$$\theta_p = \theta_p \ \exp\left(\frac{E_{\theta p}}{RT}\right) \tag{5.7.5b}$$

$$\theta_t = \{\theta_1^\circ [I_2]_0\} \ \exp\left(\frac{E_{\theta t}}{RT}\right) \tag{5.7.5c}$$

$$A(T) = C_1 - C_2(T - T_{gp})^2 \tag{5.7.5d}$$

The terms $E_{\theta p}$ and $E_{\theta t}$ are parameters to be determined from the data on gel effect.

5.8 IONIC POLYMERIZATION

As discussed earlier, ionic polymerization can be categorized according to the nature of the growing polymer centers, which yields the classifications cationic polymerization and anionic polymerization

5.8.1 Cationic Polymerization

The growth center in this class of ionic polymerizations is cationic in nature. The polymer cation adds on the monomer molecules to it sequentially, just as the polymer radical adds on the monomer in radical polymerization. The initiation of the polymerization is accomplished by catalysts that are proton donors (e.g., protonic acids such as H_2SO_4). The monomer molecules act like electron donors and react with the catalyst, giving rise to polymer ions. The successive addition of the monomer to the polymer ion is the propagation reaction. These two elementary reactions are expressed schematically as follows:

Initiation

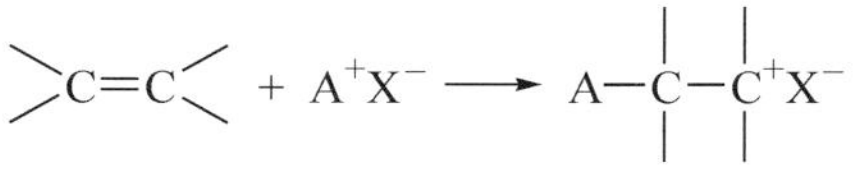

$$>\!C\!=\!C\!< \ + \ A^+X^- \longrightarrow A-\overset{|}{\underset{|}{C}}-\overset{|}{\underset{|}{C}}{}^+X^- \tag{5.8.1a}$$

Propagation

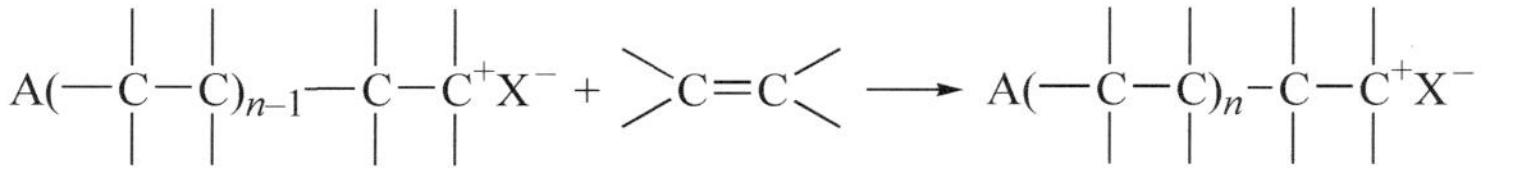

$$A(-\overset{|}{\underset{|}{C}}-\overset{|}{\underset{|}{C}})_{n-1}-\overset{|}{\underset{|}{C}}-\overset{|}{\underset{|}{C}}{}^+X^- \ + \ >\!C\!=\!C\!< \ \longrightarrow A(-\overset{|}{\underset{|}{C}}-\overset{|}{\underset{|}{C}})_n-\overset{|}{\underset{|}{C}}-\overset{|}{\underset{|}{C}}{}^+X^- \tag{5.8.1b}$$

The presence of the gegen ion in the vicinity of the growing center differentiates cationic from radical polymerization. Other common reactions in cationic polymerization include the following.

Transfer reactions. The positive charge of polymer ions is transferred to other molecules in the reaction mass. These could be impurity molecules or monomer molecules themselves. Because of these reactions, the resulting polymer has a lower molecular weight. As in the case of radical polymerization, the transfer reactions do not affect the overall reaction rate.

Chain termination. No mutual termination occurs in cationic polymerization because of the repulsion between the like charges on the two polymer ions—a phenomenon absent in radical polymerization. However, the neutralization of the polymer ion can occur by the abstraction of a proton from the polymer ion by the gegen ion, as follows:

$$A(-\overset{|}{\underset{|}{C}}-\overset{|}{\underset{|}{C}})_n-\overset{|}{\underset{|}{C}}-\overset{|}{\underset{|}{C}}{}^+X^- \longrightarrow A(-\overset{|}{\underset{|}{C}}-\overset{|}{\underset{|}{C}})_n-\overset{|}{\underset{|}{C}}=\overset{|}{\underset{|}{C}} + HX \qquad (5.8.2)$$

Such neutralization can also occur by molecules of impurities present in the reaction mass.

The true initiating species is A^+X^- (not AX), as shown in Eq. (5.8.1a). The neutral catalyst molecules must, therefore, ionize in the reaction mass before the polymer ion is formed. This implies that the initiation reaction is a two-step process:

$$AX \xrightarrow{k_i} A^+X^-$$

$$\overset{}{\underset{}{>}}C=C\overset{}{\underset{}{<}} + A^+X^- \xrightarrow{k_1} A-\overset{|}{\underset{|}{C}}-\overset{|}{\underset{|}{C}}{}^+\cdots X^- \qquad (5.8.3)$$

Either of these steps could have a lower rate of reaction and thus be the rate-determining step. If ionization is the slower of the two steps, the rate of initiation is given as

$$r_i = k_1[I_2] \qquad (5.8.4)$$

where $[I_2$ is the concentration of AX and k_i is the ionization rate constant. If the formation of the carbonium ion is the slower step, then r_i is

$$r_i = k_1[I_2][M] \qquad (5.8.5)$$

Once again, the equal reactivity hypothesis is assumed and the kinetic model is expressed as follows:

Initiation

$$AX \xrightarrow{k_i} A^+X^- \tag{5.8.6a}$$

$$A^+X^- + M \xrightarrow{k_1} P \tag{5.8.6b}$$

Propagation

$$M + P \xrightarrow{k_p} P \tag{5.8.6c}$$

Termination

$$P \xrightarrow{k_t} M_d + HX \tag{5.8.6d}$$

Transfer to monomer

$$P + M \xrightarrow{k_{trM}} M_d + P \tag{5.8.6e}$$

The rate of consumption of the monomer is given by

$$r_p = k_p[P][M] \tag{5.8.7}$$

where $[P]$ is the total concentration of the polymer ions in the reaction mass. In writing Eq. (5.8.7), the contributions of reactions (5.8.6a) and (5.8.6b) have been neglected. On application of the steady-state approximation to the polymer ion concentration,

$$\frac{d\lambda_{P0}}{dt} = -k_t\lambda_{P0} + k_j[I_2][M]^{a-1} = 0 \tag{5.8.8}$$

where $a = 1$ and $k_j = k_i$ if ionization is the rate-determining initiation step, and $a = 2$ and $k_j = k_1$ if the formation of the polymer ion is the rate-determining initiation step. This gives

$$\lambda_{P0} = \frac{k_j}{k_t}[I_2][M]^{a-1} \tag{5.8.9a}$$

$$r_p = \frac{k_j k_p}{k_t}[I_2][M]^a \tag{5.8.9b}$$

The rate of polymerization, r_p, is not affected by the transfer reaction at all, but the latter affects the kinetic chain length and the average chain length, μ_n. The kinetic chain length is given by

$$\nu = \frac{\text{Rate of propagation}}{\text{Rate of formation of the dead chains}} = \frac{k_p \lambda_{P0}[M]}{k_t \lambda_{P0} + k_{trM}[M]\lambda_{P0}} \quad (5.8.10)$$

$$\frac{1}{\nu} = \frac{k_t}{k_p[M]} + \frac{k_{trM}}{k_p} \quad (5.8.11)$$

This result shows that if the transfer reaction predominates, the average chain length, μ_n ($\mu_n = \nu$ for cationic polymerization), is independent of the monomer concentration as well as the initiator concentration.

5.8.2 Experimental Confirmation of the Model of Cationic Polymerization

Cationic polymerization is one of the least understood subjects in polymer science, and the data available are not as extensive as for radical polymerization. Normal temperatures of operation vary from $-100°C$ to $+20°C$. The appropriate temperature for any reaction is found only through experimentation and is extremely sensitive to the monomer and the catalyst chosen. Lower temperatures are preferred because they suppress several unwanted side reactions. It is necessary to have highly purified monomers and initiators because transfer reactions can easily occur with impurities, giving a polymer of very low molecular weight.

Radical and cationic polymerization differ in that, in the latter, initiation is very fast and propagation is the rate-determining step. Moreover, it has been shown experimentally that carbonium ions are much less stable than the corresponding radicals [30]. This implies that the lifetime of a polymer cation is much shorter than the corresponding polymer radical. The very rapid disappearance of the polymer cations may sometimes cause the steady-state approximation to be invalid. Hence, Eqs. (5.8.9) and (5.8.11) must be used cautiously, as they are based on the steady-state approximation.

5.8.3 Initiation in Cationic Polymerization

Cationic polymerization can be induced by initiators that release cations in the reaction mass [3]. The following are various classes of initiator systems that are commonly used:

1. *Protonic acids:* HCl, H_2SO_4, Cl_3CCOOH, $HClO_4$, and so forth
2. *Aprotonic acids:* BF_3, $AlCl_3$, $TiCl_4$, $SnBr_4$, $SbCl_3$, $SnCl_4$, $ZnCl_4$, $BiCl_3$, and so forth, with coinitiators like H_2O and organic acids

3. *Carbonium salts:* $Al(Et)_3$, $Al(Et)_2Cl$, or $Al(Et)Cl_2$ with alkyl or aryl chlorides or mineral acid coinitiators

4. *Cationogenic substances:* t-$BuClO_4$, I_2, Ph_3CCl, ionizing radiations

Let us consider protonic acids as initiators as an example. The acid must first ionize in the medium of the reaction mass before it can protonate the monomer molecule. The overall initiation reaction for HCl, for example, consists of the following three elementary reactions:

$$HCl \longrightarrow H^+ + Cl^- \quad (e_1) \tag{5.8.12a}$$

$$H^+ + C{=}C \longrightarrow H{-}C^+ \quad (-e_2) \tag{5.8.12b}$$

$$HC{-}C^+ + Cl^- \longrightarrow HC{-}C^+{-}Cl^- \quad (-e_3) \tag{5.8.12c}$$

Reaction (5.8.12a) is a simple heterolytic bond dissociation of the initiator molecule and $+e_1$ is the dissociation energy, which is always positive. The proton thus liberated attacks the monomer molecule, as shown in Eq. (5.8.12b). The energy of this reaction, $-e_2$, is a measure of the proton affinity of the monomer. Because, overall, electrical neutrality has to be maintained, the negative Cl^- ion has to move somewhere near the generated cation. This is because the energy required to keep the negative and the positive charges far apart would be very large, and the lowest-energy configuration would be obtained only when the two are at some finite distance r. The potential energy released due to the interaction of these ions is given by Coulomb's law:

$$-e_3 = \frac{e^2}{rD} \tag{5.8.13}$$

where D is the dielectric constant of the medium, e is the electric charge of the ions, and r is their distance of separation.

Equation (5.8.13) must be studied very carefully. The distance of separation between the two ions, r, depends on their relative sizes. Also, as the value of D decreases, the electrostatic energy of the interaction increases. This implies that the energy required to separate the ion pairs increases as the dielectric constant decreases. Because the polymerization progresses only by the addition of a monomer molecule to the carbonium ion, the driving force for such a process is therefore derived through the ability of the positive charge of the carbonium ion to attract the electron-rich double bond of the monomer molecule. If the carbonium ion is held with great affinity by the gegen ion (here, Cl^-), the monomer molecule will be unable to add to the carbonium ion by transferring its electron. Hence, a low dielectric constant of the medium of the reaction mass favors the formation of covalent bonds between the carbonium ion and the gegen ion. A high dielectric constant of the reaction medium, on the other hand, favors a loose association between the carbonium and gegen ions (called the solvent-separated ion pair) and promotes cationic polymerization. However, too high a value of D is also not desirable, because of thermodynamics constraints discussed next.

The total change in the energy for the initiation step can be written as a sum of the energies of the individual steps as

$$\Delta H_i = e_1 - e_2 - e_3 \tag{5.8.14}$$

and the free-energy change of initiation as

$$\Delta G_i = \Delta H_i - T\Delta S_i$$
$$= e_1 - e_2 - \frac{e^2}{rD} - T\Delta S_i \tag{5.8.15}$$

The entropy change for initiation, ΔS_i, is always negative, because the reaction moves from a less ordered state to a more ordered one. Therefore, the value of $e_1 - T\Delta S_i$ in Eq. (5.8.15) is positive.

Since ΔG_i should be negative for any process to occur, Eq. (5.8.15) shows that the initiation reaction in cationic polymerization is favored by lowering the temperature. In addition to determining e_3, the solvent or reaction medium plays an important role in influencing e_1. Usually, e_1 is a large positive number; for example, e_1 for the gaseous ionization of HCl is 130 kcal/mol. However, in the presence of suitable solvents, the dissociation energy is lowered, and in the presence of water, e_1 for HCl is as low as 25 kcal/mol. The ability of the solvent to reduce e_1 is called the solvation ability. The solvation abilities of different solvents are different, and the choice of the solvent for cationic polymerization is thus very important.

5.8.4 Propagation, Transfer, and Termination in Cationic Polymerization

The initiation reaction determines the nature of the growing polymer chain because there is always a gegen ion in the vicinity of the carbonium ion. The propagation reaction is the addition of the monomer to the growing center and it depends on the following:

1. Size and nature of the gegen ion
2. Stability of the growing center, which determines the ability to add on the monomer molecule
3. Nature of the solvent; that is, its dielectric constant and solvation ability

The propagation reaction is written schematically as

$$P_n^+ \cdots G_e^- + M \xrightarrow{k_p} P_{n+1}^+ \cdots G_e^- \tag{5.8.16}$$

where P_n^+ is the growing polymer chain ion and G_e^- is the gegen ion.

As the dielectric constant of the medium is reduced, the activation energy for the propagation reaction increases and k_p decreases considerably. As pointed out in the discussion of the initiation reaction, the lowering of the dielectric constant of the reaction mass favors the formation of a covalent bond between the carbonium and gegen ions. Therefore, the gegen ion and the propagating carbonium ion would be very tightly bound together and would not permit the monomer molecules to squeeze in. If the substituent on the carbon atom with the double bond in the monomer molecule is such that it donates electrons to the π-cloud, then the addition of the monomer to $P_n^+ \cdots G_e^-$ would be facilitated. This would increase the value of k_p.

The termination and the transfer reactions occur quite normally in cationic polymerization. The termination reaction is unimolecular—unlike in radical polymerization, where it is bimolecular. It occurs by the abstraction of a proton from the carbonium ion end of the growing polymer chain by the gegen ion, which always stays in its vicinity. How readily this occurs once again depends on the stability of the carbonium ion end of the growing polymer chain, the nature of the gegen ion, and the dielectric constant of the medium. The information on k_t is quite scanty.

One final note must be included in any discussion of the different parameters that are involved in cationic polymerization. Equations (5.8.9) and (5.8.11) for r_p and μ_n are very gross representations of the polymerization process, and they should therefore be used with caution.

5.9 ANIONIC POLYMERIZATION

Anionic polymerization is initiated by compounds that release anions in the reaction mass. Cationic and anionic polymerization are very similar in nature, except in their termination reactions. Termination reactions can occur easily in cationic polymerization, whereas they are almost absent in anionic polymerization. In both cases, there is a gegen ion adjacent to the growing center. Therefore, their initiation and propagation rates have similar characteristics.

Anionic polymerization normally consists of only two elementary reactions: initiation and propagation. In the absence of impurities, transfer and termination reactions do not occur; therefore, in this treatment, we do not discuss these reactions.

5.9.1 Initiation in Anionic Polymerization

The following are commonly used initiator systems for anionic polymerization:

1. Alkali metals and alkali metal complexes (e.g., Na, K, Li, and their

stable complexes with aromatic compounds, liquid ammonia, or ethers)
2. Organometallic compounds (e.g., butyl lithium, boron alkyl, tetraethyl lead, Grignard reagent)
3. Lewis bases (e.g., ammonia, triphenyl methane, xanthene, aniline)
4. High-energy radiation

High-energy radiation will not be discussed here because it has little commercial importance. The first system of initiation, method 1, differs from methods (2) and (3) in the process of producing growth centers. Alkali metals and alkali metal complexes initiate polymerization by transfer of an electron to the double bond of the monomer. For example, a sodium atom can attack the monomer directly to transfer an electron as follows:

$$Na + CH_2{=}CHR \rightarrow [CH_2{-}CHR]^- Na^+ \qquad (5.9.1)$$

How readily this reaction progresses in the forward direction depends on the nature of the substituents of the monomer, the nature of the gegen ion, and the ability of the alkali metal to donate the electron.

Because the sodium metal usually forms a heterogeneous reaction mass, this method of initiation is not generally preferred. On the other hand, alkali metal complexes can be prepared with suitable complexing agents. The resultant complex forms a homogeneous green solution that initiates polymerization as follows:

$$(5.9.2a)$$

$$(5.9.2b)$$

The nature of the gegen ion and that of the reaction mass control the propagation reaction, as in the case of cationic polymerization.

Initiation by organometallic compounds and Lewis bases occurs by a direct attack of these compounds on the double bond of the monomer molecule. Before the Lewis base can attack the monomer, it must ionize, and only then can a carbanion be formed. The process of initiation can be written as

$$BG \longrightarrow B^- + G^+ \qquad (+e_1) \qquad (5.9.3a)$$

$$B^- + H_2C{=}CHR \longrightarrow BCH_2{-}\underset{\underset{H}{|}}{\overset{\overset{R}{|}}{C}}{}^- \quad (-e_2) \qquad (5.9.3b)$$

$$BCH_2{-}\underset{\underset{R}{|}}{\overset{\overset{H}{|}}{C}}{}^- + G^+ BCH_2{-}\underset{\underset{R}{|}}{\overset{\overset{H}{|}}{C}} \cdots G^+ \quad (-e_3) \qquad (5.9.3c)$$

where BG is a Lewis base, e_1 is the dissociation energy, and $-e_2$ is the electron affinity. G^+ is the gegen ion, which must remain near the carbanion formed in the initiation process.

The strength of the Lewis base (measured by the pK value) required to initiate the polymerization of a particular monomer depends on the monomer itself. Monomers having substituents that can withdraw the electron from the double bond have relatively electron-deficient double bonds and can be initiated by weak Lewis bases. Because the initiation reaction consists of the ionization of the initiator and then the formation of the carbanion, the role of the solvent (which constitutes the reaction mass) in anionic polymerization would be similar to its role in cationic polymerization.

5.9.2 Propagation Reaction in Anionic Polymerization

The initiation reaction is much faster than the propagation reaction in anionic polymerization, so the latter is the rate-determining step.

The propagation reaction also depends on the nature of the gegen ion in that a monomer molecule adds to the growing chain by squeezing itself between the chain and the gegen ion. As a result of this, resonance, polar, and steric effects would be expected to play a significant role in determining k_p.

5.9.3 Kinetic Model for Anionic Polymerization [31–35]

Because the initiation reaction is much faster than the propagation reaction, we assume that all of the initiator molecules react instantaneously to give carbanions. Thus, the total number of carbanions, which is equal to the number of growing chains in the reaction mass, is exactly equal to the number of initiator molecules initially present in the reaction mass. Therefore, the molar concentration of the initiator, $[I_2]_0$, is equal to the concentration of growing chains in the reaction mass. The rate of polymerization r_p is given by

$$r_p = k_p[M][I_2]_0 \qquad (5.9.4)$$

The average chain length of the polymer formed is the ratio of the total number of monomer molecules reacted to the total number of growing polymer chains in the reaction mass. If the polymerization is carried to 100% conversion, μ_n is

$$\mu_n = \frac{[M]_0}{[I_2]} \qquad (5.9.5)$$

where $[M]_0$ is the initial concentration of the monomer.

The initiation mechanism does not directly enter into the derivation of Eqs. (5.9.4) and (5.9.5), and, therefore, these equations describe anionic polymerization only approximately. However, because little information on rates of initiation reactions is available and the initiation process is much faster than propagation, these equations serve well to describe the overall polymerization.

Anionic polymerization has found favor commercially in the synthesis of monodisperse polymers. These are found to have the narrowest molecular-weight distribution and a polydispersity index with typical values around 1.1.

One of the most important applications of anionic polymerization is to prepare a block copolymer. It may be pointed out that all monomers do not respond to this technique, which means that only limited block copolymers can, in reality, be synthesized. During the present time, as pointed out in Chapter 1, there is considerable importance placed on finding newer drugs. Therein, we also described combinatorial technique in which we showed the importance of solid supports on which chemical reactions were carried out. However, these reactions can occur provided reacting fluids can penetrate the solid support; in other words, it should be compatible with the solid supports.

One of the problems of radical polymerization is high-termination-rate constants by combination (k_{tc}) or by disproportionation (k_{td}). In view of this, polymer chains of controlled chain length cannot be formed and this technique is ill-suited for precise control of molecular structure (e.g., in star, comb, dendrimers, etc.) required for newer applications like microelectronics. The major breakthrough occurred when nonterminating initiators (which are also stable radicals) were used. Because of its nonterminating nature, this is sometimes called living radical polymerization and the first initiator that was utilized for this purpose was TEMPO (2,2,6,6-tetramethylpiperidinyl-1-oxo) [36,37]. A variation of this is atom-transfer radical polymerization (ATRP) in which, say for styrene, a mixture of 1 mol% of 1-phenyl ether chloride (R—X) and 1 mol% CuCl with two equivalents of bipyridine (bpy) is used for initiation of polymerization. Upon heating at 130°C in a sealed tube, bpy forms a complex with CuCl (bpy/CuCl), which can abstract the halide group from RX to give a radical that reacts with monomer M to give a growing radical as follows:

$$R-Cl + 2bpy/CuCl \rightarrow R^{\bullet} + 2bpy/CuCl_2 R^{\bullet} + M \rightarrow P_1^{\bullet}$$

The bpy/CuX$_2$ also complexes with growing radicals to give $P_n X$, keeping the concentration of active radicals (i.e., P_n°) small through the following equilibrium:

$$P_n-Cl + 2bpy/CuCl \rightleftharpoons P_n^{\circ} + bpy/CuCl_2$$

In the above reaction, P_n-Cl is the dormant molecule which does not give any growth of chains [38,39]. The TEMPO mediated and ATRP procedures are commonly used for controlling the architecture of the chains (comb, star, dendrite, etc.), composition of the backbone (i.e., random, gradient, or block copolymers), or inclusion of functionality (chain ends, site specific, etc.) [40].

The generation of small structures (sometimes called microfabrication) is essential to modern technologies like microelectronics and optoelectronics [41,42]. In these applications, one is interested in constructing supramolecular structures utilizing well-defined low-molecular-weight building blocks synthesized as above. For this purpose, these building blocks are first functionalized at the chain ends by cyclic pyrrolidinium salt groups and/or tetracarboxylate anions. Self-assembly is defined as spontaneous organization of molecules into a well-defined structure held together by noncovalent forces. In this case, the functionalized polymer blocks (sometimes called telechelics) are held together by electrostatic forces. On heating this self-assembly, the pyrrolodinium groups (five-ring cyclic compound) polymerize this way, giving a covalent fixation of this assembly.

5.10 ZIEGLER–NATTA CATALYSTS IN STEREOREGULAR POLYMERIZATION [43–56]

Stereoregular polymers have special properties and have therefore gained importance in the recent past. A specific configuration cannot be obtained by normal polymerization schemes (radical or ionic); special catalyst systems are required in order to produce them. The catalyst systems that give stereoregulation are called Ziegler–Natta catalysts, after their discoverers, the Nobel Prize winners Ziegler and Natta.

Ziegler–Natta catalyst systems consist of a mixture of the following two classes of compounds:

1. Compounds (normally halides) of transition elements of groups IV to VIII of the periodic table, called catalysts, such as $TiCl_3$, $TiCl_4$, $TiCl_2$, $Ti(OR)_4$, TiI_4, $(C_2H_5)_2 TiCl_2$, VCl_4, $VOCl_3$, VCl_3, vacetyl-acetonate, $ZrCl_4$, Zr tetrabenzyl, and $(C_2H_5)_2ZrCl_2$

2. Compounds (hydrides, alkyls, or aryls) of elements of groups I to IV, called cocatalysts such as $Al(C_2H_5)_3$, $Al(i\text{-}C_4H_9)_3$, $Al(n\text{-}C_6H_{13})_3$, $Al(C_2H_5)_2Cl$, $Al(i\text{-}C_4H_9)_2Cl$, $Al(C_2H_5)Cl_2$, and $Al_2(C_2H_5)_3Cl$.

Not all possible combinations of the catalysts and cocatalysts are active in stereoregulating the polymerization of a substituted vinyl monomer. Therefore, it is necessary to determine the activity of different combinations of the catalyst–cocatalyst system in polymerizing a particular monomer.

Cationic polymerizations are also known to yield stereoregular polymers, depending on experimental conditions. However, because of very low temperatures of polymerization and very stringent purity requirements of monomers and the catalyst systems, cationic polymerizations are very expensive. This is not so in the case of stereoregular polymerization, which is far less expensive and very easy to control. The only precaution that must be observed is that an inert atmosphere must be maintained in the reactor to avoid fire, because the cocatalysts are usually pyrophoric in nature.

A monomer can be in either the liquid phase or the gas phase at polymerization conditions. If monomer is a gas, a solvent medium is employed in which the Ziegler–Natta catalyst is dispersed and polymerization starts as soon as the gaseous monomer is introduced (see Fig. 5.10a for the setup). In the case of liquid monomers, a solvent is not necessary, but it is preferred because it facilitates temperature control of the reaction.

The Ziegler–Natta catalyst can either dissolve in the medium of the reaction mass or form a heterogeneous medium if insoluble. The latter is more a rule than an exception, and the commercially used Ziegler–Natta catalysts are commonly heterogeneous. The most common catalyst is $TiCl_3$, which is prepared by reducing $TiCl_4$ with hydrogen, aluminum, titanium, or $AlEt_3$, followed by activation. The catalyst is activated by grinding or milling it to a fine powder. The resultant $TiCl_3$ is crystalline, having a very regular structure. There are four crystalline modifications of $TiCl_3$ available (alpha, beta, gamma, and delta), of which the alpha form is the best known. Table 5.2 gives some of the schemes for preparing some of the important catalyst systems and the crystalline forms that result.

$TiCl_3$ is a typical ionic crystal like sodium chloride. It is a relatively nonporous material with a low specific surface area. It has a high melting point, decomposes to $TiCl_2$ and $TiCl_4$ at $450°C$, and sublimes at $830°C$ to $TiCl_4$ vapor. It is soluble in polar solvents such as alcohols and tetrahydrofuran but is insoluble in hydrocarbons. The highest specific surface area reported for these catalysts is $100\ m^2/g$, but normal values lie in the range of $10–40\ m^2/g$.

In addition to its two main components, the Ziegler–Natta system of catalysts also contains supports and inert carriers. An example of the former is $MgCl_2$, and the inert carriers include silica, alumina, and various polymers. These

TABLE 5.2 Processes for Preparing Catalyst Systems from $TiCl_4$

Ingredients (mol)	Process description	Approximate formula
$3TiCl_4 + AlEt_3$ or $3AlEt_2Cl$ or $3Al(i\text{-}Bu)_3$	On mixing, the reaction completes at 25°C	$Al_{0.3}TiCl_4$
$TiCl_4 + AlMe_2Cl$ or $AlMe_3$	The mixture is distillated. $MeTiCl_3$ is formed, which on thermal decomposition gives the catalyst.	$TiCl_3$
$\beta = Al_{0.3}TiCl_4$ $TiCl_4 + H_2$	Heat at 120–160°C for several hours. Hydrogenation reaction is carried out at 800°C.	$Al_{0.3}TiCl_4$ $TiCl_3$
$TiCl_4 + Al$, Na, Li, or Zn	Heat at 200°C. The metal is incorporated, giving the activity of the catalyst.	$Al_{0.3}\ TiCl_4$
$\alpha - Al_{0.3}TiCl_4$	Grind in ball mill, which is known to increase the activity.	
Hexane, $TiCl_4$, $AlEth_3$	Hexane, cocatalyst, and propylene are heated at 65°C and then $TiCl_4$ is added.	
$Mg(OEth)_2$ and $TiCl_4$	$Mg(OEth)_2$ is mixed with benzoyl chloride, chlorobenzene in hexane; $TiCl_4$ is added dropwise. Mixture is heated at 100°C for 3 hr.	$TiCl_4 \cdot MgCl_2 \cdot EB$ (EB: monoester base). This produces a Ti,Mg sponge having very high activity.

Source: Data from Refs. 32, 39, and 40.

differ in the way they affect the catalyst. Supports are inactive by themselves but considerably influence the performance of the catalyst by increasing the activity of the catalyst, changing the physical properties of the polymer formed, or both. Carriers do not affect the catalyst performance to any noticeable degree, but their use is warranted by technological factors. For instance, carriers dilute very active solid catalysts, make catalysts more easily transportable, and agglomerate catalysts in particles of specific shape. For example, one recipe for the catalyst for ethylene polymerization consists of dissolving $MgCl_2$ and $TiCl_4$ in a $3:1$ molar ratio in tetrahydrofuran. The solution is mixed with carrier silica powder that has already been dehydrated and treated with $Al(C_2H_5)_3$. The tetrahydrofuran is removed by drying the mixture, thus impregnating the carrier silica gel with

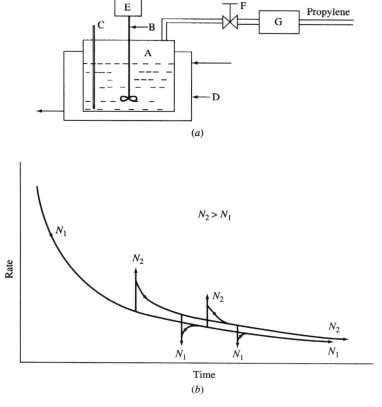

FIGURE 5.10 (a) Setup for stereoregular polymerization of propylene using $TiCl_3-AlEth_3$ catalyst in n-heptane. (b) Schematic representation of effect of stirring on polymerization for two speeds of stirring: N_1 and N_2.

MgCl$_2$ and TiCl$_4$. This product is subsequently treated with solution of Al(C$_2$H$_5$)Cl$_2$ and Al(C$_2$H$_5$)$_3$ in hexane, and the solvent is once again removed to give the final catalyst.

5.11 KINETIC MECHANISM IN HETEROGENEOUS STEREOREGULAR POLYMERIZATION

A simple system is shown in Figure 5.10a to depict heterogeneous polymerization. A gaseous monomer is continuously fed into a glass vessel. The vessel (serving as a reactor) has a suitable solvent (usually hexane for propylene) in which the catalyst–cocatalyst system is uniformly dispersed. In Figure 5.10b, the effect of stirring speed on the rate of propylene polymerization is shown schematically. These results clearly demonstrate the external mass transfer effect.

To understand the mechanism of heterogeneous polymerization, it is first necessary to understand the nature of the physical processes involved. Polymerization centers (PCs) are complexes formed by the reaction between AlEt$_3$ and TiCl$_3$ catalysts. The polymer chain is attached to these polymerization centers and grows in size by adding monomer between the PCs and the chains. Because the polymer chains coil around the catalyst particle, the PCs are buried within it. In the case of gaseous monomers, the latter must first dissolve in the medium of the reaction mass. The dissolved monomer in the reaction mass must then diffuse from the bulk and through the thin layer of polymer surrounding the PC before it reaches the catalyst surface for chemical reaction. The entire process can be written as follows:

$$
\begin{array}{ccccccc}
\text{Gaseous} & \rightleftharpoons & \text{Monomer} & \rightleftharpoons & \text{Monomer} & \rightleftharpoons & \text{Monomer} \\
\text{monomer} & & \text{in solution} & & \text{at the polymer} & & \text{at the catalyst} \\
& & & & \text{Film} \rightleftharpoons \text{Surface} & & \\
& & & & & \updownarrow & \\
& & \text{Reaction} \rightleftharpoons & & \text{Adsorption} & & \\
& & & & \text{on surface} & &
\end{array}
\tag{5.11.1}
$$

In the analysis that follows, it is assumed that the various diffusional resistances are negligible and that the reaction step in Eq. (5.11.1) is controlling. This implies that the stirring speed is very high.

Organometallic compounds used as components of Ziegler–Natta catalysts are normally liquids of a high boiling point that dissolve in aromatic hydrocarbons. Most of these exist in the following dimer form, which is stable:

Dimer of $Al(C_2H_5)_3$

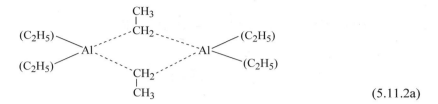

$$(5.11.2a)$$

Dimer of $Al(C_2H_5)_2Cl$

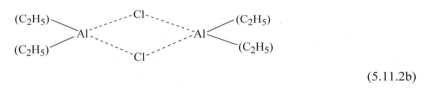

$$(5.11.2b)$$

However, there are some organometallic compounds [e.g., $Al(i-C_4H_9)_3$ and $Zr(C_2H_5)_2$] that exist in the monomeric form.

Active centers for polymerization are formed in the process of interaction between catalyst and cocatalyst systems. There is an exchange of a halogen atom between them as follows (with a $TiCl_4$ and $Al(C_2H_5)_2$ Ziegler–Natta catalyst system):

$$TiCl_4 + Al(C_2H_5)_2Cl \rightarrow Cl_3T_i-C_2H_5 + C_1-Al(C_2H_5)Cl \qquad (5.11.3)$$

This reaction is fast. The titanium–carbon bond serves as the principal constituent of the active center for polymerization because it has the ability of absorbing a monomer (vinyl or diene) molecule. These metal carbon bonds are not extremely stable and undergo several side reactions, leading to the breakage of TiCl bond. For example, the $Cl_3TiCl_2H_5$ molecule formed in Eq. (5.11.3) decomposes to give $TiCl_3$ as follows:

$$2Cl_3Ti-C_2H_5 \rightarrow 2TiCl_3 + C_2H_6 + C_2H_4 \qquad (5.11.4)$$

Even though the metal–carbon bond of $Ti-C_2H_5$ is not very stable, a significant portion of these survive under typical conditions of alkene and diene polymerization of 30–100°C and 0.5–5 hr of polymerization time. In fact, the instability of TiCl bonds strongly affects the performance of the Ziegler–Natta catalyst system and occasionally explains the reduction in its activity with time.

The transition metal–carbon bond, as stated earlier, reacts with an alkene molecule ($CH_2 = CHR$), and there is a formation of a complex, as follows:

$$Cl_3Ti-(C_2H_5) + CH_2{=}CHR \longrightarrow Cl_3Ti-C_2H_5$$

$$\uparrow$$

$$CH_2{=}CHR$$
$$\text{complex}$$

(5.11.5)

After formation of the complex, the alkene molecule is inserted in the Ti−C bond as follows:

$$Cl_3Ti-C_2H_5 \longrightarrow Cl_3Ti-CH_2-\underset{\underset{R}{|}}{CH}-C_2H_5$$

$$\uparrow$$

$$CH_2{=}CHR$$

(5.11.6)

The repeated insertion of $CH_2{=}CHR$ according to Eqs. (5.11.5) and (5.11.6) gives rise to propagation reaction, in this way forming long chain molecules.

5.12   STEREOREGULATION BY ZIEGLER–NATTA CATALYST

The most important characteristic of the Ziegler–Natta catalyst system is its ability to stereoregulate the polymer. The configuration of the resultant polymer depends on the choice of the catalyst system and its crystalline structure. Stereoregulation is believed to occur as follows:

$$Et-CH-\ldots CH_3$$
$$|$$
$$CH_2$$
$$|$$
$$CH_3-CH$$
$$(2)\quad CH_2\ldots\ldots CH-CH_3$$
$$|\qquad\qquad | \quad (1)$$
$$Ti\ldots\ldots\ldots CH_2$$

(5.12.1)

There are two kinds of interactive force existing in the activated complex shown. One is the steric hindrance between methyl groups (1) and (2) in the complex, and the other is the interaction between methyl group (1) and the chlorine ligands. Both interactions exist at any time, the relative strengths depending on the specific catalyst system. If the interactive force between the ligands and substituent (1) of the adsorbed molecule is not too large, the addition of the CH_3 group (1) of the monomer to the propagating chain occurs such that it minimizes the steric hindrance between itself and CH_3 group (2), thus giving a syndiotactic chain. If, however, the interaction between CH_3 group (1) and chlorine is large, it can compensate for the steric interaction and can lead to the formation of an isotactic chain by forcing the adsorbed molecule to approach the growing chain in a specific manner. More information on the nature of these interactions and an explanation of how the chain adds on a monomer molecule can be found elsewhere [31,49].

5.13 RATES OF ZIEGLER–NATTA POLYMERIZATION

If the diffusional resistances in Eq. (5.11.1) are neglected, the rate of polymerization, r_p, can be expressed as

$$r_p = k_p[C^*][M]_c \qquad (5.13.1)$$

where [C*] represents an active polymerization center and $[M]_c$ is the concentration of the monomer at the surface of the catalyst. If all the diffusional resistances can be neglected, $[M]_c$ can be taken as equal to the monomer concentration in the solution $[M]_s$ and can be easily determined by the vapor–liquid equilibrium conditions existing between the gaseous monomer and the liquid reaction mass. Ray et al. have used the Chao–Seader equation and Brockmeier has used the Peng–Robinson equation of state to relate $[M]_s$ to the pressure of the gas [37,38].

The size of the catalyst particle has a considerable effect on the rate of polymerization of propylene. For a constant concentration of the monomer, $[M]_0$ (i.e., at a constant propylene pressure in the gas phase), it has been found that the rate of polymerization is a function of time. For ground catalysts, a maximum is obtained, whereas for unground particles (size up to 10), the rate accelerates to approach the same asymptotic stationary value. The typical behaviors are shown schematically in Figure 5.11, which gives the different zones into which catalysts can be classified. The process has a buildup, a decay, and a stationary period. Natta has proposed the following explanation for these observations.

Like other catalyst systems, Ziegler–Natta catalyst systems have active centers, which have been amply demonstrated to be titanium atoms. In the case of unground Ziegler–Natta catalysts, there are some active sites on the outer surface where the polymerization starts immediately. As observed earlier, the polymer molecule has one of its ends attached to the site while the molecule starts growing around the catalyst particle. In this process of growth, there is a mechanical grinding on the particle by which the catalyst particles undergo fragmentation. When the larger catalyst particles break into smaller ones, additional surface is exposed and more titanium atoms are available for monomer molecules to interact. This implies that there is a generation of active sites during the process. The increase in the number of active sites leads to enhanced polymerization rates. The acceleration-type behavior is thus explained on the basis of an increase in the

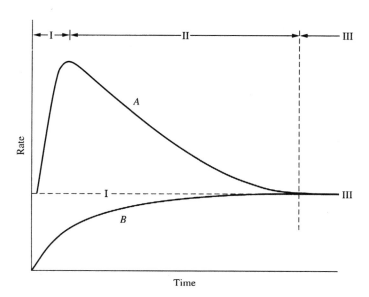

FIGURE 5.11 Typical kinetic curves obtained during propylene polymerization by TiCl₃. *A* is a decay-type curve; *B* is a buildup or acceleration-type curve. I is the buildup period; II is the decay period; III is the stationary period.

surface area with time. The smaller the particle is, the higher the mechanical energy required for further size reduction. Accordingly, the particle size approaches some asymptotic value. The stationary polymerization rate corresponds to this catalyst particle size.

5.13.1 Modeling of Stationary Rate

To determine the stationary rate of polymerization, we assume that all the sites of the Ziegler–Natta catalysts are equivalent. In the reaction mechanism given in Section 5.12, it was shown that aluminum ethylate reacts with $TiCl_3$ to form an empty ligand (and, therefore, a polymerization center). For this reaction to occur, $AlEt_3$ must first be adsorbed. We can write this schematically as

$$AlEt_3 + TI\ site \underset{adsorption}{\overset{adsorption}{\rightleftharpoons}} S^* \xrightarrow[reaction]{Chemical} PC \tag{5.13.2}$$

where S^* is a complex formed by the adsorption of the $AlEt_3$ molecule onto the Ti site. If the chemical reaction is the rate-determining step, the adsorption step in

Eq. (5.13.2) can be assumed to be at equilibrium and can be given by the following Langmuir adsorption equilibrium relation:

$$[S^*] = \frac{K[A]}{1 + K[A]} \tag{5.13.3}$$

where [A] is the concentration of aluminum ethylate in the reaction mass and K is the Langmuir equilibrium constant. If all of the S* formed is assumed to be converted to an activated polymerization center, its concentration [C*] is given by

$$[C^*] \approx [S^*] = \frac{K[A]}{1 + K[A]} \tag{5.13.4}$$

The rate of polymerization, r_∞, for large times can then be derived from Eq. (5.13.2) as follows:

$$r_\infty = k_p \frac{K[A]}{1 + K[A]} [M]_c \tag{5.13.5}$$

If the various diffusional resistances for the monomer are also neglected, $[M]_c$ can be replaced by the concentration of the monomer $[M]_s$ in the reaction mass. This may be related to the pressure P in the gas phase by using the Chao–Seader or Peng–Robinson equations of state, but for moderate pressures, Henry's law may as well be assumed, giving

$$[M]_s = K_H[P] \tag{5.13.6}$$

where K_H is Henry's law constant and P is the pressure of the gas. In these terms, r_∞ in Eq. (5.13.5) can be rewritten as

$$r_\infty = k \frac{[A]}{1 + K[A]} [P] \tag{5.13.7}$$

where

$$k = k_p K K_H \tag{5.13.8}$$

5.13.2 Modeling of the Initial Rates of Stereoregular Polymerization

In Figure 5.11, we observed that in the initial region, the rate of polymerization is a function of time. This is entirely due to the fact that the total concentration of polymerization centers, [C*], is a function of time. An expression for this can be derived with the help of Eq. (5.13.2). With this equation, the polymer formation in stereoregular polymerization can be rewritten as

$$AlEt_3 + S \rightleftharpoons S* \tag{5.13.9a}$$

$$S* + M \xrightarrow{k_c} C* \tag{5.13.9b}$$

$$C* + M \xrightarrow{k_p} C* \tag{5.13.9c}$$

where S represents an active titanium site and is the same as the first portion of Eq. (5.13.2). It has been hypothesized that a $C*$ is formed only after an $S*$ reacts with a monomer molecule and that the $C*$ shown, thus formed, is the same as a PC in Eq. (5.13.2).

It is now assumed that $[C*]_\infty$ is the total concentration of polymerization centers present at $t = \infty$ (i.e., at the stationary state). In the early stages of the reaction, $[C*]$ is expected to be less than $[C*]_\infty$. At any time, the simple mole balance is

$$[S] + [S*] + [C*] = [S]_0 \tag{5.13.10}$$

where $[S]_0$ is the total concentration of the active titanium sites. Also,

$$[S]_\infty + [S*]_\infty + [C*]_\infty = [S]_0 \tag{5.13.11}$$

Subtraction of these equations leads to

$$([S]_\infty - [S]) + ([S]_\infty - [S_\infty]) + ([C]_\infty - [C\hat{\ }*]) = 0 \tag{5.13.12}$$

$[S*]_\infty$ is zero because, at large times, monomer molecules have already reacted completely with all of the potential polymerization centers by the irreversible reaction in Eq. (5.13.9b). This is the case because $S*$ is an intermediate species in the formation of the polymerization center. $[S]$ and $[S]_\infty$ are both large numbers. Because it is assumed that only a few of the active sites participate in polymerization, $[S]_\infty - [S]$ can be neglected. Therefore,

$$[C*]_\infty - [C*] - [S*]_0 = 0 \tag{5.13.13a}$$

or

$$[S*] = [C*]_\infty - [C*] \tag{5.13.13b}$$

The rate of formation of the polymerization centers is given by

$$\frac{d[C*]}{dt} = k_c[M][S*] \tag{5.13.14}$$

With the help of Eq. (5.13.13b), this equation reduces to

$$\frac{d[C*]}{dt} = k_c[M]\{[C*]_\infty - [C*]\} \tag{5.13.15}$$

The rates of polymerization at time t and at the stationary zone are given as

$$r_p = k_p[M][C^*] \tag{5.13.16a}$$
$$r_\infty = k_p[M][C^*]_\infty \tag{5.13.16b}$$

Because [M] is constant, multiplying Eq. (5.13.15) by $k_p[M]$ gives

$$\frac{d}{dt}\{k_p[M][C^*]\} = k_c[M]\{k_p[M][C^*]_\infty - k_p[M][C^*]\}$$

or

$$\frac{dr}{dt} = k(r_\infty - r) \tag{5.13.17}$$

where $k = k_c[M]$. Equation (5.13.17) is the same as the observed empirical equation for the buildup period in the acceleration-type kinetic curve as observed in Figure 5.12. This analysis explains why a decay-type rate behavior in Figure 5.12 is observed for fine particles and not for coarse catalyst particles.

Well-ground catalysts have a larger number of active sites and, therefore, the reaction in Eq. (5.13.9) is pushed in the forward direction, the reversible reaction playing a smaller role in the early stages of reaction. As the reaction progresses, the reverse reaction starts removing the potential polymerization centers (i.e., S*) until, ultimately, the equilibrium value corresponding to the stationary zone is attained. This fact explains the maximum in [S*] and, therefore, in [C*]. This case must be differentiated from that of coarse catalyst particles, where the fragmentation of the particles occurs during the polymerization and the total number of active sites is not constant. The derivation of Eq. (5.13.17) for acceleration-type behavior is not quite correct because it is based on Eq. (5.3.10), which assumes a constant concentration of total active sites and does not account for the increase in the number of active sites by particle breakage.

Example 5.3: In the buildup period of decay-type stereoregular polymerization, the rates are found to be different when catalyst is added first from the case when gas is introduced, followed by the catalyst. Show why this happens.

Solution: Assume there are n_0 number of adsorption sites and n_A, the number of sites occupied by monomers. On addition of the catalyst first, the following adsorption equilibrium between the monomer and catalyst occurs:

$$Al_2Cl_6 + S \rightleftarrows S^* \tag{a}$$

Then,

$$r_{ads} = k_{ads}P(n_0 - n_A)$$
$$r_{des} = k_{des}n_A$$

At $K_{od}P(n_0 - n_A) = k_{des}n_A$ or

$$\theta_A = \frac{n_A}{n_0} = \frac{KP}{1 + KP}$$

where $K = k_{od}/k_{des}$

On introduction of propylene gas, the following reactions occur which are not in equilibrium:

$$S^* + M \xrightarrow{k_c} C \tag{b}$$
$$C + M \longrightarrow C \tag{c}$$

If the gas is introduced first, followed by the addition of the catalyst, reactions (a)–(c) simultaneously and consequently give different results.

5.14   AVERAGE CHAIN LENGTH OF THE POLYMER IN STEREOREGULAR POLYMERIZATION

The average chain length of the polymer at a given time can be found from the following general relation:

$$\mu_n = \frac{\text{Number of monomer molecules polymerized in time } t}{\text{Number of polymer molecules products in time } t} \tag{5.14.1}$$

The number of monomer molecules polymerized can be found from the rate of polymerization as $\int_0^t r\, dt$. The denominator can be found only if the transfer and termination rates are known. If r_t denotes the sum of these rates, then

$$\mu_n = \frac{\int_0^t r\, dt}{[C^*]_t + \int_0^t r_t\, dt} \tag{5.14.2}$$

where $[C^*]_t$ is the concentration of the polymerization centers at time t. The next step is to apply Eq. (5.14.2) for the stationary state to find μ_n. At the stationary state, r and r_t are both constant, and Eq. (5.14.2) reduces to

$$\frac{1}{\mu_n} = \frac{r_t}{r_\infty} + \frac{[C^*]_\infty}{tr_\infty} \tag{5.14.3}$$

This equation ignores the contribution to the integrals from the transition zone. For long times, the second term in Eq. (5.14.3) goes to zero and the following holds:

$$\frac{1}{\mu_n} = \frac{r_t}{r_\infty} \tag{5.14.4}$$

To be able to evaluate μ_n, termination and transfer processes must be known. These have been studied and the following termination and transfer processes for propylene have been reported.

1. Spontaneous dissociation

$$(Cat\!-\!)CH_2\!-\!\underset{\underset{CH_3}{|}}{CH}\!-\!(CH_2\!-\!\underset{\underset{CH_3}{|}}{CH})_n\!-\!R \xrightarrow{\ k_1\ }$$

$$(Cat\!-\!)H + CH_2\!=\!\underset{\underset{CH_3}{|}}{C}\!-\!(CH_2\!-\!\underset{\underset{CH_3}{|}}{CH})\!-\!R \tag{5.14.5a}$$

$$R_t = k_1[C^*] \tag{5.14.5b}$$

2. Transfer to propylene

$$(Cat\!-\!)CH_2\!-\!\underset{\underset{CH_3}{|}}{CH}\!-\!(CH_2\!-\!\underset{\underset{CH_3}{|}}{C})_n\!-\!R + CH_2 = \underset{\underset{CH_3}{|}}{CH}$$

$$\xrightarrow{\ k_1\ } (Cat\!-\!)CH_2\!-\!\underset{\underset{CH_3}{|}}{CH_2} + CH_2 = \underset{\underset{CH_3}{|}}{C}\!-\!(CH_2\!-\!\underset{\underset{CH_3}{|}}{CH})_n\!-\!R \tag{5.14.6a}$$

$$R_{t2} = k_2 [M] [C^*] \tag{5.14.6b}$$

3. Transfer to $AlEt_3$

$$(Cat\!-\!)CH_2\!-\!\underset{\underset{CH_3}{|}}{CH}\!-\!(CH_2\!-\!\underset{\underset{CH_3}{|}}{CH})_n\!-\!R = AlEt_3 \xrightarrow{\ k_3\ }$$

$$(Cat\!-\!)Et + Et_2Al\!-\!CH_2\!-\!\underset{\underset{CH_3}{|}}{CH}\!-\!(CH_2\!-\!\underset{\underset{CH_3}{|}}{CH})_n\!-\!R \tag{5.14.7a}$$

$$R_{t3} = k_3 [C^*][AlEt_3] \tag{5.14.7b}$$

The μ_n would, therefore, be given by Eq. (5.14.4) as follows:

$$\frac{1}{\mu_n} = \frac{r_t}{r_\infty}$$

$$= \frac{k_1[C^*] + k_2[C^*][M] + k_3[C^*][AlEt_3]}{k_p[C^*][M]}$$

$$= \frac{k_1 + k_2[M] + k_3[AlEt_3]}{k_p[M]} \tag{5.14.8}$$

5.15 DIFFUSIONAL EFFECT IN ZIEGLER–NATTA POLYMER [45,46,49]

As explained in Section 5.12, during polymerization the growing chain surrounds the catalyst particle. As a result, there is fragmentation of the catalyst particle, leading to increased numbers of active sites for polymerization with time. In addition, a film of polymer is formed around the polymer particle through which monomer has to diffuse. Section 5.14 has shown a semiempirical way of taking fragmentation of particles and diffusion of monomer into account, but there is a definite need to model these problems more fundamentally.

Several models have been proposed to account for both the fragmentation of catalyst particles and the diffusional resistance. The simplest is the solid-core model, in which the polymer is assumed to grow around a solid catalyst particle without any breakage. We show the model in Figure 5.12, in which the catalyst

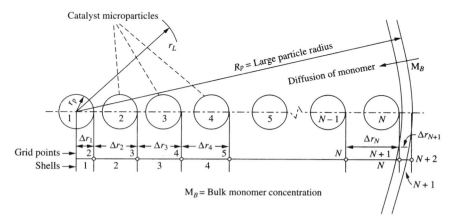

FIGURE 5.12 Schematic representation of the multigrain model for stereoregular polymerization of propylene.

particle is surrounded by a polymer shell. The dissolved monomer in the liquid phase diffuses through the accumulated polymer to the catalyst surface and reacts there. Knowing the rate of formation of the polymer at the surface, it is possible to compute the movement of the polymer shell boundary. It may be recognized that the polymerization is an exothermic reaction, which means that the heat of polymerization is liberated at the catalyst surface, which must be transported through the polymer shell by conduction. Because there is a finite resistance to transport of monomer through the shell, the temperature T and monomer concentration [M] are both dependent on radial position r and time t. This fact has been represented in Figure 5.12 by showing $T(r, t)$ and $[M](r, t)$.

As a refinement to the hard-core model just discussed, the multigrain model (recently proposed in Refs. 45 and 46) accounts for the particle breakup indirectly. This too has been depicted in Figure 5.12. A macroparticle of radius R comprises many small polymer microparticles. These particles are assumed to be lined along the macroparticle radius, touching each other. All microparticles are assumed to be spherical and of the same size. Microdiffusion in the interstices between the microparticles and microdiffusion within the particles are each assumed to exist, and the effective diffusion coefficient for the two regions need not be equal.

The microparticle diffusion is treated in the same way as in the solid-core model, and it is assumed that each of these microparticles grows independent of each other according to the existing local monomer concentration. To write the mole balance for the monomer in the macroparticle in spherical coordinates, let us define D_L as the effective diffusion coefficient for the macroparticle, r_L as the radial length, and $R(M_L, r_L)$ as the rate of consumption of monomer at r_L. The governing equation for the macroparticle can be easily derived as

$$\frac{\partial[M_L]}{\partial t} = \frac{D_L}{r_L^2}\left(\frac{\partial}{\partial r_L} r_L^2 \frac{\partial[M_L]}{\partial r_L}\right) - R([M_L], r_L) \tag{5.15.1}$$

where $[M_L]$ is the local concentration of monomer within the macroparticle. Outside this large particle, the monomer concentration is the same as the bulk concentration $[M]_{bulk}$, whereas at the center $\partial[M_L]/\partial r_L = 0$ because of the no-flux condition; that is

$$\frac{[M_L]}{r_L} = 0, \quad r_L = 0 \tag{5.15.2a}$$

$$[M_L] = [M]_{bulk}, \quad r_L = r_{poly} \tag{5.15.2b}$$

In order to solve Eq. (5.15.1), we must first derive an expression for $R(M_L)$, which can be obtained only when we solve the diffusion problem on the microparticle level. Let us define $[M_m]$ as the monomer concentration within it and D_m as the diffusivity in it. Because in this case (see Fig. 5.12) polymerization occurs within the particle at the catalyst surface, the monomer diffusion can be written as

$$\frac{\partial[M_m]}{\partial t} = D_m \frac{1}{r^2} \frac{\partial}{\partial r}\left(r^2 \frac{\partial[M_m]}{\partial r}\right) \tag{5.15.3}$$

where r is the radial length within the microparticle. If R_p is the rate of polymerization at the catalyst particle (having radius r_c), then

$$A_c D_m \frac{\partial[M_m]}{\partial r} = R_p, \quad r = r_c \tag{5.15.4a}$$

$$[M] = [M_L], \quad r = r_{\text{poly}} \tag{5.15.4b}$$

Condition (5.15.4a) arises because the surface reaction should be equal to the rate of diffusion at the catalyst surface, whereas condition (5.15.4b) arises due to continuity of monomer concentration at the boundary.

The complete rigorous solution of equations describing Ziegler–Natta polymerization is difficult. However, a numerical solution can be obtained after making several simplifying assumptions, such as quasi-steady-state for the macroparticles and equality of all macroparticles. The polymers resulting from use of Ziegler–Natta catalysts normally have a wide molecular-weight distribution and this can be explained through the analysis of this section.

5.16 NEWER METALLOCENE CATALYSTS FOR OLEFIN POLYMERIZATION [57–60]

Metallocene are group IV metals (T_1, Zr, Hf, and Rf, but commonly Zr is used) complexed with cyclopentadiene and can be activated by methyl aluminoxane (MAO) as follows:

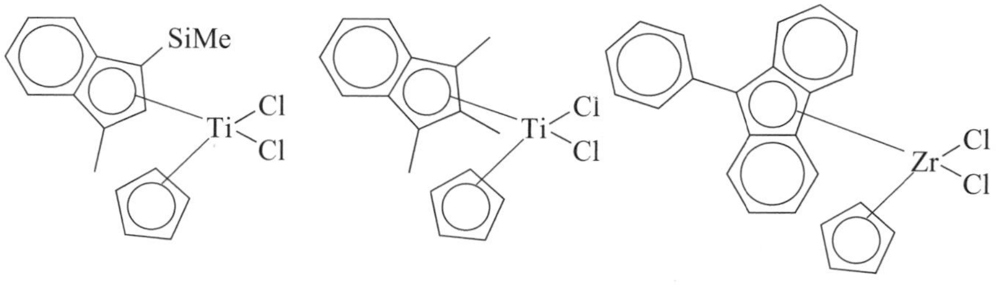

The above MAO is a reaction product of partially hydrolyzed triethyl aluminium and is mostly used in homogeneous solution. The MAO provides a cage for the cation and the pair as a whole serves as the catalyst for polymerization. Some of the metallocene complexes that produce high molecular weights of polyethylene are

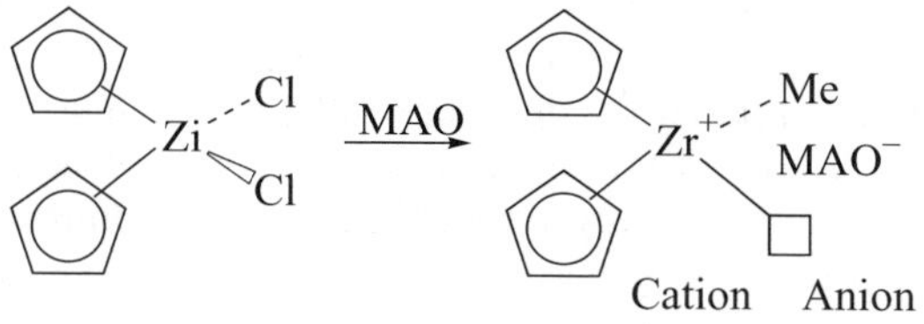

the first step in the catalyst polymerization, as discussed in Section 5.12, of the olefin to the Lewis acid metal center. The chain propagation occurs by insertion of the olefin between the metal carbon bond as follows:

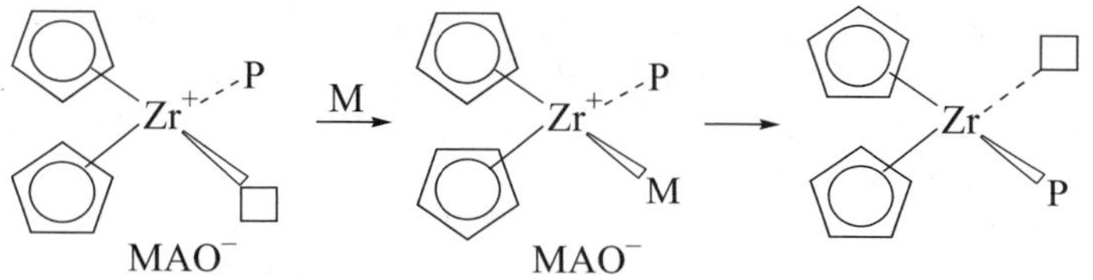

The insertion step consists of an alkyl migration to the olefin ligand. At the same time, a new free coordination site is generated at the vacant piston of the former alkyl ligand. Depending on the orientation of the monomer during insertion, the following (1,2) or (2,1) possibilities exist:

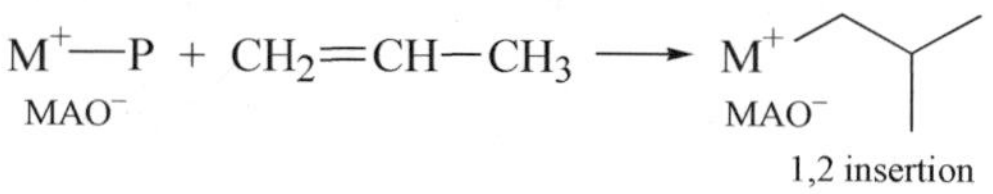

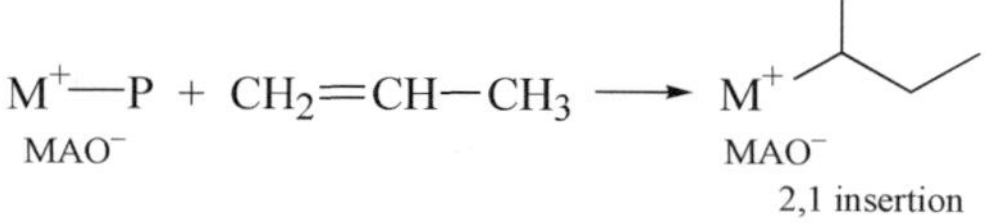

$$M^+\!-\!P + CH_2\!=\!CH\!-\!CH_3 \longrightarrow M^+$$
MAO⁻   MAO⁻
                                                      1,2 insertion

$$M^+\!-\!P + CH_2\!=\!CH\!-\!CH_3 \longrightarrow M^+$$
MAO⁻   MAO⁻
                                                      2,1 insertion

In view of this, in the polymerization of propylene, the following racemic and meso diads are formed:

Racemic diads   Meso diads

A polymer having only racemic diads gives syndiotactic polymer, whereas one having only meso diads gives isotactic polymer. The control of stereoregularity is once again by (1) catalytic site control and (2) chain end control, which is caused by the chirality of the previous monomer inserted.

Metallocene catalysts are extremely interesting because they dissolve in the reaction medium and give very high activity. The polymer formed has a polydispersity index of the order of 2, which is a considerable improvement from the usual Ziegler–Natta catalyst. The polymer thus formed has high clarity and mechanical strength. However, it has the drawback of becoming poisoned by polar comonomers. Late-metal (Hf and Pd) complexes used for ethylene homopolymerization and copolymerization. However, they have not been commercially adopted as yet. Polyethylene formed using these catalysts is highly branched and has a relatively lower molecular weight.

5.17 CONCLUSION

In this chapter, different mechanisms of chain-reaction polymerization have been discussed in detail. Based on the mechanism involved, expressions for the rate of polymerization, molecular-weight distribution, average chain lengths, and the polydispersity index can be derived.

Understanding the expressions introduced in this chapter is an essential requirement in the analysis of reactors, presented in Chapter 6.

REFERENCES

1. Odian, G., *Principles of Polymerization*, 2nd ed., McGraw-Hill, New York, 1982.
2. Flory, P. J., *Principles of Polymer Chemistry*, Cornell University Press, Ithaca, NY, 1953.
3. Kumar, A., and S. K. Gupta, *Fundamentals of Polymer Science and Engineering*, Tata McGraw-Hill, New Delhi, 1978.
4. Brandrup, J., and E. H. Immergut, *Polymer Handbook*, 2nd ed., Wiley–Interscience, New York, 1975.
5. Levenspiel, O., *Chemical Reaction Engineering*, 2nd ed., Wiley, New York, 1972.
6. Ray, W. H., On the Mathematical Modeling of Polymerization Reactors, *J. Macromol. Sci. Rev.*, C8, 1, 1972.
7. Ray, A. B., D. N. Saraf, and S. K. Gupta, Free Radical Polymerizations Associated with the Trommsdorff Effect under Semibatch Reactor Conditions: I. Modeling, *Polym. Eng. Sci.*, 35, 1290–1299, 1995.
8. Srinivas, T., S. Sivakumar, S. K. Gupta, and D. N. Saraf, Free Radical Polymerizations Associated with the Trommsdorff Effect under Semibatch Reactor Conditions: II. Experimental Responses to Step Changes in Temperature, *Polym. Eng. Sci.*, 36, 311–321, 1996.
9. Dua, V., D. N. Saraf, and S. K. Gupta, Free Radical Polymerizations Associated with the Trommsdorff Effect under Semibatch Reactor Conditions: III. Experimental Responses to Step Changes in Initiator Concentrations, *J. Appl. Polym. Sci.*, 59, 749–758, 1996.
10. Seth, V., and S. K. Gupta, Free Radical Polymerizations Associated with the

Trommsdorff Effect under Semibatch Reactor Conditions: An Improved Model, *J. Polym. Eng.*, 15, 283–326, 1995.

11. Chandran, S. R., Simulation and Optimization of Suspension PVC Batch Reactors, M.Tech. thesis, Department of Chemical Engineering, IIT, Kanpur, 1993.

12. Solomon, D. H., and G. Moad, Initiation: The Reaction of Primary Radicals, *Makromol. Chem. Makromol. Symp.*, 10/11, 109–125, 1987.

13. Kumar, A., S. K. Gupta, and R. Mohan, Effect of Shear Rate on the Polymerization of Methyl Methacrylate, *Eur. Polym. J.*, 16, 7–10, 1980.

14. Kumar, A., A. Kumar, and S. K. Gupta, Effect of Shear Rate on the Solution Polymerization of Acrylonitrile, *Polym. Eng. Sci.*, 22, 1184–1189, 1982.

15. Tirrell, M., R. Galvan, and R. L. Laurence, Polymerization Reactors, in *Chemical Reaction and Reactor Engineering*, J. J. Carberry and A. Varma (eds.), Marcel Dekker, New York, 1986.

16. Stickler, M., and E. Dumont, Free-Radical Polymerization of Methyl Methacrylate at Very High Conversions: Study of the Kinetics of Initiation by Benzoyl Peroxide, *Makromol. Chem.*, 187, 2663–2673, 1986.

17. Soh, S. K., and D. C. Sundberg, Diffusion Controlled Vinyl Polymerization: IV. Comparison of Theory and Experiment, *J. Polym. Sci. Polym. Chem. Ed.*, 20, 1345–1371, 1982.

18. Chiu, W. Y., G. M. Carratt, and D. S. Soong, A Computer Model for the Gel Effect in Free Radical Polymerization, *Macromolecules*, 16, 348–357, 1983.

19. Turner, D. T., Autoacceleration of Free Radical Polymerization: 1. The Critical Concentration, *Macromolecules*, 10, 221–226, 1977.

20. Lee, H. B., and D. T. Turner, Autoacceleration of Free-Radical Polymerization: 2. Methyl Methacrylate, *Macromolecules*, 10, 226–231, 1977.

21. Lee, H. B., and D. T. Turner, Autoacceleration of Free-Radical Polymerization: 3. Methyl Methacrylate, *Macromolecules*, 10, 231–235, 1977.

22. Cardenas, J., and K. F. O'Driscoll, High Conversion Polymerization: I. Theory and Application to Methyl Methacrylate, *J. Polym. Sci. Polym. Chem. Ed.*, 14, 883–897, 1976.

23. Marten, F. L., and A. E. Hamielec, High Conversion Diffusion Controlled Polymerization in Polymerization Reactors and Processes, *ACS Symp. Ser.*, 104, 43–90, 1979.

24. Achilias, D., and C. Kiparissides, Modeling of Diffusion Controlled Free-Radical Polymerization Reactions, *J. Appl. Polym. Sci.*, 35, 1303–1323, 1988.

25. Brooks, B. W., Kinetic Behavior and Product Formation in Polymerization Reactors Operation at High Viscosity, *Chem. Eng. Sci.*, 40, 1419–1423, 1985.

26. Stickler, M., D. Panke, and A. E. Hamielec, Polymerization of Methyl Methacrylate up to High Degrees of Conversion: Experimental Investigation of the Diffusion-Controlled Polymerization, *J. Polym. Sci. Polym. Chem. Ed.*, 22, 2243–2253, 1984.

27. Stickler, M., Free-Radical Polymerization Kinetics of Methyl Methacrylate at Very High Conversions, *Makromol. Chem.*, 184, 2563–2579, 1983.

28. Mahabadi, H. K., A Review of the Kinetics of Low Conversion Free Radical Termination Process, *Makromol. Chem. Makromol. Symp.*, 10/11, 127–150, 1987.

29. Sawada, H., *Thermodynamics of Polymerization, Marcel Dekker, New York, 1976.*

30. Szwarc, M., Living Polymers and Mechanism of Anionic Polymerization, *Adv. Polym. Sci.*, 49, 1, 1983.

31. Couso, D. A., L. M. Alassia, and G. R. Meira, Molecular Weight Distribution in a Semibatch Anionic Polymerization: I. Theoretical Studies, *J. Appl. Polym. Sci.*, 30, 3249–3265, 1985.

32. Meria, G. R., Molecular Weight Distribution Control in Continuous Polymerization Through Periodic Operations of Monomer Feed, *Polym. Eng. Sci.*, 211, 415–423, 1981.

33. Sailaja, R. R. N., and A. Kumar, Semianalytical Solution of Irreversible Anionic Polymerization with Unequal Reactivity in Batch Reactors, *J. Appl. Polym. Sci.*, 58, 1865–1876, 1996.

34. Tappe, R., and F. Bandermann, Molecular Weight Distribution of Living Polymers in Semibatch Reactors, *Makromol. Chem.*, 160, 117, 1988.

35. Chen, G. T., Kinetic Models of Homogeneous Ionic Polymerization, *J. Polym. Sci. Polym. Chem.*, 20, 2915, 1982.

36. Benoit, D., V. Chaplinski, R. Braslau, and C. J. Hawker, Development of a Universal Alkoxyamine for "Living" Free Radical Polymerizations, *J. Am. Chem. Soc.*, 121, 3904–3920, 1999.

37. Gravert, D. J., A. Datta, P. Wentworth, and K. D. Janda, Soluble Supports Tailored for Organic Synthesis: Parallel Polymer Synthesis via Sequentials Normal/Living Free Radical Processes, *J. Am. Chem. Soc.*, 120, 9481–9495, 1998.

38. Paten, T. E., J. Xia, T. Abornathy, and K. Matyjaszewski, Polymer with Very Low Polydispersities from Atom Transfer Radical Polymerization, *Science*, 272, 866–868, 1996.

39. Chen, X. P., and K. Y. Qiu, Controlled/Living Radical Polymerization of MMA via In Situ ATRP Processes, *Chem. Commun.*, 233–234, 2000.

40. Grabbs, R. B., C. J. Hawker, J. Dao, and J. M. Frechet, A Tenden Approach to Graft and Copolymers Based on Living Radical Polymerization, *Angew. Chem. Int. Ed., 36, 270–273, 1997.*

41. Xia, Y., and G. M. Whilesides, Soft Lithography, *Angew. Chem. Int. Ed.*, 37, 550–575, 1998.

42. Oike, H., H. Imaizumi, T. Mouri, Y. Yoshioka, A. Uchibori, and Y. Tezuka, Designing Unusual Polymer Topologies by Electrostatic Self Assembly and Covalent Fixation, *J. Am. Chem. Soc.*, 122, 9592–9599, 2000.

43. Keii, T., Kinetics of Ziegler–Natta Polymerization, Kodansha Scientific, Tokyo, 1972.

44. Kissin, Y. V., Principles of Polymerization with Ziegler–Natta Catalysts, in *Encyclopedia of Engineering Materials*, N. P. Cheremisnoff (ed.), Marcel Dekker, New York, 1988.

45. Keii, T., Propene Polymerization with Magnesium Supported Ziegler Catalyst: Molecular Weight Distribution, *Makromol. Chem.*, 185, 1537, 1984.

46. Yuan, H. G., T. W. Taylor, K. Y. Choi, and W. H. Ray, Polymerization of Olefins Through Heterogeneous Catalysts: I. Low Pressure Polymerization in Slurry with Ziegler–Natta Catalysis, *J. Appl. Polym. Sci., 27, 1691, 1982.*

47. Taylor, T. W., K. V. Choi, H. Yuan, and W. H. Ray, Physicochemical Kinetics of Liquid Phase Propylene Polymerization, Symposium on Transition Metal Catalyzed Polymerization, 1981.

48. Honig, J. A. J., R. P. Burford, and R. P. Caplin, Molecular Weight Phenomena at High Conversions in Ziegler–Natta Polymerization of Butadiene, *J. Polym. Sci. Polym. Chem.*, 22, 1461–1470, 1984.
49. Quirk, R. P., *Transition Metal Catalyzed Polymerization: Alkene and Dienes*, Harwood, New York, 1985.
50. Kissin, Y. V., *Isospecific Polymerization of Olefins with Heterogeneous Ziegler–Natta Catalysts*, Springer-Verlag, New York, 1985.
51. Lenz, R. W., and F. Ciardelli (eds.), *Preparation and Properties of Stereoregular Polymers*, D. Riedel, Dordrecht, 1980.
52. Chen, C. M., and W. H. Ray, Polymerization of Olefin through Heterogeneous Catalysis: XI. Gas Phase Sequential Polymerization Kinetics, *J. Appl. Polym. Sci.*, 49, 1573–1588, 1993.
53. Sau, M., and S. K. Gupta, Modeling of Semibatch Propylene Slurry Reactors, *Polymer*, 34, 4417–4426, 1993.
54. Kim, I., H. K. Choi, T. K. Han, and S. I. Woo, Polymerization of Propylene Catalyzed over Highly Active Stereospecific Catalysts Synthesized with $Mg(OEth)_2$/Benzoyl Chloride/$TiCl_4$, *J. Polym. Sci. Polym. Chem.*, 30, 2263–2271, 1992.
55. Nooijen, G. A. H., On the Importance of Diffusion of Cocatalyst Molecules Through Heterogeneous Ziegler–Natta Catalysts, *Eur. Polym. J.*, 30, 11–15, 1994.
56. McKenna, T. F., J. Dupuy, and R. Spitz, Modeling of Transfer Phenomena on Heterogeneous Ziegler Catalysts: Difference Between Theory and Experimental in Olefin Polymerization (an introduction), *J. Appl. Polym. Sci.*, 57, 371–384, 1995.
57. Keminsky, W., and M. Arndt, Metallocenes for Polymer Catalysis, *Adv. Polym. Sci.*, 127, 144–187, 1997.
58. Alt, H. G., and A. Koppel, Effect of the Nature of Metallocene Complexes of Group IV Metals on Their Performance in Catalytic Ethylene and Propylene Polymerization, *Chem. Rev.*, 100, 1205–1221, 2000.
59. Brittovsek, G. J. P., V. C. Gibson, and D. F. Wass, The Search for New Generation Olefin Polymerization Catalysts: Life Beyond Metallocenes, *Angew. Chem. Int. Ed.*, 38, 428–447, 1999.
60. Ittel, S. D., and L. K. Johnson, Late Metal Catalysts for Ethylene Homo- and Copolymerization, *Chem. Rev.*, 100, 1169–1203, 2000.

PROBLEMS

5.1. Analyze the equilibrium free-radical polymerization with the unequal reactivity in P_1 in the following propagation steps:

$$I_2 \underset{k_d'}{\overset{k_d}{\rightleftharpoons}} 2I, \qquad K_d = \frac{k_d}{k_d'}$$

$$I + M \underset{k_1'}{\overset{k_1}{\rightleftharpoons}} P_1, \qquad K_1 = \frac{k_1}{k_1'}$$

$$P_n + M \underset{k_p'}{\overset{k_p}{\rightleftharpoons}} P_{n+1} \qquad n \geq 1, \ K_p = \frac{k_p}{k_p'}$$

$$P_n + M \underset{k_{td}'}{\overset{k_{td}}{\rightleftharpoons}} M_n + M_m, \qquad n, \ m \geq 1, \ K_{td} = \frac{k_{td}}{k_{td}'}$$

Proceed the same way as in Example 5.2 and determine the following MWD of the polymer radicals:

$$P_1 = K_1 M I$$
$$P_2 = K_{P_4} M P_1$$
$$P_n = K_p M P_{n-1} = (K_p M)^{n-2} P_2, \qquad n \geq 3$$

Determine the zeroth, first, and second moments of the polymer radicals and the dead polymers.

5.2. Analyze the following equilibrium free-radical polymerization, in which P_1 reacts with itself at a different rate in the termination step:

$$I_2 \underset{k_d'}{\overset{k_d}{\rightleftharpoons}} 2I, \qquad K_d = \frac{k_d}{k_d'}$$

$$M + I \underset{k_1'}{\overset{k_1}{\rightleftharpoons}} P_1, \qquad K_1 = \frac{k_1}{k_1'}$$

$$P_1 + M \underset{k_{p1}'}{\overset{k_{p1}}{\rightleftharpoons}} P_2, \qquad K_{p1} = \frac{k_{p1}}{k_p'}$$

$$P_n + M \underset{k_p'}{\overset{k_p}{\rightleftharpoons}} P_{n+1}, \qquad n \geq 2, \ K_p = \frac{k_p}{k_p'}$$

$$P_n + P_m \underset{k_{td}'}{\overset{k_{td}}{\rightleftharpoons}} M_n + M_m, \qquad n, \ m \geq 1, \ K_{td} = \frac{k_{td}}{k_{td}'}$$

Derive the following MWD relations for the radical species as well as the dead polymer:

$$P_1 = K_1 MI$$
$$P_n = (K_p M)P_{n-1}$$

5.3. In Problem 5.2, derive expressions for the zeroth, first, and second moments of polymer radicals and the dead polymer have the following μ_n and μ_w:

$$\mu_n = \frac{1 + (R_{td} - 1)(1 - K_p M)^3}{(1 - K_p M)[1 + (R_{td} - 1)(1 - K_p M)^2]}$$

$$\mu_w = \frac{1 + K_p M + (R_{td} - 1)(1 - K_p M)^4}{(1 - K_p M)[1 + (R_{td} - 1)(1 - K_p M)^3]}$$

where $R_{td} = k_{td1}/k_{td}$.

Determine the polydispersity index of the dead polymer.

5.4. To explore as a variation of the unequal reactivity discussed in Problem 5.4, consider the following kinetic model, in which P_1 is assumed to react at a different rate with all polymer radicals:

$$I_2 \underset{k_d'}{\overset{k_d}{\rightleftharpoons}} 2I, \qquad K_d = \frac{k_d}{k_d'}$$

$$I + M \underset{k_1'}{\overset{k_1}{\rightleftharpoons}} P_1, \qquad K_1 = \frac{k_1}{k_1'}$$

$$P_n + M \underset{k_p'}{\overset{k_p}{\rightleftharpoons}} P_{n+1}, \qquad K_p = \frac{k_p}{k_p'}$$

$$P_1 + P_1 \underset{k_{td1}'}{\overset{k_{td1}}{\rightleftharpoons}} M_1 + M_1, \qquad K_{td1} = \frac{k_{td1}}{k_{td}'}$$

$$P_n + P_m \underset{k_{td}'}{\overset{k_{td}}{\rightleftharpoons}} M_m + M_n, \qquad m \geq 1, \ n \geq 2, \ K_{td} = \frac{k_{td}}{k_{td}'}$$

Find the following MWDs of the M_n and P_n species and the various moments:

$$P_1 = K_1 MI$$
$$P_n = (K_p M)^{n-1} P_1$$
$$\lambda_{M0}^2 = K_{d1} P_1 \lambda_{P0} + K_{td1} P_1(\lambda_{P0} - P_1) + K_{td}(\lambda_{P0} - P_1)^2$$
$$\lambda_{M0}^2 = K_{d1} P_1 \lambda_{P0} + K_{td1} P_1(\lambda_{P1} - P_1) + K_{td}(\lambda_{P0} - P_1)(\lambda_{P2} - P_1)$$

5.5. Polymerization of styrene has the rate constants $k_p = 145$ L/mol sec and

$k_t = 0.13 \times 10^7$ L/mol sec. The density of styrene is 0.8 g/cm^3. Benzoyl peroxide, which has a half-life of 44 hr, is used as the initiator. The polymerization of styrene uses 0.5% initiator by weight. Now, refer to the mechanism of radical polymerization, in which there is no way of measuring k_I. The only thing that is known about the mechanism is that it is the reaction between two small molecules I and M. As a result of it, $k_1 > k_p$. Assume $k_1 = 10k_p$.

Find the initiation, propagation, and termination rates under the steady-state hypothesis. Determine [P] and [I]. Find the kinetic chain length. Because termination occurs mainly by combination for styrene, find the average molecular weight of the polymer formed.

5.6. We want to polymerize styrene to a molecular weight of 10^5. To avoid the gel effect, we polymerize it in 60% toluene solution of the monomer. Find out how many grams of benzoyl peroxide should be dissolved in 1 L of the solution.

5.7. When we expose vinyl monomers to high temperatures, we find that the polymerization progresses even without an initiator. The initiation has been proposed to occur as follows:

$$CH_2{=}CH + CH_2{=}CH \xrightarrow{k_1} CH{-}\overset{\bullet}{C}H_2{-}\overset{\bullet}{C}H_2CH$$
$$\qquad\;\; | \qquad\qquad | \qquad\qquad\qquad | \qquad\qquad\; |$$
$$\qquad\;\; X \qquad\qquad X \qquad\qquad\qquad X \qquad\qquad X$$

Both of these ends can polymerize independently. Model the rate of polymerization r_p. Find the average molecular weight of the polymer.

5.8. A dilatometer is a glass bulb with a precision bore capillary; it is a convenient tool with which the rate of free-radical polymerization can be determined. A suitable initiator is dissolved in the monomer and the solution is introduced into the dilatometer through a syringe. The change in volume of the reaction mass is measured as a function of time, which can be related to the conversion of monomer through

$$\frac{\Delta[M]}{[M]_0} = \frac{\Delta V}{\rho_s V_0} \left(\frac{1}{\rho_s} - \frac{1}{\rho_{ps}} \right)^{-1}$$

where V_0 and ΔV are the initial and the change in volume; ρ_s and ρ_{ps} are the densities of the monomer and polymer (in the dissolved state), respectively; and $[M]_0$ and $\Delta[M]$ are the initial concentration and the change in the concentration of the monomer. Derive this relation.

5.9. Determine the initiator efficiency in the polymerization of styrene at 60°C in the following actual experiment. We polymerize 4.4972 cm^3 of styrene in a dilatometer with the benzoyl peroxide concentration at 6.78×10^{-3} mol/L and the concentration of 2,2-diphenyl picrylhydrazil (a strong inhibitor) as 0.3361×10^{-4} mol/L. We measure the height in the dilatometer as a function of time.

Time, min	Height, mm	Time, min	Height, mm
30	0	82	2.9
40	0.5	86	3.1
50	1.0	90	3.8
60	1.6	102	4.4
70	2.0	106	4.7

We plot these data on graph paper and extend the linear region of the plot to the abscissa to find the intercept, which is the same as the induction time. From this plot, calculate the initiator efficiency.

5.10. A dilatometric study of polymerization of styrene has been carried out. The volume of styrene is 4.4972 cm^3 and the diameter of the capillary is 1 mm. The initiator used in AZDN at a concentration of 3.87×10^{-3} mol/L. The height, h, of the monomer column in the capillary varies with time, t, as follows:

t, min	h, mm	t, min	h, mm
0	0.0	40	6.9
5	0.1	50	8.5
10	1.0	60	10.1
20	3.4	70	11.7
30	4.9		

The slope of the plot of h versus t will give r_p. Find it.

5.11. Derive an expression for the kinetic chain length in radical polymerization when a transfer reaction occurs with the monomer, the transfer agent, and the solvent. Also find the expression for μ_n.

5.12. Integrate the equation

$$r_p = k_p \left\{ \frac{fk_1[I_2]}{k_t} \right\}^{1/2} [M]$$

to find the monomer concentration as a function of time under the following assumptions:

(a) The $t^{1/2}$ of the initiator is very large, such that the concentration of the initiator is approximately constant.

(b) The initiator concentration changes following first-order kinetics.

Plot the concentration in both these cases. If this is done correctly, you will find that case (b) cannot give 100% conversion. Justify this physically, and plot μ_n as a function of time.

5.13. The kinetics of retarders (Z) are expressed as

$$P + Z \xrightarrow{k} Z_R$$

$$Z_R + M \xrightarrow{k_{td}} Z_P$$

$$Z_R + Z_R \xrightarrow{k_{zt}} \text{Nonradical product}$$

where Z_R and P are the reacted radical and polymer radicals, respectively. Z_R is a radical of lower reactivity. Supposing that a retarder Z is present in the reaction mass, the reactions shown would occur in addition to the normal ones. In this case, we neglect the reaction between Z_R and P to form an inactivated molecule. Find the rate of polymerization and DP in the presence of a known concentration of the retarder [Z]. This result is immensely important because oxygen present in a monomer even in trace amounts will retard the rate considerably, as follows:

$$P + O_2 \longrightarrow P-O-O^{\cdot}$$
$$P-O-O^{\cdot} + M \longrightarrow P-O-O-M^{\cdot}$$

$P-O-O^{\cdot}$ is the retarded radical and $P-O-O-M$ is kinetically the same as P.

5.14. Consider the following mechanism of polymerization:

$$I_2 \rightarrow 2I$$
$$I + M \rightarrow P$$
$$P + M \rightarrow P$$
$$P + P \rightarrow \text{Inactive}$$
$$P + M \rightarrow \text{Inactive}$$

In this mechanism, the monomer itself is acting as the inhibitor. Derive the expression for r_p and DP. This kind of inhibition is found in the case of polymerization of allyl monomers.

5.15. Suppose that there is some mechanism by which the termination reaction is totally removed in radical polymerization. Then, derive and plot an expression for r_p as a function of time. Note that there is no steady state existing in this case and, hence, there is no steady-state approximation.

5.16. The r_p derived in the text for radical polymerization has been done with the assumption that initiation was the rate-determining step. Find out at what state of the derivation this assumption was used. Consider the hypothetical state when propagation is the rate-determining step and derive the new r_p. Repeat the derivation when termination is the rate-determining step.

5.17. The initiator efficiency is designed to take care of the wastage of primary radicals. Find the initiator efficiency if the initiation step consists of the following reactions:

$$I_2 \underset{}{\overset{k_1}{\rightleftharpoons}} 2I$$

$$I + M \xrightarrow{k_1} P$$

$$I + S \xrightarrow{k_2} \text{Inactive species}$$

where S is a solvent molecule.

5.18. Find the initiator efficiency if the initiation step is known to consist of the following reactions:

$$I_2 \xrightarrow{k_I} 2I$$

$$I + M \xrightarrow{k_1} P$$

$$I_2 + M \xrightarrow{k_2} P + I$$

5.19. Assume that the temperature of polymerization is increased from 50°C to 100°C for pure styrene and the polymerization is carried out to completion. Do you expect the equilibrium monomer concentration to reduce or increase? Calculate the equilibrium monomer concentration as a function of temperature.

5.20. Do you expect the equilibrium monomer concentration in radical polymerization to be affected by the gel effect? Justify your claim.

5.21. The temperature of radical polymerization is increased such that the viscosity of the reaction mass remains constant. Would the gel effect occur? If yes, why do you think so?

5.22. In the buildup period of decay-type stereoregular polymerization, we find that the rate when propylene is introduced after $TiCl_3$ and $AlEt_3$ are allowed to equilibrate is different from the rate when $AlEt_3$ is added after the gas is introduced. It is assumed that the following equilibrium exists in the former case:

$$TiCl_3 + AlEt_3 \underset{\text{Equilibrium}}{\overset{\text{Adsorption}}{\rightleftharpoons}} \text{Potential PC}$$

The concentration of potential polymerization centers is given by the Langmuir equation. On the introduction of the gas, polymerization centers, C, are formed. A proper balance of C would yield the appropriate relation. Derive it.

5.23. Consider now the case in which propylene is introduced before $AlEt_3$ is added. In such a case, all of the reactions in the mechanism of polymerization occur simultaneously. Derive the following result for small extents of time:

$$r_0 = kP[a]t$$

where r_0 is the rate for small intervals of time, P is the propylene gas pressure, [A] is the concentration of $AlEt_3$, and t the time of the reaction.

5.24. Hydrogen is used as the molecular-weight regulator in stereoregular polymerization. The proposed mechanism postulates that there exists a pre-established equilibrium of dissociative adsorption of hydrogen on the $TiCl_3$ catalyst surface, as follows:

$$H_2 \rightarrow 2H_{ads}$$

It is this adsorbed hydrogen that participates in the following reaction:

$$Cat{-}P + 2H_{ads} \rightarrow Cat{-}H + PH$$

The Cat$-$H reacts with monomer molecules at a different rate in the following fashion:

$$Cat{-}H + monomer \rightarrow Cat{-}P$$

Derive the rate of polymerization.

5.25. Using the mechanism stated in Problem 5.24, find the molecular weight in the stationary zone.

6

Reaction Engineering of Chain-Growth Polymerization

6.1 INTRODUCTION

In Chapter 5, we considered the detailed mechanisms of chain-growth polymerization. Expressions for the rates and average molecular weight were developed and were shown to be sensitive to the reaction temperature, the initiator concentration, and the presence of transfer agents. In general, most polymerizations are exothermic and good temperature control can be achieved only through elaborate cooling systems. This is because most monomers and polymers have low thermal conductivity. Sometimes, the desired physical and mechanical properties cannot be obtained by a homopolymer or by blending several homopolymers. In such cases, copolymers are prepared, several of which are commonly used in industry.

This chapter discusses the reaction engineering of chain-growth polymerization. In order to form polymers of specified properties, we observe that reactor temperature is a very important variable. To find this, the energy balance equation must be solved, along with mole balance relations of various species. In the study of copolymers, the quantities of practical interest are the relative distributions of the monomers on polymer chains and the overall rates of copolymerization. With these, it is possible to carry out the reactor design.

6.2 DESIGN OF TUBULAR REACTORS [1–13]

A reactor design can be carried out if we know the following information:

1. The rate characteristics
2. The heat and mass transfer characteristics
3. External restrictions imposed by the reactor setup on fluid flow

Let us consider a tubular reactor, which is normally modeled as a plug flow reactor (PFR), shown in Figure 6.1. The composition of the reaction mass changes along the reactor length, and in the ideal reactor, all fluid elements are thought to move at the same velocity (or plug flow profile). Assuming a constant density process and no mixing, it is possible to carry out mole balance as follows:

$$F_{A0}\, dx_M = -r_p\, dV \tag{6.2.1}$$

where x_M is the conversion of monomer and r_p is the rate of consumption of monomer. The elemental volume dV can be written in terms of cross-sectional area A_c as follows:

$$dV = A_c\, dz \tag{6.2.2}$$

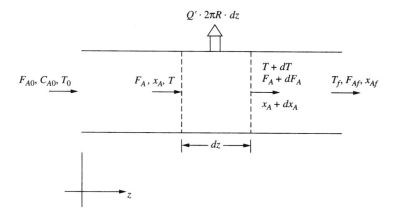

FIGURE 6.1 Schematic diagram of a plug flow reactor. F_{A0} is the molar flow rate of reactants, F_{Af} is the exit flow rate of the reactants, x_M is the conversion monomer, and Q' is the heat removed by coolants per unit surface area of the reactor.

The molar flow rate F_{A0} in Figure 6.1 can be written in terms of the velocity, v_0, at which the fluid particles are traveling, or

$$F_{A0} = A_c C_{A0} v_0 \qquad (6.2.3)$$

Substituting these in Eq. (6.2.1), we get

$$C_{A0} \frac{dx_M}{dt} = -r_p \qquad (6.2.4)$$

where

$$dt = \frac{dz}{v_0} \qquad (6.2.5)$$

In reactors used for carrying out radical polymerization, there is a continual change in the physical properties such as heat capacity, density, and viscosity. The following examples examine these effects on reactor design.

Example 6.1: Polymerization of styrene is carried out in an isothermal tubular reactor at $60°C$ up to 30% conversion. Assume average rate constants at $60°C$:

$k_p = 145 \, \text{L/mol sec}$, $k_t = 1.3 \times 10^6 \, \text{L/mol sec}$
$fk_1 = 4.4 \times 10^6 \, \text{sec}^{-1}$, $[I_2]_0 = 1.86 \times 10^{-2} \, \text{mol/L}$
$[M]_0 = 8.93 \, \text{g mol/L}$

At this temperature, styrene has a density of $0.869 \, \text{g/mL}$ and that of polystyrene is $1.047 \, \text{g/mL}$. Determine the residence time,

1. Assuming no contraction in volume.
2. Accounting for the change in density of the reaction mass.

Solution:

1. When constant molar density is assumed, Eq. (6.2.4) yields

$$\frac{d[M]}{dt} = -k_p \left\{ \frac{fk_1 [I_2]_0}{k_t} \right\}^{1/2} [M]$$

If the change in $[I_2]_0$ is neglected,

$$\ln\left(\frac{[M]_0}{[M]} \right) = -k_p \left\{ \frac{fk_1 [I_2]_0}{k_t} \right\}^{1/2} [M]$$

For 30% conversion, the residence time t is calculated to be 9.8×10^3 sec.

2. When the change in volume is to be accounted for, we write the variable value as

$$V = V_0(1 - \varepsilon x_M)$$

where ε is the incremental volume contraction due to 1% polymerization and is given by

$$\varepsilon = \frac{-1/\rho_P + 1/\rho_M}{1/\rho_M} = \frac{\rho_P - \rho_M}{\rho_P}$$

where ρ_M and ρ_P are the densities of the monomer and polymer, respectively. The monomer concentration can be written in terms of conversion as follows:

$$[M]_0 = \frac{\text{Moles of M}}{\text{Volume}} = \frac{[M]_0(1 - x_M)}{1 - \varepsilon x_M}$$

$$r_P = -\frac{d[M]}{dt} = k_p \left(\frac{fk_1}{k_t}\right)^{1/2} \frac{[I_2]_0^{1/2}[M]_0(1 - x_M)}{(1 - \varepsilon x_M)^{1/2}(1 - \varepsilon x_M)}$$

The residence time of the reactor is

$$\frac{[M]_0}{1 - \varepsilon x_M} \frac{dx_n}{dt} = r_p$$

or

$$\int dt = t = \frac{k_t^{1/2}}{k_p(fk_1)^{1/2}[I_2]_0^{1/2}} \int_0^{x_f} \frac{(1 - \varepsilon x_M)^{1/2}}{(1 - x_M)} dx_M$$

ε is determined to be 0.1704, and the integral $\int_0^{x_f}[(1 - \varepsilon x)^{1/2}/(1 - x)]\,dx$ determined numerically as 0.325 for $x_f = 0.3$. The reactor residence time is 0.95×10^5 sec.

The change in volume leads to a reduction of 3% in reactor size. It may be recalled that the polymerization rate is the slowest in the initial region, and this may represent sizable savings.

Example 6.2: Estimate the viscosity of the reaction mass in a polystyrene reactor at 1% and 10% conversion. The molecular weight of the polymer is assumed to be 3.0×10^6.

Solution: We approximate the styrene–polystyrene system by using a toluene–polystyrene system. For this, the viscosity can be estimated by

$$\frac{\eta - 0.0056}{0.0056} = 5.3\lambda_{M1} + 0.4 \times 5.3^2 \lambda_{M1} \quad \text{for } \lambda_{M1} < 5\%$$

where λ_{M1} is the concentration of polymer in g/dL and η is the viscosity of the solution (in poise). For 10% polymer, C_p is equal to 0.087 g/dL, which gives the viscosity as 0.086 P. The viscosity of 10% solution is obtained from

$$\eta = 2 \times 10^{-5} [\lambda_{M1}^{1.5} \bar{M}_w]^{3.5}$$

where $\bar{M}_w$ is the weight-average molecular weight. The calculated viscosity is in poise and is found to be 90 P, a 10,000-fold increase. This large increase in viscosity reduces the Reynolds number of the flow, which, in turn, increases the pressure drop and lowers the overall heat and mass transfer coefficients.

We have already observed in Chapter 5 that the reactor temperature plays a major role in determining the course of polymerization. This temperature can be determined if an energy balance is made over the differential element, which, at steady state, is given by

$$F\bar{C}_p(T + dT) - F\bar{C}_p T = r_p(-\Delta H_r)A_c \, dz - Q' 2\pi R \, dz \qquad (6.2.6)$$

where F is the mass flow rate, not the molar flow rate in Eq. (6.2.1), C_p is the average specific heat, and $(-\Delta H_r)$ is the heat of polymerization. Because F is in total mass units, it remains constant along the length of the tubular reactor. Rearranging Eq. (6.2.6) yields

$$\bar{C}_p \frac{dT}{dZ} = -(-\Delta H_r)r_p - Q \qquad (6.2.7)$$

where Z is the average density of the reaction mass and Q is given by

$$Q = \frac{2Q'}{R} \qquad (6.2.8)$$

It may be observed that Eq. (6.2.7) involves the specific heat, $\bar{C}_p$, which is also likely to change along the reactor length. Fortunately, though, the monomer and the polymer formed have similar $\bar{C}_p$ values. For example, styrene and polystyrene have C_p values of 0.40 and 0.41 cal/g °C at 25°C.

We discussed in Chapter 5 that, under quasi-steady-state approximation (QSSA), the net production of all intermediate species in a reaction mass can be taken as zero after a small induction time. The polymer in radical polymerization is formed only after polymer radicals are generated. At any given time, therefore, we have two distributions: one for polymer radicals (P_n, $n = 1, 2, 3$) and one for dead polymer chains (M_n, $n = 2, 3$). In Chapter 5, we assumed a QSSA and derived the zeroth moment, λ_{P0}, of the polymer radicals as follows:

$$\lambda_{P0|ss} = \left\{ \frac{f k_1 [I_2]_0}{k_t} \right\}^{1/2} \tag{6.2.9}$$

However, when the gel effect sets in, k_t decreases in value, which results in increased $\lambda_{P0|SS}$ under QSSA, thus leading to higher rates of polymerization. Because all polymerization reactions are exothermic in nature, the temperature of the polymer mass also increases for the same cooling rate, which, in turn, gives a higher rate of initiation. Hence, under the influence of the gel effect, the upward thermal drift always occurs and the QSSA is known to break down. When this happens, the mole balance relation for polymer radicals λ_{P0} is governed by a nonlinear first-order differential equation instead of the simple relation given in Eq. (6.2.9). The concentration λ_{P0} in the reaction mass is normally very low compared with the monomer or initiator concentration. For example, for styrene polymerizing at $60°C$ with benzoyl peroxide initiator, $\lambda_{P0|ss}$ [as calculated from Eq. (6.2.9)] is of the order 10^{-8} mol/L. As a consequence, in computing λ_{P0} numerically through its governing differential equation, the time increment Δt must be chosen to be very small, which would mean that the numerical determination of reactor performance requires a great amount of time for computation.

6.2.1 Moments of Radical and Dead Polymers [12,13]

In order to reduce the computational load, the following moment equations are developed as an alternative to attempting to solve a large number of coupled ordinary differential equations (governing mole balance relations for various species). Within the reaction mass, we have molecular-weight distributions of radical and dead polymers; as a result, we define moments for these separately, as follows:

$$\lambda_{Pi} = \sum_{n=1}^{\infty} n^i [P_n], \quad i = 0, 1, 2, \ldots \tag{6.2.10a}$$

$$\lambda_{Mi} = \sum_{n=1}^{\infty} n^i [M_n], \quad i = 0, 1, 2, \ldots \tag{6.2.10b}$$

Zeroth, first, and second moments ($i = 0, 1, 2$) are usually important; as a consequence, this section focuses on these moments only. In Chapter 5, we derived the mole balance for P_n; it is summarized in Table 6.1 for easy reference. The moment equations are easy to derive from the mole balance, and they are also

TABLE 6.1 Mole and Energy Balance Relations for Species in Batch (or Tubular) Reactors Used for Carrying out Radical Polymerization

---

MWD Relations

Initiator, I_2

$$\frac{d[I_2]}{dt} = -fk_1[I_2] \tag{1}$$

$$\frac{dI}{dt} = 2fk_I - [I_2] - k_1[I][M] $$

Polymer radicals, P_n

$$\frac{d[P_1]}{dt} = k_1[I][M] - k_p[M][P_1] - (k_{tc} + k_{td})\lambda_{P0}[P_1] + k_{trs}[S](\lambda_{P0} - [P_1]) \tag{2}$$

$$\frac{d[P_n]}{dt} = k_p[M]\{[P_{n-1}] - [P_n]\} - (k_{tc} + k_{td})\lambda_{P0}[P_n] + k_{trs}[S][P_n], \quad n \geq 2 \tag{3}$$

Monomer, M

$$\frac{d[M]}{dt} = -k_p[M]\lambda_{P0} \tag{4}$$

Transfer agent, S

$$\frac{d[S]}{dt} = -k_{trs}[S]\lambda_{P0} \tag{5}$$

Dead polymer, M_n

$$\frac{d[M_n]}{dt} = \frac{k_{tc}}{2}\sum_{m=1}^{n-1}[P_m][P_{n-m}] + k_{td}[P_n]\lambda_{P0} + k_{trs}[S][P_n], \quad n \geq 2 \tag{6}$$

Energy balance

$$\rho X_\pi \frac{dT}{dt} = (-\Delta H_r)k_p[M]\lambda_{P0} - \frac{4U^*}{D}(T - T_w) \tag{7}$$

Moment Relations

Radical polymers

$$\frac{d\lambda_{P0}}{dt} = k_1[I][M] - (k_{tc} + k_{td})\lambda_{P0}^2 \tag{8}$$

$$\frac{d\lambda_{P1}}{dt} = k_1[I][M] + k_p[M]\lambda_{P0} - (k_{tc} + k_{td})\lambda_{P0}\lambda_{P1} - k_{trs}(\lambda_{P1} - \lambda_{P0})[S] \tag{9}$$

$$\frac{d\lambda_{P2}}{dt} = k_1[I][M] - k_{trs}(\lambda_{P2} - \lambda_{P0})[S] + k_p[M](2\lambda_{P1} + \lambda_{P0}) - (k_{tc} + k_{td})\lambda_{P0}\lambda_{P2} \tag{10}$$

Dead polymer

$$\frac{d\lambda_{M0}}{dt} = \left(\frac{k_{tc}}{2} + k_d\right)\lambda_{P0}^2 + k_{trs}[S]\lambda_{P0} \tag{11}$$

$$\frac{d\lambda_{M1}}{dt} = (k_{tc} + k_{td})\lambda_{P0}\lambda_{P1} + k_{trs}[S]\lambda_{P1} \tag{12}$$

$$\frac{d\lambda_{M2}}{dt} = (k_{tc} + k_{td})(\lambda_{P0}\lambda_{P2}\lambda_{P1}^2) + k_{tc}(\lambda_{P1}^2 - [P_1]) + k_{trs}[S](\lambda_{P2} - [P_1]) \tag{13}$$

Note: U = overall heat transfer coefficient; T_w = coolant temperature.

given in the table. In the following, we demonstrate the technique by deriving generation relations for λ_{P2} and λ_{M2}:

$$\frac{d\lambda_{P2}}{dt} = \sum_{n=1}^{\infty} n^2 \frac{d[P_n]}{dt}$$

$$= k_1[I][M] + k_p[M]\{2^2[P_1] + 3^2[P_2] + \cdots\}$$

$$= -k_p[M]\{[P_1] + 2^2[P_2] + 3^2[P_3] + \cdots\}$$

$$\quad + k_{trs}[S]\{\lambda_{P0} - [P_1] - 2^2[P_2] - 3^2[P_3] - 4^2[P_4] - \cdots\}$$

$$\quad - (k_{tc} + k_{td})\lambda_{P0}\{[P_1] + 2^2[P_2] + 3^2[P_3] + \cdots\}$$

$$k_1 \, [I] \quad [M]$$

$$= + k_p[M] \sum_{n=1}^{\infty} \{(n+1)^2 - n^2\}[P_n] + k_{trs}[S]\{\lambda_{P0} - \lambda_{P2}\}$$

$$\quad - (k_{tc} + k_{td})\lambda_{P0}\lambda_{P2}$$

$$= k_1[I][M] - k_{trs}[S](\lambda_{P2} - \lambda_P 0) - (k_{tc} + k_{td})\lambda_{P0}\lambda_{P2}$$

$$\quad + k_p[M] \sum_{n=1}^{\infty} (2n+1)[P_n]$$

$$= k_1[I][M] - k_{trs}[S](\lambda_{P2} - \lambda_{P0}) - (k_{tc}k_{td})\lambda_{P0}\lambda_{P2}$$

$$\quad + k_p[M](2\lambda_{P_i} + \lambda_{P_n}) \tag{6.2.11}$$

which are the same as those given in Table 6.1. Similarly, for λ_{M2}, we have

$$\frac{d\lambda_{M2}}{dt} = \frac{d}{dt} \sum_{n=2}^{\infty} n^2[M_n]$$

$$= \frac{k_{tc}}{2} \sum_{n=1}^{\infty} n^2 \sum_{r=1}^{n-1} [P_r][P_{n-r}] + k_{td}\lambda_{P0} \sum_{n=2}^{\infty} n^2[P_n] + k_{trs}[S] \sum_{n=2}^{\infty} n^2[P_n] \tag{6.2.12}$$

However, we know that

$$\sum_{n=1}^{\infty} n^2 \sum_{r=1}^{n-1} [P_r][P_{n-r}] = \sum_{n=2}^{\infty} [P_n] \sum_{n=2}^{\infty} (n+r)^2[P_r]$$

$$= \sum_{n=1}^{\infty} [P_n] \sum_{r=1}^{\infty} (n^2 + r^2 + 2nr)[P_r]$$

$$= \sum_{n=1}^{\infty} [P_n](n^2\lambda_{P0} + \lambda_{P2} + 2n\lambda_{P1})$$

$$= \lambda_{P0} \sum_{n=1}^{\infty} n_2[P_n] + \lambda_{P2} \sum_{n=1}^{\infty} [P_n] + 2\lambda_{P1} \sum_{n=1}^{\infty} n[P_n]$$

$$= \lambda_{P2}\lambda_{P0} + \lambda_{P2}\lambda_{P0} + 2\lambda_{P1}^2 \tag{6.2.13}$$

Therefore, λ_{M2} is given by

$$\frac{d\lambda_{M2}}{dt} = \frac{k_{tc}}{2}(2\lambda_{P2}\lambda_{P0} + 2\lambda_{PI}^2) + k_{td}\lambda_{P0}(\lambda_{P2} - [P_1]) + k_{trs}[S](\lambda_{P2} - [P_1])$$

(6.2.14)

We have already observed that, in radical polymerization, after the monomer reacts with primary radicals I, P_1 grows very quickly (due to a large k_p) to give a high molecular weight. This means that the concentration of P_1 is always very low, and the following inequality holds:

$$\lambda_{P2} + \lambda_{PI} \gg \lambda_{P0} \qquad [P_1]$$

(6.2.15)

We subsequently assume the validity of steady-state approximation for primary radicals I, which gives the following:

$$\frac{d[I]}{dt} = 2fk_1[I_2] - k_1[M][I] = 0$$

(6.2.16)

The use of Eqs. (6.2.15) and (6.2.16) gives the moment generation relation for λ_{M2} as written in Table 6.1.

The various differential equations of Table 6.1 are nonlinear and coupled, and, in principle, they must be solved numerically, which takes excessive computational time. For isothermal reactors for time-invariant rate constants, it is possible to derive a complete analytical solution, which is given in Appendix 6.1. However, actual reactor performance is always nonisothermal; in addition, rate constants (particularly k_p and k_t) are dependent on reaction parameters in a very complex way. Tables 6.2 and 6.3 show the physical properties and rate constants for polystyrene and polymethyl methacrylate systems. Several researchers have attempted to solve for the reactor performance for these systems, and all of them have reported that the differential equations of Table 6.1 (along with the energy balance relation) take excessive computational time. The following discussion minimizes this problem by using the isothermal solution presented in Appendix 6.1.

The standard method for solving a set of ordinary differential equations is to use a fourth-order numerical technique. The total time is divided into small increments of time, Δt (the flowchart of the computer program is given in Figure 6.2). In any given time increment, Δt, the temperature (T), rate constants (k_1, k_p, and k_t), and concentration ([M]; [I$_2$]; [S]; λ_{P0}, λ_{P1}, and λ_{P2}; λ_{M0}, λ_{M1}, and λ_{M2}) are assumed constant over Δt and their next incremental values are calculated by treating the differential equations as difference equations. As a result, in the Runge–Kutta technique, the choice of incremental time is crucial and the numerical solution tends to diverge if Δt is not chosen small enough. One of the ways of overcoming this difficulty is to keep reducing Δt until the numerical solution becomes independent of it. This is called a stable solution.

TABLE 6.2   Parameters Used for Simulation of Polystyrene Reactors

$$f = -12.342396 + \frac{9577.287}{T}$$

$$\qquad\qquad k_t = \frac{d_{t0}D_0}{D_0 + k_{t0}\lambda_0\theta_t}$$

$$\quad - \frac{1743120.6}{T^2}$$

$$\qquad\qquad k_p = \frac{k_{p0}D_0}{D_0 + k_{p0} + \lambda_0\theta_p}$$

$\rho_M = 924 - 0.918\,[T\,(\text{K}) - 273.1]\,\text{g/L}$ $E_t = 31.1\,\text{kcal/mol}$

$\rho_P = 1084.8 - 0.685\,[T\,(\text{K}) - 273.1]\,\text{g/L}$ $E_p = 7.06\,\text{kcal/mol}$

$$\varepsilon = \frac{[\rho_M - \rho_P]}{\rho_P}$$

$$\qquad\qquad E_{tc} = 1.68\,\text{kcal/mol}$$

$T_{gP} = 373.1\,\text{K}$ $E_{tr} = 12.5916\,\text{kcal/mol}$

$k_d = 2.67 \times 10^{15}\,\text{sec}^{-1}$ $B = 0.02$

$k_{p0} = 1.051 \times 10^7\,\text{L/mol sec}$ $\varepsilon_1 = 0.1$

$k_{td0} = 0.0$ $\theta_p = 1.23421 \times 10^{-14}\,\exp\!\left(\dfrac{11.0614 \times 10^3}{RT}\right)$

$k_{tc0} = 1.26 \times 10^9\,\text{L/mol sec}$ $\theta_t = \left\{\dfrac{2.81437 \times 10^{-4}}{[I_2]_0}\right\}\,\exp\!\left(\dfrac{22.8488 \times 10^3}{RT}\right)$

$k_{tr0} = 2.31 \times 10^6\,\text{L/mol sec}$

$$D_0 = \frac{2.303(1 - \phi_P)}{A + B(1 - \phi_P)}$$

$$\qquad\qquad A = 0.091678 - 1.142 \times 10^{-5}\,(T - 373.1)^2$$

Source: Ref. 12.

Note that the temperature is assumed constant in the Runge–Kutta technique between any time increment Δt. Thus, instead of using a difference equation, we can easily use the integrated equations of Appendix 6.1. The flowchart of such a computer program is given in Figure 6.3. Results thus obtained are inherently stable, and there is a general insensitivity to the choice of Δt. The increment in temperature can be determined from the solution of the following energy balance equation:

$$\rho C_p \frac{dT}{dt} = (-\Delta H_r)k_p[\text{M}]\lambda_{P0} - \frac{4U}{D}(T - T_w) \qquad (6.2.17)$$

where $-\Delta H_r$ is the heat of the reaction, U is the overall heat transfer coefficient, and T_w is the coolant temperature. This can be rearranged using Eq. (6.1.26) and may be written as

$$\frac{dy}{dx} + \beta y = \alpha e^{-mX} \qquad (6.2.18)$$

TABLE **6.3** Parameters Used for Simulation of Polymethyl Methacrylate Reactors

$f = 0.58$ (AIBN)
$\rho_M = 0.9665 - 0.0011[T(K) - 273.1]\,g/cm^3$
$\rho_P = 1.2\,g/cm^3$
$\varepsilon = 0.1946 + 0.916 \times 10^{-3}[T(K) - 273.1]\,g/cm^3$
$T_{gP} = 387.1\,K$
$k_1 = 6.32 \times 10^{16}\,min$
$k_{p0} = 2.95 \times 10^7\,L/mol\,min$
$k_{td0} = 5.88 \times 10^9\,L/mol\,min$
$k_{tc0} = 0.0$
$k_{tr} = 0.0$
$E_I = 30.66\,kcal/mol$
$E_p = 4.35\,kcal/mol$
$E_{tc} = 0.701\,kcal/mol$
$B = 0.03$
$\varepsilon_1 = 1.0$
$\theta_p = 666.37 \times 10^{-16} \exp\{(24.455 \times 10^3)/1.987T\}$
$\theta_t = \dfrac{0.48139 \times 10^{-22}}{[I_2]_0} \exp\{(35.5481 \times 10^3)/1.987T\}$
$A = 0.15998 - 7.812 \times 10^{-6}\,(T - 387.1)^2$

Source: Ref. 12.

where

$$y = T - T_w \tag{6.2.19a}$$
$$X = Z_0 - Z \tag{6.2.19b}$$
$$\beta = \frac{8U}{D_\rho C_p f k_1 Z_0} \tag{6.2.19c}$$
$$\alpha = \frac{(-\Delta H_r)[M]_0 m}{\rho C_p} \tag{6.2.19d}$$

The integrating factor of Eq. (6.2.18) is $e^{\beta X}$ and, using the initial condition $T = T_0$ at $X = 0$, we obtain

$$T - T_w = \frac{\alpha e^{-m(Z_0-Z)}}{(\beta - m)} + (T_0 - T_w)e^{-\beta(Z_0-Z)} - \frac{\alpha e^{-\beta(Z_0-Z)}}{\beta - m} \tag{6.2.20}$$

This result has been built into the algorithm of Figure 6.3.

Example 6.3: It is desired to manufacture polymer of constant molecular weight in free-radical batch polymerization. How would you achieve this?

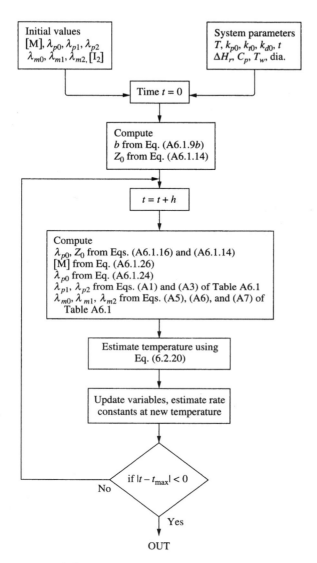

FIGURE 6.2 Flowchart of computation using an analytical scheme for free-radical polymerization.

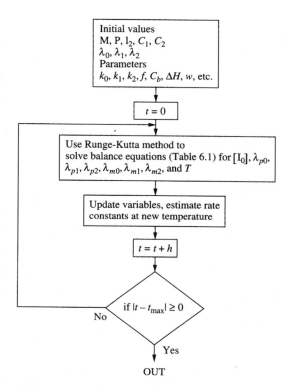

FIGURE 6.3   Runge–Kutta scheme for free-radical polymerization.

Solution: From the balance relations of Table 6.1, we have

$$\frac{d[M]}{dt} = -k_p[M]\lambda_{P0}$$

$$\rho C_p = \frac{dT}{dt} = (-\Delta H_R)k_p[M]\lambda_{P0} - \frac{4U}{D}(T - T_w)$$

$$\rho C_p = \frac{d\lambda_{P0}}{dt} = 2fk_1[I_2] - (k_{tc} + k_{td})\lambda_{P0}^2$$

$$\mu_n = \frac{k_p[M]}{2k_t\lambda_{P0}}$$

For constant μ_n, λ_{P0} is constant and is given by

$$\lambda_{P0} = \frac{k_p[M]}{2k_t\mu_n}$$

Therefore,

$$\frac{d[M]}{dt} = \frac{k_p[M]^2}{2k_t\mu_n} \overset{\Delta}{\equiv} C_1[M]^2$$

or

$$\frac{1}{[M]} - \frac{1}{[M]_0} = C_1 t$$

or

$$[M] = \frac{[M]_0}{1 + C_1[M]_0 t}$$

The energy balance equation is given by

$$\rho C_p \frac{dT}{dt} = \frac{(+\Delta H_r)k_p[M]^2}{2k_t\mu_n} - \frac{4U}{D}(T - T_w)$$

or

$$\frac{dT}{dt} = \frac{\Delta H_p}{\rho C_p} \frac{k_p}{2k_t\mu_n} \frac{[M]_0^2}{\{1 + C_1[M]_0 t\}^2} - \frac{U_A}{\rho p_D}(T - T_w)$$

which can be integrated analytically.

Example 6.4: In order to analyze equilibrium radical copolymerization of monomers A and B, we define species $N_{m,n}$ which contains m units of A and n units of B. Determine the MWD, $N_{m,n}$.

Solution: The mechanism of polymerization in terms of $N_{m,n}$ can be written as

$$1 + A \overset{K_A'}{\rightleftharpoons} N_{1,0}$$

$$1 + B \overset{K_B'}{\rightleftharpoons} N_{0,1}$$

$$N_{1,0} + A \overset{K_A}{\rightleftharpoons} N_{2,0}$$

$$N_{1,0} + B \overset{K_B}{\rightleftharpoons} N_{1,1}$$

$$N_{0,1} + A \overset{K_A}{\rightleftharpoons} N_{1,1}$$

$$N_{0,1} + B \overset{K_B}{\rightleftharpoons} N_{0,2}$$

$$N_{m,n-1} + B \overset{K_B}{\rightleftharpoons} N_{m,n}$$

We have

$$N_{1,0} = K_A'[I][A]$$
$$N_{0,1} = K_B'[I][B]$$
$$N_{2,0} = K_A'K_A[I][A]^2$$
$$N_{1,1} = (K_A'K_B + K_B'K_A)[I][A][B]$$
$$N_{0,2} = K_B'K_B[I][B]^2$$
$$N_{3,0} = K_A'K_A^2[I][A]^3$$
$$N_{2,1} = K_A(2K_A'K_B + K_B'K_A)[I][A]^2[B]$$
$$N_{1,2} = K_B(K_A'K_B + 2K_B'K_A)[I][A][B]^2$$
$$N_{0,3} = K_B'K_B[I][B]$$

By induction, one can write

$$N_{m,n} = \frac{\binom{m+n}{n}}{(m+n)}[I]K_A^{m-1}K_B^{n-1}(mK_A'K_B + nK_B'K_A)[A]^m[B]^n$$

We define

$$a = \frac{K_A'}{K_A}, \qquad b = \frac{K_B'}{K_B}$$
$$\alpha = K_A[A] \qquad \beta = K_B[B]$$

in terms of which

$$N_{m,n} = \frac{\binom{m+n}{m}}{m+n}[I](ma+nb)\alpha^m\beta^n, \qquad m+n \geq 1$$

Example 6.5: Prepare a computer program to determine the polystyrene reactor performance using the semianalytical technique [equations of Table A6.1 and Eq. (6.2.20)]. Use the data given in Table 6.2.

Solution: This is quite straightforward. Equations of Table A6.1 are in the Z domain, whereas the computations are required to be carried out in the time domain. In the program, we can choose the increment ΔZ as 0.001. Because rate constants change with the time of polymerization, the corresponding incremental time is calculated using

$$\Delta t = (t_2 - t_1) = \frac{2 \ln(Z_1/Z_2)}{fk_I}$$

The results of simulation have been plotted in Figure 6.4 for [M] and T, Figure 6.5 for λ_{P0} and λ_{M0}, and Figure 6.6 for λ_{M1} and λ_{M2}. We have also calculated results using the Runge–Kutta technique and compared the computations in these figures. We find that the results from the semianalytical technique are inherently stable, whereas those from Runge–Kutta require excessive computational time. For certain choices of Δt, there is a numerical overflow. Results from both of these techniques are identical until the thermal runaway conditions are encountered.

6.2.2   Control of Molecular Weight [8–13]

Designing a reactor for low-molecular-weight compounds involves deciding the residence time such that the desired conversion is obtained. In polymer reactors, on the other hand, it is necessary to attain not only a given conversion, but also a

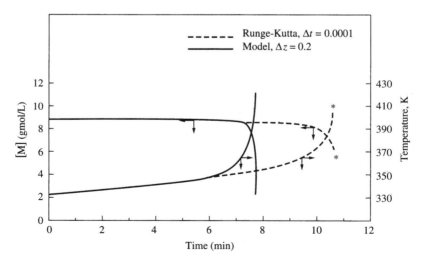

FIGURE 6.4   [M] versus time for nonisothermal polymerization with gel effect. Asterisks indicate numerical instability.

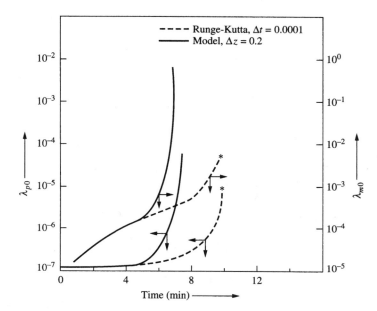

FIGURE 6.5 Total moles of polymer radicals versus time for nonisothermal polymerization with gel effect. Asterisks indicate numerical instability.

well-defined molecular weight of the polymer formed. In this context, the control of molecular weight becomes extremely important.

In Chapters 3 and 4, we noted that the rate of step-growth polymerization becomes diffusion controlled at advanced conversions and, in turn, the polymerization progresses at a slower rate. Because μ_n depends solely on conversion, the onset of the diffusional resistance delays only the attainment of the final conversion. Therefore, the control of molecular weight in step-growth polymerization implies the regulation of the final conversion in the product stream of the reactor.

As opposed to this, gel and glass effects are observed in radical polymerization. These effects lead to a marked increase in the average molecular weight of the polymer formed. Radical polymerization differs from step growth in that, in the former, the average molecular weight as well as the final conversion must be controlled individually. The control of the molecular weight in free-radical polymerization beyond the gel effect can be achieved by the following methods:

1. Increase of temperature of polymerization
2. Increase of monomer concentration, initiator concentration, or both
3. Method of weak inhibition

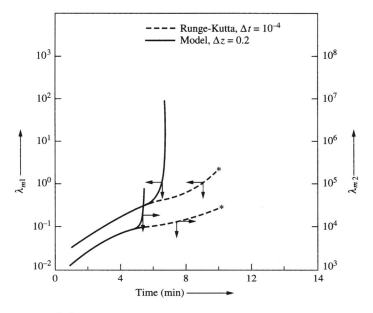

FIGURE 6.6 First and second moments of polymer formed versus time for nonisothermal polymerization with gel effect. Asterisks indicate numerical instability.

4. Use of a transfer agent

The effect of the polymerization temperature on the number-average molecular weight can be seen from the following expression:

$$\mu_n = \frac{k_p}{\{fk_t k_I\}^{1/2}} \frac{[M]}{[I_2]^{1/2}} \tag{6.2.21}$$

It has been shown that k_t reduces considerably faster than the rate constants k_p and k_I when the gel and the glass effects set in. It can be inferred from the numerical values of the activation energies that E_I is much larger than E_p and E_t. Therefore, an increase in temperature would cause an increase in k_I much greater than the corresponding increase in k_p or k_t and would lower μ_n. This is precisely what is required to control the molecular weight beyond the gel point. The increase in temperature helps in controlling the μ_n not only by increasing k_I but also by lowering the viscosity of the reaction mass, which, in turn, reduces the gel and glass effects. It is indeed possible to choose the temperature sequence in a tubular reactor such that a perfect control of the molecular weight is achieved. Finding this temperature sequence requires the following information:

1. Variation of rate constants with temperature
2. Variation of the viscosity of the reaction mass with conversion and the gel and glass effects
3. Variation of the viscosity of the reaction mass with temperature at a given conversion

Several theoretical and experimental studies have accounted for the gel effect, and, in them, the optimal temperature is determined for a given initiator concentration to minimize the residence time of a batch reactor. The optimal temperature profiles for various initiator concentrations for methyl methacrylate (MMA) polymerization, leading to a fixed μ_n^* of 5000, have been developed. It has been recommended that high isothermal temperatures be used for large concentrations of the initiator. For small concentrations, on the other hand, the temperature must be increased significantly with the time of polymerization. As the final molecular weight of the polymer (μ_n^*) is increased, it is necessary to have lower temperatures. This is reasonable because as the temperature is reduced, the molecular weight of the polymer increases even though the rate of polymerization falls.

6.3 COPOLYMERIZATION [14–25]

As pointed out earlier, several instances arise in practice where two different characteristics (e.g., the rubberlike characteristics of polybutadiene and the glasslike property of polystyrene) may be required in the final polymer. In such cases, butadiene and styrene are polymerized together, and the resultant polymer exhibits properties between those of the two homopolymers. The polymerization of two or more monomers is called copolymerization and the resultant polymer is called a copolymer.

In practice, the choice of monomers for copolymerization should be such that all of them will respond to the catalyst system used for copolymerization. In cases where any of these monomers does not respond to the catalyst, that particular monomer is not incorporated into the copolymer. Thus, a monomer (e.g., styrene) that polymerizes by the radical mechanism does not copolymerize with a monomer such as ε-caprolactam, which homopolymerizes by the step-growth mechanism.

In the analysis of copolymerization, there are two quantities of interest [14–16]. The first is the overall rate of copolymerization, and the latter is the average distribution of the monomer in the final polymer chain. Chapter 1 has already shown that monomers may be distributed randomly on the chain, may occur in blocks, or may alternate regularly [17–19]. In this chapter, we discuss the random copolymers only. The following analysis presents the copolymerization of two monomers that undergo radical polymerization. The analysis can easily be adapted for copolymerization of more than two monomers.

In general, two monomers, M_1 and M_2, homopolymerize at different rates. Thus, the rate of addition of M_1 to a polymer radical having M_2 at its end is different from the rate when it adds onto a radical with an M_1 end. In random copolymerization, polymer chain radicals are formed with M_1 and M_2 randomly distributed. Strictly speaking, all of these chain radicals are different (e.g., $M_1M_2M_2M_1$ is different from $M_1M_2M_1M_1$). The equal-reactivity hypothesis states that, for similar chains, the reactivity is independent of the chain length and, therefore, cannot be applied to the copolymerization rigorously. In general, the reactivity of a particular chain radical depends on not only the terminal monomer's unit but also the monomers attached before it. The effect of the previous monomers on the reactivity is called the penultimate effect. It is possible to figure this mathematically, but we avoid this step in order to simplify the mathematics in the analysis presented here. In the first approximation, it is assumed that the reaction rate constant is determined entirely by the terminal monomeric unit. Thus, chain radicals need to be distinguished from each other based only on the terminal units. In the copolymerization of M_1 and M_2, we need to distinguish two kinds of chain radicals having structure M_1 and M_2. These are designated as P_1 and P_2, respectively.

For radical copolymerization of M_1 and M_2, the mechanism given in Chapter 5 for homopolymerization can be extended easily for all of the elementary reactions, resulting in two initiation, four propagation, and three termination reactions, as follows:

Initiation

$$I_2 \xrightarrow{k_I} 2I \tag{6.3.1a}$$

$$M_1 + I \xrightarrow{k_{i1}} P_1 \tag{6.3.1b}$$

$$M_2 + I \xrightarrow{k_{i2}} P_2 \tag{6.3.1c}$$

Propagation

$$P_1 + M_1 \xrightarrow{k_{p11}} P_1 \tag{6.3.1d}$$

$$P_1 + M_2 \xrightarrow{k_{p12}} P_2 \tag{6.3.1e}$$

$$P_2 + M_1 \xrightarrow{k_{p21}} P_1 \tag{6.3.1f}$$

$$P_2 + M_2 \xrightarrow{k_{p22}} P_2 \tag{6.3.1g}$$

Termination

$$P_1 + P_1 \xrightarrow{k_{t11}} M_d \tag{6.3.1h}$$

$$P_1 + P_2 \xrightarrow{k_{t12}} M_d \tag{6.3.1i}$$

$$P_2 + P_2 \xrightarrow{k_{t22}} M_d \tag{6.3.1j}$$

The desired quantities, r_p, and the monomer distribution on the polymer chain can now be found as follows. The mole balance equations for M_1 and M_2 are

$$-\frac{d[M_1]}{dt} = k_{p11}[P_1][M_1] + k_{p21}[P_2][M_1] \tag{6.3.2a}$$

$$-\frac{d[M_2]}{dt} = k_{p12}[P_1][M_2] + k_{p22}[P_2][M_2] \tag{6.3.2b}$$

In writing Eqs. (6.3.2), the small consumption of the monomers by reactions (6.3.1b) and (6.3.1c) has been neglected. Dividing Eq. (6.3.2a) by Eq. (6.3.2b) gives

$$\frac{d[M_1]}{d[M_2]} = \frac{k_{p11}[P_1][M_1] + k_{p2_1}[P_2][M_1]}{k_{p12}[P_1][M_2] + k_{p2_2}[P_2][M_2]} \tag{6.3.3}$$

Whichever monomer is consumed by way of copolymerization appears on the polymer chains. Because the addition of these monomers is a random process, the relative rates of consumption [given by Eq. (6.3.3)] should be the same as the relative distribution of monomers on polymer molecules on the average. Therefore, if a differential amount, $d([M_1] + [M_2])$, of the reaction mass is polymerized in time dt, then, on the average, the number of molecules of M_1 and the number of molecules of M_2 on a small length of chain formed at that instant are in the ratio $(d[M_1]/d[M_2])$. The differential polymer composition, F_1, is defined as

$$F_1 = \frac{d[M_1]}{d\{[M_1] + [M_2]\}} = 1 - F_2 \tag{6.3.4}$$

It is possible to relate F_1 to the average composition, f_1, of the unreacted monomer in the reaction mass, defined by

$$f_1 = \frac{[M_1]}{[M_1] + [M_2]} = 1 - f_2 \tag{6.3.5}$$

In Eq. (6.3.3), $[P_1]$ and $[P_2]$ are not known. Because these are intermediate species, they can be evaluated using the steady-state approximation. The balance equations on P_1 and P_2 are

$$\frac{d[P]}{dt} = 0 = k_{i1}[I][M_1] - \{2k_{t11}[P_1]^2 + k_{t12}[P_1][P_2]$$
$$- k_{p21}[P_2][M_1] + k_{p12}[P_1][M_2]\} \tag{6.3.6a}$$

$$\frac{d[P_2]}{dt} = 0 = k_{i2}[I][M_2] - \{2k_{t22}[P_2]^2 + k_{t12}[P_1][P_2]$$
$$+ k_{p21}[P_2][M_1] - k_{p12}[P_1][M_2]\} \tag{6.36b}$$

In these equations, the propagation terms due to reactions (6.3.1e) and (6.3.1g) are much larger than the initiation and termination terms due to reactions (6.3.1a)–(6.3.1c) and (6.3.1h)–(6.3.1j). Therefore, Eqs. (6.3.6) give

$$k_{p12}[P_1][M_2] = k_{p21}[P_2][M_1] \tag{6.3.7}$$

Equation (6.3.7) can be substituted into Eq. (6.3.3) to eliminate $[P_1]/[P_2]$, as follows:

$$\frac{d[M_1]}{d[M_2]} = \frac{k_{p11}[M_1][P_1]/[P_2] + k_{p21}[M_1]}{k_{p12}[M_2][P_1]/[P_2] + k_{p22}[M_2]}$$

$$\frac{d[M_1]}{d[M_2]} = \frac{(k_{p21}/k_{p12})k_{p11}[M_1][M_1]/[M_2] + k_{p21}[M_1]}{(k_{p12}/k_{p12})[M_2][M_1]/[M_2] + k_{p22}[M_2]}$$

$$= \frac{[M_1]/[M_2]}{([M_1]/[M_2]) + r_2}\left\{r_1\frac{[M_1]}{[M]} + 1\right\}$$

where

$$r_1 = \frac{k_{p11}}{k_{p12}} \tag{6.3.9a}$$

$$r_2 = \frac{k_{p22}}{k_{p21}} \tag{6.3.9b}$$

This is sometimes called the Mayo equation. F_1 can now be written in terms of $[M_1]$ and $[M_2]$:

$$F_1 = \frac{d[M_1]}{d\{[M_1] + [M_2]\}} = \frac{d[M_1]/dp[M_2]}{1 + (d[M_1]/d[M_2])} \tag{6.3.10}$$

Substituting Eq. (6.3.8) into this equation gives the following:

$$F_1 = \frac{([M_1]/[M_2])\{1 + r_1([M_1]/[M_2])\}/\{r_2 + ([M_1]/[M_2])\}}{1 + ([M_1]/[M_2])\{1 + r_1([M_1]/[M_2])\}/\{r_2 + ([M_1]/[M_2])\}} \tag{6.3.11}$$

Equation (6.3.5) gives

$$\frac{[M_1]}{[M_2]} = \frac{f_1}{1 - f_1} = \frac{f_1}{f_2} \tag{6.3.12}$$

Substituting into Eq. (6.3.11) yields

$$F_1 = \frac{(f_1/f_2) + r_1(f_1/f_2)}{r_2 + 2(f_1/f_2) + r_1(f_1/f_2)}$$

$$= \frac{r_1 f_2^2 + f_1 f_2}{r_1 f_1^2 + 2f_1 f_2 + r_2 f_2^2} \tag{6.3.13}$$

Thus, the differential polymer composition, F_1, is related to the monomer composition, f_1, in the reaction mass at that instant. This relation has been obtained through the use of the mechanism written in Eqs. (6.3.1). The parameters r_1 and r_2 measure the relative preference that the polymer radicals P_1 and P_2 have for the two monomers M_1 and M_2. If k_{p22} and k_{p11} are zero, the resultant polymer has an alternating sequence of M_1 and M_2 on the chain. If k_{p21} and k_{p12} are zero, then there is no copolymerization, and the resulting polymer consists of two homopolymers. If, however, monomers M_1 and M_2 display equal relative preferences for P_1 and P_2—that is, if

$$\frac{k_{p11}}{k_{p12}} = \frac{k_{p21}}{k_{p22}} \tag{6.3.14}$$

then

$$r_1 r_2 = 1 \tag{6.3.15}$$

and Eq. (6.3.13) reduces to

$$F_1 = \frac{r_1 f_1^2 + f_1 f_2}{r_1 f_1^2 + 2r_1 f_1 f_2 + r_2 f_2^2} = \frac{r_1 f_1}{r_1 f_1 + f_2} \tag{6.3.16}$$

Equation (6.3.13) has been plotted in Figure 6.7 for different values of r_1 and r_2 and in Eq. (6.3.16) in Figure 6.8 for different values of r_1. The relation between F_1 and f_1, as shown in these plots, is quite similar to the vapor–liquid equilibrium relations. The vapor pressure of an ideal liquid behaves in the same way as the copolymer monomer mixture composition in Figure 6.8. Accordingly, copolymerizations with $r_1 r_2$ equal to 1 are sometimes termed ideal.

The overall balance on radicals can be written by adding Eq. (6.3.6a) and (6.3.2b):

$$k_{i1}[I][M_1] + k_{i2}[I][M_2] = 2k_{t11}[P_1]^2 + 2k_{t12}[P_2]^2 + 2k_{t12}[P_1][P_2] \tag{6.3.17}$$

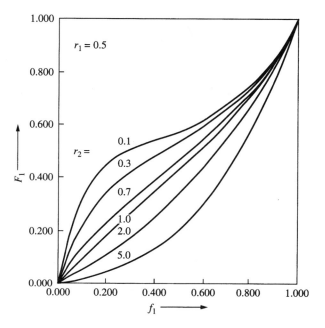

FIGURE 6.7   Plot of Eq. (6.3.13) for different values of r_1 and r_2.

The balance on primary radicals, I, reduces, under the steady-state hypothesis, to

$$2k_1[I_2] = k_{i1}[I][M_1] + k_{i2}[I][M_2] \tag{6.3.18}$$

and Eq. (6.3.17) then simplifies to

$$2k_{t11}[P_1]^2 + 2k_{t22}[P_2]^2 + 2k_{t12}[P_1][P_2] = 2k_1[I_2] \tag{6.3.19}$$

or

$$[P_1]^2 = \frac{k_1[I_2]}{k_{t11} + k_{t12}([P_2]/[P_1]) + k_{t22}([P_2]/[P_1])^2} \tag{6.3.20}$$

Equation (6.3.7) gives the following:

$$[P_1]^2 = \frac{k_1[I_2]}{k_{t11} + k_{t12}(k_{p12}[M_2]/k_{p21}[M_1]) + k_{t22}(k_{p12}^2[M_2]^2/k_{p21}^2[M_1]^2)} \tag{6.3.21}$$

Similarly, $[P_2]$ can be found. The overall rate of polymerization, r_p, can be written

$$r_p = k_{p11}[P_1][M_1] + k_{p22}[P_2][M_2 + k_{p12}[P_1][M_2] + k_{p21}[P_2][M_1] \tag{6.3.22}$$

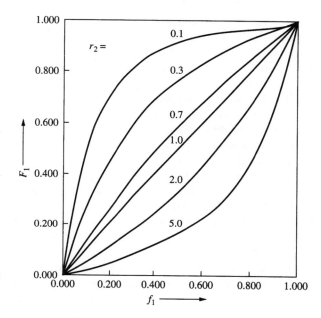

FIGURE 6.8 Incremental polymer composition F_1 versus f_1 for ideal copolymerization.

In Eq. (6.3.22), $[P_1]$ and $[P_2]$ can be eliminated from Eq. (6.3.21) and r_p (sometimes called the Melville equation) can be found:

$$r_p = \frac{\{2k_1[I_2]\}^{1/2}}{\delta_1} \frac{r_1[M_1]^2 + 2[M_1][M_2] + r_2[M_2]^2}{\{r_1^2[M_1] + 2\phi rr_1r_2(\delta_2/\delta_1)[M_1][M_2] + (\delta_2/\delta_1)(r_2[M_2])^2\}^{1/2}}$$

(6.3.23)

where

$$\delta_1 = \left\{\frac{2k_{t11}}{k_{p11}^2}\right\}^{1/2}$$

(6.3.24a)

$$\delta_2 = \left\{\frac{2k_{t22}}{k_{p22}^2}\right\}^{1/2}$$

(6.3.24b)

$$\phi = \frac{k_{t22}}{k_{p22}^{1/2}k_{t22}^{1/2}}$$

(6.3.24c)

The values of r_1 and r_2 for various monomer systems undergoing radical copolymerization have been compiled in the *Polymer Handbook* [20]. The experimental determination of r_1 and r_2 is usually tedious and sometimes requires a careful statistical analysis, but all of the measurements reported belong to short times of polymerization, where the gel effect does not exist. It is expected that r_1 and r_2 will change when the propagation rate constants begin to be affected by the glass effect [21–23].

The Melville equation for the rate of copolymerization [Eq. (6.3.23)] has been applied to several systems with ϕ as a parameter. Calculations carried out based on the molecular-orbital theory lead to a theoretical value of $\phi = 1$, which is contrary to experimental findings in which ϕ has been reported to take on any value. In the copolymerization of styrene with diethyl fumarate at 90°C shown in Figure 6.9, $\phi = 2.0$ is found to fit experimental data. The value of ϕ is also found to depend on the temperature, as shown in Figure 6.9, in which $\phi = 7.5$ is found to describe experimental data at 60°C. Furthermore, the value of ϕ depends on the monomer ratio for some systems, as has been shown for styrene–methacrylate systems in the literature. Only very few studies have reported the changes in δ_1, δ_2, and ϕ after the gel effect sets in.

If a mixture of monomers M_1 and M_2 is copolymerized from some initial concentrations $[M_1]_0$ and $[M_2]_0$, respectively, to some final concentrations $[M_1]$

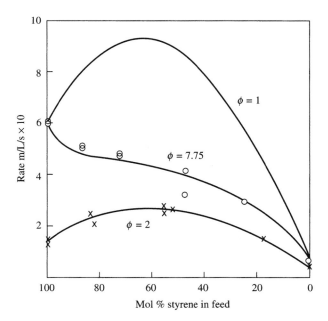

FIGURE 6.9 Copolymerization rate at different mole fractions: x, styrene and diethyl fumerate at 90°C giving $\phi = 2$; ○, styrene and diethyl fumerate at 60°C giving $\phi = 7.5$.

and $[M_2]$, respectively, then Eq. (6.3.13), as such, does not give the overall composition of monomers on the polymer chain. This is because $[M_1]$ and $[M_2]$ change with time. To obtain the composition, we proceed as follows. If $[M^*]$ is defined as

$$[M^*] = [M_1] + [M_2] \tag{6.3.25}$$

the average (i.e., the ratio of $[M_1]$ and $[M_2]$) of a polymer chain can be found when $[M^*]$ changes from $[M]_0^*$ to $[M]^*$ by making a mass balance. If f_{10} is the fraction of M_1 in the reaction mass initially, there are $f_{10}[M]_0^*N_A$ molecules of M_1 in the reaction mass (N_A is Avogadro's number) at the beginning of the copolymerization. At time t, the number of unreacted molecules of M_1 in the reaction mass is $f_1[M]^*N_A$. The total number of reacted molecules $([M]_0^* - [M]^*)N_A$ of both M_1 and M_2 would show up on the polymer chain. If F_1 is the average composition of the polymer chains at time t, then a number balance gives

$$f_{10}[M]_0^*N_A = \bar{F}_1([M]_0^* - [M]^*)N_A + f_1[M]^*N_A \tag{6.3.26}$$

$$\bar{F}_1 = \frac{f_{10} - f_1([M]/[M]_0^*)}{1 - ([M]^*/[M]_0^*)} \tag{6.3.27}$$

Equation (6.3.27) has been plotted in Figure 6.10 as a function of $(1 - [M]^*/[M]_0^*)$. Here, $\bar{F}_1$ goes to zero as $(1 - [M]^*/[M]_0^*)$ goes to 1 because styrene is more reactive ($r_1 > 1$, $r_2 < 1$), but $\bar{F}_1$ becomes 0.5 when an equimolar ratio of two monomers is copolymerized.

In Eq. (6.3.27), $[M]^*/[M]_0^*$ is the independent variable that can be determined from a plot of F_1 versus f_1 by making a number balance similar to the derivation of Eq. (6.3.27). Suppose that $[M]^*$ changes from $[M]^*$ to $[M]^* - d[M]^*$ in copolymerization. Simultaneously, f_1 changes from f_1 to $f_1 - df_1$. The quantity $f_1[M]^*N_A$ is the number of molecules of M_1 before $d[M]^*$ mol of the reaction mass copolymerize, and $(f_1 - df_1)([M]^* - d[M]^*)N_A$ is the number in the reaction mass after the copolymerization. The difference between these quantities is the number of molecules of M_1 that appear on the polymer chains. This balance can be written as

$$f_1[M]^*N_A = N_A(f_1 - df_1)([M]^* - d[M]) + (F_n \, d[M]^*N_A) \tag{6.3.28}$$

Neglecting second-order differential terms, we obtain

$$\frac{d[M]}{[M]^*} = \frac{df_1}{F_1 - f_1} \tag{6.3.29}$$

The equation can be integrated as follows:

$$\ln\left(\frac{[M]^*}{[M]_0^*}\right) = \int_{f_{10}}^{f_1} \frac{df_1}{F_1 - f_1} \tag{6.3.30}$$

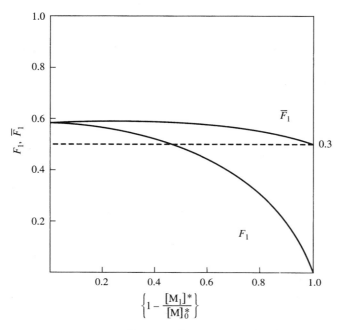

FIGURE 6.10 F_1 and $\bar{F}_1$ versus $\left\{1 - [M]^*/[M]_0^*\right\}$ for the styrene (1) and butadiene (2) systems for $r_1 = 1.39$ and $r_2 = 0.78$. Monomers are present initially in equimolar ratio.

If F_1 versus f_1 [Eq. (6.3.16)] is known, $[M]^*/[M]_0^*$ can be calculated graphically from Eq. (6.3.30) and F_1 can, therefore, be evaluated from Eq. (6.3.27). Figure 6.10 has been generated in this manner.

It may be reemphasized that some of the important variables used to control copolymerization reactors are the average composition of M_1 and M_2 and the reaction temperature. The degree of polymerization of the polymer formed is not as important—even though recent studies have attempted to focus attention on it. Incorporation of the gel effect, copolymerization in CSTRs, and the determination of optimal temperature–time histories in batch reactors are some areas that have received attention in recent studies.

Example 6.6: Anionic copolymerization involving monomers A and B have been carried out using λ_∞ mol of initiators. Determine expressions for F_1 and the rates.

Solution: In anionic polymerization, initiation is instantaneous, and in a short times, there are λ_{00} mol of growing polymer anions. Let us similarly define P_A and P_B polymer anions having monomers and A and B at the growing ends. There is no quasistate and P_A and P_B grow according to following reactions:

$$P_A + A \xrightarrow{k_{11}} P_A$$

$$P_A + B \xrightarrow{k_{12}} P_B$$

$$P_B + A \xrightarrow{k_{21}} P_A$$

$$P_B + B \xrightarrow{k_{22}} P_B$$

At times $t = 0$, let their concentrations be λ_{A00} and λ_{B00}, and at any other time, λ_{A0} and λ_{B0}. Then,

$$\lambda_{A00} + \lambda_{B00} = \lambda_{A0} + \lambda_{B0} = \lambda_{00}$$

The mole balance on λ_{A0} is given by

$$\frac{d\lambda_{A0}}{dt} = k_{12}\lambda_{A0}[B] + k_{21}\lambda_{B0}[A]$$

$$= -(k_{12}[B] + k_{21}[A])\lambda_{A0} + k_{21}[A]\lambda_{00} \overset{\Delta}{=} -C_1\lambda_{A0} + C_2\lambda_{00}$$

If we assume [A] and [B] as constants, then this equation can be easily integrated to

$$\lambda_{A0} = \frac{C_2\lambda_{00}}{C_1} + (\text{const. } \exp(-C_1 t))$$

The mole balances of monomers A and B are

$$-\frac{d[A]}{dt} = k_{11}\lambda_{0A}[A] + k_{21}\lambda_{0B}[A]$$

$$-\frac{d[B]}{dt} = k_{12}\lambda_{0A}[B] + k_{22}\lambda_{0B}[B]$$

and

$$F_1 = \frac{d[A]}{d[B]} \Big/ \left(1 + \frac{d[A]}{d[B]}\right)^{-1}$$

$$= \frac{k_{11}\lambda_{0A}[A] + k_{21}\lambda_{0B}[A]}{k_{12}\lambda_{0A}[B] + k_{22}\lambda_{0B}[B] + k_{11}\lambda_{0A}[A] + k_{21}\lambda_{0B}[A]}$$

Example 6.7: For batch reactors carrying out radical copolymerization of monomers A and B, one can use probabilistic arguments to determine average number of A and B, $\bar{N}_A$ and $\bar{N}_B$ per chain. Define relevant probabilities and determine these.

Solution: We define species $P_{m,n}$ and $Q_{m,n}$ with growing and radical groups A and B and each having m number of A and n number of B. The copolymerization reactions in terms of these are

$$P_{m,n} + A \xrightarrow{k_{p11}} P_{m+1,n}$$

$$P_{m,n} + B \xrightarrow{k_{p12}} Q_{n,m+1}$$

$$P_{m,n} + A \xrightarrow{k_{p21}} P_{n+1,n}$$

$$P_{m,n} + B \xrightarrow{k_{p22}} Q_{m,n+1}$$

We have following probabilities:

p_{AA}: probability that $P_{m,n}$ adds another A
p_{AB}: probability that $P_{m,n}$ adds another B
p_{BA}: probability that $Q_{m,n}$ adds another A
p_{BB}: probability that $Q_{m,n}$ adds another B

Then,

$$p_{AA} = \frac{k_{p11}[A]}{k_{p11}[A] + k_{p12}[B]} \stackrel{\Delta}{\equiv} \frac{r_1[A]/[B]}{1 + r_1[A]/[B]} \stackrel{\Delta}{\equiv} \frac{r_1 f_1/(1-f_1)}{1 + r_1 f_2/(1-f_1)} \stackrel{\Delta}{\equiv} \frac{\alpha}{1+\alpha}$$

$$p_{AB} = \frac{k_{p12}[B]}{k_{p11}[A] + k_{p12}[B]} = \frac{1}{1 + r_1 f_1/(1-f_1)} = \frac{1}{1+\alpha}$$

Similarly,

$$p_{BA} = \frac{1}{r_2[(1-f_1)/f_1] + 1} = \frac{1}{1+\beta}$$

and

$$p_{BB} = \frac{r_2(1-f_1)/f_1}{1 + r_2[(1-f_1)/f_1]} = \frac{\beta}{1+\alpha}$$

The probability of having exactly n units of A in a growing chain is

$$\begin{aligned} A_n &= p_{AA}^{n-1} p_{AB} \\ &= \frac{\alpha^{n-1}}{(1+\alpha)^n} \end{aligned}$$

The probability of having exactly n units of B in a growing chain is

$$B_n = p_{BB}^{n-1} p_{BA}$$

$$= \frac{\beta^{n-1}}{(i+\beta)^n}$$

$$\bar{N}_A = \sum_{n-1}^{\infty} nA_n = 1 + \alpha$$

$$\bar{N}_B = \sum_{n-1}^{\infty} nB_n = 1 + \beta$$

6.4 RECYCLING AND DEGRADATION OF POLYMERS [26–33]

With increasing industrialization, production of polymeric materials have been on the rise. Because plastics are not compatible with the environment, they do not degrade easily. Polyolefins represent the largest groups of plastics and are mainly used as packaging materials. After they are used once, they are thrown away and treated as waste. Chemical recycling of plastics is being used increasingly in recent years and is defined as the breakdown of polymer waste into materials that are reusable as fuels or chemicals. There are several applications that require the use of pure plastics and there are several other applications where mixed plastics (e.g., composites) are used. In such cases, separation and purification of these into industrially pure materials is critical to recycling of plastics [26–28].

It is desired to recover plastics waste and reprocess or use the product of the reprocessing as raw material or fuel. Currently, plastics are subjected to any one of the following options:

1. Plastics materials are dumped; however, with the increasing cost of land, this is becoming uneconomical. In addition to this, there is total wastage of the material and energy (assuming that the polymer could be burnt as fuel) [29,30].
2. Plastics are recycled; but during the recovery stage, the polymers (particularly in the case of mixed plastics such as composites) lose quality and texture due to partial degradation, which invariably occurs during the recovery stage.
3. The plastics are burned as a municipal waste in incinerators or in steel plants. In this recycle technique, the energy of combustion is recovered and utilized, but the material is lost.

There is considerable economic interest currently in the complete degradation of plastics and recover higher-valued products because of high dumping cost and stringent legal requirements of pollution. Unfortunately, monomers cannot be recovered quantitatively from these mass-produced plastics, such as polyethylene and polypropylene. Monomers have been, on the other hand, recovered from speciality polymer such as polymethyl methacrylate (PMMA), polystyrene (PS), and polytetrafluoroethylene (PTFE). Monomer recovery from PMMA is 97%, whereas those from polystyrene is only 70%. In the earlier days, the pyrolysis gave only 16% monomer recovery from PTFE, but after the advent of the present-day technology of using a fluidized bed operating at 605°C, as much as 78% PTFE has been recovered. In addition to thermal pyrolysis, alcoholysis has been used for depolymerization of polyethylene terephthalate (PET) to terephthalic acid derivates. However most of the chemical recycling processes rely heavily upon the depolymerization of polymers at higher temperatures.

A major problem faced with all of these processes is the contamination of used plastics and the uniformity of the type of the materials used. For example, polyvinyl chloride (PVC) on pyrolysis gives out HCl, which corrodes the reactor. Composites, particularly those involving metals, can also impair the reactor. Therefore, used, mixed, and contaminated plastics must be mechanically as well as thermally pretreated. In view of this, one of the possibilities of separation of the polymer has been staged pyrolysis in which different components are depolymerized one after the other with a temperature program. Below 300°C, the gaseous decomposition product of PVC is largely HCl, which autocatalyzes the dehydrochlorination and also gives polyne sequences. In commercial PVC, the molecular structure is predominantly a head-to-tail combination, but a few tail-to-tail arrangement contributes heavily to its thermal instability. In addition to this, the double bonds and branching also contributes to this instability.

Polymer degradation is an old subject and was studied mainly because people wanted to know the limiting factor (loss of mechanical strength) in different environments (the pH, temperature, pressure). Thermal degradation as a technique was also utilized to determine the molecular structure of copolymers, but now it has gained considerable importance because of recycling of polymers.

Degradation of polymer by heat in the absence of oxygen depends not only on the temperature but also the molecular structure of the polymer. In Chapter 2, we gave the mechanism of thermal degradation and showed that the polymer degradation could occur from either (or a combination) of any other mechanisms [31].

1. Elimination of low-molecular-weight compounds (such as HCl in PVC)
2. Unzipping (sometimes called depolymerization) of monomers (as in PMMA)
3. Cyclization (as in polyacrylonitrile)
4. Random scission of polymer chains (as in polyethylene)
5. Formation of specific molecular-weight compounds, sometimes called specific depolymerization.

The fifth mode of degradation is similar to the second one, except that in the former, the compound formed is not the monomer. For example, in the case of PS, the specific compounds formed are C_8H_8 (styrene), $C_8H_8 \cdot C_3H_6O$, and $C_8H_8(C_3H_6O)_2$. In addition to this, PS undergoes random scission (or mode 4). Also, the polymer radicals thus formed react with each other either through combination or disproportionation reactions. If the radical center per polymer chain is only one, linear (if it is at the chain end) or branched (if it is in the middle of the chain) dead polymers are formed. However, if it is more than one per chain, the final polymer is always a network.

Recently, there have been great efforts to find catalysts which would lead to specific depolymerization. In this regard, polyethylene was depolymerized in the presence of NO, O_2, and N_2 (275 kPa NO, 690 kPa O_2, and 3170 kPa N_2) to a mixture of benzoic acid, 4-nitrobenzoic acid, and 3-nitrobenzoic acid [32]. In an alternate work [33], zirconium hydride supported on silica alumina catalyst has been reported, which, in presence of hydrogen, cleaves the C−C bonds of polyethylene and polypropylene. The end products of the hydrogenolysis of these polymers have been diesel and lower alkanes and is still a subject of vigorous research.

6.5 CONCLUSION

In this chapter, we have considered the reaction engineering of chain-growth polymerization. In order to manufacture polymers of desired physical and mechanical properties, the performance of the reactors must be closely controlled. To do this, various transport equations governing their performance must be established, which, in principle, can be solved numerically. The usual Runge–Kutta technique takes considerable computational time and, at times, gives numerical instability. To overcome all of these problems, a semianalytical approach can be used.

There are certain applications in which homopolymers or their blends are not adequate. In such cases, copolymers are synthesized. In this chapter, we have presented the analysis of copolymerization. By developing relations, the rate of polymerization and the monomer distribution on polymer chains can be determined.

APPENDIX 6.1 SOLUTION OF EQUATIONS DESCRIBING ISOTHERMAL RADICAL POLYMERIZATION

The mole balances for radical and dead polymers for batch (or tubular) reactors are given in Table 6.1. We define concentration of polymer radicals, λ_{P0}, as

$$\lambda_{P0} = \sum_{n=1}^{\infty} [P_n] \qquad (A6.1.1)$$

With the help of Eq. (2) of Table 6.1, we can derive an expression for the time variation of λ_{P0}, as follows:

$$\frac{d\lambda_{P0}}{dt} = 2fk_I[I_2] - k_t\lambda_{P0}^2 \tag{A6.1.2}$$

When the gel and thermal effects are present in radical polymerization, the rate constants k_I and k_t are dependent on temperature and monomer conversion and, therefore, the above equation cannot be integrated. We show that Eq. (A6.1.2) has a solution for constant values of k_I and k_t and then show that these results can be naturally adopted in the presence of gel and thermal effects.

A6.1.1 Solution for λ_{P0}

The mole balance for the initiator given in Eq. (1) of Table 6.1 can be integrated for time-invariant fk_I as follows:

$$[I_2] = [I_2]_0 e^{-f_I t} \tag{A6.1.3}$$

where $[I_2]_0$ is the concentration of the initiator at $t = 0$. Let us now transform λ_{P0} in Eq. (A6.1.2):

$$\lambda_{P0} = \frac{1}{k_t} \frac{dy/dt}{y} \tag{A6.1.4}$$

Substituting Eq. (A6.1.3) for $[I_2]$ in Eq. (8) of Table 6.1 gives

$$\frac{d^2y}{dt^2} = 2fk_dk_t[I_2]_0 e^{-fk_I t} \tag{A6.1.5}$$

Further, we define x related to time of polymerization as

$$x = 2fk_dk_I[I_2]_0 e^{-fk_I t} y \tag{A6.1.6}$$

This gives

$$\frac{dx}{dt} = -fk_I x \tag{A6.1.7a}$$

$$\frac{dy}{dt} = \frac{dy}{dx}\frac{dx}{dt} = -fk_I x \frac{dy}{dx} \tag{A6.1.7b}$$

$$\frac{d^2y}{dt^2} = \frac{d}{dx}\left(\frac{dy}{dt}\right)\frac{dx}{dt} \tag{A6.1.7c}$$

$$= (fk_I)^2 x \frac{dy}{dx} + (fk_I)^2 x^2 \frac{d^2y}{dx^2} \tag{A6.1.7d}$$

In terms of these, Eq. (A6.1.5) becomes

$$(fk_I)^2 x^2 \frac{d^2y}{dx^2} + (fk_I)^2 x \frac{dy}{dx} - xy = 0 \tag{A6.1.8}$$

We further transform x by

$$Z = 2\sqrt{bx} \tag{A6.1.9}$$

where

$$b = \left(\frac{1}{fk_I}\right)^2 \tag{A6.1.10}$$

in terms of which Eq. (A6.1.8) reduces to

$$Z^2 \frac{d^2y}{dZ^2} + Z \frac{dy}{dZ} - (Z^2 - 0)y = 0 \tag{A6.1.11}$$

For this equation, y has the solution

$$y = C_1 I_0(Z) + C_2 K_0(Z) \tag{A6.1.12}$$

where $I_0(Z)$ and $K_0(Z)$ are the modified zeroth-order Bessel functions and C_1 and C_2 are the constants of integration. With the help of Eq. (A6.1.4), we can derive λ_{P0} as

$$\lambda_{P0} = \frac{fk_I Z}{2k_t} \frac{1}{y} \frac{dy}{dZ} = \left\{ \frac{fk_I Z}{2k_t} \frac{C_2 K_1(Z) - C_1 I_I(Z)}{C_2 K_0(Z) + C_1 I_0(Z)} \right\}$$

$$= \frac{fk_d Z}{2k_t} \frac{K_I(Z) - CI_1(Z)}{K_0(Z) + CI_0(Z)}$$

where $C = C_1/C_2$, which is to be determined by the initial conditions.

Let us assume that at $t = 0$, the concentration of polymer radicals is λ_{P00}. In the Z plane, time $t = 0$ corresponds to Z_0, given by

$$Z_0 = \sqrt{\frac{2k_t[I_2]_0}{fk_I}} \tag{A6.1.14}$$

and C in Eq. (A6.1.13) is given by

$$C = \frac{K_1(Z_0) - (2k_t[P]_0/fk_I Z_0)K_0(Z_0)}{I_1(Z_0) + (2k_t[P]_0/fK_I Z_0)I_0(Z_0)}$$

$$= \frac{K_I(Z_0)/I_1(Z_0) - \lambda_{P0}^*[K_0(Z_0)/I_1(Z_0)]}{1 + \lambda_{P0}^*[I_0(Z_0)/I_1(Z_0)]} \tag{A6.1.15}$$

where

$$\lambda_{\text{P0}}^{*} = \frac{\lambda_{\text{P0}}}{\lambda_{\text{P0s}}} = \lambda_{\text{P000}}\sqrt{\frac{k_t}{2fk_1[I_2]_0}} \tag{A6.1.16}$$

Here, λ_{P0s} is the concentration of polymer radicals assuming the steady-state approximation; it is given by Eq. (6.2.9).

The variation in the monomer concentration [M] is governed by Eq. (4) of Table 6.1. We substitute Eq. (A6.1.9) into it to obtain

$$\frac{d[\text{M}]}{dt} = \frac{d[\text{M}]}{dz}\frac{dz}{dt} = -\frac{fk_1Z}{2}\frac{d[\text{M}]}{dZ} = -k_p[\text{M}]\lambda_{\text{P0}}$$

or

$$\frac{fk_1Z}{2}\frac{d[\text{M}]}{dZ} = k_p[\text{M}]\left(\frac{fk_1Z}{2k_t}\right)\frac{K_1(Z) - CI_1(Z)}{K_0(Z) + CI_0(Z)} \tag{A6.1.17}$$

$$\frac{d[\text{M}]}{dZ} = [\text{M}]\frac{k_p}{k_t}\left(\frac{K_1(Z) - CI_1(Z)}{K_0(Z) - CI_0(Z)}\right)$$

We substitute

$$u = K_0(Z) + CI_0(Z) \tag{A6.1.18}$$

which, on differentiating with respect to Z, gives

$$\frac{du}{dz} = -K_1(Z) + CI_1(Z) \tag{A6.1.19}$$

Comparison of Eqs. (A6.1.17) and (A6.1.19) gives

$$\frac{du}{dz} = -K_1(Z) + CI_1(Z) \tag{A6.1.20}$$

which, on integration, leads to

$$-\frac{d[\text{M}]}{[\text{M}]} = \frac{k_p}{k_t}\frac{du}{u}$$

$$[\text{M}] = [\text{M}]_0\left[\frac{K_0(Z) + CI_0(Z)}{K_0(Z_0) + CI_0(Z_0)}\right]^{-k_p/k_t} \tag{A6.1.21}$$

where $[\text{M}]_0$ is the monomer concentration at $t = 0$ (or $Z = Z_0$) and Z_0 is defined in Eq. (A6.1.14). The magnitude of Z, as defined in Eq. (A6.1.9), is very large. For example, for methyl methacrylate polymerizing at $60°\text{C}$ with AIBN, $[I_2]_0$ is equal to a 0.0258-mol/L initiator; the rate constants $k_1 k_p$, and k_t are $0.475 \times 10^{-3}\,\text{min}^{-1}$, $0.4117 \times 10^5\,\text{L/mol min}$, and $0.20383 \times 10^{10}\,\text{L/mol min}$, respectively. Taking the initiator efficiency f to be 0.58, at time $t = 0$ we get b as defined by Eq. (A6.1.9) and x_0 as defined by Eq. (A6.1.6) to be 1.3175×10^7 and

2.8756×10^4, respectively. Therefore, Z at time $t = 0$ is given by Eq. (A6.1.14) and is equal to 0.1231×10^7. Hence, the asymptotic expansion of Bessel functions valid for large arguments ($Z > 5$) may be used. Neglecting terms involving Z in the denominator, we get

$$K_1(Z) = K_0(Z) = \sqrt{\frac{\pi}{2Z}} e^{-z} \tag{A6.1.22a}$$

$$I_1(Z) = I_0(Z) = \frac{e^Z}{\sqrt{2\pi Z}} \tag{A6.1.22b}$$

Making use of these approximations for Bessel functions, C as defined by Eq. (A6.1.15) is

$$C = \psi \pi e^{-2Z_0} \tag{A6.1.23}$$

where

$$\psi = \frac{1 - \lambda_{P0}^*}{1 + \lambda_{P0}^*}$$

Equation (A6.1.13) for λ_{P0} may be written as

$$\lambda_{P0} = \left(\frac{fk_1}{2kZ}\right) \frac{[K_1(Z)/(I_0(Z)] - C[I_1(Z)/I_0(Z)]}{2k_t[K_0(Z)/I_0(Z)] + C}$$

After substituting for C and Bessel functions, we get an expression for the time variation of [P]:

$$\lambda_{P0} = \left(\frac{fk_1 Z}{2k_i}\right) \frac{fk_1 Z}{2k_t} \frac{1 - \psi e^{-2(Z_0-Z)}}{1 + \psi e^{-2(Z_0-Z)}} \tag{A6.1.24}$$

The quantities $CI_0(z)$ and $CI_0(Z_0)$ in Eq. (A6.1.21) for [M] are given by

$$CI_0(Z) = \psi \pi e^{-Z_0} \frac{e^{-(Z_0-Z)}}{\sqrt{2\pi Z}} \tag{A6.1.25}$$

These may be taken as zero, because both Z_0 and Z are of the order 10^6. Hence, Eq. (A6.1.21) for [M] simplifies to

$$[M] = [M]_0 e^{-m(Z_0-Z)} \tag{A6.1.26a}$$

where

$$m = \frac{k_p}{k_t} \tag{A6.1.26b}$$

By making use of these expressions for [M] and λ_{P0}, we can obtain analytical expressions for various moments of radical and dead polymer distributions. The final results are summarized in Table A6.1.

TABLE A6.1 Analytical solution of radical and dead polymer moments

For $k_{trs} = 0$,

$$\lambda_{p1} = C_1\left(\frac{[M]}{[M]_0}\right) + C_2\left(\frac{[M]}{[M]_0}\right)^{1/m} + \frac{fk_1}{2k_t}(Z+1) \tag{A1}$$

where
$$C_1 = \frac{-[M]_0}{1 - 1/m} \quad \text{and} \quad C_2 = \lambda_{p10} + \frac{[M]_0}{1 - 1/m} - \frac{fk_1}{2k_t}(Z_0 + 1) \tag{A2}$$

$$\begin{aligned}\lambda_{p2} = {}&\frac{fk_1(Z+1)}{2k_t} - \frac{4k_p[M]_0 C_1}{fk_1 Z_0(2m-1)}\left(\frac{[M]}{[M]_0}\right)^2 - \frac{4k_p[M]_0 C_2}{fk_1 Z_0 m}\left(\frac{[M]}{[M]_0}\right)^{1+(1/m)}\\&- \frac{2k_p[M]_0}{k_t(m-1)}\left(\frac{[M]}{[M]_0}\right) - \frac{2k_p[M]_0}{k_t Z_0(m-1)}\left(\frac{[M]}{[M]_0}\right) - \frac{[M]_0}{1-1/m}\left(\frac{[M]}{[M]_0}\right) + C^*\left(\frac{[M]}{[M]_0}\right)^{1/m}\end{aligned} \tag{A3}$$

where C^* is a constant evaluated using the initial condition $\lambda_{p2} = \lambda_{20}$ at $t = 0$. Therefore, we get

$$\begin{aligned}C^* = {}&\lambda_{p20} - \frac{fk_1(Z_0+1)}{2k_t} + \frac{4k_p[M]_0 C_1}{fk_1 Z_0(2m-1)} + \frac{4k_p[M]_0 C_2}{fk_1 Z_0 m} + \frac{2k_p[M]_0}{k_t(m-1)}\\&+ \frac{2k_p[M]_0}{k_t Z_0(m-1)} + \frac{[M]_0}{1-1/m}\end{aligned} \tag{A4}$$

$$\lambda_{m0} - \lambda_{m00} = \frac{(0.5k_{tc} + k_{td})}{4k_p k_t}m \cdot fk_1 Z_0\{2(Z_0 - Z) - 2\ln(1 + \psi) + 2\ln[1 + (\psi e)^{-2(Z_0 - Z)}]\} \tag{A5}$$

$$\lambda_{m1} - \lambda_{m10} = \frac{k_t}{k_p}\left[C_1\left(1 - \frac{[M]}{[M]_0}\right) + mC_2\left\{1 - \left(\frac{[M]}{[M]_0}\right)^{1/m}\right\}\right] + \frac{fk_1}{2k_t}\left[\frac{1}{2}(Z_0^2 - Z^2) + (Z_0 - Z)\right] \tag{A6}$$

$$\begin{aligned}\lambda_{m2} - \lambda_{m20} = {}&\frac{2k_{tc}}{fk_1}\left[\frac{C_1^2}{2mZ_0}\left\{1 - \left(\frac{[M]}{[M]_0}\right)^2\right\} + \frac{C_2^2}{2Z_0}\left\{1 - \left(\frac{[M]}{[M]_0}\right)^{2/m}\right\}\right.\\&+ \left(\frac{fk_1}{2k_t}\right)^2\left[\frac{1}{2}(Z_0^2 - Z^2) + 2(Z_0 - Z) + \ln\left(\frac{Z_0}{Z}\right)\right]\\&+ \frac{2C_1 C_2}{(m+1)Z_0}\left\{1 - \left(\frac{[M]}{[M]_0}\right)^{1+(1/m)}\right\} + \frac{C_1}{m}\frac{fk_1}{k_t}\left\{1 - \left(\frac{[M]}{[M]_0}\right)\right\}\\&+ \frac{C_1}{mZ_0}\left(\frac{fk_1}{k_t}\right)\left\{1 - \left(\frac{[M]}{[M]_0}\right)\right\} + \frac{C_2 fk_1}{k_t}\left\{1 - \left(\frac{[M]}{[M]_0}\right)^{1/m}\right\}\\&\left.+ \frac{C_2 fk_1}{k_t Z_0}\left\{1 - \left(\frac{[M]}{[M]_0}\right)^{1/m}\right\}\right]\\&+ \frac{k_t}{k_p}\left(\frac{mfk_1}{2k_t}\left[\frac{1}{2}(Z_0^2 - Z^2) + (Z_0 - Z)\right] - \frac{2k_p[M]_0 C_1}{fk_1 Z_0(2m-1)}\left\{1 - \left(\frac{[M]}{[M]_0}\right)^2\right\}\right.\\&- \frac{4k_p[M]_0 C_2}{fk_1 Z_0(m+1)}\left\{1 - \left(\frac{[M]}{[M]_0}\right)^{1+(1/m)}\right\} - \frac{2k_p[M]_0}{k_t(m-1)}\left\{1 - \left(\frac{[M]}{[M]_0}\right)\right\}\\&- \frac{2k_p[M]_0}{k_t Z_0(m-1)}\left\{1 - \left(\frac{[M]}{[M]_0}\right)\right\} - \frac{[M]_0}{(1-[1/m])}\left(1 - \frac{[M]}{[M]_0}\right)\\&\left.+ mC^*\left\{1 - \left(\frac{[M]}{[M]_0}\right)^{1/m}\right\}\right)\end{aligned} \tag{A7}$$

REFERENCES

1. Brooks, B. W., Kinetic Behavior and Product Formation in Polymerization Reactors' Operation at High Viscosity, *Chem. Eng. Sci.*, 40(8), 1419–1423, 1985.
2. Biesenberger, J. A., R. Capinpin, and J. C. Yang, A Study of Chain Addition Polymerizations with Temperature Variations: II. Thermal Runaway and Instability—A Computer Study, *Polym. Eng. Sci.*, 16, 101–116, 1976.
3. Stolin, A. M., A. G. Merzhanov, and A. Y. Malkin, Non-Isothermal Phenomena in Polymer Engineering and Science, A Review: Part I—Non-Isothermal Polymerization, *Polym. Eng. Sci.*, 19, 1065–1073, 1979.
4. Agarwal, S. S., and C. Kleinstreuer, Analysis of Styrene Polymerization in a Continuous Flow Tubular Reactor, *Chem. Eng. Sci.*, 41(12), 3101–3093, 1986.
5. Hamer, J. W., and W. H. Ray, Continuous Tubular Polymerization Reactors: I. A Detailed Model, *Chem. Eng. Sci.*, 41(12), 3083, 1986.
6. Tien, N. K., E. Flaschel, and A. Renken, Bulk Polymerization of Styrene in a Static Mixer, *Chem. Eng. Commun.*, 36, 25, 1985.
7. Baillagou, P. E., and S. D. Soong, Free-Radical Polymerization of Methyl Methacrylate in Tubular Reactors, *Polym. Eng. Sci.*, 25, 212, 1985.
8. Louie, B. M., and D. S. Soong, Optimization of Batch Polymerization Processes—Narrowing the MWD: I. Model Simulation, *J. Appl. Polym. Sci.*, 30, 3707–3749, 1985.
9. Biesenberger, J. A., Thermal Runaway in Chain-Addition Polymerizations and Copolymerizations, in *Polymerization Reactors and Processes*, J. N. Henderson and T. C. Bouton (eds.), ACS Symposium Series No. 104, American Chemical Society, Akron, OH, 1979.
10. Noronha, J. A., M. R. Juba, H. M. Low, and E. J. Schiffhaner, High Temperature Free-Radical Polymerization in Viscous Systems, in *Polymerization Reactors and Processes*, J. N. Henderson and T. C. Bouton (eds.), ACS Symposium Series No. 104, American Chemical Society, Akron, OH, 1979.
11. Venkateswaran, G., and A. Kumar, Solution of Free Radical Polymerization, *J. Appl. Polym. Sci.*, 45, 187–215, 1992.
12. Venkateswaran, G., Simulation and Experimental Validation of Radical Polymerization with AIBN Initiator in Presence of Shear, Department of Chemical Engineering, I.I.T. Kanpur, India, 1989.
13. Kumar, A., and S. K. Gupta, *Fundamentals of Polymer Science and Engineering*, Tata McGraw-Hill, New Delhi, 1978.
14. Wittmer, P., Kinetics of Copolymerization, *Makromol. Chem.* 3(Suppl.), 129, 1979.
15. Leal, H. P., P. M. Reilly, and K. F. O'Driscoll, On Estimation of Reactivity Ratio, *J. Polym. Sci. Polym. Chem.*, 18, 219, 1980.
16. Fukuda, T., Y. D. Ma, and H. Inagaki, Reexamination of Free Radical Copolymerization Kinetics, *Makromol. Chem.* 12(Suppl.) 125, 1985.
17. Fukuda, T., Y. D. Ma, and H. Inagaki, New Interpretation for the Propagation Rate Versus Composition Curve, *Makromol. Chem. Rapid Commun.*, 8, 495–499, 1987.
18. Hill, D. J. T., and J. H. O'Donnel, Evaluation of Alternative Models for the Mechanism of Chain-Growth Copolymerization, *Macromol. Chem. Macromol. Symp.*, 10/11, 375, 1987.

19. Ebdon, J. R., C. R. Towns, and K. Dodson, On the Role of Monomer–Monomer Complexes in Alternating Free Radical Copolymer: The Case of Styrene and Maleic Anhydride, *JMS—Rev. Macromol. Chem. Phys.*, C26, 523, 1986.

20. Brandrup, J., and E. H. Immergut, *Polymer Handbook*, 2nd ed., Wiley–Interscience, New York, 1975.

21. Dionisio, J. M., and K. F. O'Driscoll, High Conversion Copolymerization of Styrene and Methyl Methacrylate, *J. Polym. Sci. Polym. Lett. Ed.*, 17, 701–707, 1979.

22. Zilberman, E. N., Relative Monomer Reactivities in the Copolymerization at Large Conversions, *Polym. Sci. USSR*, 25, 2198–2204, 1980.

23. Kartavykh, V. P., Y. N. Barantsevice, V. A. Lavrov, and S. S. Ivanchev, The Effect of the Change in Volume of the Reaction System on the Kinetics of Free-Radical Polymerization, *Polym. Sci. USSR*, 22, 1319–1325, 1980.

24. Tirrell, M., and K. Gromley, Composition Control of Batch Copolymerization Reactors, *Chem. Eng. Sci.*, 36, 367–375, 1981.

25. Johnson, A. F., B. Khaligh, J. Ramsay, and K. F. O'Driscoll, Temperature Effects in Free Radically Initiated Copolymerization, *Polym. Commun.*, 24, 35, 1983.

26. Brandrup, J., *Recycling and Recovery of Plastics*, Hanser/Gardener, Munchen, 1996.

27. Rader, C. P., S. D. Baldwin, D. D. Cornell, G. D. Sadler, and R. F. Stockel, *Plastics, Rubber and Paper Recycling*, American Chemical Society, Washington, DC, 1995.

28. Andrews, G. D., and P. H. Subramanian, *Emerging Technologies in Plastics Recycling*, American Chemical Society, Washington, DC, 1992.

29. Chevassus, F., and R. D. Broutelles, *Stabilisation of PVC*, 2nd ed., Arnolds, London, 1993.

30. McNeill, I. C., *Thermal Degradation in Comprehensive Polymer Science*, G. Allen and J. C. Bevington (eds.), Pergamon, London, 1989, Vol. 6, pp. 451–500.

31. Wang, M., J. M. Smith, and B. J. McCoy, Continuous Kinetics for Thermal Degradation of Polymer in Solution, *AIChE J.*, 41, 1521–1533, 1995.

32. Pifer, A., and A. Sen, Chemical Recycling of Plastics to Useful Organic Compounds by Oxidative Degradation, *Angew. Chem. Int. Ed.*, 37, 3306–3308, 1998.

33. Defaud, V., and J. M. Basset, Catalytic Hydrogenolysis at Low Temperature and Pressure of Polyethylene and Polypropylene to Diesel or Lower Alkanes by a Zirconium Hydride Supported on Silica–Alumina: A Step Toward Polyolefin Degradation by Microscopic Reverse of Ziegler–Natta Polymerization, *Angew. Chem. Int. Ed.*, 37, 806–810, 1998.

PROBLEMS

6.1. Suppose that two monomers responding to radical initiators are copolymerized in the presence of a transfer agent, S. Can you use Eq. (6.3.11) for F_1? If not, derive the corrected result. Find the overall rate of polymerization.

6.2. If an inhibitor is used instead of a transfer agent in Problem 6.1, what happens to the equation derived therein?

6.3. Derive the equations that are parallel to Eq. (6.3.11) for cationic as well as stereoregular polymerizations.

6.4. In writing Eqs. (6.3.1a)–(6.3.1j), it was assumed that the effect of penultimate groups on the reactivity is negligible. Suppose that this effect cannot be neglected. In that case, we define species $P_1(AA)$, $Q_1(BA)$, $P_2(AB)$, and $Q_2(BB)$. Now, write the kinetic mechanism of copolymerization involving these species.

6.5. Find the size of the CSTR as well as the PFR for free-radical polymerization of styrene to achieve a 60% conversion. Assume for styrene, $k_p = 145 \, \text{L/mol sec}$, $fk_I = 4.4 \times 10^{-6} \, \text{sec}^{-1}$, $[M]_0 = 8.93 \, \text{mol/L}$, and $k_t = 1.3 \times 10^6 \, \text{L/mol sec}$.

6.6. It is well known that the flow is never a plug type in a reaction that is carried out in a flow reactor. Instead, we find a parabolic flow profile. Find the correction in the calculation when the flow is assumed to be a plug type. What is the effect of the velocity profile on μ_n?

6.7. An engineer suggests a sequence of two CSTRs such that the polymer is separated from the monomer at the intermediate point. The monomer is the ARB type. Would the separation of the polymer make any difference on the total residence time? Would the separation affect the μ_n? If so, by how much? Use the same numerical values as given in Problem 6.5.

6.8. Carry out the analysis in Problem 6.7 for the following cases also: (1) two PFRs in sequence and (2) a CSTR and a PFR in sequence.

6.9. The text has assumed so far that the initiator concentration is constant in the design of a CSTR (as shown here) for radical polymerization. In general, this would not be so.

$[M]_0[I_2]_0$ → ⟶ $[M],[I_2]$

Find $[I_2]$, $[M]$, and μ_n.

6.10. Consider the free-radical polymerization of a monomer in the presence of a weak inhibitor (Z). The mechanism of the polymerization is given by the following:

$$I_2 \xrightarrow{k_I} 2I$$

$$I + M \xrightarrow{k_1} P_1$$

$$P_n + M \xrightarrow{k_p} P_{n+1}$$

$$P_n + P_m \xrightarrow{k_t} M_n + M_m + M_{m+n}$$

$$P + Z \xrightarrow{k_z} M_n - Z$$

Find an expression for r_p^* where r_p^* and r_p are the polymerization rates in the presence and absence of inhibitor. When the gel effect sets in, k_t decreases and r_p, in turn, increases explosively. However, due to the presence of inhibitor, the rate remains within bounds. Derive an expression of μ_n and confirm through similar arguments that μ_n also remains within bounds.

6.11. It is necessary to use the weak inhibitor method for controlling the molecular weight of polymethyl methacrylate in batch bulk polymerization. At $60°C$ k_p is $580 \, L/mol \, sec$, whereas the value of k_z for terephthonitrile is 31.2. Assume that $[I_2] = 10^{-2} \, mol/L$ and find λ_{P0}^* in this problem.

6.12. In a CSTR, a vinyl monomer is polymerized at $60°C$ in the presence of a weak inhibitor. Develop relevant equations when an inhibitor is used for controlling the molecular weight of the polymer formed.

6.13. When the conversion approaches 100% in radical polymerization, we can neglect the mutual termination compared with primary termination. Then, the mechanism of polymerization is as follows and we cannot assume a steady-state approximation:

$$I_2 \xrightarrow{k_1} 2I$$

$$I + M \xrightarrow{k_1} P_1$$

$$P_n + M \xrightarrow{k_p} P$$

$$P + I \xrightarrow{k_t} M_n$$

The results have to be calculated in the time domain; what would be the molecular-weight distribution of the polymer formed?

6.14. From the mechanism of stereoregular polymerization, find the molecular-weight distribution of the polymer formed in a batch reactor.

6.15. In this chapter, only λ_0, λ_1, and λ_2 have been related to the generating function. Find the general expression of λ_k in terms of $G(s, t)$.

6.16. Note in Example 6.4 that the moles of initiator consumed would be equal to the total moles of polymer. If $[I]_0$ is the initial concentration of the initiator, find its concentration at the equilibrium.

6.17. For the $N_{m,n}$ distribution in Example 6.4, evaluate $\sum_{m,n} N_{m,n}$, $\sum_{m,n} mN_{m,n}$, $\sum_{m,n} nN_{m,n}$, $\sum_{m,n} m^2 N_{m,n}$, $\sum_{m,n} n^2 N_{m,n}$, and $\sum_{m,n} mnN_{m,n}$. Then, determine the following:

(a) Average amount of monomer A in copolymer: $\bar{P}_A = (\sum_{m,n} mN_{m,n})/(\sum_{m,n} N_{m,n})$.

    (b) Average amount of monomer B in copolymer: $\bar{P}_B = (\sum_{m,n} nN_{m,n})/(\sum_{m,n} N_{m,n})$.

    (c) Number-average chain length: $\bar{P} = \bar{P}_A + \bar{P}_B$.

6.18. Continuing as in Problem 6.17, go on to find the following:

    (a) Number-average molecular weight: $\bar{M}_n = \bar{P}_A M_A + \bar{P}_B M_B$, where M_A and M_B are the molecular weights of monomers A and B.

    (b) Weight-average molecular weight M_w:

$$\bar{M}_w = \bar{P}\frac{\bar{P}_A(1+\alpha-\beta)M_A^2 + 2(\alpha\bar{P}_B + \beta\bar{P}_A)M_A M_B + P_B(1-\alpha-\beta)M_B^2}{(\bar{P}_A M_A + \bar{P}_B M_B)^2}$$

    (c) The mole fraction of A and B in copolymer, F_A and F_B:

$$F_A = \frac{\bar{P}_A}{\bar{P}} = 1 - F_B$$

    (d) The weight fractions of A and B in copolymer, F_{AW} and F_{BW}:

$$F_{AW} = \frac{\bar{P}_A M_A}{\bar{P}_A M_A + \bar{P}_B M_B} = 1 - F_{BW}$$

6.19. For the reversible anionic polymerization, the initiation step is fast, and the following kinetic model may represent the equilibrium better:

$$A \underset{}{\overset{K'_A}{\rightleftharpoons}} N_{1,0}$$

$$B \underset{}{\overset{K'_B}{\rightleftharpoons}} N_{0,1}$$

$$N_{1,0} + A \underset{}{\overset{K_A}{\rightleftharpoons}} N_{2,0}$$

$$N_{1,0} + B \underset{}{\overset{K_B}{\rightleftharpoons}} N_{1,1}$$

$$N_{0,1} + A \underset{}{\overset{K_A}{\rightleftharpoons}} N_{1,1}$$

$$N_{0,1} + B \underset{}{\overset{K_B}{\rightleftharpoons}} N_{0,2}$$

$$N_{m-1,n} + A \underset{}{\overset{K_A}{\rightleftharpoons}} N_{m,n}$$

$$N_{m,n-1} + B \underset{}{\overset{K_B}{\rightleftharpoons}} N_{m,n}$$

Prove that

$$N_{m,n} = \frac{\binom{m+n}{m}}{(m+n)}(ma + nb)\alpha^m\beta^n, \quad m+n \geq 1$$

where a, b, α, and β are defined as in Example 6.4.

6.20. For the MWD in Problem 6.19, find the information sought in Problems 6.17 and 6.18.

6.21. A variation of the equilibrium copolymerization model is as follows:

$$A + A \underset{}{\overset{K'_A}{\rightleftharpoons}} N_{2,0}$$

$$B + A \underset{}{\overset{K'_A}{\rightleftharpoons}} N_{1,1}$$

$$A + B \underset{}{\overset{K'_B}{\rightleftharpoons}} N_{1,1}$$

$$B + B \underset{}{\overset{K'_B}{\rightleftharpoons}} N_1$$

$$N_{2,0} + A \underset{}{\overset{K_A}{\rightleftharpoons}} N_{3,0}$$

$$N_{2,0} + B \underset{}{\overset{K_B}{\rightleftharpoons}} N_{2,1}$$

$$N_{1,1} + A \underset{}{\overset{K_A}{\rightleftharpoons}} N_{2,1}$$

$$N_{1,1} + B \underset{}{\overset{K_B}{\rightleftharpoons}} N_{1,2}$$

$$N_{m-1,n} + A \underset{}{\overset{K_A}{\rightleftharpoons}} N_{m,n}$$

$$N_{m-1,n} + B \underset{}{\overset{K_B}{\rightleftharpoons}} N_{m,n}$$

Prove that the MWD is now given by

$$N_{m,n} = \left\{ \frac{a}{K_A} \binom{m+n-2}{n} \left(\frac{a}{K_A} + \frac{b}{K_A} \right) \binom{m+n-2}{n-1} \right.$$
$$\left. + \frac{b}{K_A} \binom{m+n-2}{m} \right\} \alpha^m \beta^n, \quad m+n \geq 2$$

6.22. For the MWD of Problem 6.20, find all of the information sought in Problems 6.17 and 6.18.

7

Emulsion Polymerization

7.1 INTRODUCTION

Emulsion polymerization is a technique of polymerization where polymer formation occurs in an inert medium in which the monomer is sparingly soluble (not completely insoluble). Traditionally, water is the inert medium and the initiator is chosen such that it is water soluble. Monomers undergoing step-growth reaction do not require any initiation and are not polymerized by this method. Emulsion polymerization is commonly used for vinyl monomers undergoing addition polymerization and, even among these, those that polymerize by the radical mechanism are preferably polymerized by this method. Water-based emulsions for ionic polymerizations are uncommon because of high-purity requirements. This discussion is therefore restricted to the polymerization of monomers following the radical mechanism only.

The water-soluble initiator commonly used is potassium or sodium persulfate, and the usual recipe for emulsion polymerization is 200 parts by weight of water, 100 parts by weight of the monomer, and 2–5 parts by weight of a suitable emulsifier [1,2]. The monomer should be neither totally soluble nor totally insoluble in the water medium and must form a separate phase. The emulsifier is necessary to ensure that the monomer is dispersed uniformly as in a true emulsion [3–8]. The polymer that is formed from emulsion polymerization is in the form of small particles having an average diameter around 5 µm. The particles form a stable emulsion in water. Their separation can be effected only through the

299

evaporation of water, and once the water is evaporated, these particles coalesce to a solid mass. The rate of emulsion polymerization, r_p, is found experimentally to be much greater than that for the corresponding bulk polymerization, and the average molecular weight of the polymer formed is simultaneously very high—a property that is not achieved in bulk polymerization.

7.2 AQUEOUS EMULSIFIER SOLUTIONS

Emulsifiers are known to play a very important role in emulsion polymerization. To appreciate the role of an emulsifier, we must understand the physicochemical properties of emulsifier solutions. When an emulsifier is dissolved in water, several physical properties of the solution (e.g., osmotic pressure, conductivity, relative viscosity, and surface tension) change. Figure 7.1 shows these changes as a function of the molar concentration of the emulsifier. Beyond a particular level of concentration, there is a sudden change in the slope of these physicochemical properties, as shown in the figure. This concentration is called the critical micelle concentration (CMC).

An emulsifier molecule consists of a long hydrocarbon chain, which is hydrophobic in nature, and a small hydrophilic end, as shown in Figure 7.2. For very small concentrations of the emulsifier, molecules of the latter arrange themselves on the free surface of water such that the hydrophobic ends point

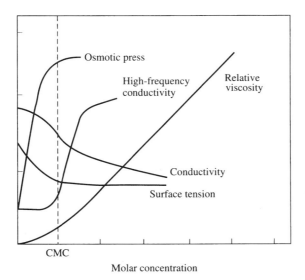

FIGURE 7.1 Changes in physical properties of water as a function of the concentration of the sodium dodecyl sulfate emulsifer.

outward and the hydrophilic ends are buried in the water. In this way, the total free energy of the system is minimized. When more molecules of the emulsifier are present than necessary to form a monolayer on the free surface, they tend to form aggregates, called micelles, so as to minimize the energy of interaction. This aggregate formation starts when the emulsifier concentration increases above the critical micellar concentration. The idealized lamellae and spherical aggregates are shown in Figure 7.2. Beyond the CMC, the emulsifier molecules stay primarily in micellar form. These micelles are responsible for the changes in the physical properties that can be observed in Figure 7.1.

When an emulsifier (or detergent) is added to water and a sparingly soluble monomer is dissolved, the solubility of the monomers is found to increase. The apparently higher solubility is attributed to the presence of micelles, which really become a kind of reservoir for the excess monomer, as shown in Figures 7.2b and 7.2c. At the beginning of the emulsion polymerization, therefore, an emulsifier acts as the solubilizer of the monomer, thus giving a higher rate of emulsion polymerization.

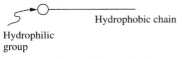

Hydrophobic chain

Hydrophilic group

(*a*) An emulsifier molecule

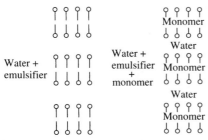

(*b*) Lammellar micelles in water and monomer solution

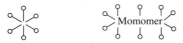

(*c*) Spherical micelles in water and monomer solution

FIGURE 7.2 Schematic representation of micelle formation in emulsion polymerization.

7.2.1 Polymerization in Water Emulsion

It may be recognized that the water-soluble initiator (say, sodium persulfate) decomposes to give a hydrophilic radical that cannot enter the monomer phase due to thermodynamic constraints. We have already pointed out that the monomer is sparingly soluble in the aqueous phase, which means that once initiation occurs in water, the polymer radical begins to grow in chain length. If the polymer radical grows to a certain critical chain length (say, n^*), it nucleates into primary particles, but in the meantime there is a finite probability of being trapped by monomer-swollen micelles, already-existing polymer particles (through coagulation), or monomer droplets. If the radical is trapped in a monomer droplet, the monomer–water equilibrium is totally disturbed, and there is a tendency to eject it and form a smaller monomer droplet and primary particles, as shown in Figure 7.3. Primary particles can similarly be formed from the monomer-swollen micelles.

When emulsion polymerization is started, physical changes occur in the medium as the reaction progresses. For example, it is found that beyond about 5% conversion, the surface tension increases suddenly. Because micelle formation involves a sharp decrease in surface tension, emulsifier molecules are not present in the form of micelles beyond about 5% conversion. The region before this point is referred to as the first stage of polymerization, whereas the region beyond is referred to as the second stage. The third stage of the emulsion polymerization is the stage in which monomer is not available as droplets. Whatever the monomer is present in the reaction mass, it is available in (monomer) swollen polymer particles. These stages are depicted in Figure 7.4.

Figure 7.4, a summary of various studies, reveals that in the second stage, fewer primary particles are formed and polymerization occurs essentially by the growth of polymer particles. As the propagation continues, monomer molecules from the emulsified monomer droplets (see Fig. 7.5) diffuse toward the propagating chains within the polymer particles. The diffusion of the monomer to the polymer particles continues at a fairly rapid rate to maintain constant monomer concentration in the polymer particles.

We will first discuss the modeling of the second stage of emulsion polymerization, because most of the polymerization occurs in this stage. One of the simplest (and oldest) models existing is that of Smith and Ewart. This model was the first to explain gross experimental observations. It may be added that the Smith and Ewart theory assumes that primary radicals (SO_4^{-2}) can enter into the polymer particles. Although we have already explained that this is not possible because of thermodynamic constraints, it is an important simplifying assumption of this theory.

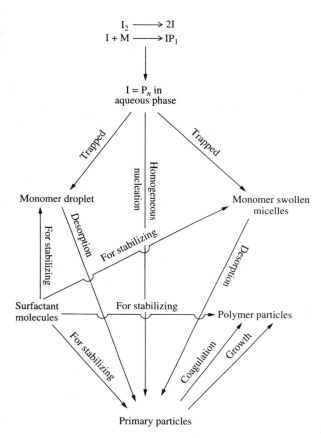

FIGURE 7.3 The overall polymerization process in emulsion polymerization.

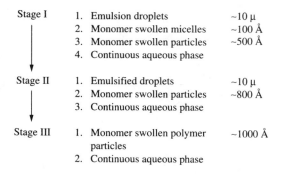

FIGURE 7.4 Various phases in different stages of emulsion polymerization.

7.3 SMITH AND EWART THEORY FOR STATE II OF EMULSION POLYMERIZATION [1,2]

This theory is based on the observation that no new polymer particles are formed in the second stage of emulsion polymerization. As depicted in Figure 7.5, monomer-swollen polymer particles exist in this stage in the form of a stable emulsion. It is assumed that initiator radicals formed in the water phase can enter into these particles to start or stop polymerization. Thus, polymer radicals lie only within these polymer particles.

The total number of polymer particles per unit volume of emulsion in the second stage of the emulsion polymerization is assumed to be N_t. Out of these, N_0 particles are assumed to have no polymer radicals, N_1 to have one polymer radical each, N_2 to have two polymer radicals, and so forth. Therefore,

$$N_t = \sum_{i=0}^{\infty} N_i = N_0 + N_1 + N_2 + \cdots \qquad (7.3.1)$$

If n_t is the total number of polymer radicals per unit volume in the reaction mass, then

$$n_t = \sum_{i=0}^{\infty} iN_i = N_1 + 2N_2 + 3N_3 + \cdots \qquad (7.3.2)$$

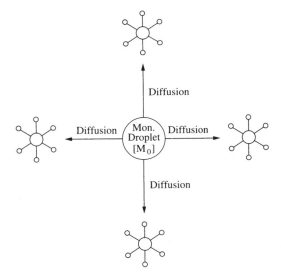

FIGURE 7.5 Representation of physical processes in emulsion polymerization in stage 2. Monomer concentration within polymer particles is maintained constant through diffusion.

The upper limit on these summations would theoretically be infinity. It is assumed that ρ primary radicals are generated in the aqueous phase per unit time per unit volume. These primary radicals enter the polymer particles and generate polymer radicals therein. These polymer radicals can also diffuse out of the particle or undergo a mutual termination with other polymer radicals in the particle. Because primary radicals are generated in the aqueous phase at a rate of ρ radicals per unit volume per unit time and there are N_t polymer particles per unit volume, ρ/N_t primary radicals per unit time (on average) enter a given particle. Polymer radicals can diffuse out of a particle at a rate r_1, given by

$$r_1 = -k_0 a \left(\frac{n}{v} \right) \tag{7.3.3}$$

where a is the surface area of the polymer particle, v is its volume, k_0 is the mass transfer coefficient, and n is the number of polymer radicals in it. It is assumed that all particles are of equal size. The polymer radicals are also destroyed by mutual termination and the rate, r_2, at which this happens is given by

$$r_2 = -\frac{2k_2 n(n-1)}{v} \tag{7.3.4}$$

This form of the equation appears because the polymer radical cannot react with itself. The term $n(n-1)/v$ is proportional to the total number of collisions between the radicals per unit volume.

The rate of generation of particles having n radicals can now be written as follows. A particle having n polymer radicals is formed when one primary radical enters into a particle having $n-1$ polymer radicals, one radical diffuses out from a particle having $n+1$ radicals, or two polymer radicals undergo a mutual termination in a particle having $n+2$ radicals. Similarly, the population of particles having n radicals is reduced when a primary radical enters into them or any one radical diffuses out or there is a mutual termination. At steady state, the rate of increase of N_n equals the rate of decrease of N_n; that is,

$$N_{n-1}\left(\frac{\rho}{N_t} \right) + N_{n+1} k_0 a \left(\frac{n+1}{v} \right) + N_{n+2} k_2 \frac{(n+2)(n+1)}{v}$$
$$= N_n \left(\frac{\rho}{N_t} \right) + N_n k_0 \left(\frac{n}{v} \right) + N_n k_2 n \frac{n(n-1)}{v} \tag{7.3.5}$$

This equation is valid if the number of particles is large. The factor of 2 does not appear with k_2 because only one particle having n radicals is formed from one having $n+2$ radicals. Equation (7.3.5) is the basic recursion relation for emulsion polymerization derived by Smith and Ewart. Many refinements to this equation have been suggested by several authors, but their net result is similar to that of Smith and Ewart. It is difficult to solve Eq. (7.3.5) in its general form; the following discussion develops a few simplified forms.

7.3.1 Case I: Number of Free Radicals per Polymer Particle Is Small Compared with Unity

It is clear that polymer radicals of one polymer particle cannot undergo a mutual termination with polymer radicals of another polymer particle. Therefore, when the number of polymer radicals per particle is on average less than 1, there would not be any mutual termination and the corresponding term in Eq. (7.3.5) should be dropped; that is,

$$N_{n-1}\left(\frac{\rho}{N_t}\right) + N_{n+1}k_0a\left(\frac{n+1}{v}\right) = N_n\left(\frac{\rho}{N_t}\right)N_nk_0a\left(\frac{n}{v}\right) \qquad (7.3.6)$$

This equation is valid only when the diffusion of free radicals out of the particle is much higher than the mutual termination of radicals within it. In general, there are some particles having more than one polymer radical, but such cases are rare because as soon as more radicals are formed, they are transported out of the particle. As a good approximation, therefore, Eq. (7.3.1) can be modified for this case to

$$N_t \cong N_0 + N_1 \qquad (7.3.7)$$

where it is assumed that the number of particles having more than one polymer radical is small (i.e., $N_2 = N_3 = N_4 = \cdots \cong 0$). Then,

$$N_1k_0\frac{a}{v} = N_0\left(\frac{\rho}{N_t}\right) \qquad (7.3.8)$$

from which N_1 can be determined as follows:

$$N_1 = \frac{N_t}{1 + (k_0a/\rho_v)N_t} \qquad (7.3.9)$$

If N_1 is small, $N_0 = N_t$ and Eqs. (7.3.8) and (7.3.9) give

$$N_1 \cong \frac{\rho_v}{k_0a} \qquad (7.3.10)$$

Therefore, the overall rate of polymerization, r_p, per unit volume of the aqueous phase is given by

$$r_p = k_p[M]N_1$$
$$= \frac{k_p[M]\rho v}{k_0a} \qquad (7.3.11)$$

where [M] is the local monomer concentration within the polymer particles. Procedures to obtain this value are discussed later.

7.3.2 Case II: No Transfer of Polymer Radicals out of the Particle Through Diffusion Combined with a High Termination Rate

This occurs when polymer radicals are intertwined in the particles and $k_0 = 0$. In this case, the recursion relation, Eq. (7.3.5), becomes

$$N_{n-1}\left(\frac{\rho}{N_t}\right) + N_{n+2}k_2\frac{(n+2)(n+1)}{v} = N_n\frac{\rho}{N_t} + N_nk_2\frac{n(n-1)}{v} \tag{7.3.12}$$

which can be written as

$$\frac{N_{n-1}}{N_n} + \frac{N_{n+2}}{N_n}\beta(n+2)(n+1) = 1 + \beta n(n-1) \tag{7.3.13}$$

where

$$\beta = \frac{k_2N_t}{v\rho} \tag{7.3.14}$$

The following can now be defined:

$$x = \frac{N_{n+1}}{N_n}$$

$$x_1 = \frac{N_n}{N_{n+1}} \tag{7.3.15}$$

$$x_2 = \frac{N_{n+1}}{N_{n+2}}$$

as a result of which Eq. (7.3.13) can be written as follows

$$x = 1 + \beta n(n-1)\left(1 - \frac{1}{x_1x_2}\frac{(n+2)(n+1)}{n(n-1)}\right) \tag{7.3.16}$$

If we examine the convergence of the series $\sum_{n=0}^{\infty} N_n$ and $\sum_{n=0}^{\infty} nN_n$, it is clear that for n_t and N_t to be finite, these series must be such that x, x_1, and x_2 are each greater than 1. Because β and n can take on only positive values, the following expression can be deduced from Eq. (7.3.16) to be valid for $x > 1$:

$$1 - \frac{1}{x_1x_2}\frac{(n+2)(n+1)}{n(n-1)} > 0 \tag{7.3.17}$$

If this difference is defined as Δ,

$$\Delta \equiv 1 - \frac{1}{x_1 x_2} \frac{(n+2)(n+1)}{n(n-1)} > 0 \tag{7.3.18}$$

where Δ is a positive quantity; then, Eq. (7.3.16) becomes

$$x = 1 + \beta n(n-1)\Delta \tag{7.3.19}$$

For any arbitrary positive value of Δ, as n or β tends to very large values, x also tends to a very large value. Consequently, x_1 and x_2 also take on large values. Therefore, in the limit of large β or large n, the factor $(1/x_1 x_2) \times [(n+2)(n+1)]/n(n-1)$ goes to zero and the following approximation can be made:

$$x \equiv \frac{N_{n-1}}{N_n} \approx 1 + \beta n(n-1) \tag{7.3.20}$$

which is true for large β or large n. From Eq. (7.3.20),

$$\frac{N_0}{N_1} = 1 \qquad \text{for } n = 1 \text{ (and large } \beta) \tag{7.3.21}$$

$$\frac{N_1}{N_2} = 1 + 2\beta \qquad \text{for } n = 2 \tag{7.3.22}$$

$$\frac{N_2}{N_3} = 1 + 6\beta \qquad \text{for } n = 3 \tag{7.3.23}$$

and so on. From these results the values of N_1, N_2, and so forth can be solved as follows:

$$N_1 = N_0 \tag{7.3.24a}$$

$$N_2 = \frac{N_0}{1 + 2\beta} \tag{7.3.24b}$$

$$N_3 = \frac{N_0}{(1 + 2\beta)(1 + 6\beta)} \tag{7.3.24c}$$

The average number of polymer radicals per polymer particle can now be determined as follows:

$$\frac{n_t}{N_t} = \frac{N_1 + 2N_2 + 3N_3 + \cdots}{N_0 + N_1 + N_2 + \cdots} \tag{7.3.25}$$

N_1, N_2, and so forth are now substituted from Eq. (7.3.24) into Eq. (7.3.25), giving

$$\frac{n_t}{N_t} = \left(N_0 + \frac{2N_0}{1+2\beta} + \frac{3N_0}{(1+2\beta)(1-6\beta)} + \cdots\right)$$
$$\times \left(N_0 + N_0 + \frac{N_0}{1+2\beta} + \frac{N_0}{(1+2\beta)(1+6\beta)} + \cdots\right)^{-1} \qquad (7.3.26)$$

For large β, we have

$$\frac{n_t}{N_t} \simeq \frac{N_0}{2N_0} = \frac{1}{2} \qquad (7.3.27)$$

The rate of polymerization, r_p, in the second stage can now be written as follows:

$$r_p = k_p[M]n_t = k_p[M]\frac{N_t}{2} \qquad (7.3.28)$$

where [M] is, again, the local concentration of the monomer in the polymer particles. In this equation, N_1 is constant because—according to Harkin's observations—it does not change in the second stage of emulsion polymerization. N_t can be found as discussed next. The degree of polymerization of the polymer formed can easily be written as follows:

$$\mu_n = \frac{\text{Rate of propagation}}{\text{Rate of termination}} = \frac{r_p}{\rho/2} = k_p\frac{[M]N_t}{\rho} \qquad (7.3.29)$$

Equation (7.3.27) can also be derived using statistical arguments as given in Ref. 2. Each of the polymer particles can be imagined as having at most one growing polymer radical at a given time. Because k_t is large, as soon as an initiator radical enters a polymer particle, it terminates the polymer radical if there is any present. If there are no polymer radicals, the entry of any initiator radical into the polymer particle starts the propagation step once again. Thus, there can be either one or zero polymer radicals in the particle under this condition. On an average, then, of the N_t polymer particles, half will have no radical and the other half will have one radical.

Example 7.1: Consider the diffusion of species A through a stationary medium B around a stationary particle. Determine k_0 in Eq. (7.3.3).

Solution: This particular problem is solved in Ref. 9 and the rate of diffusion, n_A, is given by

$$\dot{n}_A = -D_w 4\pi r^2 C\left(\frac{dx_A}{dt}\right)_{R=r}$$

where D_w is the diffusivity of species A, C is the total molar concentration, and x_A is the mole fraction of species A. If Δ is the thickness of the stationary B layer through which the diffusion occurs, then n_A can be derived as follows:

$$\dot{n}_A = -D_w 4\pi r^2 C \frac{r+\delta}{\delta}(1-x_a)\ln\left(\frac{1-x_a}{1-x_w}\right)$$

where x_a and x_w are the values of x_A at the particle surface and in the bulk, respectively. If x_A is small, we can write $1 - x_A \approx 1$, which gives

$$\dot{n}_A = -D_w 4\pi r r \frac{r+\delta}{\delta}(Cx_w - Cx_a)$$

If the size of the particle is considerably smaller than δ, n_A reduces to

$$\dot{n}_A = -4\pi r D_w(C_w - C_a)$$

where C_w and C_a are the concentrations of A in the bulk and at the surface of the particle, respectively. On comparison with Eq. (7.3.3), we find the following [8]:

$$k_0 a = 4\pi r D_w$$

Example 7.2: The Smith and Ewart theory for emulsion polymerization yields the average number of radicals per particle as 0.5. Experimental results for vinyl chloride give the following [10]:

$$[M]_p = 6 \frac{mol}{L}$$

$$k_p = 3.6 \times 10^7 \frac{L}{mol\ h}$$

Burnett has calculated n in Eq. (7.3.28) to be in the range of $(0.1-5) \times 10^{-4}$, which implies that the theory of Smith and Ewart does not work, and he attributes this to the presence of transfer and termination in the aqueous phase. Modify the analysis to account for this special case of low n.

Solution: Let us assume that ρ_w is the rate of radical production in water and that the polymer radicals desorbed from polymer particles can be terminated in the water with the rate constant k_{tw} [as opposed to k_2 in Eq. (7.3.4) within the polymer particles]. We define the following:

N_w = number of latex particles per liter H_2O
N_0 = number of latex particles having no radicals
N_1 = number of particles having one radical
N_2 = number of particles having two radicals
N_t = sum of the number of radicals in the latex particles and in the water phase per liter of H_2O

We assume that

$$N_w \gg N_1 \gg N_2 \quad \text{or} \quad N_0 \approx N_w - (N_1 + N_2) \tag{7.3.30}$$

which is based on the fact that there are very few latex particles that have growing radicals. If $[I]_w$ is the concentration of radicals in the water phase, the rate of adsorption of radicals by N_0 particles is given by

$$\text{Rate of adsorption} = k_a[I]_w \frac{N_0}{N_w} \approx k_a[I]_w \frac{N_w - (N_1 + N_2)}{N_w}$$

We can now make a count balance on N_1, N_2, and n_t very easily. We observe that N_1 is formed when N_0 receives a radical through adsorption or when there is desorption in N_2. Conversely, N_1 is depleted when it receives a radical or there is desorption in it; that is,

$$\frac{dN_1}{dt} = k_a[I]_w \frac{N_w - (N_1 + N_2)}{N_w} + 2k_dN_2 - k_dN_1 - k_a[I]_w \frac{N_1}{N_w} \tag{7.3.31}$$

We can similarly make balances for N_2 and n_t as follows:

$$\frac{dN_2}{dt} = k_a[I]_w \frac{N_1}{N_w} - 2k_dN_2 - 2\frac{k_{tp}}{v}N_2 = 0 \tag{7.3.32}$$

$$\frac{dn_1}{dt} = \rho_w - \frac{4k_{tp}}{v}N_2 - 2k_{tw}[I]_w^2 \tag{7.3.33}$$

If we assume the existence of a steady state, we get

$$\frac{dN_1}{dt} = \frac{dN_2}{dt} = \frac{dn_1}{dt} = 0$$

which, with the help of Eq. (7.3.30), gives the following:

$$[I]_w = \frac{k_d}{k_aN_1} \tag{7.3.34a}$$

$$N_2 = \frac{k_dN_1^2N_w}{2k_d + 2k_{tp}/v} \tag{7.3.34b}$$

On substituting these into Eq. (7.3.33), we obtain N_1 as follows:

$$N_1 = \rho_w^{1/2} \left(\frac{(V_pk_d + N_wk_{tp})k_a^2}{2k_{tp}k_dk_a^2 + 2k_{tw}k_d^2(V_pk_a^2 + N_wk_{tp})} \right)^{1/2} \tag{7.3.35}$$

As $N_1 \gg N_2$, the total number of particles containing one radical (i.e., N_1) can be set equal to the total number of radicals. The rate of polymerization, r_p, is then given by

$$R_p = \frac{k_p[\mathrm{M}]_p}{N_A}(\rho_w)^{1/2}\left(\frac{(v_p k_d + N_w k_{tp})k_a^2}{2k_{tp}k_d k_a^2 + 2k_{tw}k_d^2(V_p k_d + N_w k_{tp})}\right)^{1/2} \qquad (7.3.36)$$

If the termination in the aqueous phase (i.e., k_{tw}) is small, or

$$2k_{tp}k_d k_a^2 \gg 2k_{tw}k_d^2(V_p k_d + N_w k_{tp}) \qquad (7.3.37)$$

then the rate is given by

$$r_p = \frac{k_p[\mathrm{M}]_p}{N_A}\rho_w\left(\frac{V_p}{2k_{tp}} + \frac{N_w}{2k_d}\right)^{1/2}$$

Example 7.3: In the Smith and Ewart theory, we define state i of the polymer particles the same as the number of polymerizing free radicals within it. Experimentally, it has been demonstrated that the polymer particles have a particle size distribution (PSD), which cannot be obtained through this theory. Assume that within the reaction mass, particles have either 0 or 1 growing radicals and develop necessary relations that give the PSD [11].

Solution: Let us define particle size distribution $n_i(v, t)$ as the number of particles n_i in state i having volume v at time t. Because N_i represents particles of all sizes, we have the following general relation:

$$N_i(t) = \int_0^\infty n_i(v, t)\, dv$$

We have stated that only N_0 and N_1 exist; therefore,

$$n_i(v, t) = [n_0(v, t), n_1(v, t)]$$

Let us divide the general PSD into various boxes, each having an elemental volume ΔV. Subsequently, we define n_0^j and n_1^j as the number concentrations of n_0 and n_1 particles, respectively, in the jth box. We further observe that the number of nongrowing particles n_0 is increased whenever a radical is captured or desorbed by n_1. If Δn_0^j is the change, then

$$\Delta n_0^j = \rho n_1 \Delta t - \rho n_0 \Delta t + k n_1 \Delta t \qquad (7.3.38)$$

where ρ and k have the same meaning as in the discussion of the Smith and Ewart theory. Because we have assumed that there can be at most one growing radical in any polymer particle, it is evident that the mutual termination of radicals cannot occur. We can make a similar balance for n_1, except that now we must account for the growth of the particle. We define a dimensionless number f as the fraction of growing n_1 in an elemental box Δv that leaves the box in time Δt, or

$$f = K \frac{\Delta t}{\Delta v} \tag{7.3.39}$$

where K is a constant. In these terms,

$$\Delta n_1^j = \rho n_0 \Delta t - \rho n_1 \Delta t - k n_1 \Delta t - f n_1^j + f n_1^{j-1} \tag{7.3.40}$$

Dividing Eqs. (7.3.38) and (7.3.40) by Δt and using Eq. (7.3.39) gives

$$\frac{\Delta n_0^j}{\Delta t} = (\rho + k) n_i^j - \rho_{n0}^j$$

$$\frac{\Delta n_1^j}{\Delta t} = \rho_{n0}^j (\rho + k) n_i^j - K \frac{n_1^j - n_1^{j-1}}{\Delta v}$$

These reduce to the following partial differential equations in the limit when Δt and Δv both are allowed to go to zero:

$$\frac{\partial n_0}{\partial t} = (\rho + k) n_i - \rho_{n0}$$

$$\frac{\partial n_1}{\partial t} = \rho_{n0} (\rho + k) n_i - K \frac{\partial n_1}{\partial v}$$

7.4 ESTIMATION OF THE TOTAL NUMBER OF PARTICLES, N_t

The total number of polymer particles in the second stage is estimated from the fact that all of the emulsifier molecules form a monolayer over the particles at the beginning of this stage. According to Figure 7.5, some emulsifier molecules are used up in stabilizing monomer droplets, but these are assumed to be negligible in number.

In the first stage of polymerization, the number of polymer particles N changes with time. We assume that the overall rate of polymerization is given by the same equation as for the second stage when $N = N_t$ (a constant); that is, the polymerization rate per particle is $k_p([M]/2)$. The volumetric rate of growth of the particles, μ, would then be proportional to this rate and is given as

$$\mu = k_1 \left(\frac{k_p}{2} \right) [M] \tag{7.4.1}$$

where k_1 is the proportionality constant.

We now consider the emulsion polymerization in stage 1 at time t from the beginning of the reaction. If a polymer particle is born (by entry of the first initiator radical in a monomer-swollen micelle) after time τ from the beginning, its age at time t is $t - \tau$. The volume, V_t, of this particle at time t is

$$V_{t,\tau} = \mu(t - \tau) \tag{7.4.2}$$

assuming that the size of the particle at birth is negligible. The term $a_{t,\tau}$ is the surface area of this particle at time t and is given by

$$a_{t,\tau} = 4\pi \left(\frac{3\mu(t - \tau)}{4\pi}\right)^{2/3} = [(4\pi)^{1/2}3\mu(t - \tau)]^{2/3} \tag{7.4.3}$$

To determine the total area, A_t, of all polymer particles at time t, the rate at which polymer particles are formed must be known. Knowing this rate allows us to specify the age distribution of polymer particles at time t. In general, primary radicals generated in the aqueous phase are either caught by monomer-swollen micelles or by monomer-swollen polymer particles existing at the time considered. To simplify the analysis, it is assumed that primary radicals are trapped by monomer-swollen micelles only. In that case, the rate of generation of particles, $dN/d\tau$, is given as follows

$$\frac{dN}{d\tau} = \rho \tag{7.4.4}$$

Therefore, A_t can be found:

$$A_t = \int a_{t,\tau} \, dN = [(4\pi)^{1/2}3\mu]^{2/3} \int_{\tau=0}^{t} (t - \tau)^{2/3}\rho \, d\tau \tag{7.4.5}$$

which, on integration, gives

$$A_t = \frac{3}{5}[(4\pi)^{1/2}3\mu]^{2/3}\rho t^{5/3} \tag{7.4.6}$$

The emulsifier is assumed to form a monolayer around the polymer particles when the polymerization just enters the second stage. It is assumed that t_1 is the time when the emulsion polymerization enters the second stage. If $[S]_0$ is the initial concentration of the emulsifier and a_e is the area occupied by a unit mole of the emulsifier for a monolayer, then

$$A_{t1} = [S]_0 a_e \tag{7.4.7}$$

This value of A_{t1} is attained at time t_1, given by Eq. (7.4.6):

$$t_1 = \frac{5^{3/5}\mu^{-2/5}}{3(4\pi)^{1/5}} \left(\frac{[S]_0 a_e}{\rho}\right)^{3/5} = 0.53\left(\frac{[S]_0 a_e}{\rho}\right)^{3/5} \mu^{-2/5} \tag{7.4.8}$$

The total number of particles generated by the time t_1 can now be found by integrating Eq. (7.4.5) as follows:

$$N(t = t_1) = N_t = 0.53 \left(\frac{[S]_0 a_e}{\rho} \right)^{3/5} \mu^{-2/5} \rho$$

$$= 0.53[S]_0 a_e^{3/5} \left(\frac{\rho}{\mu} \right)^{2/5} \tag{7.4.9}$$

7.5 MONOMER CONCENTRATION IN POLYMER PARTICLES, [M]

As shown in Figure 7.5, the concentration [M] of the monomer in the polymer particles remains almost constant with time because of diffusion from the monomer droplets. The monomer is held in the polymer particles by surface tension forces, and the polymer in these particles is in the swollen state. If the monomer is a good solvent for the polymer formed, it is possible to derive the monomer concentration in the particle through fundamental principles. The concentration is given by

$$RT \left[\ln \phi_m + \left(1 - \frac{1}{\mu_n} \right) \phi_p + X_{\text{FH}} \phi_p^2 \right] + \frac{2 V_m \gamma}{r} = 0 \tag{7.5.1}$$

where ϕ_m and ϕ_p are the volume fractions of monomer and polymer in the particle, respectively, X_{FH} is the Flory–Huggins interaction constant (whose value is known for a monomer–polymer system), V_m is the partial molar volume of the monomer, γ is the interfacial tension, and r is the radius of the polymer particle. The first term in the brackets in Eq. (7.5.1) is the chemical potential of the monomer in the absence of surface tension effects, which results from the Flory–Huggins theory, discussed later. The second term represents the contribution to the chemical potential of the monomer due to surface tension effects. Thus, Eq. (7.5.1) represents equilibrium between surface tension effects and solubility effects. Equation (7.5.1) has been plotted in Figure 7.6, where the saturation swelling has been plotted as a function of ϕ_m and the Flory–Huggins parameter. The volume fraction ϕ_m can easily be converted to [M] using additivity of volumes.

In deriving Eq. (7.5.1), it has been assumed that the monomer-swollen particle is homogeneous. This assumption, however, does not give a correct physical picture. Based on thermodynamic considerations, the core-and-shell theory [12–16] on the other hand, proposes that the particle has a polymer-rich core with relatively little monomer and an outer shell around it, consisting of practically pure monomer, as shown in Figure 7.7. This means that [M] in Eq. (7.5.1) is the molar density of the pure monomer. It has been argued that Eq.

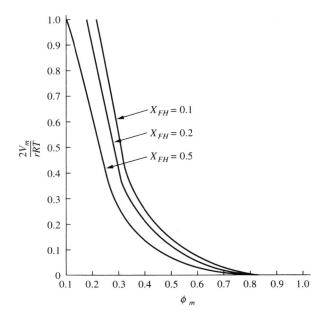

FIGURE 7.6   Saturation swelling of polymer particles by monomer.

(7.5.1) predicts reasonable results when the monomer is a good solvent for the polymer, as in the styrene–polystyrene system [12–16]. However, when the monomer is a poor solvent for the polymer, as for the vinyl acetate–polyvinyl acetate and vinylidene chloride–polyvinylidene chloride systems, the core-and-shell theory gives better results.

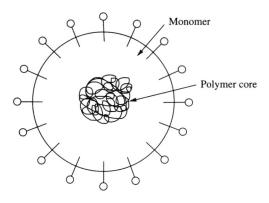

FIGURE 7.7   Core-and-shell model of polymer particles.

7.5.1 Experimental Results

The results derived for region II using the Smith and Ewart theory have been confirmed abundantly. Figure 7.8 gives the conversion in emulsion polymerization as a function of time for isoprene using a potassium laurate emulsifier. On integration of Eq. (7.3.28) with [M] constant, it is found that the plot of the yield versus time should be a straight line. In Figure 7.8, the plot is observed to be essentially linear except near the beginning of the emulsion polymerization (i.e., in stage I), which is attributed to the variation of N with time. The tapering of the curve in stage III, on the other hand, is attributed to the disappearance of the emulsified monomer droplets and the consequent decrease in [M] within polymer particles.

One of the assumptions in the Smith and Ewart theory is that no new polymer particles are generated in the second stage of emulsion polymerization. The rate of polymerization per particle has been experimentally measured in the

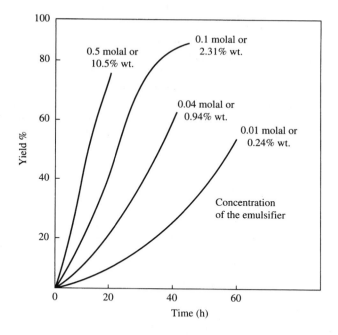

FIGURE 7.8 Polymerization of isoprene in aqueous emulsion with 0.3 g $K_2S_2O_8$ initiator per 100 g monomer and potassium laurate emulsifier at 50°C. (Reprinted with permission from Harkins, J. A., *J. Am. Chem. Soc.*, 69, 1428, 1943. Copyright 1943 American Chemical Society.)

second stage and is found to be flat for N_t between 10^{12} and 10^{14} particles per milliliter of water solution. However, for larger concentrations of particles, N_t has been observed to change with conversion [8]. It was believed earlier that the nucleation of particles in stage I of emulsion polymerization occurred solely in micelles. However, it has been shown more recently that the nucleation can occur equally well in the aqueous phase. The change in N_t in the second stage (in the range of large concentrations of particles) has been explained by the coagulation of particles. When this occurs, emulsifier molecules are released in the water phase, which can generate fresh particles even at high conversions. As can be seen, this fresh nucleation has considerable ramifications on the rate of polymerization as well as on the molecular-weight distribution (MWD) of the polymer formed by emulsion polymerization. In addition, in some systems, the gel effect manifests itself in the third stage of emulsion polymerization.

Comparison of Eqs. (7.4.9) and (7.3.28) reveals that the rate of polymerization in stage II of emulsion polymerization is proportional to $[I]^{0.40}$. This proportionality has also been experimentally confirmed. Moreover, if the polymerization has entered the second stage, and then some more initiator is added, the rate should not change, according to the Smith and Ewart theory. This is because N_t attains a constant value [according to Eq. (7.4.9)] corresponding to the initial concentration $[I_2]_0$. Addition of the initiator after the second stage is reached does not alter N_t. This has also been confirmed by Bovey and Kolthoff [1].

Finally, according to Eq. (7.4.9), the total number of particles, N_t, should be proportional to $[S]_0^{0.60}$. This proportionality has been confirmed for styrene, but for other monomers, deviations have been found. Table 7.1 gives empirical correlations for different emulsifiers and monomers [4].

TABLE 7.1 Empirical Correlations for Emulsion Polymerization with Sodium Lauryl Sulfate Emulsifier and Potassium Persulfate Initiator

Monomer	Empirical correlations
Styrene	$r \propto N_t$ $N_t \propto [S]_0^{5/7}[I_2]_0^{2/7}$
Methyl methacrylate	$N_t \propto [S]_0^{0.6}[I_2]_0^{0.4}$ μ_n decreases with $[I_2]_0$
Acrylonitrile	r_p increases with time $r_p \propto [S]_0^{1/6}[I_2]_0^{1/2}$ μ_n decreases with conversion
Vinyl acetate	$r_p \propto N_t^{0.14-0.2}[I_2]_0^{0.8-1.0}$
Vinyl chloride	$r_p \propto [I_2]_0^{0.5} N_t^{0.05-0.15}$

7.6 DETERMINATION OF MOLECULAR WEIGHT IN EMULSION POLYMERIZATION [17,18]

We have already observed that emulsion polymerization has three phases (i.e., water, polymer particle, and micelles). In each of these, the following radical polymerization reactions occur:

Propagation

$$P_n + Mk_p \xrightarrow{k_p} P_n + 1 \tag{7.6.1a}$$

Monomer transfer

$$P_n + M \xrightarrow{k_{pm}} M_n + P_1 \tag{7.6.1b}$$

Polymer transfer

$$P_n + M_s \xrightarrow{k_{pf}} M_n + P_s \tag{7.6.1c}$$

Terminal double bond

$$P_r + M_x \xrightarrow{k_p^*} P_r + x \tag{7.6.1d}$$

Termination by disproportionation

$$P_r + P_s \xrightarrow{k_{td}} M_r + M_s \tag{7.6.1e}$$

Termination by combination

$$P_r + P_s \xrightarrow{k_{tc}} M_r + s \tag{7.6.1f}$$

Note that in Eq. (7.6.1), the transfer reaction can occur anywhere on the polymer chain. As a result, when the polymer radical P_s grows, it gives rise to branched chains, as seen in Figure 7.9. It is also observed that the disproportionation reaction [Eq. (7.6.1e)] gives rise to dead polymer chains, which can react as in Eq. (7.6.1d). This also leads to the formation of branched polymers, as shown in Figure 7.9.

Most of the polymer is present in polymer particles and, in view of this, we must analyze the polymer formation therein if we want to determine the average chain length and the polydispersity index. We have already noted that initiator radicals are absorbed on the surface of polymer particles, initiating the formation of polymer radicals. These polymer radicals grow and terminate according to Eq. (7.6.1) or get desorbed. For the analysis presented in this section, we assume that dead polymer chains do not get desorbed, even though polymer radicals (i.e., P_n)

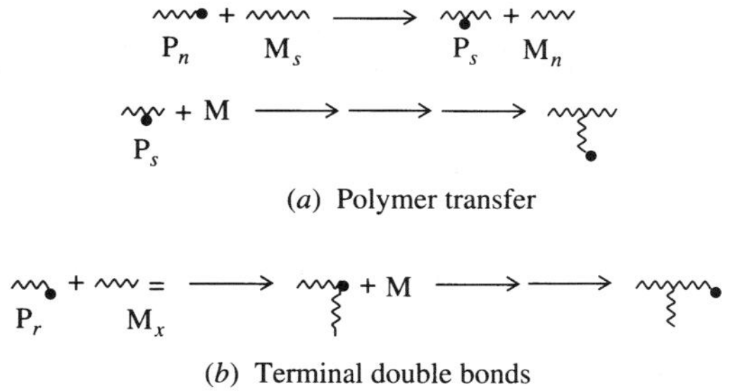

P_n
M_s
P_s
M_n
+ M
P_s
(a) Polymer transfer
+ M
P_r
M_x
(b) Terminal double bonds

FIGURE 7.9 Formation of branched polymer in the emulsion polymerization of vinyl acetate and vinyl chloride.

could. If V_p is the total volume of polymer particles within the reaction mass, the rate of formation of polymer molecules, M_n, within the particle is given by

$$\begin{aligned}\frac{1}{V_p}(V_p[M]_n]) &= k_{fm}[M]_p(1-x)[P_n]\sum_{r=1}^{\infty} r[M_r] \\ &= k_{fpn}[M_n]\sum_{n=1}^{\infty}[P_n] - k_p^*[M_n]\sum_{r=1}^{\infty}[P_r]\end{aligned} \tag{7.6.2}$$

where x is the conversion of the monomer and $(1-x)[M]_{p0}$ gives the monomer concentration. Also, $\sum r[M_r]$ refers to the total moles of repeat units of dead polymer, which is equal to the moles of monomer reacted. In other words,

$$\sum r[M_r] = x[M]_0 \tag{7.6.3}$$

In writing Eq. (7.6.2), expressions for termination and initiation have not been included. Within a given particle, these reactions are small due to a sufficiently slow rate of polymerization and, therefore, have negligible influence on the molecular weight distribution.

We can similarly make a mole balance equation within polymer particles for polymer radicals P_r and derive the following reaction, assuming that the steady-state approximation is valid:

$$\begin{aligned}\frac{1}{V_p}\frac{d}{dt}(V_p[P_n]) &= 0 \\ &= k_p[M]_{p0}(1-x)[P_{n-1}] - \{k_p[M]_{p0}(1-x) + k_{fm}(1-x)[M]_{p0} \\ &\quad + k_{fp}[M]_{p0}x\}[P_n] - k_p^*\left(\sum_{r=1}^{\infty}[M_r]\right)[P_n] \\ &\quad + k_{fp}(n[M_n])\sum_{r=1}^{\infty}[P_r] + k_p^*\sum[M_{n-r}][P_r]\end{aligned} \tag{7.6.4}$$

Within particles, we have a distribution of dead polymer molecules M_r as well as polymer radicals. Consequently, we have the following two kinds of moment:

$$\lambda_{M_n} = \sum_{r=1}^{\infty} r^n [M_r], \quad n = 0, 1, 2 \tag{7.6.5a}$$

$$\lambda_{P_n} = \sum_{r=1}^{\infty} r^n [P_r], \quad n = 0, 1, 2 \tag{7.6.5b}$$

In these terms, Eqs. (7.6.2) and (7.6.5) can be used to derive the moment generation relations as follows:

$$\frac{1}{V_p} \frac{d(V_p \lambda_{M_N})}{dt} = [M]_{p0}\{k_{fm}(1-x) + k_{fp}x\}\lambda_{P_n}$$
$$- k_{fp}\lambda_{M_{(n+1)}}\lambda_{p0} - k_p^* \lambda_{M_n}\lambda_{p0} \tag{7.6.6a}$$

$$\lambda_{pn}\{k_{fm}(1-x)[M]_{p0} + k_{fp}x[M]_{p0} + k_p^*\lambda_{M_0}\}$$
$$= nk_p(1-x)[M]_{p0}\lambda_{P_{(n-1)}} + k_{fp}\lambda_{p0}\lambda_{M_{n+1}} + k_p^*F_n \tag{7.6.6b}$$

where

$$F_n = \sum_{r=2}^{\infty} r^n \sum_{s<r}^{\infty}[M_{r-s}][P_s] = \sum_{r=1}^{\infty}\sum_{s=1}^{\infty}(r+s)^n[M_r][P_s] \tag{7.6.7a}$$

$$\sum r^n[M_{r-1}] = \sum_r r^n[M_r] + n\sum r^{n-1}[M_r] \tag{7.6.7b}$$

In order to get the number- and weight-average chain lengths (μ_n and μ_w), the first three moments of both these distributions are needed. For $n = 0, 1,$ and 2, Eq. (7.6.6a) yields

$$\frac{1}{V_p}\frac{d}{dt}(V_p\lambda_{M0}) = z\lambda_{p0} - k_{fp}\lambda_{M1}\lambda_{p0} - k_p^*\lambda_{M0}\lambda_{p0} \tag{7.6.8a}$$

$$\frac{1}{V_p}\frac{d}{dt}\{V_p\lambda_{M0}\} = z\lambda_{p1} - k_{fp}\lambda_{M2}\lambda_{p0} - k_p^*\lambda_{M1}\lambda_{p0} \tag{7.6.8b}$$

$$\frac{1}{V_p}\frac{d}{dt}\{V_p\lambda_{M2}\} = z\lambda_{p2} - k_{fp}\lambda_{M3}\lambda_{p0} - k_p^*\lambda_{M2}\lambda_{p0} \tag{7.6.8c}$$

where

$$z = \{k_{fm}(1-x) + xk_{fp}\}[M]_{p0} \tag{7.6.9}$$

We further observe the following:

$$F_1 = \sum_r \sum_s (r+s)[M_r][P_s]$$
$$= \lambda_{M1}\lambda_{p0} + \lambda_{M0}\lambda_{p1} \tag{7.6.10}$$
$$F_2 = \sum_r \sum_s (r^2 + 2rs + s^2)[M_r][P_s]$$
$$= \lambda_{M2}\lambda_{p0} + 2\lambda_{M1}\lambda_{p1} + \lambda_{M0}\lambda_{p2} \tag{7.6.11}$$

In these terms, Eq. (7.6.6b) gives

$$z\lambda_{p1} = k_p(1-x)[M]_{p0}\lambda_{p0} + k_{fp}\lambda_{M2}\lambda_{p0} + k_p^*\lambda_{M1}\lambda_{p0} \tag{7.6.12a}$$
$$z\lambda_{p2} = 2k_p(1-x)[M]_{p0}\lambda_{p1} + k_{fp}\lambda_{M3}\lambda_{p0} + k_p^*(\lambda_{M2}\lambda_{p0} + 2\lambda_{M1}\lambda_{p1}) \tag{7.6.12b}$$

On eliminating $z\lambda_{p1}$ and $z\lambda_{p2}$ in Eq. (7.6.8b) and with the help of Eqs. (7.6.12a) and (7.6.12b), the following are obtained:

$$\frac{1}{V_p}\frac{d(V_p\lambda_{M1})}{dt} = k_p(1-x)[M]_{p0}\lambda_{p0} \tag{7.6.13a}$$

$$\frac{1}{V_p}\frac{d(V_p\lambda_{M2})}{dt} = 2\{k_p(1-x)[M]_{p0} + k_p^*\lambda_{M1}\}\lambda_{p1} \tag{7.6.13b}$$

We have already observed that there is an equilibrium during stage II of emulsion polymerization between the monomer-swollen polymer particles and the separate monomer phase. As a result of this equilibrium, the ratio of monomer and polymer within them remains time invariant. Equations (7.6.8a), (7.6.13a), and (7.6.13b) during stage II reduce to the following:

$$\frac{\lambda_{M0}}{V_p}\frac{dV_p}{dt} = (z - k_{fp}\lambda_{M1} - k_p^*\lambda_{M1})\lambda_{p0} \tag{7.6.14a}$$

$$\frac{\lambda_{M1}}{V_p}\frac{dV_p}{dt} = k_p(1-x)[M]_{p0}\lambda_{p0} \tag{7.6.14b}$$

$$\frac{\lambda_{M2}}{V_p}\frac{dV_p}{dt} = 2\{k_p(1-x)[M]_{p0} + k_p^*\lambda_{M1}\}\lambda_{p1} \tag{7.6.14c}$$

It may be mentioned that the net effect of polymerization in stage II is the growth of polymer particles. If Eqs. (7.6.14a) and (7.6.14b) are divided, we get the average chain length of the polymer within the particles as follows:

$$\mu_n = \frac{\lambda_{M1}}{\lambda_{M0}} = \frac{k_p(1-x)[M]_{p0}}{(z - k_{fp}\lambda_{M1} - k_p^*\lambda_{M0})} \tag{7.6.15}$$

Dividing Eqs. (7.6.14b) and (7.6.14c) and eliminating $\lambda P_1 / \lambda P_0$ with the help of Eq. (7.6.12a) gives the weight-average chain length μ_w:

$$\mu_w = \frac{\lambda_{M2}}{\lambda_{M1}} = \frac{2[k_p(1-x)[M]_{p0} + k_p^* \lambda_{M1} k_p (1-x)[M]_{p0} + k_{fp} \lambda_{M2} + k_p^* \lambda_{M1}}{k_p(1-x)[M]_{p0}}$$

(7.6.16)

Note that the first moment of monomer, λ_{M_1}, can be directly determined by the conversion of the monomer

$$\lambda_{M1} = x[M]_{p0} \tag{7.6.17}$$

and once this is known, Eqs. (7.6.15) and (7.6.16) can be conveniently used to determine μ_n and μ_w.

Example 7.4: Polymerizing systems such as vinyl chloride and vinyl acetate give branched polymers in emulsion. Find the extent of branching in the second stage. Also, find the relations governing μ_n and μ_w in the third stage of emulsion polymerization.

Solution: Whenever reactions occur through Eqs. (7.6.1c) and (7.6.1d) within the polymer particles, branched polymers are formed. Let us denote by [B] the concentration of branch points per unit volume of the particles. A balance on [B] is given by

$$\frac{1}{V_p} \frac{d}{dt}(V_p[B]) = k_{fp}\left(\sum r_1[M_r]\right)\left(\sum P_n\right) + k_p^*\left(\sum P_n\right)\left(\sum M_n\right)$$

$$= k_{fp} \lambda_{M1} \lambda_{p0} + k_p^* \lambda_{M0} \lambda_{p0}$$

For the second stage, if we assume that [B] is constant, we get

$$\frac{[B]}{V_p} \frac{dV_p}{dt} = (k_{fp}\lambda_{M1} + k_p^*\lambda_{M0})\lambda_{p0}$$

During stage III of emulsion polymerization, the various moments within the polymer particles may change. Assuming that changes in volume of the particles are considerably slower compared to changes in moments, Eq. (7.6.8) reduces to the following:

$$\frac{d\lambda_{M0}}{dt} = (z - k_{fp}\lambda_{M1} - k_p^*\lambda_{M1})\lambda_{p0}$$

$$\frac{d\lambda_{M1}}{dt} = k_p(1 - x)[M]_{p0}\lambda_{p0}$$

$$\frac{d\lambda_{M_2}}{dt} = 2\{k_p(1 - x)[M]_{p0} + k_p^*\lambda_{M1}\}\lambda_{p1}$$

7.7 EMULSION POLYMERIZATION IN HOMOGENEOUS CONTINUOUS-FLOW STIRRED-TANK REACTORS [8]

Operation of homogeneous continuous-flow stirred-tank reactors (HCSTRs) for emulsion polymerization offers several advantages over batch reactors, the most important of which are high production rate and better polymer quality. In commercial batch reactors, it has been found that there is always a small variation from batch to batch in monomer conversion, particle number and size, molecular weight, and polymer branching. HCSTRs, on the other hand, eliminate this problem and also give a narrower MWD of the polymer. We have already observed that steady-state HCSTRs operate at the exit concentrations, which can be chosen by the designer. This provides flexibility during operation, and the reactor can be conveniently controlled at optimal conditions.

Figure 7.10 shows an HCSTR carrying out emulsion polymerization. The feed consists of a monomer dispersed in aqueous phase that already has initiator and emulsifier in it. During polymerization within the reactor, polymer particles are generated. These particles are macroscopic in size, and we cannot assume that all particles stay within the reactor for the same residence time. The reasoning for this is as follows. We first observe that the feed has no polymer particles; they are later nucleated within the reactor at different times. Because the particles grow and attain their final size in the remainder of the time within the reactor, there must be a size distribution of the particles. As a result, for the specified flow conditions, the particles of different size move toward the reactor exit at different velocities.

In analyzing an HCSTR, we rarely analyze the flow conditions existing within it, but assume (based on experiments) the following particle age distribution:

$$f(\tau) = \theta^{-1} \exp\left(-\frac{\tau}{\theta}\right) \tag{7.7.1}$$

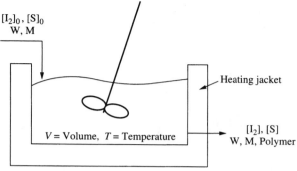

W = Water
M = Monomer
I_2 = Initiator
S = Emulsifier

FIGURE 7.10   Schematic diagram of an HCSTR carrying out emulsion polymerization.

where θ represents the reactor residence time and $f(\tau)$ represents the fraction of particles that stays in the reactor for time τ. As a result of this age distribution, there is a particle size distribution, which can be determined as follows. Let us consider a single particle operating during the second stage of emulsion polymerization, when the rate r_{pp} in a given particle is given by

$$r_{pp} = \frac{k_p[M]_p \bar{n}}{N_A} \tag{7.7.2}$$

where $\bar{n}$ is the average number of free radicals per particle. If the excess monomer is available within the reactor (which is the case for second stage), the monomer concentration, $[M]_p$, within the particle remains constant. The particles grow in size due to polymerization and the reacted monomer is replenished by diffusion from the droplets. Under these conditions, the particle growth (radius r and volume v_p) can be represented by

$$\frac{dv_p}{dt} = 4\pi r^2 \frac{dr}{d\tau} = \frac{k_p[M]_p \bar{n}}{\rho_p N_{AV}} \equiv K_1[M]_p \bar{n} \tag{7.7.3}$$

where K_1 is a constant and ρ_p is the density of the particle. If we assume case II of the Smith–Ewart theory, n is equal to 0.5 and Eq. (7.7.3) can be easily obtained as follows:

$$v = 0.5K_1[M]_p \tau \tag{7.7.4a}$$

$$r^3 = \frac{3K_1[M]_p \tau}{8\pi} \tag{7.7.4b}$$

where we assume that particles have negligible volume after stage I. We now eliminate τ in Eq. (7.7.1) using Eq. (7.7.4b) to obtain particle size distribution $F(r)$ as

$$F(r) = f(\tau) \left(\frac{dr}{dt}\right)^{-1}$$
$$= \frac{8\pi r^2}{K_1[M]_p \theta} \exp\left(-\frac{8\pi r^3}{3K_1[M]_p \theta}\right) \tag{7.7.5}$$

If the inlet concentration of the emulsifier is $[S]_0$ and that in the exit stream is $[S]$, then it is possible to estimate the total number of polymer particles, N_p. It is observed that the total surface area of these particles is $N_p \int_0^\infty 4\pi r^2 F(r)/dr$ and that

$$[S]_0 - [S] = \frac{4\pi N_p}{a_s} \int_0^\infty \frac{r^2 F(r)}{dr}$$

where a_s is the surface area coverage offered by a single surfactant molecule.

7.8 TIME-DEPENDENT EMULSION POLYMERIZATION

There are several situations described in the literature where the phenomenon of transience in emulsion polymerization must be considered. Some of these are reactor start-up or shutdown [19,20], reactor stability [21,22], and reactor controls [23,24] There are also a few applications in which the amount of emulsifier used is such that it gives a concentration less than the CMC. The particle sizes that result lie in the range 0.5–10 µm and are larger than those obtained from the usual emulsion polymerization described earlier. The product obtained in this condition is not a true emulsion from which polymer particles precipitate out when diluted with water. This is known as dispersion polymerization, to which the analysis developed earlier is not applicable [25–27].

The time-dependent behavior of emulsion polymerization arises due to variation in monomer concentration, changes in the number of polymer particles N_t, or both. We have already observed that N_t changes due to nucleation in stage I of emulsion polymerization and this normally ends at about 10–15% conversion. However, when the monomer-to-water ratio (M/W) is high or the monomer is more than "sparingly" soluble, the constancy of N_t cannot be assumed up to conversions as large as 50%. If the monomer droplets are sufficiently small, they also become the loci of particle formation and, in such circumstances, the Smith–Ewart theory is inadequate to explain the experimental phenomena. We now present the outline of a mathematical model of emulsion polymerization that is based on kinetic theory and thus overcomes the inadequacy of the Smith–Ewart theory.

The overall emulsion polymerization can be described in terms of reactions as follows. The initiator molecule (I_2) is present in the aqueous phase, where it undergoes thermal decomposition to give initiator radical (I). These radicals are hydrophilic and cannot enter into hydrophobic monomer droplets (see Fig. 7.3). Consequently, they can react only with monomer dissolved in water, giving the following:

$$I_2 \xrightarrow{k_{da}} 2I \tag{7.8.1a}$$

$$I + {}^aM \xrightarrow{k_{ia}} {}^aP_1 \tag{7.8.1b}$$

$$ {}^aP_1 + {}^aM \xrightarrow{k_{pa}} {}^aP_{i+1} \tag{7.8.1c}$$

These polymer radicals (${}^aP_i$) in the aqueous phase can grow up to a certain critical length (say, n^*), after which they precipitate to form a primary particle. Here, a polymer particle is represented P_{aj}, where the index j denotes the number of times it has been initiated. This means that P_{a1} represents the primary particle. Before reaching the critical length n^*, the polymer radicals ${}^aP_i$ can terminate as they do in the homogeneous polymerization discussed in Chapter 5. This means that the following reactions occur:

$$ {}^aP_{n-1}^* + {}^aM \xrightarrow{k_{pa}} {}^aP_{n^*} \xrightarrow{\text{Nucleation}} P_{a1} \tag{7.8.2a}$$

$$ {}^aP_i + {}^aP_j \xrightarrow{k_{ta}} {}^aP_{i+j} \quad \text{as long as } i+j < n^* \tag{7.8.2b}$$

$$ {}^aP_i + {}^aP_j \xrightarrow[ta]{k} {}^aP_{i+j} \xrightarrow{\text{Nucleation}} P_{a1} \quad \text{when } i+j > n^* \tag{7.8.2c}$$

The polymer radicals can also enter into micelles (indicated by subscript or superscript c), monomer droplets (indicated by subscript or superscript d), or polymer particles. As soon as one of the former two happens, the micelle (MC) and the droplet (MD) become particles with one radical in them:

$$ {}^aP_i + MD \xrightarrow{K_{md}} Pa_1 \tag{7.8.3a}$$

$$ {}^aP_i + MC \xrightarrow{K_{mc}} Pa_1 \tag{7.8.3b}$$

$$ {}^aP_i + P_{aj} \xrightarrow{K_c} Pa_{j+1} \tag{7.8.3c}$$

In writing Eq. (7.8.3c), we assume that, on average, a polymer particle has mostly dead chains and any radical entry into it amounts to initiating polymerization therein. Radical desorption from the particles can be written as

$$Pa_j K_i \xrightarrow{k_{ei}} Pa_j + {}^aP_i \qquad \text{for all } j \qquad (7.8.4)$$

The coagulation of polymer particles [Eq. (7.8.5a)], micelle disappearance to cover newly formed particle surface [Eq. (7.8.5b)], and the coalescence and breakage of monomer droplets [Eq. (7.8.5c)] can be represented as follows:

$$Pa_i + Pa_j \xrightarrow{K_{fij}} Pa_{i+j} \qquad \text{for all } i \text{ and } j \qquad (7.8.5a)$$

$$MC + Pa_j \xrightarrow{R_{mp}} Pa_j \qquad \text{for all } j \qquad (7.8.5b)$$

$$MD + MD \underset{R_{md1}}{\overset{R_{mp1}}{\rightleftharpoons}} MD \qquad (7.8.5c)$$

With the model presented in Eqs. (7.8.1)–(7.8.5) it is possible to model the transience of emulsion polymerization. As an example, let us derive the rate of formation of polymer particles as follows. If N_t represents the total number of particles in the reaction mass, then

$$\frac{1}{N_A}\frac{dN_t}{dt} = [\text{Nucleation in aqueous phase by Eq. (7.8.2a)}] + [\text{Nucleation}$$
$$\text{in micelles by Eq. (7.8.3c)}] + [\text{Nucleation in monomer}$$
$$\text{droplets by Eq. (7.8.3a)}] - [\text{Coalescence of particles by}$$
$$\text{Eq. (7.8.5a)} + [\text{Precipitation of oligomers from the} \qquad (7.8.6)$$
$$\text{aqueous phase by termination by Eq. (7.8.2c)}]$$

$$= k_p[{}^aM][{}^aP_{n^*-1}] + K_{mc}[MC]\sum_{i=1}^{n^*-1}[{}^aP_i] + K_{md}[MD]\sum_{i=1}^{n^*-1}[{}^aP_i]$$

$$-\frac{1}{2}\sum_{i=1}^{n_c^*}\sum_{j=1}^{n^*}K_{fij}[Pa_i][Pa_j] = K_{ta}\sum_{i=1}^{n^*-1}\sum_{k=1}^{i}[{}^aP_i][{}^aP_{n^*-k}]$$

We can establish mole balance equations for each species present in different phases (i.e., aqueous, micelle, droplet, and particle phases) and solve these simultaneously. Song and Pohlein have solved these sets of differential equations and have finally arrived at the following analytical solution of Eq. (7.8.6) [28].

$$\frac{N_t}{N_s} = \frac{1 - e^{-t/\tau}}{1 + e^{-t/\tau}/A_2} \qquad (7.8.7)$$

where N_s represents particle concentration in number per volume of the aqueous phase at the steady state. Here, A_2 and τ are the model parameters, which can be estimated from the basic kinetic parameters and reaction conditions.

Example 7.5: In an effort to graft gelatin with polymethyl methacrylate (PMMA), 2 g of potassium persulfate and 20 g of gelatin are dissolved in water. This is added to 40 g of methyl methacrylate (MMA), and the reaction mass is made up to 500 cm^3. This recipe does not contain any surfactant and the polymerization at 70°C is found to give stable emulsion polymenzation. Experimental analyses of samples show a copious formation of gelatin grafts, which suppress the homopolymerization of MMA [29]. Explain this phenomenon through a kinetic model.

Solution: Because there are no surfactants in the reaction mass, there are only two phases: aqueous and monomer droplet phases. Gelatin has hydroxyl ($-$OH) functional groups on its chain that can react with a persulfate radical easily and give growing polymer radicals. Such a reaction would lead to the formation of grafted gelatin, which is the intention of the experiment described. Accordingly, we define the following symbols for species in the reaction mass:

I_2 = initiator molecules
I = initiator radicals
M = monomer
P_H = homopolymer radicals
P_g = graft polymer radicals
OH_G = gelatin
M_G = dead graft homopolymer molecules
M_H = dead homopolymer molecules
Superscript w = aqueous phase
Superscript d = droplet phase

The aqueous phase consists of I_2, gelatin (OH_G^w), dissolved methyl methacrylate monomer (M^w), primary radicals (SO_4^-), and desorbed homopolymers (M_H^w). The grafting of gelatin occurs mostly in the aqueous phase but as the length of the graft chain increases, it begins to migrate to the droplet surface due to thermodynamic considerations. We have therefore grafted gelatin in the aqueous phase (OH_G^w) as well as at the interface of the droplets surrounding it (OH_G^d), with the graft dangling inside.

The various reactions occurring in the system are all shown schematically in Figure 7.11, and the rate constants for these are given in Table 7.2. It is possible to establish balance equations for various species; these are given in Table 7.3. Note that all polymer radicals, such as P_H and P_G, are intermediate species and their rates of formation could be taken as zero after the steady-state approximation is made.

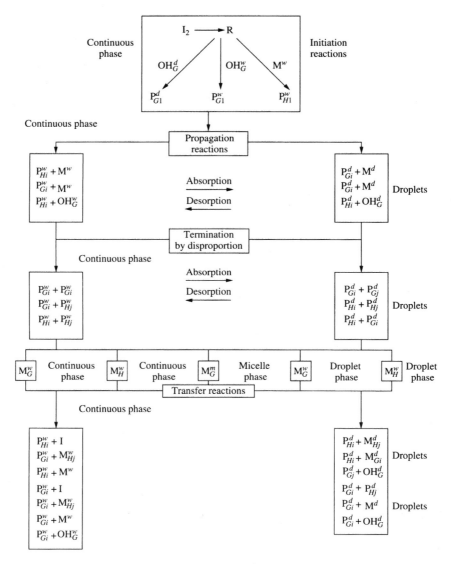

FIGURE 7.11   Various reactions giving grafting of gelatin for Example 7.5.

TABLE 7.2 Equations of the Model of Example 7.5

Initiation reactions

$$I^w \xrightarrow{K_1} 2R^{-w}$$

$$R^{-w} + M^w \xrightarrow{K_2} P_{H1}^w$$

$$R^w + OH_G^w \xrightarrow{K_3} P_{G1}^w$$

$$R^w + OH_G^d \xrightarrow{K_4} P_{G1}^w$$

Propagation of MMA and branched gelatin in water phase

$$P_{Hi}^w + M^w \xrightarrow{K_5} P_{Hi+1}^w, \quad i = 1, 2, \ldots$$

$$P_{Gi}^w + M^w \xrightarrow{K_6} P_{Gi+1}^w, \quad i = 1, 2, \ldots$$

$$P_{Hi}^w + OH_G^w \xrightarrow{K_7} P_{Gi}^w$$

Termination by disproportionation in water phase

$$P_{Gi}^w + P_{Gj}^w \xrightarrow{K_8} M_{Gi}^w + M_{Gj}^w$$

$$P_{Gi}^w + P_{Hj}^w \xrightarrow{K_9} M_{Gi}^w + M_{Hj}^w$$

$$P_{Hi}^w + P_{Hj}^w \xrightarrow{K_{10}} M_{Hi}^w + M_{Hj}^w$$

Absorption–desorption reactions between water and droplet phase

$$M_{Gi}^w \underset{K_{-33}}{\overset{K_{33}}{\rightleftharpoons}} M_{Gi}$$

$$M_{Hi}^w \underset{K_{-34}}{\overset{K_{34}}{\rightleftharpoons}} M_{Hi}$$

$$P_{Hi}^w \underset{K_{-11}}{\overset{K_{11}}{\rightleftharpoons}} P_{Hi}^d$$

$$P_{Gi}^w \underset{K_{-12}}{\overset{K_{12}}{\rightleftharpoons}} P_{Gi}^d$$

Propagation of MMA and branched gelatin in the droplet phase

$$P_{Hi}^d + M^d \xrightarrow{K_{13}} P_{Hi+1}^d, \quad i = 1, 2, \ldots$$

$$P_{Gi}^d + M^d \xrightarrow{K_{14}} P_{Gi+1}^d, \quad i = 1, 2, \ldots$$

$$P_{Hi}^d + OH_G^d \xrightarrow{K_{15}} P_{Gi}^d$$

Termination reactions in the droplet phase

$$P_{Gi}^d + P_{Gj}^d \xrightarrow{K_{16}} M_{Gi}^d + M_{Gj}^d$$

$$P_{Hi}^d + P_{Hj}^d \xrightarrow{K_{17}} M_{Hi}^d + M_{Hj}^d$$

$$P_{Hi}^d + P_{Gj}^d \xrightarrow{K_{18}} M_{Hi}^d + M_{Gj}^d$$

$$M_{Hi}^d + OH_G^d \xrightarrow{K_{19}} M_{Gi}^d$$

Transfer reactions with P_{Hi} in water phase

$$P_{Hi}^w + I \xrightarrow{K_{20}} M_{Hi}^w + P_{Hi}^w$$

$$P_{Hi}^w + M_{Hj}^w \xrightarrow{K_{21}} M_{Hi}^w + P_{H1}^w$$

$$P_{Hi}^w + M^w \xrightarrow{K_{22}} M_{Hi}^w + P_{H1}^w$$

Transfer reaction with P_{Hi} in droplet phase

$$P_{Hi}^d + M_{Hj}^d \xrightarrow{K_{23}} M_{Hi}^d + P_{Hj}^d$$

$$P_{Hi}^d + M^d \xrightarrow{K_{24}} M_{Hi}^d + P_{H1}^d$$

$$P_{Hi}^d + OH_G^d \xrightarrow{K_{25}} M_{Hi}^d + P_{G1}^d$$

Transfer reactions with P_{Gi} in water phase

$$P_{Gi}^w + I \xrightarrow{K_{26}} M_{Gi}^w P_{H1}^w$$

$$P_{Gi}^w + M_{Hj}^w \xrightarrow{K_{27}} M_{Gi}^w P_{Hj}^w$$

$$P_{Gi}^w + M^w \xrightarrow{K_{28}} M_{Gi}^w P_{H1}^w$$

$$P_{Gi}^w + OH_G^w \xrightarrow{K_{29}} M_{Gi}^w P_{G1}^w$$

Transfer reaction with P_{Gi} in droplet phase

$$P_{Gi}^d + M_{Hj}^d \xrightarrow{K_{30}} M_{Gi}^d + P_{Hj}^d$$

$$P_{Gi}^d + M^d \xrightarrow{K_{31}} M_{Gi}^d + p_{H1}^d$$

$$P_{Gi}^d + OH_G^d \xrightarrow{K_{32}} M_{Gi}^d + p_{G1}^d$$

Source: Ref. 29.

TABLE 7.3 Mole Balance of Various Species in Example 7.5

$$\frac{d[P_G]^w}{dt} = V_x(-K_6[P_G]^w M^w + K_7[P_H]^w OH_G^w - K_8[P_G]^w - K_9[P_G]^w[P_H]^w - K_{12}[P_G]^w$$
$$- K_{26}P_G]^w I^w - K_{27}[P_G]^w[M_H]^w - K_{28}[P_G]^w M^w - K_{29}[P_G]^w OH_G^w)$$
$$+ (1 - V_x)K_{-12}[P_H]^d + V_x K_3[R]^w OH_G^w = 0 \tag{1}$$

$$\frac{d[P_G]^d}{dt} = (1 - V_x)(-K_{-12}[P_G]^d - K_{14}[P_G]^d M^d + K_{15}[P_H]^d OH_G^d$$
$$- K_{16}[P_G^d]^2 - K_{18}[P_H]^d[P_G]^d - K_{30}[P_G]^d[M_H]^d - K_{31}[P_G]^d M^d - K_{32}[P_G]^d OH_G^d)$$
$$+ V_x K_{12}[P_G]^w + (1 - V_x)K_4[R]^w OH_G^d = 0 \tag{2}$$

$$\frac{d[P_H]^w}{dt} = V_x(-K_5[P_H]^w - K_{11}[P_H]^w - K_{20}[P_H]^w I^w - K_{21}[P_H]^w[M_H]^w - K_{22}[P_H]^w M^w$$
$$- K_{10}[P_H]^w)(1 - V_x)K_{-11}[P_H]^d + V_x K_2[R]^w M^w = 0 \tag{3}$$

$$\frac{d[P_H]^d}{dt} = (1 - V_x)(-K_{-11}[P_H]^d - K_{13}[P_H]^d M^d + K_{15}[P_H]^d OH_G^d$$
$$- K_{17}[P_H^d]^2 - K_{18}[P_H]^d[P_g]^d$$
$$- K_{23}[P_H]^d[M_H]^d - K_{24}[P_H]^d M^d - K_{25}[P_H]^d OH_G^d) + V_x K_{11}[P_H]^w = 0 \tag{4}$$

$$\frac{d[I]}{dt} = V_x(-2K_1 f[I] - K_{20}[P_h]^w I - K_{26}[P_g]^w I)$$

$$\frac{d[M]^w}{dt} = V_x(-K_2[R]M^w - K_5[P_H]^w M^w - K_6[P_G]^w M^w - K_{22}[P_H]^w M^w - K_{28}[P_G]^w M^w \tag{5}$$

$$\frac{d[M]^d}{dt} = (1 - V_x)(-K_{13}[P_H]^d M^d - K_{14}[P_G]^d M^d - K_{24}[P_H]^d M^d - K_{31}[P_G]^d M^d) \tag{6}$$

$$\frac{d[OH_G]^w}{dt} = V_x(-K_3[R]OH_G^w - K_7[P_H]^w OH_G^w - K_{29}[P_G]^w OH_G^w) \tag{7}$$

$$\frac{d[OH_G]^d}{dt} = (1 - V_x)(-K_4[R]OH_G^d - K_{15}[P_H]^d OH_G^d - K_{19}[M_H]^d OH_G^d$$
$$- K_{25}[P_H]^d OH_G^d - K_{32}[P_G]^d OH_G^d) \tag{8}$$

$$\frac{d[M_G]^w}{dt} = V_x(2K_8[P_G^w]^2 + K_9[P_G]^w + [P_H]^w - K_{33}[M_G]^w - K_{26}[P_G]^w$$
$$+ K_{27}[P_G]^w[M_H]^w + K_{28}[P_G]^w M^w + K_{29}[P_G]^w OH_G^w) + (1 - V_x)K_{-33}[M_G]^d \tag{9}$$

$$\frac{d[M_G]^d}{dt} = (1 - V_x)\{-K_{-33}[M_G]^d + K_{16}[P_G^d]^2 + K_{19}[M_H]^d OH_G^d + K_{30}[P_G]^d[M_H]^d$$
$$+ K_{31}[P_G]^d M^d + K_{32}[P_G]^d OH_G^d\} + V_x K_{33}[M_G]^w \tag{10}$$

$$\frac{d[M_H]^d}{dt} = V_x(K_9[P_G]^w[P_H]^w + 2K_{10}[P_H^w]^2 - K_{34}[M_H]^w$$
$$+ K_{34}[M_H]^w + K_{20}[P_H]^w I^w + K_{22}[P_H]^w M^w) + (1 - V_x)K_{-34}[M_H]^d \tag{11}$$

$$\frac{d[M_H]^d}{dt} = (1 - V_x)(-K_{-34}[M_H]^d + 2[P_H^d]^2 K_{18}[P_H]^d[P_G]^d - K_{19}[M_H]^d OH_G^d$$
$$+ K_{24}[P_H]^d M^d + K_{25}[P_H]^d H_G^d) + V_x K_{34}[M_H]^w \tag{12}$$

The kinetic model of Table 7.2 is very complex and can be simplified as follows:

1. The dissociation rate constant K_1 for the potassium persulfate initiator in water has been reported in the literature as 3.42×10^{-3} mL/g hr [30]. The same value is used here.
2. The initiation rate constant for gelatin in water (K_3) and on the surface of the droplet phase (K_4) have been assumed to be equal. Because there is no information on K_2 and K_3, we have treated these as parameters and, through curve fitting, we have obtained $K_2 = 1250$ mL/mg hr at 60°C and $K_3 = 4.5 \times 10^{-1}$ mL/g hr.
3. The propagation rate constants in water (K_5, K_6, K_7) and in droplet (K_{13}, K_{14}, K_{15}) are assumed to be equal.

$$K_5 = K_6 = K_7 = K_p^w \text{ (say)}; \qquad K_{13} = K_{14} = K_{15} = K_p^d \text{ (say)}$$

K_p^d is available from the *Polymer Handbook* [30] and the average propagation rate constant K_p is taken as $K_p = V_x K_p^w + (1 - V_x) K_p^d$. The term V_x is the volume fraction of water phase in the reaction mass. The values taken in the simulation are as follows:

$$V_x = 0.75 \quad K_p^d = 2.06 \times 10^7 \text{ mL/g hr}$$
$$K_p^w = \text{(assumed)} 13.18 \times 10^6 \text{ mL/g hr}$$
$$K_p = 10.4 \times 10^7 \text{ mL/g hr}$$

4. The termination rate constants for MMA and graft polymer in water (K_8, K_9, K_{10}) and those in the droplet phase $(K_{16}, K_{17}, K_{18}, K_{19})$ are assumed to be equal. The following weighted average termination rate constant is defined as

$$K_t = V_x K_t^w + (1 - V_x) K_t^d$$

The value K_t^d is known from homopolymerization data and K_t^w has been assumed to be a quantity of the same order of magnitude. At 60°C, K_t^w is assumed to be 38.55×10^{11} mL/g hr, K_t^p is 4.28×10^{11} mL/g hr, and K_t is 29.98×10^{11} mL/g hr.
5. The absorption rate constants K_{11}, K_{12}, K_{33}, and K_{34} for homopolymer, graft copolymer radicals, and dead molecules from water to droplet phase have been assumed to be equal. These can be estimated [10] using $K_c = 4\pi r_p D_w N_A$, where r_p is the particle radius ($\sim 2.1 \times 10^{-7}$), D_w is the diffusion coefficient ($\sim 2.86 \times 10^{-8}$ cm²/sec), and N_A is Avogadro's number (6.023×10^{23}). This gives K_{11}, K_{12}, K_{33}, and K_{34} as 1.634×10^{12} mL/g hr. In addition, we have assumed that the desorption rate constants are zero; that is, $K_{-11} = K_{-12} = K_{-33} = K_{-34} = 0.0$.

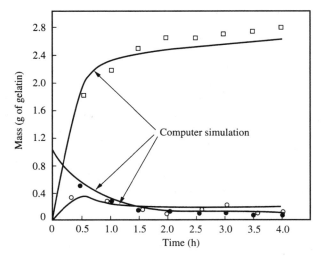

FIGURE 7.12 Numerical solution of differential equations of Table 7.3 and the experimental results of Example 7.5. (From Ref. 29.)

6. All of the transfer reaction rate constants K_{20} to K_{32} have been assumed to be equal; that is, $K_{20} = K_{21} = K_{22} = K_{32} = K_{tr}$ (say). The value of K_{tr} that gives good fit of the data at 60°C is 1060 mL/g hr.

    The ordinary differential equations of Table 7.3 can be solved numerically and the results have been plotted in Figure 7.12. Note the decrease in the amount of homopolymer PMMA for large times and a rapid increase in the formation of graft copolymer. This result could not be predicted by the Smith–Ewart theory.

7.9 CONCLUSIONS

The molecular weight of the polymer formed in emulsion polymerization is normally considerably higher than that formed in corresponding radical polymerization. As might be clear from the discussion of this chapter, this is largely because of the presence of micelles present in the water where the polymerization has been hypothesized to occur. The micelles offer a flexible cavity which serves as a cage where polymer radicals can grow but cannot terminate due to thermodynamic constraints. Similar results have been obtained when polymerization is carried out in more rigid cages of lipid bylayers, liquid crystals, organic crystals, inclusion complexes, microporous zeolites, and mesoporous materials [31].

In Chapters 1–7, readers have been introduced to formulate polymers for a given end use. The purpose of the remainder of the chapters is to make the readers familiar with subjects relevant to polymer physics, such as characterization, thermodynamics rheology and so forth.

REFERENCES

1. Bovey, F. A., I. M. Kolthoff, A. I. Medalia, and E. J. Meehan, *Emulsion Polymerization*, Wiley–Interscience, New York, 1955.
2. Blackley, D., *Emulsion Polymerization*, Applied Science, London, 1975.
3. Flory, P. J.., *Principles of Polymer Chemistry*, Cornell University Press, Ithaca, NY, 1953.
4. Min, K. W., and W. H. Ray, On the Mathematical Modelling of Emulsion Polymerization Reactors, *J. Macromol. Sci. Rev. Macromol. Chem.*, C11, 177, 1974.
5. Lissant, K. J. (ed.), *Emulsion and Emulsion Technology, Part II*, Marcel Dekker, New York, 1974.
6. Kumar, A., and S. K. Gupta, *Fundamentals of Polymer Science and Engineering*, Tata McGraw-Hill, New Delhi, 1978.
7. Bassett, D. R., and A. E. Hamielec (eds.), *Emulsion Polymers and Emulsion Polymerization*, ACS Symp. Series, Vol. 165, American Chemical Society, Washington, DC, 1982.
8. Piirma, I., *Emulsion Polymerization*, Academic Press, New York, 1982.
9. Bird, R. B., W. E. Stewart, and E. N. Lightfoot, *Transport Phenomena*, Wiley, New York, 1960.
10. Ugelstad, J., P. C. Mork, P. Dahl, and P. Rangnes, A Kinetic Investigation of Emulsion Polymerization of Vinyl Chloride, *J. Polym. Sci.*, C27, 49, 1969.
11. Lichti, G., R. G. Gilbert, and D. H. Napper, Growth of Polymer Colloids, *J. Polym. Sci. Polym. Chem.*, 15, 1957–1971, 1977.
12. Reichert, K. H., and H. U. Moritz, Continous Free Radical Polymerization in Disperse System, *Makromol. Chem. Macromol. Symp.*, 10/11, 571, 1987.
13. Uranek, C. K., Molecular Weight Control of Elastomers Prepared by Emulsion Polymerization, *Rubber Chem. Technol.*, 49, 610–649, 1976.
14. Ugelstad, J., and F. K. Hansen, Kinetics and Mechanism of Emulsion Polymerization, *Rubber Chem. Technol.*, 49, 536–609, 1976.
15. Bataille, P., B. T. Van, and Q. B. Pham, Emulsion Polymerization of Styrene: I. Review of Experimental Data and Digital Simulation, *J. Polym. Sci. Polym. Chem.*, 20, 795, 1982.
16. Hansen, F. K., and J. Ugelstad, Particle Nucleation in Emulsion Polymerization: Ia. Theory of Homogeneous Nucleation, *J. Polym. Sci. Polym. Chem.*, 16, 1953, 1978; II. Nucleation in Emulsifier-Free Systems Investigated by Seed Polymerization; III. Nucleation in Systems with Anionic Emulsifier Investigated by Seeded and Unseeded Polymerization; IV. Nucleation in Monomer Droplets, *J. Polym. Sci. Polym. Chem.*, 17, 3033–3046, 3047–3068, 3069, 1979.
17. Graessley, W. W., H. M. Hehauser, and R. Maramba, MWD in Free Radical Polymers, Effect of Branching on Distribution Breadth and Prediction of Gel Point in Diene Polymerization, *Makromol. Chem.*, 86, 129–138, 1965.
18. Friis, N., and A. E. Hamielec, Kinetics of Vinyl Chloride and Vinyl Acetate in Emulsion Polymerization, *J. Appl. Polym. Sci.*, 19, 97–113, 1975.

19. Chiang, A. S. T., and R. W. Thompson, Modelling of Transient Behaviour of Continuous Emulsion Polymerization Reactors, *AIChE J.*, 25, 552–554, 1979.
20. Badder, E. E., and B. W. Brooks, Startup Procedure for Continous Flow Emulsion Polymerization Reactors, *Chem. Eng. Sci.*, 39, 499, 1984.
21. Rawlings, J. B., and W. H. Ray, Stability of Continuous Emulsion Polymerization Reactors: A Detailed Model Analysis, *Chem. Eng. Sci.*, 42, 2767–2777, 1987.
22. Rawlings, J. B., and W. H. Ray, Emulsion Polymerization Reactor Stability: Simplified Model Analysis, *AIChE J.*, 33, 1663, 1987.
23. Louie, B. M., W. Y. Chiu, and D. S. Soong, Control of Free-Radical Emulsion Polymerization of Methyl Methacrylate by Oxygen Injection: I. Modeling Study, *J. Appl. Polym. Sci.*, 30, 3189–3223, 1985.
24. Louie, B. M., T. Franaszek, T. Pho, W. Y. Chiu, and D. S. Soong, Control of Free-Radical Emulsion Polymerization of Methyl Methacrylate by Oxygen Injection: II. Experimental Study, *J. Appl. Polym. Sci.*, 30, 3841–3856, 1985.
25. Fitch, R. M. (ed.), *Polymer Colloids*, Plenum, New York, 1971.
26. Barret, K. E. J., and H. R. Thomas, The Preparation of Polymer Dispersion in Organic Liquids, in *Dispersion Polymerization in Organic Media*, K. E. J. Barrett (ed.), Wiley, London, 1975.
27. Almog, Y., and M. Levy, Effect of Initiator on Molecular Weight Distribution in Dispersion Polymerization of Styrene, *J. Polym. Sci. Polym. Chem.*, 18, 1–11, 1980.
28. Song, Z., and G. W. Pohlein, Particle Formation in Emulsion Polymerization: Transient Particle Concentration, *J. Macromol. Sci. Chem.*, 25, 403–443, 1988.
29. Roy, P., and A. Kumar, Modelling of Grafting of Gelatin with Polymethyl Methacrylate in Aqueous Medium, *Polym. Eng. Sci.*, 31, 1001–1008, 1991.
30. Brandrup, J., and E. H. Immergut, *Polymer Handbook*, 2nd ed., Wiley, New York, 1975.
31. Tajima, K., and T. Aida, Controlled Polymerization with Constrained Geometries, J. *Chem. Soc., Chem. Commun.*, 2399–2412, 2000.

PROBLEMS

7.1. Take k_p for styrene as 500 L/mol sec and [M] as 5.2 mol/L at 50°C. Let there be 10^{14} particles/cm³ of water. Let primary radicals be generated in water at the rate of 10^{12} radicals/sec (cm³ of water) $(= \rho)$. Calculate r_p. Calculate μ_n using the following relation:

$$\mu_{n,\text{emulsion}} = \frac{k_p N_t [\text{M}]}{\rho}$$

Compare this rate and μ_n for bulk polymerization at the same monomer concentration and ρ.

7.2. Suppose that there is a chain transfer agent that is water soluble and does not dissolve in the monomer at all. Would the expression for μ_n given in Problem 7.1 be valid? What would happen to the overall rate?

7.3. Write the elementary reactions when styrene is polymerized in emulsion with a transfer agent that is water insoluble but styrene soluble.

7.4. In the derivation of the total number of particles, N_t, it was assumed that ρ was constant. Supposing that the decomposition of initiator follows first-order kinetics, determine the time at which the polymerization enters the second stage.

7.5. In Problem 7.4, find out how the rate and DP would vary with time.

7.6. Within latex particles, the concentration of the polymer is much higher and considerably restricts the motion of macroradicals. When the polymerization temperature, T, is less than the glass transition temperature, T_g, of the reaction mass, a limiting conversion x_1 is approached. The free-volume theory gives this to be

$$x_1 = \frac{(\rho_p/\rho_m)\phi_p}{1 - (1 - \rho_p/\rho_m)\phi_p}$$

$$T = T_g = \frac{\alpha_p \phi_p T_{gp} + \alpha_m (1 - \phi_p) T_{gm}}{\alpha_p \phi_p + \alpha_m (1 - \phi_p)}$$

where ϕ_p is the volume fraction of the polymer, T_{gp} and T_{gm} are the T_g values of the pure polymer and monomer, respectively, α_p and α_m are the differences in coefficients of the volumetric expansion of polymer (or monomer) in the melt and glass state, respectively, and ρ_p and ρ_m are the densities of the polymer and monomer, respectively. Using these equations, plot T versus x_1 for styrene, for which $T_{gp} = 92.5°C$, $T_{gm} = -106°C$, $\alpha_p = 0.48 \times 10^{-3}$ C^{-1}, and $\alpha_m = 1.0 \times 10^{-3}$ C^{-1}.

7.7. Consider the seed polymerization in which seeds of polymer particles (each of volume v) are each charged to the reaction mass having monomer and initiator. When desorption of radicals and water-phase termination of radicals are neglected, the average number of growing polymer radicals n can be derived:

$$\bar{n} = \frac{(a/4)I_0(a)}{I_1(a)}$$

where $a^2 = 8\rho v/(k_t N)$, and N is the total number of particles. Derive an expression for conversion as a function of time.

7.8. It is evident that k_t and k_p within polymer particles change with time and have been modeled using the following:

$$\frac{k}{k_0} = \frac{D}{D_0}$$

Determine the variation of k_t and k_p within the particle and write the appropriate particle size and MWD equations.

7.9. The emulsion polymerization of vinyl acetate has been modeled as consisting of the initiation of radicals in the water phase, the initiation of particles in water phase, the entry of radicals into polymer particles, the termination of particles, the chain transfer to monomer in polymer particles, and the escape of radicals in polymer particles. Write the unsteady-state balance for the total number of particles N_T and the number of polymer radicals having 1 and 2 radicals (N_1 and N_2).

7.10. Latexes prepared by mixing methyl methacrylate and ionogenic monomers containing carboxylic acid groups constitute a special class of latexes, which are used in carpet backing, adhesives, surface coating, and paper coating. List a few monomers that can be used for this purpose. Their solubility improves by increasing pH.

7.11. To prepare a 340-nm sphere, the following recipe has been suggested: 19.9 g of methyl methacrylate, 3.5 g of methacrylic acid, 10.5 g of hydroxy-ethyl methacrylate, 1.1 g of ethylene glycol dimethacrylate, and 0.1 g of emulsifier sodium dodecyl sulfate (SDS) are added to 64.9 g of distilled water. Polymerization is carried out for 1 hr at 98°C in a sealed tumbling container. Write down the approximate molecular structure of the surface and indicate whether the polymer particle would dissolve in any solvent.

7.12. The polymer particles formed in Problem 7.11 are suspended in water adjusted to pH 10.5 with NaOH and activated at 25°C with 10 mg of cyanogen bromide per milliliter of suspension. After 15 min, it is diluted with equal volume of cold 0.1 M borate buffer at pH 8.5 and 4°C. The immunolatex conjugates thus prepared are known to bind antigens. Explain why this occurs.

7.13. In the carbodiimide method, the latex particles of Problem 7.11 are reacted in an aqueous medium with water-soluble diimide. This can be reacted to antibodies. Write down the chemical reactions.

7.14. Suppose two monomers in a 1 : 1 ratio (responding to radical initiators) are mixed and copolymerized in an emulsion. These monomers dissolve in water in the same proportion. Find the rate of polymerization.

7.15. In Problem 7.14, the resultant polymer is a copolymer. We know the values of r_1 and r_2 from the study of bulk copolymerization. Can we use the same values for emulsion copolymerization? Write down all the equations that you would use for this case.

7.16. Suppose we want to copolymerize styrene and acrylonitrile in an emulsion. The solubility of acrylonitrile is 7.4% and that of styrene is 0.04% by weight in water. Let S (volume of styrene), w (volume of water), and a (volume of acrylonitrile) be mixed together. Find the concentrations of monomers in micelles (or polymer particles) at equilibrium. From the results developed in Problem 7.14, calculate F_s, the fraction of styrene in the resultant polymer.

7.17. We can carry out rigorous modeling of the molecular weight of polymer in emulsion polymerization as follows. The state of the particle is defined by the number of polymerization radicals in it. The following additional variables are defined:

N_i = number of polymer particles in state i
X_i = number of dead chains within particles in state i
W_i = total weight of live chains in state i
Y_i = total weight of dead chains in state i
x_i = mean number of dead chains within particle in state i
y_i = mean total weight of dead chains per particle in state i
w_i = mean weight of live chains in state i
λ_i = mean weight of dead chains in state i

Write the contribution to W, X_i, N_i, and Y_i from polymerization, radical entry, radical exit, chain transfer, and bimolecular termination.

7.18. Derive overall expressions for X_i, N_i, Y_i, and W_i. The mean dead weight per particle in state i is given by

$$\lambda_i = \frac{Y_i}{X_i}$$

and the grand mean is given by

$$\lambda = \frac{\sum_{i=0}^{\infty} Y_i}{\sum_{i=0}^{\infty} X_i}$$

7.19. In inverse emulsion polymerization of acrylamide, we carry out polymerization in iso-octane using nonionic emulsifier pentaerithritol myristate (PEM) with oil-soluble azoinitiators (e.g., AIBN). The standard recipe (see page 209 of Reichert and Geiseler) is as follows:

Oil phase	Weight (g)	Water phase	Weight (g)
Iso-octane	649.64	Water	195.15
PEM	13.86	Acrylamide	53.31
AIBN	0.164		

Show the schematic model of the inverse polymerization along with various reactions occurring in different phases.

7.20. Acrylonitrile and methyl acrylate are mixed in water with a suitable water-soluble initiator and emulsifier. Determine the composition of the polymer.

8

Measurement of Molecular Weight and Its Distribution

8.1 INTRODUCTION

A solid polymer is a mosaic of structures. For a crystallizable homopolymer, for example, we can vary the amount and nature of crystallinity and the shape and size of the crystals. In addition, we can vary the orientation of the polymer chains in both the crystalline and amorphous phases. This variation can be brought about either by changing material variables or process conditions. The former include the chemical structure, the molecular weight and its distribution, the extent of chain branching, and the bulkiness of the side groups. The latter include the temperature and the deformation rate. It is the interplay within this multitude of variables that leads to the physical structure visible in the finished product. This structure, in turn, determines the properties of the solid polymer. In this chapter, we examine the methods of measuring the polymer's molecular weight and its distribution. These quantities were defined in Chapter 1, and knowledge thereof can be helpful to the process engineer in optimizing desired polymer properties. These properties include mechanical properties such as impact strength, flow properties such as viscosity, thermal properties such as the glass transition temperature, and optical properties such as clarity.

There are several other reasons why we might want to measure the molecular weight. The molecular weight and its distribution determine the viscous and elastic properties of the molten polymer. This affects the processi-

340

bility of the melt and also the behavior of the resulting solid material (see also Chapter 12). To cite a specific example [1], a resin suitable for extrusion must have a high viscosity at low shear rates so that the extrudate maintains its integrity. To be suitable for injection molding, however, the same resin must have a low viscosity at high shear rates so that the injection pressure not be excessive. Both of these requirements can be satisfied by a proper adjustment of the molecular-weight distribution. More often, though, different grades of the same polymer are marketed for different products that are fabricated via different polymer processing operations; the resin used for making polycarbonate water bottles, for example, differs significantly in molecular weight from the poly-carbonate that goes into compact disks. Differences in molecular-weight distribu-tion also influence the extent of polymer chain entanglement and the amount of melt elasticity, as measured by phenomena such as extrudate swell. The effect of swell shows up during processing, wherein flow results in different amounts of chain extension and orientation, which remain frozen within the solidified part. As a consequence, two chemically similar polymers, processed identically, that have the same molecular weight but different molecular-weight distributions may result in products that show significantly different shrinkages, tensile properties, and failure properties [2]. For this very important reason, it is advantageous to know the molecular weight and molecular-weight distribution of the polymers used. Furthermore, because polymers can mechanically degrade during proces-sing and during use (polymers such as nylon can also increase in molecular weight), a second measurement of the molecular weight can reveal the extent of chain scission or postcondensation. These measurements are also useful in verifying that the various kinetic schemes postulated for polymer synthesis in Chapters 3–7 do, indeed, produce the molecular-weight distributions predicted theoretically. Other situations where the molecular weight and its distribution directly influence results include phase equilibrium and crystallization kinetics.

A variety of methods are available for molecular-weight determination and they are applicable in different ranges of molecular weight. Also, they provide different amounts and kinds of information. Thus, end-group analysis and colligative property measurements yield the number-average molecular weight. Light scattering, on the other hand, furnishes the weight-average molecular weight and the size of the polymer in solution. Intrinsic viscosity supplies neither number-average molecular weight ($\bar{M}_n$) nor weight-average molecular weight ($\bar{M}_w$); it gives a viscosity-average molecular weight. The entire distribution can be obtained using either ultracentrifugation or size-exclusion chromatography. However, the former technique is an absolute one, whereas the latter is indirect and requires calibration. All of these methods mandate that the polymer be in solution. Other, less commonly encountered methods are described elsewhere [3].

8.2 END-GROUP ANALYSIS

The simplest conceptual method of measuring polymer molecular weight is to count the number of molecules in a given polymer sample. The product of the sample weight and Avogadro's number when divided by the total number of molecules gives the number-average molecular weight. This technique works best with linear molecules having two reactive end groups that each can be titrated in solution. Consequently, linear condensation polymers made by step-growth polymerization and possessing carboxyl, hydroxyl, or amine chain ends are logical candidates for end-group analysis.

Nylon 66, a polyamide and one of the earliest polymers to be synthesized, contains amine and carboxyl end groups. The number of amine groups in a sample can be determined by dissolving the polymer in a phenol–water solvent [4]. Typically, ethanol and water are added to this solution and the mixture is titrated to a conductometric end point with hydrochloric acid in ethanol. Because the number of amine end groups may not equal the number of carboxyl end groups, the acid groups are counted separately by dissolving the nylon in hot benzyl alcohol, and titration is carried out with potassium hydroxide in benzyl alcohol to a phenolphthalein end point. Finally, assuming that the reaction goes to completion and that each nylon molecule has two titratable ends, it is possible to calculate

$$\bar{M}_n = \frac{2}{[NH_2] + [COOH]} \tag{8.2.1}$$

where the quantities in square brackets are concentrations of end groups in moles per gram of polymer. Results are comparable in magnitude to those obtained using osmotic pressure and vapor-pressure osmometry in the range of applicability of these techniques [5].

In addition to polyamides, end-group analysis has been used with polyesters, polyurethanes, and polyethers. Besides titration, counting methods that have been employed include spectroscopic analyses and radioactive labeling. Because the number of chain ends for a given mass of sample reduces with increasing molecular weight, the method becomes less and less sensitive as the size of the polymer molecules increases. The molecular weight of most condensation polymers, however, is less than 50,000, and in this range, end-group analysis works fine [6]. Note also that the amount of polymer needed for end-group analysis is relatively small.

Example 8.1: In order to determine the number of carboxyl end groups in a sample of polyethylene terephthalate, Pohl dissolved 0.15 g of the polymer in hot benzyl alcohol, to which some chloroform was subsequently added [7]. This solution, when titrated with 0.105 N sodium hydroxide, required 35 µL of the alkali. If a blank solution of the benzyl alcohol plus chloroform required 5 µL of the base, how many carboxyl end groups were contained in the polymer sample?

Solution: Because 30 µL of 0.105 gram equivalent per liter of the base reacted with the polymer, the concentration of gram equivalents of end groups was

$$\frac{(30)(10^{-6})(0.105)}{0.15} = 21 \times 10^{-6} \text{ equivalents per gram}$$

8.3 COLLIGATIVE PROPERTIES

It is easily observed that dissolving a nonvolatile solute in a liquid results in a depression of the freezing point; that is, the temperature at which a solid phase is formed from solution is lower than the temperature at which the pure solvent freezes. This is the principle at work in an ice cream maker and in snow removal when salt is used to melt and thereby remove snow and ice from roads. Besides lowering the freezing point, the addition of a nonvolatile solute also reduces the vapor pressure at a given temperature, with the consequence that the solution boils at a higher temperature than the pure solvent does. Furthermore, a solution can develop a large osmotic pressure (explained later), which can be measured with relative ease. These four effects—depression of freezing point, elevation of boiling point, lowering of solvent vapor pressure, and development of an osmotic pressure—are called *colligative properties* and they depend only on the number concentration of the solute in solution in the limit of infinite dilution. Thus, beginning with a known mass of solute, a knowledge of any of these colligative properties reveals the total number of molecules in solution, which, in turn, allows computation of the number-average molecular weight. However, the relative magnitude of these effects is such that as the molecular weight of the solute increases and the number of molecules in a given sample mass decreases, not all four colligative properties can be measured with equal accuracy or ease; indeed, membrane osmometry is the method of choice for measuring the number-average molecular weight of high polymers.

Phase equilibrium is the basic principle used to obtain expressions for the magnitude of the different colligative properties. It is known from thermodynamics that when two phases are in equilibrium, the fugacity, $\hat{f}$, of a given component is the same in each phase. Thus, if, as shown in Figure 8.1, pure vapor A is in equilibrium at temperature T and pressure P with a liquid mixture of A and B, where B is a nonvolatile solute,

$$f_A^v(T, P) = \hat{f}_A^L(T, P, x_A) \qquad (8.3.1)$$

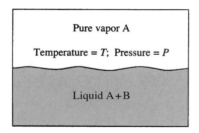

FIGURE 8.1 Pure solvent vapor in equilibrium with a polymer solution.

where the superscripts v and L denote vapor phase and liquid phase, respectively, and x_A is the mole fraction of A in the liquid phase. Also, a "hat" $(\hat{\ })$ on f_A signifies a component in solution as opposed to a pure component.

If the mixture of A and B is sufficiently dilute, it will behave as an ideal solution, for which the following holds [8]:

$$\hat{f}_A^L(T,\ P,\ x_A) = f_A^L(T,\ P)x_A \qquad (8.3.2)$$

and the result is known as the Lewis and Randall rule. Consequently,

$$f_A^v(T,\ P) = f_A^L(T,\ P)x_A \qquad (8.3.3)$$

If at the same pressure P, pure A boils at temperature T_b, then it is obvious that

$$f_A^v(T_b,\ P) = f_A^L(T_b,\ P) \qquad (8.3.4)$$

Dividing the left-hand side of Eq. (8.3.4) by the left-hand side of Eq. (8.3.3), taking the natural logarithm, and equating the result to the logarithm of the ratio of the corresponding right-hand sides gives the following:

$$\ln\left[\frac{f_A^v(T_b,\ P)}{f_A^v(T,\ P)}\right] = \ln\left[\frac{f_A^L(T_b,\ P)}{f_A^L(T,\ P)}\right] - \ln x_A \qquad (8.3.5)$$

For a pure material,

$$\left(\frac{d\ln f}{dT}\right)_P = \frac{h^0 - h}{RT^2} \qquad (8.3.6)$$

where h is the specific enthalpy at temperature T and pressure P, and h^0 is the same quantity at temperature T but at a low enough pressure that the material behaves as an ideal gas.

Integrating Eq. (8.3.6) from temperature T_b to temperature T at constant pressure and noting that $T_b \approx T$ and, therefore, $TT_b \approx T_b^2$,

$$\ln\left[\frac{f(T,\ P)}{f(T_b,\ P)}\right] = \frac{h^0 - h}{R}\left[\frac{1}{T_b} - \frac{1}{T}\right] = \frac{h^0 - h}{RT_b^2}(T - T_b) \qquad (8.3.7)$$

Applying Eq. (8.3.7) to pure A in the vapor phase and then to pure A in the liquid phase and introducing the results in Eq. (8.3.5) gives

$$-\left[\frac{h^0 - h^v}{RT_b^2}\right]\Delta T_b = -\left[\frac{h^0 - h^L}{RT_b^2}\right]\Delta T_b - \ln x_A \qquad (8.3.8)$$

where ΔT_b equals $T - T_b$, the elevation in boiling point.

Because $\ln x_A$ equals $\ln(1 - x_B)$, which for small x_B is the same as $-x_B$, we see the following:

$$\Delta T_b = \frac{(RT_b^2 x_B)}{\Delta h_v} \qquad (8.3.9)$$

in which Δh_v equals $h^v - h^L$, the molar latent heat of vaporization of pure solvent A. From the definition of the mole fraction,

$$\begin{aligned} x_B &= \frac{\text{Moles of B}}{\text{Total moles in mixture}} \\ &= \frac{\text{Mass of B}}{(\text{Mol. wt. of B})(\text{Total moles})} \frac{\text{Mixture volume}}{\text{Mixture volume}} \\ &\approx \frac{cv_A}{\bar{M}_n} \end{aligned} \qquad (8.3.10)$$

where c is the mass concentration of B, $\bar{M}_n$ is the number-average molecular weight of B, and v_A is the molar volume of the solvent. Finally, we can derive

$$\frac{\Delta T_b}{c} = \frac{RT_b^2 v_A}{\Delta h_v \bar{M}_n} \qquad (8.3.11)$$

which allows us to compute $\bar{M}_n$ from a measurement of ΔT_b. Note that one typically extrapolates the left-hand side to $c = 0$ in order to ensure ideal solution behavior. This technique of molecular-weight measurement is also known as *ebulliometry*.

If we consider the situation depicted in Figure 8.2 instead of that shown in Figure 8.1, then an analysis similar to the one carried out earlier leads to an expression for the depression in freezing point, which is identical to Eq. (8.3.11) except that ΔT is now $T_f - T$, where T_f and T are the freezing points of the pure solvent and the solution, respectively. Also, Δh becomes the molar latent heat of fusion of the pure solvent, and T_b is replaced by T_f. This measurement is known as *cryoscopy*.

For an ideal solution, the vapor pressure p_A of the solvent in solution is given by Raoult's law as follows [8]:

$$p_A = x_A P_A \qquad (8.3.12)$$

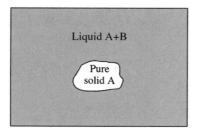

FIGURE 8.2   Pure solid solvent in equilibrium with a polymer solution.

where P_A is the vapor pressure of the pure solvent at temperature T. This lowering in vapor pressure is utilized for the measurement of molecular weight in the technique known as *vapor-pressure osmometry*.

Figure 8.3 shows a schematic diagram of a vapor-pressure osmometer. Two thermistor probes are positioned inside a temperature-controlled cell that is saturated with solvent vapor. If syringes are used to introduce drops of pure solvent on the thermistor probes, then at thermal equilibrium, the temperature of the two probes is the same and an equal amount of solvent vaporizes and condenses at each probe. If, however, one of the solvent drops is replaced by a drop of solution, there is an initial imbalance in the amount of solvent condensing and vaporizing at that probe. Because of the lowering in vapor pressure, less solvent vaporizes than condenses, which leads to a rise in temperature due to the additional heat of vaporization. When equilibrium is reestablished, the temperature T at this probe is higher than the temperature T_S at the other probe which is in contact with the drop of pure solvent. Under these conditions, the vapor pressure of pure solvent at temperature T_S equals the vapor pressure of the solvent in solution at temperature T, and the situation is analogous to ebulliometry. Therefore, we can use Eq. (8.3.11) again if we define ΔT as $T - T_S$. The temperature difference itself is measured as a difference in electrical resistance by

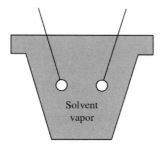

FIGURE 8.3   Schematic drawing of a vapor-pressure osmometer.

making the two thermistors be the two arms of a Wheatstone bridge. Commercial instruments can, at best, measure ΔT down to about 5×10^{-5}°C. Because of heat losses and solution nonidealities, the measured ΔT does not equal the value calculated based on Eq. (8.3.11). It is necessary to calibrate the instrument using a material of known molecular weight. The range of commercial vapor pressure osmometers is from 40 to 50,000 g/mol, with the lower limit being set by solute volatility [6].

For polymer molecular weights of 100,000 and greater, the temperature differences predicted by Eq. (8.3.11) for a dilute polymer solution in a typical organic solvent are about 10^{-5}°C (see Table 8.1). Such small changes in temperature are very difficult to measure with any degree of precision. Consequently, when working with high-molecular-weight polymers, we turn to other techniques of molecular-weight determination, especially osmotic pressure.

When a polymer solution is separated from the pure solvent by a semipermeable membrane that allows passage of the solvent but not the solute, then (as shown in Fig. 8.4) the tendency to equalize concentrations results in flux of the solvent across the membrane and into the solution. As mass transfer proceeds, a pressure head builds up on the solution side, tending to slow down and ultimately stop the flow of solvent through the membrane. At equilibrium, the liquid levels in the two compartments differ by h units; the difference in pressure π is known as the osmotic pressure of the solution. Note that if additional pressure is applied to the solution, solvent can be made to flow back to the solvent side from the solution side; this is known as *reverse osmosis*. As the following analysis demonstrates, osmotic pressure can be employed to measure the number-average molecular weight of a polymeric solute.

If we designate solvent properties by subscript 1 and solute properties by subscript 2, then the following relations hold at thermodynamic equilibrium, using the condition of phase equilibrium:

$$f_1(T, P) = \hat{f}_1(T, P + \pi, x_1) \approx x_1 f_1(T, P + \pi) \qquad (8.3.13)$$

TABLE 8.1 Colligative Properties of Polystyrene-in-Toluene Solutions at a Mass Concentration of 0.01 g/cm³

$\bar{M}_n$	ΔT_b (K)	ΔT_f (K)	π (cm of solvent at 25°C)
50,000	7.8×10^{-4}	8.5×10^{-4}	5.8
100,000	3.9×10^{-4}	4.25×10^{-4}	2.9
500,000	7.8×10^{-5}	8.5×10^{-5}	0.58
5,000,000	7.8×10^{-6}	8.5×10^{-6}	0.058

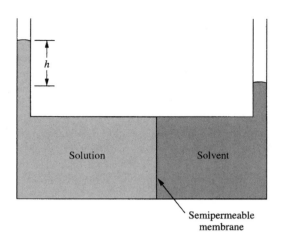

FIGURE 8.4   Osmosis through a semipermeable membrane.

where the second equality follows from the Lewis and Randall rule. Consequently,

$$\ln\left(\frac{f_1(T, P)}{f_1(T, P + \pi)}\right) = \ln x_1 \tag{8.3.14}$$

For a pure material, however, laws of thermodynamics give

$$\left(\frac{d \ln f}{dP}\right)_T = \frac{v_1}{RT} \tag{8.3.15}$$

where v_1 is the molar volume of the solvent. Integrating Eq. (8.3.15) between P and $P + \pi$ at constant temperature yields

$$\ln\left[\frac{f(T, P + \pi)}{f(T, P)}\right] = \frac{v_1 \pi}{RT} \tag{8.3.16}$$

Comparing Eqs. (8.3.14) and (8.3.16) reveals

$$-\ln x_1 = \frac{v_1 \pi}{RT} \tag{8.3.17}$$

Because it is possible to write the left-hand side as x_2 for dilute solutions, a further use of Eq. (8.3.10) converts Eq. (8.3.17) to

$$\frac{\pi}{c} = \frac{RT}{\bar{M}_n} \tag{8.3.18}$$

where c is the mass concentration of the solute. Again, we typically extrapolate π/c to $c = 0$ to ensure that ideal solution behavior is obtained and Eq. (8.3.18) holds. Expected values of the osmotic pressure for dilute solutions of polystyrene in toluene are listed in Table 8.1.

A typical plot of experimental data for aqueous solutions of polyethylene oxide at 20°C is shown in Figure 8.5 [9]; these data are extremely easy to obtain even though 2 days are required for equilibrium to be reached. It is seen that the plot has a nonzero slope, and significant error can occur if extrapolation to zero concentration is not carried out. This nonzero slope can be theoretically predicted if real solution theory is used instead of assuming ideal solution behavior. For instance, if we employ the Flory–Huggins theory (considered in detail in Chapter 9) and equate the fugacities (or, equivalently, the chemical potentials) of the solvent on both sides of the membrane, the use of Eq. (9.3.30) along with the known dependence of the chemical potential on temperature leads to the following result (see Chapter 9):

$$\pi = -\frac{RT}{v_1}\left[\ln \phi_1 + \phi_2\left(1 - \frac{1}{m}\right) + \chi_1 \phi_2^2\right] \tag{8.3.19}$$

where ϕ_1 and ϕ_2 are the volume fractions of the two components, respectively, m is the ratio of the molar volume of the solute to the molar volume of the solvent, and χ_1 is the *interaction parameter*.

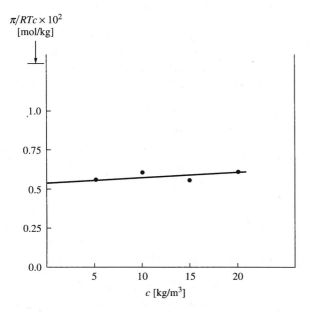

FIGURE 8.5 Osmotic pressure of aqueous polyethylene oxide solutions at 20°C. (From Ref. 9.)

On expanding $\ln \phi_1$ in a Taylor series about $\phi_2 = 0$ and noting that the mass concentration c of the polymer equals $\bar{M}_n \phi_2 / m v_1$, Eq. (8.3.19) becomes

$$\frac{\pi}{c} = \frac{RT}{\bar{M}_n}\left[1 + \left(\frac{1}{2} - \chi_1\right)m\phi_2 + \frac{m}{3}\phi_2^2 + \cdots\right] \tag{8.3.20}$$

which can be written in terms of the polymer density ρ because ϕ_2 equals c/ρ and m equals $\bar{M}_n / \rho v_1$. Finally,

$$\frac{\pi}{c} = \frac{RT}{\bar{M}_n} + \frac{RT(\frac{1}{2} - \chi_1)c}{\rho^2 v_1} + \frac{RTc^2}{3\rho^3 v_1} + \cdots \tag{8.3.21}$$

and it is seen that Eq. (8.3.18) is obtained by letting c tend to zero in Eq. (8.3.21). As discussed in Chapter 9, the latter equation can be used to estimate the interaction parameter if the polymer number-average molecular weight is known.

Because the Flory–Huggins theory is not strictly valid at low polymer concentrations, it is common practice to rewrite Eq. (8.3.21) in the form of a virial equation (as is done in thermodynamics):

$$\frac{\pi}{c} = RT\left[\frac{1}{\bar{M}_n} + A_2 c + A_3 c^2 + \cdots\right] \tag{8.3.22}$$

in which A_2 and A_3 are known as the second and third virial coefficients, respectively.

Commercial membrane osmometers are designed to reduce the time of measurement from a few hours to a few minutes. This is done by reducing the cell volume and increasing the membrane surface area. In addition, pressure transducers are used to detect solvent flow, and external pressure is applied to the solution to help achieve rapid equilibrium. Molecular weights between 10^3 and 10^6 can be measured at temperatures exceeding $100°C$. The lower limit on the molecular weight is set by solute permeability, whereas the upper limit is governed by the sensitivity and accuracy of the pressure measuring system. Additional details may be found in the literature [6,10].

8.4 LIGHT SCATTERING

A beam of light is a transverse wave made up of sinusoidally varying electric and magnetic field vectors that are perpendicular to each other and also to the direction of propagation of the wave. Such a wave contains energy that is measured in terms of the wave intensity I, defined as the power transmitted per unit area perpendicular to the direction in which the wave is traveling. Using the principles of physics [11], it is easy to show that the average intensity or the power averaged over one cycle is proportional to the square of the wave amplitude. When such a beam travels through a polymer solution, it can go through unaltered, but, more commonly, it is either absorbed or scattered. Absorption occurs only if the wave frequency is such that the energy of radiation exactly equals the energy gap between, say, the electronic or vibrational energy levels of the molecules making up the liquid medium; this phenomenon is the basis of methods such as infrared and nuclear magnetic resonance spectroscopy. Scattering, on the other hand, involves attenuation of the incident beam with simultaneous emission of radiation in all directions by the scattering molecules due to the presence of induced instantaneous dipole moments. In this case, the energy of the incident beam equals the sum of the energies of the transmitted beam and all of the scattered beams. Here, scattered light has the same frequency as the incident light, and the process is called *elastic light scattering*. Sometimes, though, scattered light has a different frequency, which is called *inelastic* or *Raman scattering*. From an observation of the time-averaged intensity of elastically scattered light (called *static light scattering*), we can get information about the weight-average molecular weight, the second virial coefficient, and the size or radius of gyration (root-mean-square distance of chain elements from the center of gravity of the molecule) of macromolecules. Instantaneous scattering intensity or dynamic light scattering reveals, in addition, the translational diffusion coefficient [12]. Note that, in recent times, dynamic light scattering has also been applied to, among other things, studies of bulk polymers, micelles, microemulsions, and polymer gels [13]. Here, however, we consider the application of static light scattering to the determination of polymer molecular weight under conditions where absorption effects are not important.

The theory of light scattering was developed many years ago; a review [14], excellent books [15,16], and an elementary treatment [17] on the subject are available. The essential features of elastic light scattering can be understood with reference to Figure 8.6, which shows an unpolarized beam of light of intensity I_0 and wavelength λ passing through a cylindrical sample cell of unit volume. The intensity I_θ of the scattered beam is measured at a distance r from the cell and at an angle θ with the direction of the transmitted beam. If the cell contains N noninteracting, identical particles of an ideal gas (polymer solutions are consid-

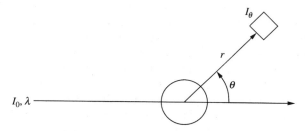

FIGURE 8.6 Schematic diagram of static light scattering.

ered later) and if the size of the scattering particles is small compared to the wavelength of the incident light, then we have the following [15]:

$$\frac{I_\theta}{I_0} = \frac{2\pi^2(1+\cos^2\theta)(dn/dc)^2 Mc}{N_A \lambda^4 r^2} \tag{8.4.1}$$

in which n is the refractive index of the gas, c is the mass concentration, N_A is Avogadro's number, and M is the molecular weight of the particles. According to this equation, which is known as the Rayleigh equation, if N is fixed, the scattering intensity is proportional to the square of the molecular weight because c equals NM/N_A. Thus, if there were a mixture of two kinds of particles, with one kind being much larger than the other, the contribution of the larger particles to the scattered light intensity would be the dominant one. This fact is used to great advantage in determining the molecular weight of polymeric solutes in solution. In this situation, for light scattering from an ideal polymer solution, Eq. (8.4.1) is modified to read as follows [15]:

$$\frac{I_\theta}{I_0} = \frac{2\pi^2(1+\cos^2\theta)n_0^2(dn/dc)^2 c}{N_A \lambda^4 r^2 / M} \tag{8.4.2}$$

in which n_0 is the refractive index of the solvent and n now becomes the refractive index of the solution, whereas M is the molecular weight of the polymer and c its mass concentration.

Equation (8.4.2) is valid only at infinite dilution. For finite concentrations, the use of a virial expansion of the type introduced in Eq. (8.3.22) leads to

$$\frac{I_\theta}{I_0} = \frac{2\pi^2(1+\cos^2\theta)n_0^2(dn/dc)^2 c}{N_A \lambda^4 r^2 (1/M + 2A_2 c + 3A_3 c^2 + \cdots)} \tag{8.4.3}$$

and Eq. (8.4.3) properly reduces to Eq. (8.4.2) when c tends to zero.

For most polymer molecules, the limitation that the particle size be much smaller than the wavelength of light, which, in practice, means that all molecular dimensions should be less than $\lambda/20$, is too restrictive. When the particle size becomes comparable to the wavelength of the incident beam, scattering occurs from different parts of the same molecule, resulting in interference due to phase differences. This tends to progressively reduce I_θ as θ increases. The result can be seen in Table 8.2, which lists data for polystyrene-in-toluene solutions. However, because Eq. (8.4.3) still holds when θ is zero, we can either make measurements at different θ values and extrapolate data to θ equal to zero, or make measurements at very small values of θ. The latter situation is practical because the use of lasers as light sources allows us to conduct experiments at θ values as small as 4°. Note that light scattering at nonzero θ values depends on the geometric shape of the scattering particle; it is from the deviation of the data from the predictions of Eq. (8.4.3) that we estimate the radius of gyration of the polymer molecule.

TABLE 8.2 Light-Scattering Data, I_θ, in Arbitrary Units for Solutions of Polystyrene in Toluene at 20°C

Scattering angle, θ (deg)	Concentration (g/cm³)			
	0.0002	0.0004	0.001	0.002
25.8	3.49	5.82	7.86	8.65
36.9	2.98	4.88	7.38	8.02
53.0	2.19	3.82	6.37	7.41
66.4	1.74	3.12	5.58	6.88
90.0	1.22	2.25	4.42	5.95
113.6	0.952	1.80	3.73	5.35
143.1	0.763	1.48	3.15	4.79

Source: Ref. 18.

It is common practice to define a quantity R_θ (called the *Rayleigh ratio*) as follows:

$$R_\theta = \frac{I_\theta r^2}{I_0(1 + \cos^2 \theta)} \tag{8.4.4}$$

so that Eq. (8.4.3) takes a form similar to Eq. (8.3.22):

$$\frac{K_c}{R_\theta} = \frac{1}{M} + 2A_2 c + 3A_3 c^2 + \cdots \tag{8.4.5}$$

where K is a optical constant given by

$$K = \frac{2\pi^2 n_0^2 (dn/dc)^2}{N_A \lambda^4} \tag{8.4.6}$$

and n_0 and dn/dc are usually measured using a refractometer and a differential refractometer, respectively [6].

On measuring R_θ as a function of concentration at a low θ value, K_c/R_θ is plotted versus c. The intercept of such a plot represents extrapolation to zero concentration, and from Eq. (8.4.5), we have the following:

$$\lim_{c \to 0} R_\theta = MKc \tag{8.4.7}$$

which yields the polymer molecular weight. Also, the slope of the plot allows us to compute the second virial coefficient. If measurements are made at several constant temperatures, the temperature value at which A_2 equals zero is the θ temperature.

If the polymer sample is polydisperse, then R_θ can be written as a sum $\sum R_i$ over all of the molecular-weight fractions; so, Eq. (8.4.7) becomes

$$R_\theta = K \sum M_i c_i = K \sum M_i \frac{w_i}{v} \tag{8.4.8}$$

because the mass concentration of each molecular-weight fraction equals the ratio of the respective mass divided by the solution volume. The total mass concentration c, however, is

$$c = \sum \frac{w_i}{v} \tag{8.4.9}$$

so that

$$\frac{R_\theta}{c} = K \frac{\sum M_i w_i}{\sum w_i} = K \bar{M}_w \tag{8.4.10}$$

and it is clear that, despite the similarities between Eqs. (8.3.22) and (8.4.5), light scattering yields the weight-average molecular weight.

The usual range of molecular weights that can be measured by light scattering is from a few thousand to a few million. A photomultiplier is used as a detector; measurements can be made in either aqueous or organic solvents and can be made in the presence of salts or buffers. However, care must be taken to exclude dust particles, which can influence the results and introduce errors. Light-scattering measurements are typically time-consuming, and the equipment is significantly more expensive than that needed for colligative property measurement. Nonetheless, light scattering is a powerful technique, especially when coupled with other techniques such as gel permeation chromatography; the combination of the two can give the complete molecular-weight distribution. Additional applications of classical light scattering have been discussed in the literature [19].

8.5 ULTRACENTRIFUGATION

In order to understand the theory of the ultracentrifuge, let us first consider an analogous situation, that of a single sphere settling under gravity in a long tube filled with a Newtonian liquid, as shown in Figure 8.7. If the sphere of mass m and volume V is dropped from a state of rest, it initially accelerates, but soon reaches a constant velocity, known as the *terminal velocity*, at which point the vector sum of all the forces acting on the sphere is exactly zero. As long as the tube radius is large compared to the sphere radius, the forces that act on the

FIGURE **8.7** Settling of a sphere in a Newtonian liquid.

sphere are gravity, buoyancy, and the viscous drag of the liquid F_d tending to slow down the sphere. At equilibrium, therefore,

$$\rho_s Vg - \rho Vg - F_d = 0 \qquad (8.5.1)$$

where ρ_s is the density of the sphere, ρ is the density of the liquid, and g is the acceleration due to gravity.

As shown later in Section 13.4 of Chapter 13, the drag force on an isolated sphere can be written in terms of the Stokes–Einstein equation as

$$F_d = \frac{kTv}{D} \qquad (8.5.2)$$

in which k is Boltzmann's constant, T is the absolute temperature, v is the terminal velocity of the sphere, and D is the sphere diffusion coefficient. Introducing Eq. (8.5.2) into Eq. (8.5.1), solving for the sphere volume, and multiplying the result by the sphere density gives the following:

$$m = \frac{kTv}{Dg[1 - (\rho/\rho_s)]} \qquad (8.5.3)$$

On multiplying both sides of Eq. (8.5.3) by Avogadro's number, the molecular weight M of the sphere can be derived as

$$M = \frac{RTv}{Dg[1 - (\rho/\rho_s)]} \qquad (8.5.4)$$

where R is the universal gas constant. It can be seen that a measurement of the terminal velocity makes it possible to compute the molecular weight if the other quantities in Eq. (8.5.4) are known.

If, instead of a single particle, a large number of particles are dropped into the tube, then, in the absence of particle–particle interactions, the mass flux of spheres at any cross section is given by

$$\text{Flux} = vc \tag{8.5.5}$$

where c is the local mass concentration of spheres.

As time proceeds, spheres build up at the bottom of the tube, and the tendency to equalize concentrations causes a diffusive flux of spheres upward in the tube. The magnitude of the flux is given by Fick's law (see Chapter 13) as follows:

$$\text{Flux} = D\,\frac{dc}{dz} \tag{8.5.6}$$

where D is the same diffusion coefficient appearing in Eq. (8.5.2) and z is the distance measured along the tube axis. For a steady state to be reached in the sphere concentration, the fluxes given by Eqs. (8.5.5) and (8.5.6) have to be equal in magnitude. Equating these two quantities and replacing the terminal velocity by an expression obtained with the help of Eq. (8.5.4) gives

$$\frac{dc}{dz} = \frac{cMg}{RT}\left[1 - \frac{\rho}{\rho_s}\right] \tag{8.5.7}$$

Separating the variables and integrating the result yields

$$\ln c = \frac{Mg}{RT}\left(1 - \frac{\rho}{\rho_s}\right)z + \text{constant} \tag{8.5.8}$$

and the slope of the straight-line plot of $\ln c$ versus z again allows for the determination of the molecular weight. The advantage of using Eq. (8.5.8) instead of Eq. (8.5.4) is that the value of the diffusion coefficient is not needed.

If we try to apply the foregoing theoretical treatment to the determination of polymer molecular weight from the sedimentation of a dilute polymer solution, we discover that, in practice, polymer molecules do not settle. This is the case because the equivalent sphere radii are so small that colloidal forces [not accounted for in Eq. (8.5.1)] predominate over gravitational forces and keep the polymer molecules from settling. However, the situation is not irredeemable. If the polymer solution is placed in a horizontal, pie-shaped cell and the cell is rotated at a large angular velocity ω about a vertical axis as shown in Figure 8.8, the centrifugal force that develops can exceed the force of gravity by a factor of a few hundred thousand. Indeed, the centrifugal force can and does cause sedimentation of polymer molecules in the direction of increasing r. Because

Axis of
rotation

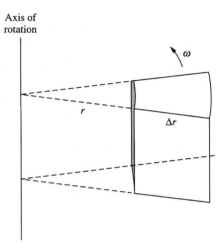

FIGURE 8.8 Schematic diagram of an ultracentrifuge.

the centrifugal acceleration equals $\omega^2 r$, we can replace g by this quantity in Eqs. (8.5.4) and (8.5.7). Also, v and z are replaced by dr/dt and r, respectively, where r is the radial distance from the axis of rotation. With these modifications, Eq. (8.5.4) is known as the Svedberg equation, and the quantity $(dr/dt)/(\omega^2 r)$ is known as the *sedimentation coefficient*, S. Also, Eq. (8.5.8) becomes

$$\ln c = \frac{M\omega^2}{2RT}\left[1 - \frac{\rho}{\rho_s}\right]r^2 + \text{constant} \tag{8.5.9}$$

and it is now necessary to plot $\ln c$ versus r^2.

The equipment used to measure polymer molecular weight according to either the Svedberg equation or Eq. (8.5.9) is an ultracentrifuge, which can rotate a horizontal cell in an evacuated chamber at tens of thousands of revolutions per minute (rpm). These high rotational speeds are needed to give rise to measurable sedimentation velocities. The sedimentation velocity itself is measured using either ultraviolet (UV) absorption or with the help of Schlieren optics [20]; in the latter technique, it is possible to determine the location of a change in concentration by measuring the refractive index gradient as a function of position. Much lower rotational speeds, of the order of 10,000 rpm, are needed to achieve sedimentation equilibrium. However, the time needed to attain equilibrium can easily be a couple of days. Molecular weights can be measured over a wide range up to about 40×10^6.

These techniques work well provided that the polymer solution is such that the assumptions made in deriving either the Svedberg equation or Eq. (8.5.9) remain valid. This is possible for biological molecules (such as proteins and nucleic acids) that act like relatively compact and rigid spheres in solution. They are also monodisperse. Even so, solvation effects make the effective sphere density lower than the polymer density, and the large pressure developed due to the centrifugal force alters both the solvent viscosity and density. In addition, it is necessary to extrapolate data to infinite dilution; otherwise, the drag force becomes concentration dependent due to polymer–polymer interactions.

The ultracentrifuge is rarely used to measure the molecular weight of synthetic polymers, because these are permeable to the solvent and their size changes easily depending on the process conditions. Furthermore, polydispersity introduces both theoretical and experimental difficulties. During sedimentation, for example, we no longer observe sharp concentration boundaries. Also, a plot of $\ln c$ versus r^2 is a curve rather than the straight line expected on the basis of Eq. (8.5.9). In principle, though, data can be analyzed to yield the weight-average molecular weight. These and other details are available elsewhere [6,15,20].

Example 8.2: Sedimentation data on aqueous solutions of hydroxypropyl guar, a biopolymer, have been reported in the literature [21]. At low concentrations and a rotational speed of 40,000 rpm, the sedimentation coefficient is 5.4×10^{-13} sec at 20°C. If the measured diffusion coefficient is 0.32×10^{-7} cm^2/sec and $1 - \rho/\rho_s$ equals 0.377, what is the molecular weight?

Solution: When applied to the ultracentrifuge, Eq. (8.5.4) is

$$M = \frac{RTS}{D[1 - (\rho/\rho_s)]}$$

where S and D are evaluated in the limit of infinite dilution. Inserting numbers yields the following:

$$M = \frac{8.314 \times 10^7 \text{ (erg/mol K)} \times 293 \text{ (K)} \times 5.4 \times 10^{-13} \text{ (sec)}}{0.32 \times 10^{-7} \text{ (cm}^2/\text{sec)} \times 0.377}$$
$$= 1.09 \times 10^6 \text{ g/mol}$$

8.6 INTRINSIC VISCOSITY

It is an experimental fact that the viscosity of a polymer solution is generally much larger than that of the solvent alone even at low polymer concentrations, and it increases with increasing molecular weight at a fixed mass concentration. The measurement of solution viscosity can, therefore, be used to estimate polymer molecular weights. Indeed, a large number of sophisticated viscometers exist for the accurate measurement of solution viscosity and its variation with concentration, shear rate, and temperature. Details of these instruments and methods of data analysis are discussed at length in Chapter 14. For molecular-weight measurements, however, it is customary to employ dilute polymer solutions and to use glass capillary viscometers of the type illustrated in Figure 8.9. This particular instrument is known as a *suspended level Ubbelhode viscometer*. In use, the bulb A is filled with a solution of known concentration. A volume V of this solution is then transferred to completely fill bulb C between marks E and F by closing arm N and applying a pressure down arm L. On simultaneously opening N and releasing the pressure in L, excess liquid drains back into A, leaving bulb C filled. At this stage, the pressure at point B at the bottom end of the capillary is atmospheric. Further draining of liquid out of bulb C is prevented by closing arm M, and the viscometer is transferred to a thermostatted bath. Once thermal equilibrium is reached, the polymer solution is allowed to flow under gravity through the capillary, and the time taken for the liquid level to move from mark F to mark E is recorded. The process is then repeated for the pure solvent and also for the polymer solution at different concentrations. In each instance, the efflux time is noted.

If the radius of the capillary is R, its length is L, and the viscosity of the solution (assumed Newtonian) is η, then according to the well-known Hagen–Poiseuille equation [22], the volumetric flow rate Q through the capillary is given by

$$Q = \frac{\pi R^4 \Delta p}{8\eta L} \tag{8.6.1}$$

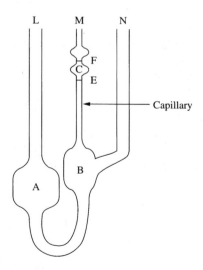

FIGURE 8.9 An Ubbelhode viscometer.

where Δp is the difference in dynamic pressure across the capillary and represents the combined effect of static pressure and the gravitational force.

The efflux time t is given by

$$t = \frac{V}{Q} \qquad (8.6.2)$$

and the ratio of the efflux time t of the solution to t_s corresponding to that of the solvent (because Δp is essentially the same in both cases if we neglect the minor difference between the densities of the solution and the solvent) is

$$\frac{t}{t_s} = \frac{\eta}{\eta_s} \qquad (8.6.3)$$

so that we may calculate η knowing η_s, provided that non-Newtonian effects such as shear thinning can be neglected. Therefore, we want to work with zero-shear-rate viscosities. For this and other reasons explained later, it is essential to use low polymer concentrations or to extrapolate data to infinite dilution. Note that the ratio η/η_s is generally known as the *relative viscosity* or the *viscosity ratio* and denoted η_R.

Relating measured viscosity to molecular weight is generally done by first relating viscosity to some measure of the size of polymer molecules in solution and then relating the size to the molecular weight. The process begins by appealing to the behavior of a dilute suspension of spheres in a Newtonian liquid for which the relative viscosity is given by the Einstein result [23]:

$$\eta_R = 1 + 2.5\phi \qquad (8.6.4)$$

wherein ϕ is the volume fraction of spheres.

If we consider each polymer molecule in dilute solution to be an isolated random coil of spherical shape and volume v_e, we may apply Eq. (8.6.4) with η_R given by Eq. (8.6.3) and the volume fraction of polymer by

$$\phi = \frac{n_2 v_e}{V} \qquad (8.6.5)$$

where n_2 is the number of polymer molecules in a solution of volume V.

Multiplying and dividing the right-hand side of Eq. (8.6.5) by $M N_A$, the product of the polymer molecular weight and Avogadro's number, yields

$$\phi = \frac{v_e c N_A}{M} \qquad (8.6.6)$$

in which c is the mass concentration. Introducing this result into Eq. (8.6.4) and rearranging gives

$$\eta_R - 1 = \frac{2.5 v_e c N_A}{M} \qquad (8.6.7)$$

where the left-hand side is known as the *specific viscosity*, η_{sp}. The ratio of the specific viscosity to the mass concentration of polymer is called the *reduced viscosity* or the *viscosity number*. Thus,

$$\frac{\eta_{sp}}{c} = \frac{2.5v_e N_A}{M} \tag{8.6.8}$$

The viscosity number ought to be independent of polymer concentration. However, the Einstein equation is valid only for noninteracting spheres; this situation prevails as the concentration tends to zero. Consequently, we can extrapolate data to infinite dilution, and the result is known as the *intrinsic viscosity* or *limiting viscosity number* $[\eta]$. In the past, this quantity was measured in units of deciliters per gram; recent practice has been to use milliliters per gram. Data can generally be represented in terms of the Huggins equation,

$$\frac{\eta_{sp}}{c} = [\eta] + k[\eta]^2 c \tag{8.6.9}$$

where k is known as the Huggins constant. Alternatively, the Kraemer equation can be used:

$$\frac{\ln \eta_R}{c} = [\eta] + k'[\eta]^2 c \tag{8.6.10}$$

in which the left-hand side is known as the *inherent viscosity* or the *logarithmic viscosity number*. That both of these methods of data representation yield the same value of the intrinsic viscosity is demonstrated in Figure 8.10 using data on solutions of nylon 66 in formic acid [24]. If the radius of each polymeric sphere is taken to be proportional to the root-mean-square radius of gyration $\langle s^2 \rangle^{1/2}$, then, using Eq. (8.6.8), we can derive

$$\lim_{c \to 0} \frac{\eta_{sp}}{c} = [\eta] \alpha \frac{\langle s^2 \rangle^{3/2}}{M} = \left[\frac{\langle s^2 \rangle}{M}\right]^{3/2} M^{1/2} \tag{8.6.11}$$

If a linear polymer molecule is represented as a freely jointed chain having n links each of length l, then, as shown in Chapter 10,

$$\langle s^2 \rangle \propto l^2 n \tag{8.6.12}$$

provided that there are no excluded volume effects that are long-range interactions due to attraction and repulsion forces between widely separated chain segments or between polymer segments and solvent molecules. This happens at what is known as the *theta condition*.

Because the polymer molecular weight is also proportional to n, the ratio $\langle s^2 \rangle / M$ must be independent of chain length or molecular weight. As a consequence,

$$[\eta] \propto M^{1/2} \tag{8.6.13}$$

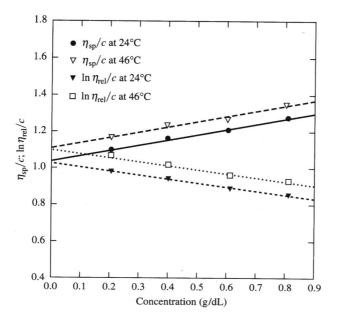

FIGURE 8.10   Reduced viscosity and inherent viscosity of nylon 66 in 90% formic acid. (From Ref. 24.)

and a plot of $\log[\eta]$ as a function of $\log M$ should be a straight line of slope 0.5.

In general, theta conditions do not occur, and polymer–solvent interactions lead to coil expansion with the following result [25]:

$$\langle s^2 \rangle = \alpha^2 \langle s^2 \rangle_0 \tag{8.6.14}$$

where α is a linear coil expansion factor that depends on n, and the subscript 0 denotes theta conditions. Because of this complication, it is difficult to modify Eq. (8.6.13) theoretically. However, by experiment, we can find that

$$[\eta] = KM^a \tag{8.6.15}$$

which is known as the Mark–Houwink equation. The values of the constants K and a are determined experimentally using monodisperse polymer fractions, and these may be found in standard handbooks [26]. The value of the exponent a typically varies from 0.5 at the theta temperature to 0.8 for good solvents, although values exceeding unity have been measured for extended chain polymers [27]. Once the intrinsic viscosity has been related to the molecular weight, we can also use the experimental data to relate the relative viscosity at a fixed concentration to the molecular weight. Correlations of this kind are often used in industrial practice [28] and they may lead to a significant reduction in experimental work.

Thus far, we have assumed that polymer molecules are monodisperse. For a polydisperse sample, we introduce the Mark–Houwink equation into Eq. (8.6.7) and write

$$\eta_{sp} = K \sum c_i M_i^a \qquad (8.6.16)$$

Dividing both sides of this equation by the mass concentration of polymer and noting that

$$c = \sum c_i = \sum \frac{n_i M_i}{N_A V} \qquad (8.6.17)$$

where n_i is the number of molecules of molecular weight M_i, N_A is Avogadro's number, and V is the sample volume, we find the following:

$$\frac{\eta_{sp}}{c} = \frac{K \sum n_i M_i^{a+1}}{\sum n_i M_i} \qquad (8.6.18)$$

Accordingly, it is clear that the use of the Mark–Houwink equation gives an average molecular weight $\bar{M}_v$ known as the viscosity-average molecular weight and defined by

$$\bar{M}_v = \left[\frac{\sum n_i M_i^{a+1}}{\sum n_i M_i} \right]^{1/a} \qquad (8.6.19)$$

which is generally intermediate between $\bar{M}_n$ and $\bar{M}_w$ but much closer to the latter quantity.

Despite the fact that intrinsic viscosity measurements result in neither the mass-average nor the weight-average molecular weight, the viscosity-average molecular weight is a very commonly encountered quantity due to the ease of measurement and also the simplicity and low cost of the viscometer. Regarding the experimental determination of intrinsic viscosity, it is preferable to extrapolate data to low concentrations rather than attempt measurements on very dilute solutions. This is because errors associated with the dilution process become so important that the overall result becomes worse following each dilution.

Example 8.3: If the Mark–Houwink exponent for nylon 66 dissolved in 90% formic acid is 0.72, calculate $\bar{M}_n$, $\bar{M}_w$, and $\bar{M}_v$ for a sample that contains 50% by weight of a fraction having molecular weight 10,000 and 50% by weight of a fraction of molecular weight 20,000.

Solution:

$$\bar{M}_n = \frac{1}{0.5/10{,}000 + 0.5/20{,}000} = 13{,}333$$

$$\bar{M}_w = (10{,}000 \times 0.5) + (20{,}000 \times 0.5) = 15{,}000$$

$$\bar{M}_v = \left[\frac{0.5}{10{,}000}(10{,}000)^{1.72} + \frac{0.5}{20{,}000}(20{,}000)^{1.72}\right]^{1.39} = 14{,}876$$

In closing this section, we note that, from Eq. (8.6.8), $[\eta]M/N_A$ is the volume of a polymer molecule multiplied by 2.5, whereas cN_A/M is obviously the number of polymer molecules per unit volume. As a consequence, $[\eta]c$, which is the product of these two quantities, represents the volume fraction of polymer multiplied by 2.5. If this number is small compared to unity, the polymer solution is considered to be dilute; if it is of the order of unity, the solution is considered moderately concentrated with a near certainty of intermolecular interactions.

8.7 GEL PERMEATION CHROMATOGRAPHY

The simplest conceptual method of determining the molecular weight distribution of a polymer sample is to separate the polydisperse sample into its constituent fractions and then measure the molecular weight of each fraction using any of the techniques discussed so far. This is exactly what used to be done until the commercialization in the mid-1960s of the procedure known as *gel permeation chromatography* (GPC) or *size-exclusion chromatography*. In the old method, a polymer in solution was fractionated either by the sequential addition of nonsolvents or by the progressive lowering of temperature (see Chapter 9 for the theory of polymer–polymer phase equilibrium). However, this was a very tedious and time-consuming process that was obviously ill-suited to routine laboratory procedures. The new method of GPC uses the fact that large polymer molecules are excluded from the small channels in a porous gel, with the result that different molecular weight fractions travel down a column packed with the porous medium at different rates, leading to separation based on size.

A schematic diagram of a GPC setup is shown in Figure 8.11. Solvent is made to flow at a low but constant flow rate of about 1 mL/min through a packed column with the help of a pump. The solvent is typically an organic liquid such as tetrahydrofuran or toluene for room-temperature work and methyl ethyl ketone for high-temperature separations. Temperatures as high as 150°C can be achieved, and these are needed while working with crystalline polymers, which can often only be dissolved in supercritical solvents. The column is maintained at a constant temperature and is packed with beads made from a cross-linked polymer gel that can be swollen by the solvent used to dissolve the polymer being assayed. In the case of organic solvents, the packing is usually cross-linked styrenedivi-

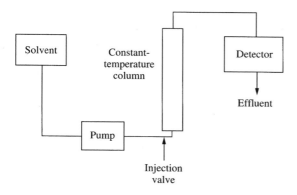

FIGURE 8.11   Schematic diagram of a gel permeation chromatograph.

nylbenzene; for water-soluble polymers, cross-linked methyl methacrylate is used. For analytical work, the column is about 1 cm in diameter and 30–50 cm long. Frequently, the column walls are flexible so that the packing material digs into the side of the wall, which prevents channeling of fluid along the wall.

Once solvent flow has been established, a sample of polymer (less than 1 mL of 0.1–1% solution) is injected upstream of the column. Depending on the pore size in the packing (anywhere from 10^3 to 10^6 Å), molecules above a certain size are completely excluded from the beads and continue flowing with the mobile solvent. By contrast, very small molecules below a critical size are free to enter even the smallest channel and tend to dissolve in the pure solvent that is immobilized there; these, therefore, travel slowly through the column. Molecules between these two size extremes travel at intermediate speeds and emerge from the column at different times, resulting in separation based on molecular weight. The mass concentration of the solute leaving the column is generally detected with the help of a differential refractometer that measures the refractive index difference between the solution and the solvent; a chromatogram might look like the one shown in Figure 8.12. Instead of time, it is common practice to use as the abscissa an equivalent quantity called the *elution volume*, which is the volume of solvent emerging from the column from the instant of sample injection. If the volume of the mobile solvent in the column is V_m and the volume of the stationary solvent is V_s, then, assuming that equilibrium between the two phases is achieved instantly, the elution volumes in Figure 8.11 will range from V_m to $V_m + V_s$. These two extremes correspond to the times taken for the largest and smallest molecules, respectively, to flow out through the column.

Although we know that the polymer sample has been fractionated due to passage through the column, we do not, in general, know the molecular weight corresponding to a particular elution volume; therefore, Figure 8.12 is useful only

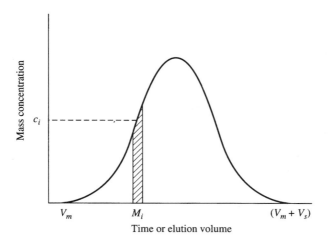

FIGURE 8.12 A typical GPC chromatogram.

for comparative purposes. For quantitative work, we need a separate calibration curve of the type shown in Figure 8.13, which relates the molecular weight to the elution volume (or travel time through the column); sometimes, it is necessary to use more than one column in series if the molecular-weight range of a single column is inadequate for a sample at hand.

The calibration curve is best prepared using monodisperse samples of the same polymer in the same solvent and at the same temperature. Unfortunately,

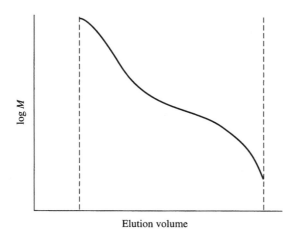

FIGURE 8.13 Characteristic shape of a GPC calibration curve.

extremely few polymers are available in narrow molecular-weight fractions, and it is necessary to examine other possibilities for calibration. Because polymers such as polystyrene, polyethylene oxide, polyethylene glycol, and polyacrylic acid are commercially available over a wide range of narrow molecular weights, we use these as calibration standards. Note that polystyrene and polyethylene are also available as standard reference materials from the National Institute of Standards and Technology. The hydrodynamic size of a polymer molecule in solution, however, depends on both the temperature and the thermodynamic quality of the solvent used. Consequently, a given polymer when dissolved in different solvents or in the same solvent at different temperatures will have a different radius of gyration in each case; the situation is similar for different polymers of the same molecular weight that are dissolved in the same solvent. To get around this problem and construct a universal calibration curve, we can make use of Eq. (8.6.11), according to which the following holds:

$$[\eta]M \propto \langle s^2 \rangle^{3/2} \tag{8.7.1}$$

and because $\langle s^2 \rangle^{3/2}$ is proportional to the volume of a polymer molecule in solution, this equation says that $[\eta]M$ is a surrogate quantity for the hydrodynamic volume. As a result, for a given column and specified set of operating conditions, a plot of $[\eta]M$ as a function of elution volume should be a universal curve independent of polymer type. This is, indeed, found to be the case, and such a curve is shown in Figure 8.14.

A knowledge of the elution volume (taken together with Fig. 8.14) therefore yields the product $[\eta]M$. The molecular weight is, in turn, obtained from a measurement of the intrinsic viscosity. It is for this reason that modern GPCs come equipped with instrumentation for viscosity measurement. Even if we do not actually measure the intrinsic viscosity, we can still estimate the molecular weight of each fraction with the help of Eq. (8.6.15) if the Mark–Houwink constants are known. Thus, if the calibration curve is prepared using polystyrene, we know the product $[\eta]_p M_p$ corresponding to polystyrene. If the intrinsic viscosity of the unknown polymer sample is $[\eta]$, then the molecular weight M of the unknown sample is

$$M = \frac{[\eta]_p M_p}{[\eta]} = \frac{[\eta]_p M_p}{K M^a} \tag{8.7.2}$$

or

$$M = \left[\frac{[\eta]_p M_p}{K} \right]^{1/(1+a)} \tag{8.7.3}$$

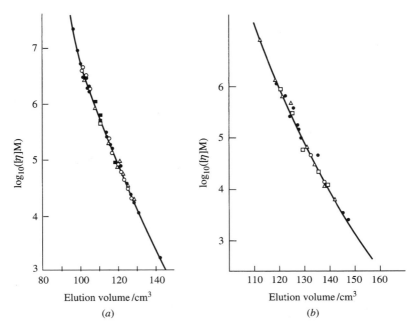

FIGURE 8.14   Examples of the universal calibration curve for GPC. (a) Universal calibration for tetrahydrofuran solution at ambient temperature: ●, linear polystyrene; □, polybutadiene; ○, branched polystyrene; ×, polymethyl methacrylate; △, styrene/ methyl methacrylate copolymer; ■, poly(phenyl siloxane). (b) Universal calibration for various types of polyethylene. Solvent, *o*-dichlorobenzene at 130°C. ●, linear polyethylene fractions; open symbols, branched polyethylene fractions. (From Ref. 6.)

in which all the quantities are known. Of course, the entire calibration problem is avoided if light scattering is employed to determine the absolute molecular weight of each molecular-weight fraction leaving the column.

Using any of these methods, we can easily convert the abscissa of Figure 8.12 to molecular weight. If we now consider a vertical slice as shown in Figure 8.12, the shaded area equals the product of the mass concentration c_i and the volume Δv. This equals the mass w_i of the polymer contained in the fraction of molecular weight M_i. By repeating this process for similar slices, we obtain the entire molecular-weight distribution. Note that the upper limit on the molecular weight for GPC is almost 10^7 for commercial instruments. Note also that these data allow for the determination of the polydispersity index (PDI). As mentioned in earlier chapters, the PDI can be close to unity for anionic polymerization, whereas for step-growth polymerization, it approaches 2 for 100% conversion; for chain-growth polymerization, the PDI can range between 2 and 5.

8.8 CONCLUSION

In this chapter, we have introduced a variety of techniques that can be utilized to measure either a particular molecular-weight average or the entire molecular-weight distribution of a given sample. Although each method has its advantages and disadvantages, the two techniques that are encountered most commonly are intrinsic viscosity and GPC. These are both easy to use and are employed for routine quality control purposes. GPC has proved to be a rapid and precise method of molecular-weight determination, often requiring as little as a half hour per sample. One increasingly frequent industrial application that calls for measurement of molecular-weight distribution is polymer recycling. Mixtures of virgin and reground material rarely have the same molecular-weight distribution or even the same average molecular weight if the recycled material comes from unknown sources. Therefore, it is necessary to ensure that the molecular-weight distribution of the mixture does not change significantly from batch to batch, or else the properties of the fabricated part will also vary, sometimes in an unacceptable manner [29]. For this and the other reasons enumerated in Section 8.1, it is essential that the practicing polymer engineer be thoroughly familiar with methods of molecular-weight measurement.

Finally, we mention that a number of technologically important polymers, especially fluoropolymers, are insoluble in suitable solvents, and their molecular-weight distributions cannot be determined using the methods outlined in this chapter. Consequently, indirect methods are needed, and one such method is based on the theory of linear melt viscoelasticity [30]; this is explained in Chapter 14. We also mention that commercial polymers normally contain additives that serve to minimize changes in molecular weight during processing and use. In the case of polyolefins, molecular-weight changes are the result of the generation of free radicals. This process can be arrested by the addition of phenolics that can donate a hydrogen atom; these phenolics are known as primary antioxidants. Also commonly added are phosphites, called secondary antioxidants, because they act in concert with the phenolics to retard the formation of free radicals.

REFERENCES

1. Hagan, R. S., and J. R. Thomas, Key Indicators for Plastics Performance in Consumer Products, *Polym. Eng. Sci.*, 14, 177–188, 1974.
2. Thomas, D. P., and R. S. Hagan, The Influence of Molecular Weight Distribution on Melt Viscosity, Melt Elasticity, Processing Behavior, and Properties of Polystyrene, *Polym. Eng. Sci.*, 9, 164–171, 1969.
3. Ezrin, M. (ed.), *Polymer Molecular Weight Methods*, Advances in Chemistry Series No. 125, American Chemical Society, Washington, DC, 1973.
4. Waltz, J. E., and G. B. Taylor, Determination of Molecular Weight of Nylon, *Anal. Chem.*, 19, 448–451, 1947.

5. Burke, J. J., and T. A. Orofino, Nylon 66 Polymers: I. Molecular Weight and Compositional Distribution, *J. Polym. Sci.*, A2(7), 1–25, 1969.

6. Billingham, N. C., *Molar Mass Measurements in Polymer Science*, Wiley, New York, 1977.

7. Pohl, H. A., Determination of Carboxyl End Groups in a Polyester, Polyethylene Terephthalate, *Anal. Chem.*, 26, 1614–1616, 1954.

8. Smith, J. M., H. C. Van Ness, and M. Abbott, *Introduction to Chemical Engineering Thermodynamics*, 5th ed., McGraw-Hill, New York, 1996.

9. Cherutich, C. K., Osmotic Pressure of Polymer Solutions in the Presence of Fluid Deformation, M.S. thesis, Chemical Engineering, State University of New York, Buffalo, 1990.

10. Young, R. J., and P. A. Lovell, *Introduction to Polymers*, 2nd ed., Chapman & Hall, London, 1991.

11. Resnick, R., and D. Halliday, *Physics—Part I*, Wiley, New York, 1966.

12. Berne, B. J., and R. Pecora, *Dynamic Light Scattering*, Wiley, New York, 1976.

13. Pecora, R. (ed.), *Dynamic Light Scattering*, Plenum, New York, 1985.

14. Oster, G., The Scattering of Light and Its Application to Chemistry, *Chem. Rev.*, 43, 319–365, 1948.

15. Tanford, C., *Physical Chemistry of Macromolecules*, Wiley, New York, 1961.

16. Kerker, M., *The Scattering of Light and Other Electromagnetic Radiation*, Academic Press, New York, 1969.

17. Hiemenz, P. C., *Polymer Chemistry*, Marcel Dekker, New York, 1984.

18. Zimm, B. H., Apparatus and Methods for Measurement and Interpretation of the Angular Variation of Light Scattering; Preliminary Results on Polystyrene Solutions, *J. Chem. Phys.*, 16, 1099–1116, 1948.

19. Kratochvil, P., Advances in Classical Light Scattering from Polymer Solutions, *Pure Appl. Chem.*, 54, 379–393, 1982.

20. Hiemenz, P. C., *Principles of Colloid and Surface Chemistry*, Marcel Dekker, New York, 1977.

21. Budd, P. M., and S. Chakrabarti, Ultracentrifugal Studies of the Degradation of a Fracturing Fluid Polymer: Hydroxypropyl Guar, *J. Appl. Polym. Sci.*, 42, 2191–2196, 1991.

22. Bird, R. B., W. E. Stewart, and E. N. Lightfoot, *Transport Phenomena*, 2nd ed., Wiley, New York, 2002.

23. Einstein, A., Eine neuve bestimmung der molekuldimension, *Ann. Phys.*, 19, 289–306, 1906.

24. Walia, P. S., Influence of Polymeric Additives on the Melting and Crystallization Behavior of Nylon 6,6, Ph.D. thesis, Chemical Engineering, West Virginia University, Morgantown, 1998.

25. Kurata, M., and W. H. Stockmayer, Intrinsic Viscosities and Unperturbed Dimensions of Long Chain Molecules, *Fortschr. Hochpolym.-Forsch.*, 3, 196–312, 1963.

26. Brandrup, J., and E. H. Immergut, *Polymer Handbook*, 3rd ed., Wiley, New York, 1989.

27. Yang, H. H., Aramid Fibers, in *Fibre Reinforcements for Composite Materials*, A. R. Bunsell (ed.), Elsevier, Amsterdam, 1988.

28. Kohan, M. I. (ed.), *Nylon Plastics Handbook*, Hanser, Munich, 1995.
29. Liang, R., and R. K. Gupta, Rheological and Mechanical Properties of Recycled Polycarbonate, *Soc. Plast. Eng. 58th Annual Tech. Conf.*, Vol XLVI, 2903–2907, 2000.
30. Wu, S., Dynamic Rheology and Molecular Weight Distribution of Insoluble Polymers: Tetrafluoroethylene–Hexafluoropropylene Copolymers, *Macromolecules*, 18, 2023–2030, 1985.
31. Han, L., Study of the Rheological Properties of Nomex Fibrids, M.S. thesis, Chemical Engineering, West Virginia University, Morgantown, 2001.
32. Gelman, R. A., and H. G. Barth, Viscosity Studies of Hydrophobically Modified (Hydroxyethyl) Cellulose, Advances in Chemistry Series No. 213, American Chemical Society, Washington, DC, 1986, pp. 101–110.
33. Wagner, H. L., The Polymer Standard Reference Materials Program at the National Bureau of Standards, Advances in Chemistry Series No. 125, American Chemical Society, Washington, DC, 1973, pp. 17–24.

PROBLEMS

8.1. If the polymer sample in Example 8.1 contains 1% by weight of an impurity having a molecular weight of 500 and containing one carboxyl group per molecule, calculate the percentage error that would result in the determination of carboxyl end groups in the polymer due to the presence of the impurity.

8.2. Starting from first principles, derive the expression for ΔT_f for cryoscopy measurements.

8.3. If you had the choice of using either water or camphor as the solvent for cryoscopy, which one would you prefer? Justify your answer by doing some theoretical calculations.

8.4. If a polymer were soluble in both water and toluene, which of the two would be the preferred solvent for vapor pressure osmometry? Why?

8.5. Calculate the colligative properties of aqueous polyethylene oxide solutions and compare the results with the polystyrene–toluene system data given in Table 8.1.

8.6. Repeat Problem 8.5 for aqueous sodium chloride solutions.

8.7. If the room temperature varies by $10°C$ over the course of osmotic pressure measurements, what is the maximum percentage error that is likely to result?

8.8. In the Zimm method of representing light-scattering data [18], C/I_θ is plotted against $\sin^2(\theta/2) + kc$, where k is an arbitrary constant picked to yield a reasonable spread of points. Show that a grid of points is obtained if the data given in Table 8.2 are plotted in this manner. Join points corresponding to a given angle and extrapolate to zero concentration. Next, use the extrapolated points to show that in the limit of zero concentration and zero scattering angle, C/I_θ equals 3.47×10^{-5} in this case.

8.9. Instead of measuring the intensity of scattered light, it is possible to measure the intensity I of transmitted or unscattered light. If the fractional change in light intensity due to unit distance traveled through the sample equals the product of the number of scattering particles per unit volume N and their cross-sectional area (csa), how might one experimentally determine the turbidity τ defined as $N \times$ csa? If $I \approx I_0$, show that τ equals $(I_s/I_0)/l$, where I_s is the total scattered light intensity and l is the sample thickness.

8.10. According to Stokes' law, the drag force F_d on an isolated sphere of diameter D and moving through a Newtonian liquid of viscosity η is

$$F_d = 3\pi\eta Dv$$

where v is the velocity of the sphere. Use this relationship to calculate the terminal velocity of a 1-mm glass sphere of $2.5\,g/cm^3$ density in a concentrated sugar syrup of $1.4\,g/cm^3$ density and $100\,P$ viscosity. What is the corresponding value of the diffusion coefficient at a temperature of $25°C$?

8.11. When the polymer in Example 8.2 was mechanically sheared, it tended to degrade with a reduction in molecular weight. If, after shearing, the measured sedimentation and diffusion coefficients were 3.82×10^{-13} sec and 1.27×10^{-7} cm^2/sec, respectively, what was the percent reduction in molecular weight?

8.12. Show that $[\ln \eta_R]$ equals η_{sp} when η_R is only slightly larger than unity and, thus, demonstrate that Eqs. (8.6.9) and (8.6.10) must necessarily yield the same value for intrinsic viscosity.

8.13. Han [31] measured efflux times of neutrally buoyant glass spheres suspended in a hydraulic oil, and his results are listed below.

Volume fraction of spheres	Efflux time (sec)
0.0047	454.3
0.0094	459.6
0.0141	465.2
0.0188	470.5

If the efflux time of the oil alone is 449.3 sec, do these results validate Eq. (8.6.4)?

8.14. Listed below are values of the intrinsic viscosity as a function of the degree of polymerization for solutions of hydrophobically modified (hydroxyethyl) cellulose in 0.1% sodium oleate [32]. Determine the Mark–Houwink parameters.

Degree of polymerization	Intrinsic viscosity (dL/g)
630	3.16
1250	5.62
2510	10.5
4000	15.8

8.15. The cumulative molecular-weight distribution for a polyethylene sample as obtained using gel permeation chromatography is given below [33]. Determine $\bar{M}_n$, $\bar{M}_w$, and the polydispersity index.

$\log M$	Wt%	$\log M$	Wt%	$\log M$	Wt%
2.800	0.0	4.014	15.2	5.065	90.7
2.865	0.005	4.070	18.1	5.113	92.2
2.929	0.020	4.126	21.5	5.161	93.7
2.992	0.052	4.182	25.2	5.209	94.8
3.056	0.105	4.237	29.3	5.256	95.8
3.119	0.185	4.292	33.7	5.303	96.6
3.181	0.343	4.346	38.5	5.349	97.3
3.243	0.475	4.440	43.4	5.395	97.9
3.305	0.706	4.454	48.5	5.440	98.4
3.366	0.999	4.507	53.5	5.485	98.7
3.427	1.38	4.560	58.3	5.530	99.1
3.488	1.88	4.612	62.9	5.574	99.3
3.548	2.51	4.664	67.3	5.618	99.5
3.607	3.30	4.715	71.4	5.662	99.7
3.667	4.28	4.766	75.1	5.705	99.8
3.725	5.46	4.817	78.15	5.789	99.9
3.784	6.87	4.868	81.6	5.87	100.0
3.842	8.56	4.918	84.4		
3.900	10.50	4.967	86.7		
3.957	12.7	5.016	88.9		

8.16. Use the data given in Problem 8.15 to plot the mole fraction distribution and the weight fraction distribution as a function of the logarithm of the degree of polymerization.

9

Thermodynamics of Polymer Mixtures

9.1 INTRODUCTION

As with low-molecular-weight substances, the solubility of a polymer (i.e., the amount of polymer that can be dissolved in a given liquid) depends on the temperature and pressure of the system. In addition, however, it also depends on the molecular weight. This fact can be used to separate a polydisperse polymer sample into narrow molecular-weight fractions in a conceptually easy, albeit tedious, manner. It is obvious that any help that thermodynamic theory could afford in selecting solvent and defining process conditions would be quite useful for optimizing polymer fractionation. Such polymers having a precise and known molecular weight are needed in small quantities for research purposes. Although today we use gel permeation chromatography for polymer fractionation, a working knowledge of polymer solution thermodynamics is still necessary for several important engineering applications [1].

In the form of solutions, polymers find use in paints and other coating materials. They are also used in lubricants (such as multigrade motor oils), where they temper the reduction in viscosity with increasing temperature. In addition, aqueous polymer solutions are pumped into oil reservoirs for promoting tertiary oil recovery. In these applications, the polymer may witness a range of temperatures, pressures, and shear rates, and this variation can induce phase separation. Such a situation is to be avoided, and it can be, with the aid of thermodynamics. Other situations in which such theory may be usefully applied are devolatilization of polymers and product separation in polymerization reactors. There are also instances in which we want no polymer–solvent interactions at all, especially in cases where certain liquids come into regular contact with polymeric surfaces.

In addition, polymer thermodynamics is very important in the growing and commercially important area of selecting components for polymer–polymer blends. There are several reasons for blending polymers:

1. Because new polymers with desired properties are not synthesized on a routine basis, blending offers the opportunity to develop improved materials that might even show a degree of synergism. For engineering applications, it is generally desirable to develop easily processible polymers that are dimensionally stable, can be used at high temperatures, and resist attack by solvents or by the environment.

2. By varying the composition of a blend, the engineer hopes to obtain a gradation in properties that might be tailored for specific applications. This is true for miscible polymer pairs such as polyphenylene oxide and polystyrene that appear and behave as single-component polymers.

3. If one of the components is a commodity polymer, its use can reduce the cost or, equivalently, improve the profit margin for the more expensive blended product.

Although it is possible to blend two polymers by either melt-mixing in an extruder or dissolving in a common solvent and removing the solvent, the procedure does not ensure that the two polymers will mix on a microscopic level. In fact, most polymer blends are immiscible or incompatible. This means that the mixture does not behave as a single-phase material. It will, for example, have two different glass transition temperatures, which are representative of the two constituents, rather than a single T_g. Such incompatible blends can be homogenized somewhat by using copolymers and graft polymers or by adding surface-active agents. These measures can lead to materials having high impact strength and toughness.

In this chapter, we present the classical Flory–Huggins theory, which can explain a large number of observations regarding the phase behavior of concentrated polymer solutions. The agreement between theory and experiment is, however, not always quantitative. Additionally, the theory cannot explain the phenomenon of phase separation brought about by an increase in temperature. It is also not very useful for describing polymer–polymer miscibility. For these reasons, the Flory–Huggins theory has been modified and alternate theories have been advanced, which are also discussed.

9.2 CRITERIA FOR POLYMER SOLUBILITY

A polymer dissolves in a solvent if, at constant temperature and pressure, the total Gibbs free energy can be decreased by the polymer going into solution. Therefore, it is necessary that the following hold:

$$\Delta G_M = \Delta H_{mix} - T \Delta S_{mix} < 0 \qquad (9.2.1)$$

For most polymers, the enthalpy change on mixing is positive. This necessitates that the change in entropy be sufficiently positive if mixing is to occur. These changes in enthalpy and entropy can be calculated using simple models; these calculations are done in the next section. Here, we merely note that Eq. (9.2.1) is only a necessary condition for solubility and not a sufficient condition. It is possible, after all, to envisage an equilibrium state in which the free energy is still lower than that corresponding to a single-phase homogeneous solution. The single-phase solution may, for example, separate into two liquid phases having different compositions. To understand which situation might prevail, we need to review some elements of the thermodynamics of mixtures.

A partial molar quantity is the derivative of an extensive quantity M with respect to the number of moles n_i of one of the components, keeping the temperature, the pressure, and the number of moles of all the other components fixed. Thus,

$$\bar{M}_i = \left(\frac{\partial M}{\partial n_i}\right)_{T,P,n_j} \qquad (9.2.2)$$

It is easy to show [2] that the mixture property M can be represented in terms of the partial molar quantities as follows:

$$M = \sum_i \bar{M}_i n_i \qquad (9.2.3)$$

For an open system at constant temperature and pressure, however,

$$dM = \sum_i \bar{M}_i \, dn_i \qquad (9.2.4)$$

but Eq. (9.2.3) gives

$$dM = \sum_i \bar{M}_i \, dn_i + \sum_i n_i \, d\bar{M}_i \qquad (9.2.5)$$

so that

$$\sum_i n_i \, d\bar{M}_i = 0 \qquad (9.2.6)$$

which is known as the Gibbs–Duhem equation.

Let us identify M with the Gibbs free energy G and consider the mixing of n_1 moles of pure component 1 with n_2 moles of pure component 2. Before mixing, the free energy of both components taken together, G_{comp}, is

$$G_{comp} = \sum_{i=1}^{2} g_i n_i \qquad (9.2.7)$$

where g_i is the molar free energy of component i. After mixing, the free energy of the mixture, using Eq. (9.2.3), is as follows:

$$G_{mixture} = \sum_{i=1}^{2} \bar{G}_i n_i \qquad (9.2.8)$$

Consequently, the change in free energy on mixing is

$$\Delta G_M = \sum_{i=1}^{2} (\bar{G}_i - g_i) n_i \qquad (9.2.9)$$

and dividing both sides by the total number of moles, $n_1 + n_2$, yields the corresponding result for 1 mol of mixture,

$$\Delta g_m = \sum_{i=1}^{2} (\bar{G}_i - g_i) x_i \qquad (9.2.10)$$

where x_i denotes mole fraction.

It is common practice to call the partial molar Gibbs free energy $\bar{G}_i$ the *chemical potential* and write it as μ_i. Clearly, g_i is the partial molar Gibbs free energy for the pure component. Representing it as μ_i^0, we can derive from Eq. (9.2.10) the following:

$$\Delta g_m = x_1 \Delta \mu_1 + x_2 \Delta \mu_2 \qquad (9.2.11)$$

where $\Delta \mu_1 = \mu_1 - \mu_1^0$ and $\Delta \mu_2 = \mu_2 - \mu_2^0$. Because $x_1 + x_2$ equals unity, Eq. (9.2.11) can be written

$$\Delta g_m = \Delta \mu_1 + x_2 (\Delta \mu_2 - \Delta \mu_1) \qquad (9.2.12)$$

Differentiating this result with respect to x_2 gives

$$\frac{d\Delta g_m}{dx_2} = \frac{d\mu_1}{dx_2} + (\Delta \mu_2 - \Delta \mu_1) + x_2 \left(\frac{d\mu_2}{dx_2} - \frac{d\mu_1}{dx_2} \right) \qquad (9.2.13)$$

$$= (\Delta \mu_2 - \Delta \mu_1) + x_2 \frac{d\mu_2}{dx_2} + x_1 \frac{d\mu_1}{dx_2}$$

From Eq. (9.2.6), however, $\sum_{1}^{2} x_i \, d\mu_i$ equals 0. Therefore, Eq. (9.2.13) becomes

$$\frac{d\Delta g_m}{dx_2} = \Delta \mu_2 - \Delta \mu_1 \qquad (9.2.14)$$

and solving Eq. (9.2.14) simultaneously with Eq. (9.2.11) yields

$$\Delta\mu_1 = \Delta g_m - x_2 \frac{d\Delta g_m}{dx_2} \tag{9.2.15}$$

$$\Delta\mu_2 = \Delta g_m + x_1 \frac{d\Delta g_m}{dx_2} \tag{9.2.16}$$

Thus, if Δg_m can be obtained by some means as a function of composition, the chemical potentials can be computed using Eqs. (9.2.15) and (9.2.16). The chemical potentials are, in turn, needed for phase equilibrium calculations.

Let us now return to the question of whether a single-phase solution or two liquid phases will be formed if the ΔG_M of a two-component system is negative. This question can be answered by examining Figure 9.1, which shows two possible Δg_m versus x_2 curves; these two curves may correspond to different temperatures. It can be reasoned from Eqs. (9.2.15) and (9.2.16) that the chemical potentials at any composition x_2 can be determined simply by drawing a tangent to the Δg_m curve at x_2 and extending it until it intersects with the $x_2 = 0$ and $x_2 = 1$ axes. The intercept with $x_2 = 0$ gives $\Delta\mu_1$, whereas that with $x_2 = 1$ gives $\Delta\mu_2$.

Following this reasoning, it is seen that the curve labeled T_1 has a one-to-one correspondence between $\Delta\mu_1$ and x_2 or, for that matter, between $\Delta\mu_2$ and x_2. This happens because the entire curve is concave upward. Thus, there are no two composition values that yield the same value of the chemical potential. This implies that equilibrium is not possible between two liquid phases of differing compositions; instead, there is complete miscibility. At a lower temperature T_2, however, the chemical potential at x_2' equals the chemical potential at x_2''. Solutions of these two compositions can, therefore, coexist in equilibrium. The points x_2' and x_2'' are called *binodal points*, and any single-phase system having a composition between these two points can split into these two phases with relative

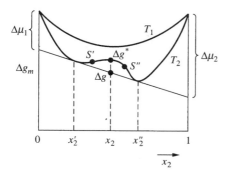

FIGURE 9.1 Free-energy change of mixing per mole of a binary mixture as a function of mixture composition.

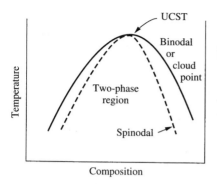

FIGURE 9.2 Temperature–composition diagram corresponding to Figure 9.1.

amounts of each phase determined by a mass balance. Phase separation occurs because the free energy of the two-phase mixture denoted by the point marked Δg is less than the free energy Δg^* of the single-phase solution of the same average composition. Points S' and S'' are inflection points called *spinodal points*, and between these two points the Δg_m curve is concave downward. A solution having a composition between these two points is unstable to even the smallest disturbance and can lower its free energy by phase separation. Between each spinodal point and the corresponding binodal point, however, Δg_m is concave upward and, therefore, stable to small disturbances. This is called a *metastable region*; here, it is possible to observe a single-phase solution—but only for a limited period of time.

The presence of the two-phase region depends on temperature. For some solutions, at a high enough temperature called the upper critical solution temperature, the spinodal and binodal points come together and only single-phase mixtures occur above this temperature. This situation is depicted in Figure 9.2 on a temperature–composition diagram. Here, the locus of the binodal points is called the *binodal curve* or the *cloud point curve*, whereas the locus of the spinodal points is called the *spinodal curve*. Next, we direct our attention to determining the free-energy change on mixing a polymer with a low-molecular-weight solvent.

9.3 THE FLORY–HUGGINS THEORY

The classical Flory–Huggins theory assumes at the outset that there is neither a change in volume nor a change in enthalpy on mixing a polymer with a low-molecular-weight solvent [3–5]; the influence of non-athermal ($\Delta H_{mixing} \neq 0$) behavior is accounted for at a later stage. Thus, the calculation of the free-energy

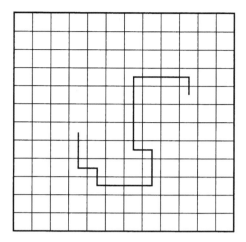

FIGURE 9.3 Schematic diagram of a polymer molecule on a two-dimensional lattice.

change on mixing at a constant temperature and pressure reduces to a calculation of the change in entropy on mixing. This latter quantity is determined with the help of a lattice model using formulas from statistical thermodynamics.

We assume the existence of a two-dimensional lattice with each lattice site having z nearest neighbors, where z is the coordination number of the lattice; an example is shown in Figure 9.3. Each lattice site can accommodate a single solvent molecule or a polymer segment having a volume equal to a solvent molecule. Polymer molecules are taken to be monodisperse, flexible, initially disordered, and composed of a series of segments the size of a solvent molecule. The number of segments in each polymer molecule is m, which equals V_2/V_1, the ratio of the molar volume of the polymer to the molar volume of the solvent. Note that m is not the degree of polymerization.

We begin with an empty lattice and calculate the number of ways, Ω, of arranging n_1 solvent molecules and n_2 polymer molecules in the $n_0 = n_1 + mn_2$ lattice sites. Because the heat of mixing has been taken to be zero, each arrangement has the same energy and is equally likely to occur. The only restriction imposed is by the connectivity of polymer chain segments. It must be ensured that two segments connected to each other lie on the nearest neighboring lattice sites. Once Ω is known, the entropy of the mixture is given by $k \ln \Omega$, where k is Boltzmann's constant.

9.3.1 Entropy Change on Mixing

In order to calculate the entropy of the mixture, we first arrange all of the polymer molecules on the lattice. The identical solvent molecules are placed thereafter. If j

polymer molecules have already been placed, the number of lattice sites still available number $n_0 - jm$. Thus, the first segment of the $(j + 1)$st molecule can be arranged in $n_0 - jm$ ways. The second segment is connected to the first one and so can be placed only in one of the z neighboring sites. All of these may, however, not be vacant. If the polymer solution is relatively concentrated so that chain overlap occurs, we would expect that, on average, the fraction of neighboring sites occupied (f) would equal the overall fraction of sites occupied. Thus, $f = jm/n_0$. As a result, the second segment of the $(j + 1)$st molecule can be placed in $z(1 - f)$ ways. Clearly, the third segment can be placed in $(z - 1)(1 - f)$ ways, and similarly for subsequent segments. Therefore, the total number of ways Ω_{j+1} in which the $(j + 1)$st polymer molecule can be arranged is the product of the number of ways of placing the first segment with the number of ways of placing the second segment and the number of ways of placing each subsequent segment. Thus,

$$\Omega_{j+1} = (n_0 - jm)z(1 - f) \prod_{3}^{m} (z - 1)(1 - f) \tag{9.3.1}$$

where the symbol $\prod$ denotes product. As a consequence,

$$\begin{aligned}
\Omega_{j+1} &= (n_0 - jm)z(z - 1)^{m-2}(1 - f)^{m-1} \\
&\cong (n_0 - jm)(z - 1)^{m-1}(1 - f)^{m-1} \\
&= (n_0 - jm)(z - 1)^{m-1}\left(1 - \frac{jm}{n_0}\right)^{m-1} \\
&= (n_0 - jm)^m \left(\frac{z - 1}{n_0}\right)^{m-1}
\end{aligned} \tag{9.3.2}$$

The total number of ways of arranging all of the n_2 polymer molecules, Ω_p, is the product of the number of ways of arranging each of the n_2 molecules in sequence. This fact and Eq. (9.3.2) yield

$$\Omega_p = \prod_{j=0}^{n_2-1} \left[(n_0 - jm)^m \left(\frac{z - 1}{n_0}\right)^{m-1} \right] \tag{9.3.3}$$

where the index only goes up to $n_2 - 1$ because $j = 0$ corresponds to the first polymer molecule. The development so far assumes that all of the polymer molecules are different. They are, however, identical to each other. This reduces the total number of possible arrangements by a factor of $n_2!$, and it is therefore necessary to divide the right-hand side of Eq. (9.3.3) by $n_2!$.

Having arranged all of the polymer molecules, the number of ways of fitting all of the indistinguishable solvent molecules into the remaining lattice sites is exactly one. As a result, Ω_p equals Ω, the total number of ways of placing all the polymer and solvent molecules on to the lattice. Finally, then,

$$S_{\text{mixture}} = k \ln \Omega \qquad (9.3.4)$$

and using Eq. (9.3.3) properly divided by $n_2!$:

$$\frac{S_{\text{mixture}}}{k} = -\ln(n_2!) + m \sum_{j=0}^{n_2-1} \ln(n_0 - jm) + (m-1) \sum_{j=0}^{n_2-1} \ln\left(\frac{z-1}{n_0}\right) \qquad (9.3.5)$$

Because j does not appear in the last term on the right-hand side of Eq. (9.3.5), that term adds up to $(m-1)n_2 \ln[(z-1)/n_0]$. Also, the first term can be replaced by Stirling's approximation:

$$(n_2!) = n_2 \ln n_2 - n_2 \qquad (9.3.6)$$

Now, consider the summation in the second term:

$$\sum_{j=0}^{n_2-1} \ln(n_0 - jm) = \sum_{j=0}^{n_2-1} \ln\left[m\left(\frac{n_0}{m} - j\right)\right]$$

$$= n_2 \ln m + \sum_{j=0}^{n_2-1} \ln\left(\frac{n_0}{m} - j\right) \qquad (9.3.7)$$

Furthermore,

$$\sum_{j=0}^{n_2-1} \ln\left(\frac{n_0}{m} - j\right)$$

$$= \ln\left(\frac{n_0}{m}\right) + \ln\left(\frac{n_0}{m} - 1\right) + \cdots + \ln\left(\frac{n_0}{m} - n_2 + 1\right)$$

$$= \ln\left[\left(\frac{n_0}{m}\right)\left(\frac{n_0}{m} - 1\right)\left(\frac{n_0}{m} - 2\right)\cdots\left(\frac{n_0}{m} - n_2 + 1\right)\right]$$

$$= \ln\left\{\frac{\left(\frac{n_0}{m}\right)\left(\frac{n_0}{m} - 1\right)\cdots\left(\frac{n_0}{m} - n_2 + 1\right)\left(\frac{n_0}{m} - n_2\right)\cdots 1}{\left(\frac{n_0}{m} - n_2\right)\cdots 1}\right\}$$

$$= \ln\left[\frac{(n_0/m)!}{(n_0/m - n_2)!}\right] \qquad (9.3.8)$$

Combining all of these fragments and again using Stirling's approximation in Eq. (9.3.8) yields

$$
\begin{aligned}
\frac{S_{\text{mixture}}}{k} = {} & -n_2 \ln n_2 + n_2 + m\left[n_2 \ln m + \left(\frac{n_0}{m}\right) \ln \left(\frac{n_0}{m}\right) - \frac{n_0}{m} \right. \\
& \left. - \left(\frac{n_0}{m} - n_2\right) \ln \left(\frac{n_0}{m} - n_2\right) + \left(\frac{n_0}{m} - n_2\right) \right] \\
& + (m-1)n_2 \ln \left(\frac{z-1}{n_0}\right)
\end{aligned}
\tag{9.3.9}
$$

which, without additional tricks, can be simplified to the following:

$$
\begin{aligned}
\frac{S_{\text{mixture}}}{k} = {} & -n_2 \ln \left(\frac{n_2}{n_0}\right) + n_2 - mn_2 - n_1 \ln \left(\frac{n_1}{n_0}\right) \\
& + (m-1)[n_2 \ln(z-1)]
\end{aligned}
\tag{9.3.10}
$$

Adding to and subtracting $n_2 \ln m$ from the right-hand side of Eq. (9.3.10) gives the result

$$
\begin{aligned}
\frac{S_{\text{mixture}}}{k} = {} & -n_2 \ln \left(\frac{mn_2}{n_0}\right) - n_1 \ln \left(\frac{n_1}{n_0}\right) \\
& + n_2[(m-1)\ln(z-1) + (1-m) + \ln m]
\end{aligned}
\tag{9.3.11}
$$

The entropy of the pure polymer S_2 can be obtained by letting n_1 be zero and n_0 be mn_2 in Eq. (9.3.11):

$$
\frac{S_2}{k} = n_2[(m-1)\ln(z-1) + (1-m) + \ln m]
\tag{9.3.12}
$$

Similarly, the entropy of the pure solvent S_1 is obtained by setting n_2 equal to zero and n_1 equal to n_0:

$$
\frac{S_1}{k} = 0
\tag{9.3.13}
$$

Using Eqs. (9.3.11)–(9.3.13),

$$
\begin{aligned}
\Delta S_{\text{mixing}} &= \Delta S_{\text{mixture}} - S_1 - S_2 \\
&= -k\left[n_1 \ln \left(\frac{n_1}{n_0}\right) + n_2 \ln \left(\frac{mn_2}{n_0}\right) \right]
\end{aligned}
\tag{9.3.14}
$$

From the way that m and n_0 have been defined, it is evident that n_1/n_0 equals ϕ_1, the volume fraction of the solvent, and mn_2/n_0 equals ϕ_2, the volume fraction of the polymer. As a result,

$$
\Delta S = -k[n_1 \ln \phi_1 + n_2 \ln \phi_2]
\tag{9.3.15}
$$

which is independent of the lattice coordination number z. The change in entropy on mixing n_1 moles of solvent with n_2 moles of polymer will exceed by a factor of Avogadro's number the change in entropy given by Eq. (9.3.15); multiplying the right-hand side of this equation by Avogadro's number gives

$$\Delta S = -R[n_1 \ln \phi_1 + n_2 \ln \phi_2] \tag{9.3.16}$$

where R is the universal gas constant and n_1 and n_2 now represent numbers of moles. Note that if m were to equal unity, ϕ_1 and ϕ_2 would equal the mole fractions and Eq. (9.3.16) would become identical to the equation for the change in entropy of mixing ideal molecules [2]. Note also that Eq. (9.3.16) does not apply to dilute solutions because of the assumption that f equals jm/n_0 and is independent of position within the lattice.

Example 9.1: One gram of polymer having molecular weight 40,000 and density $1\,\text{g/cm}^3$ is dissolved in $9\,\text{g}$ of solvent of molecular weight 78 and density $0.9\,\text{g/cm}^3$.

(a) What is the entropy change on mixing?
(b) How would the answer change if a monomer of molecular weight 100 were dissolved in place of the polymer?

Solution:

(a) $n_1 = 9/78 = 0.115$; $n_2 = 2.5 \times 10^{-5}$; $\phi_1 = (9/0.9)/[(9/0.9) + 1] = 0.909$; $\phi_2 = 0.091$. Therefore, $\Delta S = -R[0.115 \ln 0.909 + 2.5 \times 10^{-5} \ln 0.091] = 0.011R$.

(b) In this case, $\Delta S = -R[n_1 \ln x_1 + n_2 \ln x_2]$, with $n_2 = 0.01$, $x_1 = 0.92$, and $x_2 = 0.08$ so that $\Delta S = 0.035R$.

9.3.2 Enthalpy Change on Mixing

If polymer solutions were truly athermal, ΔG of mixing would equal $-T\Delta S$, and, based on Eq. (9.3.16), this would always be a negative quantity. The fact that polymers do not dissolve very easily suggests that mixing is an endothermic process and $\Delta H > 0$. If the change in volume on mixing is again taken to be zero, ΔH equals ΔU, the internal energy change on mixing. This latter change arises due to interactions between polymer and solvent molecules. Because intermolecular forces drop off rapidly with increasing distance, we need to consider only nearest neighbors in evaluating ΔU. Consequently, we can again use the lattice model employed previously.

Let us examine the filled lattice and pick a polymer segment at random. It is surrounded by z neighbors. Of these, $z\phi_2$ are polymeric and $z\phi_1$ are solvent. If the energy of interaction (a negative quantity) between two polymer segments is represented by e_{22} and that between a polymer segment and a solvent molecule by e_{12}, the total energy of interaction for the single polymer segment is

$$z\phi_2 e_{22} + z\phi_1 e_{12}$$

Because the total number of polymer segments in the lattice is $n_0\phi_2$, the interaction energy associated with all of the polymer segments is

$$\frac{z}{2}n_0\phi_2(\phi_2 e_{22} + \phi_1 e_{12})$$

where the factor of $\frac{1}{2}$ has been added to prevent everything from being counted twice.

Again, by similar reasoning, the total energy of interaction for a single solvent molecule picked at random is

$$z\phi_1 e_{11} + z\phi_2 e_{12}$$

where e_{11} is the energy of interaction between two solvent molecules. Because the total number of solvent molecules is $n_0\phi_1$, the total interaction energy is

$$\frac{zn_0\phi_1}{2}(\phi_1 e_{11} + \phi_2 e_{12})$$

For the pure polymer, the energy of interaction between like segments before mixing (using a similar lattice) is

$$\frac{n_0\phi_2 z e_{22}}{2}$$

For pure solvent, the corresponding quantity is

$$\frac{n_0\phi_1 z e_{11}}{2}$$

From all of these equations, the change in energy on mixing, ΔU, is the difference between the sum of the interaction energy associated with the polymer and solvent in solution and the sum of the interaction energy of the pure components. Thus,

$$
\begin{aligned}
\Delta U &= \frac{z}{2}n_0\phi_2(\phi_2 e_{22} + \phi_1 e_{12}) + \frac{zn_0\phi_1}{2}(\phi_1 e_{11} + \phi_2 e_{12}) \\
&\quad - \frac{n_0\phi_2 z e_{22}}{2} - \frac{n_0\phi_1 z e_{11}}{2} \\
&= \frac{zn_0}{2}[2\phi_1\phi_2 e_{12} - \phi_1\phi_2 e_{11} - \phi_1\phi_2 e_{22}] \\
&= \Delta e z n_0 \phi_1 \phi_2
\end{aligned}
\tag{9.3.17}
$$

where $\Delta e = (1/2)(2e_{12} - e_{11} - e_{22})$, and the result is found to depend on the unknown coordination number z. Because z is not known, it makes sense to lump Δe along with it and define a new unknown quantity χ_1, called the *interaction parameter*:

$$\chi_1 = \frac{z\Delta e}{kT} \tag{9.3.18}$$

whose value is zero only for athermal mixtures. For endothermic mixing, χ_1 is positive (the more common situation), whereas for exothermic mixing, it is negative. Combining Eqs. (9.3.17) and (9.3.18) yields

$$\Delta H_M = \Delta U_M = kT\chi_1 n_0 \phi_1 \phi_2 \tag{9.3.19}$$
$$= kT\chi_1 n_1 \phi_2$$

and the magnitude of χ_1 has to be estimated by comparison with experimental data.

9.3.3 Free-Energy Change and Chemical Potentials

If we assume that the presence of a nonzero ΔH_M does not influence the previously calculated ΔS_M, a combination of Eqs. (9.2.1), (9.3.15), and (9.3.19) yields

$$\Delta G_M = kT[n_1 \ln \phi_1 + n_2 \ln \phi_2 + \chi_1 n_1 \phi_2] \tag{9.3.20}$$

Because volume fractions are always less than unity, the first two terms in brackets in Eq. (9.3.20) are negative. The third term depends on the sign of the interaction parameter, but it is usually positive. From Eq. (9.3.18), however, χ_1 decreases with increasing temperature so that ΔG_M should always become negative at a sufficiently high temperature. It is for this reason that a polymer–solvent mixture is warmed to promote solubility. Also note that if one increases the polymer molecular weight while keeping n_1, ϕ_1, ϕ_2, and T constant, n_2 decreases because the volume per polymer molecule increases. The consequence of this fact, from Eq. (9.3.20), is that ΔG_M becomes less negative, which implies that a high-molecular-weight fraction is less likely to be soluble than a low-molecular-weight fraction. This also means that if a saturated polymer solution containing a polydisperse sample is cooled, the highest-molecular-weight component will precipitate first. In order to quantify these statements, we have to use the thermodynamic phase equilibrium criterion [2]

$$\mu_i^A = \mu_i^B \tag{9.3.21}$$

where $i = 1, 2$ and A and B are the two phases that are in equilibrium. In writing Eq. (9.3.21), it is assumed that the polymer, component 2, is monodisperse. The effect of polydispersity will be discussed later.

The chemical potentials required in Eq. (9.3.21) can be computed using Eq. (9.3.20), the definition of the chemical potential as a partial molar Gibbs free energy, and the fact that

$$\Delta G_M = G_{mixture} - G_1 - G_2 \tag{9.3.22}$$

so that

$$G_{mixture} = n_1 g_1 + n_2 g_2 + RT[n_1 \ln \phi_1 + n_2 \ln \phi_2 + \chi_1 n_1 \phi_2] \tag{9.3.23}$$

where n_1 and n_2 now denote numbers of moles rather than numbers of molecules, and g_1 and g_2 are the molar free energies of the solvent and polymer, respectively. Differentiating Eq. (9.3.23) with respect to n_1 and n_2, in turn, gives the following:

$$\mu_1 = \frac{\partial G_{mixture}}{\partial n_1}$$

$$= g_1 + RT\left[\ln \phi_1 + \frac{n_1}{\phi_1}\frac{\partial \phi_1}{\partial n_1} + \frac{n_2}{\phi_2}\frac{\partial \phi_2}{\partial n_1} + \chi_1 \phi_2 + \chi_1 n_1 \frac{\partial \phi_2}{\partial n_1}\right] \tag{9.3.24}$$

$$\mu_2 = \frac{\partial G_{mixture}}{\partial n_2}$$

$$= g_2 + RT\left[\frac{n_1}{\phi_1}\frac{\partial \phi_1}{\partial n_2} + \ln \phi_2 + \frac{n_2}{\phi_2}\frac{\partial \phi_2}{\partial n_2} + \chi_1 n_1 \frac{\partial \phi_2}{\partial n_2}\right] \tag{9.3.25}$$

Recognizing that

$$\phi_1 = \frac{n_1}{n_1 + mn_2} \quad \text{and} \quad \phi_2 = \frac{mn_2}{n_1 + mn_2}$$

gives the following:

$$\frac{\partial \phi_1}{\partial n_1} = \frac{\phi_2}{n_1 + mn_2} \tag{9.3.26}$$

$$\frac{\partial \phi_1}{\partial n_2} = -\frac{m\phi_1}{n_1 + mn_2} \tag{9.3.27}$$

$$\frac{\partial \phi_2}{\partial n_1} = -\frac{\phi_2}{n_1 + mn_2} \tag{9.3.28}$$

$$\frac{\partial \phi_2}{\partial n_2} = \frac{m\phi_1}{n_1 + mn_2} \tag{9.3.29}$$

Introducing these results into Eqs. (9.3.24) and (9.3.25) and simplifying gives

$$\frac{\mu_1 - \mu_1^0}{RT} = \ln(1 - \phi_2) + \phi_2\left(1 - \frac{1}{m}\right) + \chi_1\phi_2^2 \tag{9.3.30}$$

$$\frac{\mu_2 - \mu_2^0}{RT} = (1 - \phi_2)(1 - m) + \ln\phi_2 + \chi_1 m(1 - \phi_2)^2 \tag{9.3.31}$$

in which g_1 and g_2 have been relabeled μ_1^0 and μ_2^0, respectively. The preceding two equations can now be used for examining phase equilibrium.

9.3.4 Phase Behavior of Monodisperse Polymers

If we mix n_1 moles of solvent with n_2 moles of polymer having a known molar volume or molecular weight (i.e., a known value of m), the chemical potential of the solvent in solution is given by Eq. (9.3.30). If we fix χ_1, we can easily plot $(\mu_1 - \mu_1^0)/RT$ as a function of ϕ_2. By changing χ_1 and repeating the procedure, we get a family of curves at different temperatures, because there is a one-to-one correspondence between χ_1 and temperature. Such a plot is shown in Figure 9.4 for m equaling 1000, taken from the work of Flory [3,5]. Note that increasing χ_1 is equivalent to decreasing temperature.

By examining Figure 9.4, we find that for values of χ_1 below a critical value χ_c, there is a unique relationship between μ_1 and ϕ_2. Above χ_c, however, the plots are bivalued. Because the same value of the chemical potential occurs at two different values of ϕ_2, these two values of ϕ_2 can coexist at equilibrium. In other words, two phases are formed whenever $\chi_1 > \chi_c$. To calculate the value of χ_c, note that at $\chi_1 = \chi_c$, there is an inflection point in the μ_1 versus ϕ_2 curve. Thus, we can obtain χ_c by setting the first two derivatives of μ_1 with respect to ϕ_2 equal to zero. Using Eq. (9.3.30) to carry out these differentiations,

$$\frac{\partial\mu_1}{\partial\phi_2} = -\frac{1}{1 - \phi_2} + \left(1 - \frac{1}{m}\right) + 2\chi_1\phi_2 \tag{9.3.32}$$

$$\frac{\partial^2\mu_1}{\partial\phi_2^2} = -\frac{1}{(1 - \phi_2)^2} + 2\chi_1 \tag{9.3.33}$$

At $\chi_1 = \chi_c$ and $\phi_2 = \phi_{2c}$, these two derivatives are zero. Solving for χ_c from each of the two equations yields the following:

$$\chi_c = \frac{1}{2\phi_{2c}(1 - \phi_{2c})} - \left(1 - \frac{1}{m}\right)(2\phi_{2c})^{-1} \tag{9.3.34}$$

$$\chi_c = \frac{1}{2(1 - \phi_{2c})^2} \tag{9.3.35}$$

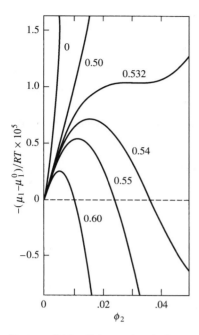

FIGURE 9.4   Solvent chemical potential as a function of polymer volume fraction for $m = 1000$. The value of χ_1 is indicated on each curve. (Reprinted from Paul J. Flory, Principles of Polymer Chemistry. Copyright ©1953 Cornell University and copyright © 1981 Paul J. Flory. Used by permission of the Publisher, Cornell University Press.)

Equating the right-hand sides of the two previous equations gives

$$\phi_{2c} = \frac{1}{1 + \sqrt{m}} \tag{9.3.36}$$

which means that

$$\chi_c = \frac{1}{2} + \frac{1}{\sqrt{m}} + \frac{1}{2m} \tag{9.3.37}$$

and $\chi_c \to \frac{1}{2}$ as m becomes very large. Thus, knowing m allows us to derive χ_c or, equivalently, the temperature at which two liquid phases first begin to appear; this is the upper critical solution temperature (UCST) shown in Figure 9.2. The corresponding UCST for polymer of infinite molecular weight is known as the Flory temperature or theta temperature, and it is higher than the UCST of polymer having a finite molecular weight. It is clear, however, that if the theory is valid, complete solubility should be observed for $\chi_1 \le 0.5$. It is also desirable to plot the binodal or the temperature–composition curve separating the one- and two-phase regions. The procedure for doing this is deferred until after we discuss the method of numerically relating χ_1 to temperature.

9.3.5   Determining the Interaction Parameter

The polymer–solvent interaction parameter χ_1 can be calculated from Eq. (9.3.30) in conjunction with any experimental technique that allows for a measurement of the chemical potential. This can be done via any one of several methods, including light scattering and viscosity, but most commonly with the help of vapor-pressure or osmotic pressure measurements [1–6]. Let us examine both.

If we consider a pure vapor to be ideal, then the following is true at constant temperature:

$$d\mu = dg = v \, dP = \frac{RT}{P} \, dP \tag{9.3.38}$$

where g and v are the molar free energy and molar volume, respectively. Integrating from a pressure P_0 to pressure P gives

$$\mu(T, \ P) - \mu(T, \ P_0) = RT \ln\left(\frac{P}{P_0}\right) \tag{9.3.39}$$

The equivalent expression for a component, say 1, in a mixture of ideal gases with mole fraction y_1 is given by the following [2]:

$$\mu_1(T, \ P, \ y_1) - \mu_1(T, \ P_0) = RT \ln\left(\frac{Py_1}{P_0}\right) \tag{9.3.40}$$

If the vapor is in equilibrium with a liquid phase, the chemical potential of each component has to be the same in both phases. Also, for a pure liquid at equilibrium, P equals the vapor pressure P_1^0. Thus, denoting as μ_1^0 the pure liquid 1 chemical potential, we can derive the following, using Eq. (9.3.39):

$$\mu_1^0 = \mu_1(T, \ P_0) + RT \ln\left(\frac{P_1^0}{P_0}\right) \tag{9.3.41}$$

Similarly, for component 1 in a liquid mixture in equilibrium with a mixture of gases, the liquid-phase chemical potential is, from Eq. (9.3.40),

$$\mu_1 = \mu_1(T, \ P_0) + RT \ln\left(\frac{Py_1}{P_0}\right) \tag{9.3.42}$$

Subtracting Eq. (9.3.41) from Eq. (9.3.42) to eliminate $\mu_1(T, \ P_0)$ gives the following [7]:

$$\mu_1 - \mu_1^0 = RT \ln\left(\frac{Py_1}{P_1^0}\right) \tag{9.3.43}$$

but Py_1 is the partial pressure P_1 of component 1 in the gas phase. Combining Eqs. (9.3.30) and (9.3.43) gives

$$\ln\left(\frac{P_1}{P_1^0}\right) = \ln(1 - \phi_2) + \phi_2\left(1 - \frac{1}{m}\right) + \chi_1\phi_2^2 \tag{9.3.44}$$

where the left-hand side is also written as a_1, in which a_1 is the solvent activity. Thus, measurements of P_1 as a function of ϕ_2 can be used to obtain χ_1 over a wide range of concentrations.

The situation with osmotic equilibrium is shown schematically in Figure 8.4, and it has been discussed previously in Chapter 8. At equilibrium, the chemical potential of the solvent is the same on both sides of the semipermeable membrane. Thus,

$$\mu_1(T, P) = \mu_1(T, P + \pi, x_1) \tag{9.3.45}$$

where π is the osmotic pressure and x_1 is the mole fraction of solvent in solution. From elementary thermodynamics, however,

$$\mu_1(T, P + \pi, x_1) = \mu_1(T, P, x_1) + \int_P^{P+\pi} \bar{V}_1 \, dP \tag{9.3.46}$$

in which $\bar{V}_1$ is the partial molar volume. The term $\mu_1(T, P)$ is the same as what we have been calling μ_1^0; therefore, Eqs. (9.3.45) and (9.3.46) imply that

$$\mu_1 - \mu_1^0 = -\int_P^{P+\pi} \bar{V}_1 \, dP \cong -v_1 \, d\pi \tag{9.3.47}$$

because the partial molar volume is not too different from the molar volume of the solvent.

Using the Flory–Huggins expression for the difference in chemical potentials in Eq. (9.3.47) gives

$$\pi = -\left(\frac{RT}{v_1}\right)\left[\ln(1 - \phi_2) + \left(1 - \frac{1}{m}\right)\phi_2 + \chi_1\phi_2^2\right] \tag{9.3.48}$$

which can be rewritten in a slightly different form if we expand $1 - \phi_2$ in a Taylor series about $\phi_2 = 0$. Retaining terms up to ϕ_2^3, we get

$$\pi = \left(\frac{RT}{v_1}\right)\left[\frac{\phi_2}{m} + \left(\frac{1}{2} - \chi_1\right)\phi_2^2 + \frac{\phi_2^3}{3} + \cdots\right] \tag{9.3.49}$$

which can again be used to evaluate χ_1 using experimental data. A comparison of Eq. (9.3.49) with Eq. (8.3.22) shows that the second virial coefficient is 0 at the theta temperature because χ_1 equals 0.5 at that condition.

Typical data for χ_1 as a function of ϕ_2 obtained using these methods are shown in Figure 9.5 [5]. It is found that although solutions of rubber in benzene

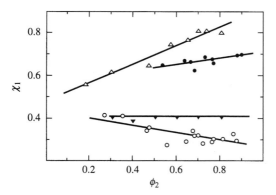

FIGURE 9.5 Influence of composition on the polymer–solvent interaction parameter. Experimental values of the interaction parameter χ_1 are plotted against the volume fraction ϕ_2 of polymer. Data for polydimethylsiloxane ($M = 3850$) in benzene ($\triangle$), polystyrene in methyl ethyl ketone ($\bullet$), and polystyrene in toluene ($\bigcirc$) are based on vapor-pressure measurements. Those for rubber in benzene ($\blacktriangledown$) were obtained using vapor-pressure measurements at higher concentrations and isothermal distillation equilibration with solutions of known activities in the dilute range. (Reprinted from Paul J. Flory, Principles of Polymer Chemistry. Copyright © 1953 Cornell University and copyright © 1981 Paul J. Flory. Used by permission of the Publisher, Cornell University Press.)

behave as expected, most systems are characterized by a concentration-dependent interaction parameter [5,8]. In addition, χ_1 does not follow the expected inverse temperature dependence predicted by theory [5]. This suggests that ΔH_M is not independent of temperature. To take the temperature dependence of ΔH_M into account, Flory uses the following expression for χ_1 that involves two new constants, θ and ψ [5]:

$$\chi_1 = \frac{1}{2} - \psi \left(1 - \frac{\theta}{T} \right) \tag{9.3.50}$$

One way of determining these constants is to first determine the upper critical solution temperature, T_c, as a function of polymer molecular weight. At T_c, χ_1 is equal to χ_c. Equations (9.3.37) and (9.3.50) therefore yield

$$\frac{1}{2} + \frac{1}{\sqrt{m}} + \frac{1}{2m} = \frac{1}{2} - \psi \left(1 - \frac{\theta}{T_c} \right) \tag{9.3.51}$$

or, upon rearrangement,

$$\frac{1}{T_c} = \frac{1}{\theta} \left[1 + \frac{1}{\psi} \left(\frac{1}{2m} + \frac{1}{\sqrt{m}} \right) \right] \tag{9.3.52}$$

TABLE 9.1   UCST Data for Solutions of PS in DOP

Molecular weight ($\times 10^{-5}$)	UCST (°C)	Molar volume ratio ($\times 10^{-3}$)
2.00	5.9	0.456
2.80	7.4	0.639
3.35	8.0	0.770
4.70	8.8	1.072
9.00	9.9	2.069
18.00	12.0	4.131

Source: Ref. 10.

so that a plot of $1/T_c$ versus $[(1/2m) + (1/\sqrt{m})]$ should be a straight line with a slope of $1/\theta\psi$ and an intercept of $1/\theta$. These are, in fact, the results obtained by Schultz and Flory [9], and this allows for easy determination of ψ and θ. Clearly, χ_1 equals 0.5 when T equals θ and, therefore, the parameter θ is the theta temperature referred to earlier and is the maximum in the cloud point curve for an infinite-molecular-weight polymer. It can be shown that at the theta temperature, the effect of attraction between polymer segments exactly cancels the effect of the excluded volume and the random coil described in the next chapter exactly obeys Gaussian statistics. Also, the Mark–Houwink exponent equals $\frac{1}{2}$ under theta conditions.

The value of the interaction parameter is often used as a measure of solvent quality. Solvents are normally designated as "good" if $\chi_1 < 0.5$ and "poor" if $\chi_1 > 0.5$; an interaction parameter value of exactly 0.5 denotes an ideal solvent or a theta solvent.

Example 9.2: Listed in Table 9.1 are data for the upper critical solution temperature of six polystyrene (PS)-in-dioctylphthalate (DOP) solutions as a function of molecular weight [10]. Also given is the corresponding ratio of molar volumes. Determine the temperature dependence of the interaction parameter.

Solution: The data of Table 9.1 are plotted in Figure 9.6 according to Eq. (9.3.52). From the straight-line graph, we find that $\psi = 1.45$ and $\theta = 288$ K. This value of the theta temperature is bracketed by similar values estimated by viscometry and light-scattering techniques [10].

9.3.6   Calculating the Binodal

Once the interaction parameter in the form of Eq. (9.3.50) has been determined, the entire temperature–composition phase diagram or the binodal curve can be calculated using the conditions of phase equilibrium. At a chosen temperature, let the two polymer compositions in equilibrium with each other be ϕ_2^C and ϕ_2^D. Let

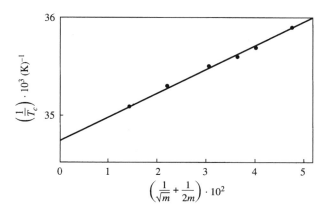

FIGURE 9.6 Plot of the reciprocal of the critical precipitation temperatures $(1/T_c)$ against $[1/\sqrt{m} + 1/(2m)]$ for six polystyrene fractions in DOP. (From Ref. 10.)

the corresponding chemical potentials be μ_2^C and μ_2^D. Because the latter two values must be equal to each other, Eq. (9.3.31) implies the following:

$$\ln \phi_2^C - (m-1)(1-\phi_2^C) + \chi_1 m(1-\phi_2^C)^2$$
$$= \ln \phi_2^D - (m-1)(1-\phi_2^D) + \chi_1 m(1-\phi_2^D)^2 \qquad (9.3.53)$$

This equation can be solved to give χ_1 in terms of ϕ_2^C and ϕ_2^D. Another expression for χ_1 in terms of ϕ_2^C and ϕ_2^D can be obtained by using Eq. (9.3.30) to equate the chemical potentials of the solvent in the two phases. These two expressions for χ_1 can be used to obtain a single equation relating ϕ_2^C to ϕ_2^D. Thereafter, we simply pick a value of ϕ_2^C and solve for the corresponding value of ϕ_2^D. By picking enough different values of ϕ_2^C, we can trace the entire binodal curve because the value of χ_1 and, therefore, T is known for any ordered pair ϕ_2^C, ϕ_2^D. Approximate analytical expressions for the resulting compositions and temperature have been provided by Flory [5], and sample results for the polyisobutylene-in-diisobutyl ketone system are shown in Figure 9.7 [5,9]. Although the theoretical predictions are qualitatively correct, the critical point occurs at a lower than measured concentration. Also, the calculated binodal region is too narrow. Tompa has shown that much more quantitative agreement could be obtained if χ_1 were made to increase linearly with polymer volume fraction [11]. We shall, however, not pursue this aspect of the theory here.

In closing this subsection, we note that the phase equilibrium calculation for polydisperse polymers is conceptually straightforward but mathematically tedious. Each polymer fraction has to be treated as a separate species with its own chemical potential given by an equation similar to Eq. (9.3.31). The interaction parameter, however, is taken to be independent of molecular weight. It is

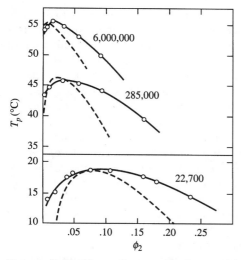

FIGURE 9.7 Phase diagram for three polyisobutylene fractions (molecular weights indicated) in diisobutyl ketone. Solid curves are drawn through the experimental points. The dashed curves have been calculated from theory. (Reprinted with permission from Shultz, A. R., and P. J. Flory: "Phase Equilibria in Polymer-Solvent Systems," J. Am. Chem. Soc., vol. 74, pp. 4760–4767, 1952. Copyright 1952 American Chemical Society.)

necessary to again equate chemical potentials in the two liquid phases and carry out proper mass balances to obtain enough equations in all of the unknowns. Details are available elsewhere [12]. The procedure can be used to predict the results of polymer fractionation [13].

9.3.7 Strengths and Weaknesses of the Flory–Huggins Model

The Flory–Huggins theory, which has been described in detail in this chapter, is remarkably successful in explaining most observations concerning the phase behavior of polymer–solvent systems. For a binary mixture, this theory includes the prediction of two liquid phases and the shift of the critical point to lower concentrations as the molecular weight is increased (see Fig. 9.7). In addition, the theory can explain the phase behavior of a three-component system—whether it is two polymers dissolved in a common solvent or a single polymer dissolved in two solvents. The former situation is relevant to polymer blending [14], whereas the latter is important in the formation of synthetic fibers [15] and membranes [16] by phase inversion due to the addition of nonsolvent. Computation of the phase diagram is straightforward [5], and results are represented on triangular diagrams. Note, though, that the index i in Eq. (9.3.21) ranges from 1 to 3 and, in general, we have three separate interaction parameters relating the three different components. We may also use the theory to interpret the swelling equilibrium of cross-linked polymers brought into contact with good solvents [5]. Because a cross-linked polymer cannot dissolve, it imbibes solvent in a manner similar to that in osmosis. As with osmosis, the process is again self-limiting because swelling causes polymer coil expansion, generating a retractile force (see Chapter 10) that counteracts further absorption of the solvent. The extent of swelling can be used to estimate the value of the polymer–solvent interaction parameter. A technological application of this phenomenon is in the synthesis of porous polymer sorbents as replacements for activated carbon used in the removal of volatile organic compounds from wastewater streams. In this process, a nonporous polymer is lightly cross-linked and then made to swell with the help of an appropriate solvent [17]. Further cross-linking in the swollen state gives a material having a very high degree of porosity.

The Flory–Huggins theory has weaknesses, however. Although some quantitative disagreement between the observed and predicted size of the binodal region has already been noted, the major failure has to do with the inability to predict phase separation above a critical temperature, known as the *lower critical solution temperature*. Freeman and Rowlinson have found that even nonpolar polymers that do not interact with the solvent would demix with increasing temperature [18]. Because the ΔS of mixing is always positive in the Flory–Huggins theory and because χ_1 always decreases with increasing temperature, such a phase separation is totally inexplicable. The resolution of this enigma is discussed in the next section. We close this section by also noting that the Flory–Huggins theory fails for very dilute solutions due to the breakdown of the spatially uniform polymer concentration assumption. The actual entropy change on mixing is found to be less than the predicted theoretical value because polymer molecules in dilute solution exist as isolated random coils whose sizes are a function of the molecular weight. This makes χ_1 a function of the polymer chain length [5,19].

Note that the Flory–Huggins theory applies to flexible macromolecules only. Rodlike particles can be treated in an analogous manner [20] and the results can be used to explain the behavior of polymeric liquid crystals.

9.4 FREE-VOLUME THEORIES

A basic assumption in the Flory–Huggins theory is the absence of a change in volume on mixing. This, however, is not exactly true. As Patterson explains in his very readable review [21], the free volume of the polymer differs markedly from the solvent free volume. (See Chapter 13 for an extensive discussion about the free volume.) The solvent is much more "expanded" due to its larger free volume. When mixing occurs, the solvent loses its free volume and there is a net decrease in the total volume. This result is analogous to, but not the same as, the process of condensation of a gas; in a condensation process, latent heat is evolved and there is an increase in order. Thus, both ΔH and ΔS are negative. This happens even when the polymer and the solvent are chemically similar. Both of these contributions need to be included in the free-energy change on mixing. As the free-volume dissimilarity between the polymer and the solvent increases with increasing temperature, the free-volume effect is likely to be more important at elevated temperatures. One way of accounting for this effect is to consider the interaction parameter χ_1 to be composed of an entropic part in addition to the enthalpic part. Thus,

$$\chi_1 = \chi_H + \chi_S \tag{9.4.1}$$

Indeed, Eq. (9.3.50) already does this, with χ_H being $\psi\theta/T$ and χ_S being $(\frac{1}{2} - \psi)$. Now, we also have to add the free-volume contributions. This is done using an equation of state that allows for a calculation of the volume, enthalpy, and entropy change on mixing from a knowledge of the pure-component properties and a limited amount of solution data. Qualitatively, though, we expect the χ_1 contribution arising from free-volume effects to increase with increasing tempera- ture. This is shown in Figure 9.8. When this free-volume contribution is added to the interaction parameter given by Eq. (9.3.50), the result is a minimum in the χ_1 versus temperature curve. Because phase separation originates from a large

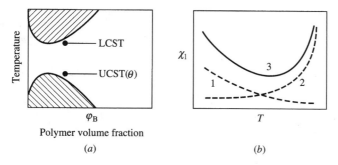

FIGURE 9.8 (a) Phase diagram of a polymer solution showing the phase separation occurring at high temperatures above the lower critical solution temperature (LCST). (b) The temperature dependence of the χ_1 parameter: curve 3, total χ_1; curve 2, contribution to χ_1 due to free-volume dissimilarity between polymer and solvent; curve 1, contribution to χ_1 due to contact-energy dissimilarity between polymer and solvent. (Reprinted with permission from Patterson, D.: "Free volume and Polymer Solubility: A Qualitative View," Macromolecules, vol. 2, pp. 672–677, 1969. Copyright 1969 American Chemical Society.)

positive value of χ_1 and because this can now happen at both low and high temperatures, the phenomenon of a lower critical solution temperature is easily understood. Note that the critical value of χ_1 is still given by Eq. (9.3.37), but, as seen from Figure 9.8, it now corresponds to two different temperatures—a lower critical solution temperature and an upper critical solution temperature. Except for this change, the phase boundaries are again computed using the procedure outlined in Section 9.3.6. A description of the actual procedure for computing the modified χ_1 versus temperature curve shown schematically in Figure 9.8 is beyond the scope of this book, but details are available in the literature [22–24]. Note that using the free-volume theory permits us to explain the existence of an interaction parameter that depends on both temperature and concentration in a manner that logically leads to the prediction of a lower critical solution temperature.

9.5 THE SOLUBILITY PARAMETER

The solubility parameter of Hildebrand [25], generally denoted δ, is a useful alternative to the interaction parameter χ_1 in many situations. It is used to estimate the endothermic heat of mixing that accompanies the dissolution of an amorphous polymer by a low-molecular-weight solvent. The technique has been used extensively in the paint and rubber industries [26]. In the former application, the parameter is used for identifying appropriate solvents, and in the latter, it is used for preventing the swelling of volcanized rubber by solvents. As will be seen here, the major argument in favor of using the solubility parameter is that solution properties are not required; all necessary information can be obtained from data on pure components.

For purposes of motivation, let us consider the mixing of n_1 molecules of a low-molecular-weight species with n_2 molecules of another low-molecular-weight species having the same volume v per molecule. Then, using the same argument enunciated in Section 9.3.2 [setting m as unity in Eq. (9.3.17)], the following can be derived:

$$\Delta H_M = zn\phi_1\phi_2[e_{12} - \tfrac{1}{2}(e_{11} + e_{22})] \tag{9.5.1}$$

in which n equals $n_1 + n_2$ and the e_{ij} terms are all negative quantities. Denoting Avogadro's number by N_A and the total mixture volume by V, Eq. (9.5.1) can be rewritten as

$$\Delta H_M = \frac{V\phi_1\phi_2}{N_A v}\left[-N_A z\,|e_{12}| + \frac{N_A z}{2}\,|e_{11}| + \frac{N_A z}{2}\,|e_{22}|\right] \tag{9.5.2}$$

in which v is the volume per molecule. To make further progress, we assume that

$$|e_{12}| = \sqrt{|e_{11}||e_{22}|} \tag{9.5.3}$$

This assumption is reasonable when there are no specific interactions among molecules, as is true for nonpolar molecules [26], and it allows us to replace a mixture property in terms of pure component properties. Introducing Eq. (9.5.3) into Eq. (9.5.2) yields

$$\Delta H_M = \frac{V\phi_1\phi_2}{N_A v}\left[\left(\frac{N_A z}{2}|e_{11}|\right)^{1/2} - \left(\frac{N_A z}{2}|e_{22}|\right)^{1/2}\right]^2 \tag{9.5.4}$$

If one goes back to the lattice model, $N_A z/2$ represents the total number of interactions among 1 mol of molecules. Multiplication with $|e_{ii}|$ yields the molar internal energy change, ΔU_i, for vaporizing species i. Thus,

$$\Delta H_M = \frac{V\phi_1\phi_2}{N_A v}(\Delta U_1^{1/2} - \Delta U_2^{1/2})^2$$

$$= V\phi_1\phi_2\left[\left(\frac{\Delta U_1}{N_A v}\right)^{1/2} - \left(\frac{\Delta U_2}{N_A v}\right)^{1/2}\right]^2$$

$$= V\phi_1\phi_2(\delta_1 - \delta_2)^2 \tag{9.5.5}$$

where the solubility parameters δ_1 and δ_2 are defined by the above equation, Eq. (9.5.5). The quantity δ^2 is usually called the *cohesive energy* density. Its value $(\Delta U/N_A v)$ is obtained by dividing the molar energy of vaporization by the molar volume. It is obvious that a material with a high cohesive energy density prefers its own company. It is, therefore, more difficult to disperse than a material with a low cohesive energy density.

We can extend the concept of the solubility parameter to macromolecules by defining the solubility parameter of a polymer as

$$\delta_p = \left(\frac{N_A z |e_{pp}|}{2V_p}\right)^{1/2} \tag{9.5.6}$$

where $|e_{pp}|$ is the energy of interaction between two polymer segments and V_p is the volume of 1 mol of polymer segments. Because polymers generally decompose on heating, δ_p cannot be obtained using data on the energy of vaporization, and an indirect method is needed.

An examination of Eq. (9.5.5) shows that ΔH_M vanishes when the solubility parameters of the two components equal each other. Because the theory assumes a positive ΔS_M, the free-energy change on mixing is the most negative when the solubility parameters are matched. A cross-linked polymer would, therefore, swell the most when its solubility parameter equaled that of the solvent. This suggests that one ought to slightly cross-link the polymer whose solubility parameter is sought to be measured and allow it to swell in various solvents having known solubility parameters. The unknown solubility parameter is then equal to the solubility parameter of the liquid that gives rise to the maximum amount of swelling. The same logic also provides the reason for tabulating solubility parameters in the first place: We can select the best solvent for a given polymer simply by finding a liquid with a solubility parameter of the same value.

If we equate the right-hand side of Eq. (9.3.19) to the right-hand side of Eq. (9.5.5), we can relate the interaction parameter to the solubility parameter as follows:

$$\chi_1 = \frac{V_p(\delta_1 - \delta_2)^2}{RT} \tag{9.5.7}$$

Small values of χ_1 promote polymer solubility. Because δ is a surrogate for χ_1, it can be used in much the same way as the interaction parameter. Naturally, it suffers from the same drawbacks, as well as those resulting from the assumption embodied in Eq. (9.5.3). Values of the solubility parameter for selected polymer and nonpolar solvents are listed in Table 9.2. These may also be estimated using the method of group contributions [27]. For a mixture of solvents, the solubility parameter may be taken to be a weighted average of the solubility parameters of the constituents; weighting is done with respect to the volume fraction of the components.

TABLE 9.2 Representative Solubility Parameter Values for Nonpolar Liquids and Amorphous Polymers

Compound	δ (cal/cm^3)$^{1/2}$
Liquid	
n-Pentane	7.1
n-Hexane	7.3
1-Hexene	7.3
n-Octane	7.5
n-Hexadecane	8.0
Toluene	8.9
Benzene	9.2
Styrene	9.3
Carbon disulfide	10
Polymer	
Polytetrafluoroethylene	6.2
Polyethylene	7.9
Polyisobutylene	8.1
Polybutadiene	8.6
Polystyrene	9.1

Source: Ref. 22.

Example 9.3: Based on the solubility parameter concept, which solvent should be used for polystyrene?

Solution: According to Table 9.2, the solubility parameter of polystyrene is 9.1. This suggests the use of either benzene (which has a solubility parameter of 9.2) or toluene (which has a solubility parameter of 8.9). Indeed, these two solvents have been used extensively with polystyrene.

Hansen [28] has suggested one way of modifying the solubility parameter to account for the presence of specific interactions between the polymer and solvent. In this approach, the solubility parameter is considered to be a vector made up of three components: one due to hydrogen-bonding, another due to dipole interactions, and a third due to dispersive forces. Values of each of the three components for different polymers and different solvents have been determined based on experimental observations as well as on theoretical modeling, and these have been tabulated in books [28]. A polymer is found to be soluble in a liquid when the magnitude of the vector difference between the two vectors representing the Hansen solubility parameters of the polymer and the liquid is less than a certain amount. This method has found widespread application in the paint industry.

9.6 POLYMER BLENDS

As explained in Section 9.1, it is much more cost-effective to blend polymers of known properties than it is to try to synthesize new polymers having properties as yet unknown. Therefore, the driving force is the same as the one for the development of composite materials. Indeed, we can look upon immiscible polymer blends as composites on a microscopic scale. It is for this reason that miscibility, by itself, is not the paramount criterion for utility [26]. A specific example in which immiscibility is beneficial is the impact modification of (relatively brittle) polystyrene by rubber; energy absorption results from crazing (see Chapter 12) of the polystyrene matrix in the region between the rubber particles. On the other hand, miscibility is important in applications where segregation of the constituents could lead to deleterious mechanical properties, such as might happen at a weld line in injection molding (see Chapter 15). In addition, because polymers can be processed only between the glass transition temperature and the temperature at which chemical degradation sets in, the addition of a lower-T_g miscible component can often open a processing window whose size depends on the proportion of the material added. An example of this is the commercial blend of polystyrene and high-T_g poly(2,6-dimethyl-1, 4-phenylene oxide) (PPO). Miscible blends of the two materials have an intermediate value of T_g, which suggests that PPO can be processed at temperatures lower than would otherwise be possible. This intermediate value is given by the Flory–Fox equation [26].

$$\frac{1}{T_g} = \frac{w_1}{T_{g1}} + \frac{w_2}{T_{g2}} \qquad (9.6.1)$$

where T_{gi} is the T_g of component i and w_i is its mass fraction. Extensive tabulations of commercial blends (both miscible and immiscible), their properties, and their applications are available in the literature [26,29,30]. Mixing rules for predicting blend properties are also available [31].

In order to predict polymer–polymer miscibility, we might turn to the Flory–Huggins theory, where each lattice site has an interacting segment volume v_s. Dividing both sides of Eq. (9.3.20) by the total mixture volume V and using the definition of the interaction parameter given by Eq. (9.3.18) yields

$$\frac{\Delta G_M}{V} = kT\left[\frac{n_1 v_1}{v_1 V}\ln\phi_1 + \frac{n_2 v_2}{v_2 V}\ln\phi_2\right] + \frac{\Delta e z n_0 \phi_1 \phi_2}{V} \qquad (9.6.2)$$

where v_1 and v_2 are the volume per molecule of the two polymers, respectively. From the definitions of v_s and the volume fractions, we have

$$\frac{\Delta G_M}{V} = kT\left[\frac{\phi_1}{v_1}\ln\phi_1 + \frac{\phi_2}{v_2}\ln\phi_2\right] + \frac{\Delta e z}{v_s}\phi_1 \phi_2 \qquad (9.6.3)$$

Because both v_1 and v_2 are substantially greater than v_s, the first two terms on the right-hand side of Eq. (9.6.2) are negligible compared to the third term. As a consequence, $\Delta G_M \cong \Delta H_M$ and miscibility depends entirely on the energetics of intermolecular interactions. In other words, a negative value of Δe or, equivalently, of the interaction parameter is needed to assure polymer–polymer miscibility.

Example 9.4: If the 1 g of polymer of Example 9.1 is dissolved in 9 g of a different polymer of molecular weight 80,000, what would be the entropy change on mixing? Assume that the density of the two polymers is the same.

Solution: According to Eq. (9.3.16),

$$\Delta S = -R\left[\frac{\ln 0.1}{40,000} + \frac{9}{80,000}\ln 0.9\right] = 6.94 \times 10^{-5}R$$

This number is almost three orders of magnitude smaller than those calculated in Example 9.1.

There is no general theory that might predict *a priori* as to which polymer pairs are likely to be miscible with each other. However, if the solubility parameters of two polymers are matched, then any favorable interactions between the two different kinds of macromolecules are likely to make the enthalpy of mixing be negative; these interactions might include [14] hydrogen-bonding, as in the case of polyvinyl chloride and polyester, and electron donor–electron acceptor molecular complex formation, as in the case of PPO and polystyrene. An indication of whether two polymers may show exothermic mixing may be gained by examining their low-molecular-weight analogs. If these mix in an exothermic manner, the polymers might do as well; if the enthalpy change on mixing is endothermic, then the corresponding polymers will certainly not interact in a favorable manner. Modifications of the Flory–Huggins theory that account for specific interactions, especially hydrogen-bonding, have been discussed at length by Coleman et al. [32].

9.7 CONCLUSION

In this chapter, we have seen how classical and statistical thermodynamics coupled with simple ideas of lattice theory can be used to predict the phase behavior of polymer solutions. For polymers dissolved in low-molecular-weight solvents, the Flory–Huggins theory and its various modifications can adequately explain data obtained for quiescent solutions. More recently, the theory has been applied to predict the shift in the binodal under the influence of an imposed shear deformation [33]. For macromolecular solvents, however, development of the theory has not reached the same stage of maturity as for low-molecular-weight solvents. This remains an area of current and active research.

REFERENCES

1. Blanks, R. F., Engineering Applications of Polymer Solution Thermodynamics, *Chemtech*, 396–401, June 1976.
2. Smith, J. M., H. C. Van Ness, and M. Abbott, *Introduction to Chemical Engineering Thermodynamics*, 5th ed., McGraw-Hill, New York, 1996.
3. Flory, P. J., Thermodynamics of High Polymer Solutions, *J. Chem. Phys.*, 10, 51–61, 1942.
4. Huggins, M. L., Thermodynamic Properties of Solutions of Long-Chain Compounds, *Ann. NY Acad. Sci.*, 43, 1–32, 1942.
5. Flory, P. J., *Principles of Polymer Chemistry*, Cornell University Press, Ithaca, NY, 1953.
6. Sheehan, C. J., and A. L. Bisio, Polymer/Solvent Interaction Parameters, *Rubber Chem. Tech.*, 39, 149–192, 1966.
7. Murrell, J. N., and E. A. Boucher, *Properties of Liquids and Solutions*, Wiley, Chichester, 1982.
8. Huggins, M. L., Properties of Rubber Solutions and Gels, *Ind. Eng. Chem.*, 35, 216–220, 1943.

9. Shultz, A. R., and P. J. Flory, Phase Equilibria in Polymer-Solvent Systems, *J. Am. Chem. Soc.*, 74, 4760–4767, 1952.

10. Rangel-Nafaile, C., and J. J. Munoz-Lara, Analysis of the Solubility of PS in DOP Through Various Thermodynamic Approaches, *Chem. Eng. Commun.*, 53, 177–198, 1987.

11. Tompa, H., *Polymer Solutions*, Butterworth Scientific, London, 1956.

12. Flory, P. J., Thermodynamics of Heterogeneous Polymers and Their Solution, *J. Chem. Phys.*, 12, 425–438, 1944.

13. Cantow, M. R. J., *Polymer Fractionation*, Academic Press, New York, 1967.

14. Barlow, J. W., and D. R. Paul, Polymer Blends and Alloys—A Review of Selected Considerations, *Polym. Eng. Sci.*, 21, 985–996, 1981.

15. Han, C. D., and L. Segal, A Study of Fiber Extrusion in Wet Spinning: II. Effects of Spinning Conditions on Fiber Formation, *J. Appl. Polym. Sci.*, 14, 2999–3019, 1970.

16. Tsay, C. S., and A. J. McHugh, Mass Transfer Modeling of Asymmetric Membrane Formation by Phase Inversion, *J. Polym. Sci. Polym. Phys.*, 28, 1327–1365, 1990.

17. Davankov, V. A., and M. P. Tsyurupa, Structure and Properties of Hypercrosslinked Polystyrene—The First Representative of a New Class of Polymer Networks, *React. Polym.*, 13, 27–42, 1990.

18. Freeman, P. I., and J. S. Rowlinson, Lower Critical Points in Polymer Solutions, *Polymer*, 1, 20–26, 1960.

19. Krigbaum, W. R., and P. J. Flory, Statistical Mechanics of Dilute Polymer Solutions: IV. Variation of the Osmotic Second Coefficient with Molecular Weight, *J. Am. Chem. Soc.*, 75, 1775–1784, 1953.

20. Flory, P. J., Phase Equilibria in Solutions of Rod-Like Particles, *Proc. Roy. Soc. London*, A234, 73–89, 1956.

21. Patterson, D., Free Volume and Polymer Solubility: A Qualitative View, *Macromolecules*, 2, 672–677, 1969.

22. von Tapavicza, S., and J. M. Prausnitz, Thermodynamics of Polymer Solutions: An Introduction, *Int. Chem. Eng.*, 16, 329–340, 1976.

23. Cassasa, E. F., Thermodynamics of Polymer Solutions, *J. Polym. Sci. Polym. Symp.*, 54, 53–83, 1976.

24. Carpenter, D. K., Solution Properties, in *Encyclopedia of Polymer Science Engineering*, 2nd ed., H. F. Mark, N. M. Bikales, C. G. Overberger, and G. Menges (eds.), Wiley, New York, 1989, Vol. 15, p. 481.

25. Hildebrand, J. H., and R. L. Scott, *The Solubility of Nonelectrolytes*, 3rd ed., Reinhold, New York, 1950.

26. Olabisi, O., L. M. Robeson, and M. T. Shaw, *Polymer–Polymer Miscibility*, Academic Press, New York, 1979.

27. Small, P. A., Some Factors Affecting the Solubility of Polymers, *J. Appl. Chem.*, 3, 71–80, 1953.

28. Hansen, C. M., *The Three Dimensional Solubility Parameter and Solvent Diffusion Coefficient*, Danish Technical Press, Copenhagen, 1967.

29. Paul, D. R., and J. W. Barlow, Polymer Blends (or Alloys), *J. Macromol. Sci. Macromol. Chem.*, C18, 109–168, 1980.

30. Utracki, L. A., *Polymer Alloys and Blends*, Hanser, Munich, 1990.

31. Nielson, L. E., *Predicting the Properties of Mixtures*, Marcel Dekker, New York, 1978.

32. Coleman, M. M., J. F. Graf, and P. C. Painter, *Specific Interactions and the Miscibility of Polymer Blends*, Technomic, Lancaster, PA, 1991.

33. Rangel-Nafaile, C., A. B. Metzner, and K. F. Wissbrun, Analysis of Stress-Induced Phase Separations in Polymer Solutions, *Macromolecules*, 17, 1187–1195, 1984.

PROBLEMS

9.1. What is the value of m, the ratio of the molar volume of polymer to the molar volume of solvent, for polystyrene of 250,000 molecular weight dissolved in toluene? Thus, determine χ_c and the polymer volume fraction corresponding to the upper critical solution temperature. Assume that the density of polystyrene is $1.1 \, \text{g/cm}^3$ and that of toluene is $0.86 \, \text{g/cm}^3$.

9.2. By noting the significance of intrinsic viscosity and by examining the data given in Figure 8.10, determine the lower limit of polymer concentration at which you might expect the Flory–Huggins theory to apply to the polymer solution at 24°C. Note that concentration is measured in units of grams per deciliter.

9.3. Use Eq. (9.3.30) to obtain an equation similar to Eq. (9.3.53). Then, use the data given in Figure 9.7 to compute the interaction parameter for polyisobutylene in diisobutyl ketone at 15°C for the 22,700 molecular-weight polymer. Assume that $1/m$ is negligible compared to unity.

9.4. Show that it is possible to determine the solvent activity needed in Eq. (9.3.44) from measurements of the boiling point elevation. In particular, show that

$$-\ln a_1 = \frac{\Delta h_1^v}{RT_b^2} \Delta T_b$$

9.5. How will the slope of the straight-line plot in Figure 8.5 change as the solvent becomes a progressively better solvent? In other words, how does the second virial coefficient depend on solvent quality?

9.6. A polymer sample dissolves in toluene but not in ethyl acetate. Is the polymer likely to be polyisobutylene or polystyrene?

9.7. Why does vinyl upholstery become less and less flexible with use?

9.8. What is one likely to observe if a solution of polystyrene in a mixture of dichloromethane and diethyl ether is added dropwise to a beaker containing water?

9.9. What simplification occurs in Eq. (9.3.44) as the polymer molecular weight increases? Further, if the solvent volume fraction is small, show that

$$P_1 = P_1^0 \phi_1 e^{1+\chi_1}$$

9.10. Use the procedure outlined in Section 9.3.6 to predict the solubility diagram for the polystyrene–DOP system using data for the 200,000 molecular-weight polymer listed in Table 9.1.

9.11. Use the Flory–Huggins theory to calculate the free-energy change on mixing polystyrene of 200,000 molecular weight with DOP at a concentration of $0.05\,g/cm^3$ of solution. The molecular weight of DOP is 390 and its density is $0.98\,g/cm^3$.

      (a) Would you expect to see a single-phase solution or two phases at a temperature of 1°C? At 15°C?

      (b) Are the results obtained in part (a) consistent with those of Problem 9.6?

9.12. Under what conditions does a polymer solution in a mixed solvent act as if the polymer were dissolved in a single solvent?

9.13. A good solvent can be looked upon as one that promotes polymer–solvent interactions leading to a larger size of the polymer coil in solution as compared to the corresponding coil size in a theta solvent. Based on this reasoning, how do you expect the viscosity of a solution of polyisobutylene in n-hexadecane to vary as increasing amounts of carbon disulfide are added to the solution? Assume that the viscosity of each of the two solvents is the same.

9.14. Use literature values of the molar energy of vaporization and the molar volume to estimate the solubility parameter of water at 25°C.

9.15. The glass transition temperature of poly(ether ether ketone) (PEEK), a semicrystalline polymer, is 145°C, whereas that of poly(ether imide) (PEI), an amorphous polymer miscible with PEEK, is 215°C. What will be the T_g of a blend of these two materials containing 10% by weight PEI? Speculate on what the presence of the PEI in the blend might do to the rate of crystallization of PEEK.

9.16. Would a polymer that hydrogen-bonds with itself be more likely or less likely to form miscible blends with other polymers compared with a polymer that does not hydrogen-bond with itself?

9.17. Use the results of Example 9.2 and the data given in Table 9.2 to estimate the value of the solubility parameter of DOP at 20°C.

10

Theory of Rubber Elasticity

10.1 INTRODUCTION

As mentioned in Chapter 2, all polymers are stiff, brittle, glassy materials below their glass transition temperature, T_g. However, they soften and become pliable once above T_g and, ultimately, flow at still higher temperatures. For crystalline polymers, the flow temperature is slightly above the crystalline melting point. In this chapter, we examine the mechanical behavior of solid polymers above T_g, whereas polymer crystallization is considered in Chapter 11, and the deformation and failure properties of glassy polymers are presented in Chapter 12. The stress-versus-strain behavior of amorphous polymers above T_g is similar to that of natural rubber at room temperature and very different from that of metals and crystalline solids. Although metals can be reversibly elongated by only a percent or so, rubber can be stretched to as much as 10 times its length without damage. Furthermore, the stress needed to achieve this deformation is relatively low. Thus, polymers above T_g are soft elastic solids; this property is known as *rubberlike elasticity*. Other extraordinary properties of rubber have also been known for a long time. Gough's experiments in the early 1800s revealed that, unlike metals, a strip of rubber heats up on sudden elongation and cools on sudden contraction [1]. Also, its modulus increases with increasing temperature. These properties are lost, however, if experiments are performed in cold water. Explaining these remarkable observations is useful not only for satisfying intellectual curiosity but also for the purpose of generating an understanding that is beneficial for tailoring the properties of rubberlike materials (called *elastomers*) for specific applications. Recall that rubber (whether natural or synthetic) is used to manufacture tires, adhesives, and footwear, among other products. Note also that because polymer properties change so drastically around T_g, the use temperature of most polymers is either significantly below T_g (as in the case of plastics employed for structural applications) or significantly above T_g (as in the case of elastomers).

Chemically, rubber is *cis*-1,4-polyisoprene, a linear polymer, having a molecular weight of a few tens of thousands to almost four million, and a wide molecular-weight distribution. The material collected from the rubber tree is a latex containing 30–40% of submicron rubber particles suspended in an aqueous protein solution, and the rubber is separated by coagulation caused by the addition of acid. At room temperature, natural rubber is really an extremely viscous liquid because it has a T_g of $-70°C$ and a crystalline melting point of about $-5°C$. It is the presence of polymer chain entanglements that prevents flow over short time scales.

In order to explain the observations made with natural rubber and other elastomers, it is necessary to understand the behavior of polymers at the microscopic level. This leads to a model that predicts the macroscopic behavior. It is surprising that in one of the earliest and most successful models, called the *freely jointed chain* [2,3], we can entirely disregard the chemical nature of the polymer and treat it as a long slender thread beset by Brownian motion forces. This simple picture of polymer molecules is developed and embellished in the sections that follow. Models can explain not only the basics of rubber elasticity but also the qualitative rheological behavior of polymers in dilute solution and as melts. The treatment herein is kept as simple as possible. More details are available in the literature [1–7].

10.2 PROBABILITY DISTRIBUTION FOR THE FREELY JOINTED CHAIN

One of the simplest ways of representing an isolated polymer molecule is by means of a freely jointed chain having n links each of length l. Even though real polymers have fixed bond angles, such is not the case with the idealized chain. In addition, there is no correspondence between bond lengths and the dimensions of the chain. The freely jointed chain, therefore, is a purely hypothetical entity. Its behavior, however, is easy to understand. In particular, as will be shown in this section, it is possible to use simple statistical arguments to calculate the probability of finding one end of the chain at a specified distance from the other end when one end is held fixed but the other end is free to move at random. This probability distribution can be coupled with statistical thermodynamics to obtain the chain entropy as a function of the chain end-to-end distance. The expression for the entropy can, in turn, be used to derive the force needed to hold the chain ends a particular distance apart. This yields the force-versus-displacement relation for the model chain. If all of the molecules in a block of rubber act similarly to each other and each acts like a freely jointed chain, the stress–strain behavior of the rubber can be obtained by adding together contributions from each of the chains. Because real polymer molecules are not freely jointed chains, the final results cannot be expected to be quantitatively correct. The best that we can hope for is that the form of the equation is correct. This equation obviously involves the chain parameters n and l, which are unknown. If we are lucky, all of the unknown quantities will be grouped as one or two constants whose values can be determined by experiment. This, then, is our working hypothesis.

To proceed along this path, let us conduct a thought experiment. Imagine holding one end of the chain fixed at the origin of a rectangular Cartesian coordinate system (as shown in Fig. 10.1) and observe the motion of the other end. You will find that the distance r between the two ends ranges all the way from zero to nl even though some values of the end-to-end distance occur more frequently than others. In addition, if we use spherical coordinates to describe the location of the free end, different values of θ and ϕ arise with equal frequency. As a consequence, the magnitude of the projection on any of the three axes x, y, and z of a link taken at random will be the same and equal to $l/\sqrt{3}$.

To determine the probability distribution function for the chain end-to-end distance, we first consider a freely jointed, one-dimensional chain having links of length $l_x = l/\sqrt{3}$, which are all constrained to lie along the x axis. What is the probability that the end-to-end distance of this one-dimensional chain is ml_x? The

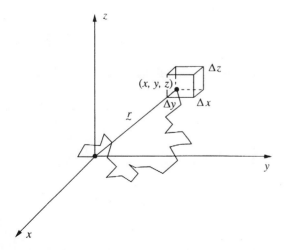

FIGURE 10.1 The unconstrained freely jointed chain.

answer to this question can be obtained by analyzing the random walk of a person who starts out from the origin and takes n steps along the x axis; n_+ of these steps are in the positive x direction and n_- are in the negative x direction, and there is no relation between one step and the next one. Clearly, m equals $(n_+ - n_-)$.

From elementary probability theory, the probability that an event will occur is the ratio of the number of possible ways in which that event can occur to the total number of events. As a consequence, the probability, $p(m)$, of obtaining an end-to-end distance of ml_x is the number of ways in which one can take n_+ forward steps and n_- backward steps out of n steps divided by the total number of ways of taking n steps. The numerator, then, is the same as the number of ways of putting n objects (of which n_+ are of one kind and n_- are of another kind) into a container having n compartments. This is $n!/(n_+!n_-!)$. Because any given step can either be a forward step or a backward step, each step can be taken in two ways. Corresponding to each way of taking a step, the next step can again be taken in two ways. Thus, the total number of ways of taking n steps is 2^n, which gives us

$$p(m) = \frac{n!}{2^n n_+! n_-!} \tag{10.2.1}$$

We can rewrite n_+ and n_- as follows:

$$n_+ = \frac{1}{2}(n + m) \tag{10.2.2}$$

$$n_- = \frac{1}{2}(n - m) \tag{10.2.3}$$

For large n we can use Stirling's formula:

$$n! = \frac{\sqrt{2\pi}n^{(2n+1)/2}}{e^n} \tag{10.2.4}$$

In Eq. (10.2.1), introduce Eqs. (10.2.2) and (10.2.3) in the result and simplify to obtain the following:

$$p(m) = \sqrt{\frac{2}{n\pi}} \left[\left(1 + \frac{m}{n}\right)^{(n+m+1)/2} \left(1 - \frac{m}{n}\right)^{(n-m+1)/2} \right]^{-1} \tag{10.2.5}$$

Taking the natural logarithm of both sides of Eq. (10.2.5) and recognizing that

$$\ln\left(1 + \frac{m}{n}\right) \cong \frac{m}{n} - \frac{m^2}{2n^2} \tag{10.2.6}$$

provided that m/n is small,

$$\ln p(m) = \frac{1}{2}\ln\left(\frac{2}{n\pi}\right) - \frac{m^2}{2n} + \frac{m^2}{2n^2} \tag{10.2.7}$$

Neglecting the very last term in Eq. (10.2.7),

$$p(m) = \left(\frac{2}{n\pi}\right)^{1/2} e^{-m^2/2n} \tag{10.2.8}$$

which is known as a Gaussian or Normal distribution. Note that for all of these relations to be valid, n has to be large and m/n has to be small.

Equation (10.2.8) represents a discrete probability distribution and is the probability that x lies between ml_x and $(m+2)l_x$. This is because if n_+ increases by 1, n_- has to decrease by 1 and m increases by 2. Simultaneously, the distance between the chain ends goes up by $2l_x$. To obtain the continuous probability distribution $p(x)\,dx$, which is the probability that the end-to-end distance ranges from x to $x+dx$, we merely multiply $p(m)$ by $dx/(2l_x)$. Furthermore, because m equals x/l_x,

$$p(x)\,dx = (2n\pi l_x^2)^{-1/2} e^{-x^2/2nl_x^2}\,dx \tag{10.2.9}$$

In order to extend the one-dimensional results embodied in Eq. (10.2.9) to the three-dimensional case of practical interest, we use the law of joint probability. According to this law, the probability of a number of events happening simultaneously is the product of the probabilities of each of the events occurring individually. Thus, the probability, $p(r)\,dr$, that the unconstrained end of the freely jointed chain lies in a rectangular parallelepiped defined by x, y, z, $x+dx$, $y+dy$, and $z+dz$ (see Fig. 10.1) is the product $p(x)\,dx\,p(y)\,dy\,p(z)\,dz$, where $p(y)\,dy$ and $p(z)\,dz$ are defined in a manner analogous to $p(x)\,dx$. Therefore,

$$p(r)\,dr = (2n\pi)^{-3/2}(l_x^2 l_y^2 l_z^2)^{-1/2} \exp\left[-\frac{1}{2n}\left(\frac{x^2}{l_x^2}+\frac{y^2}{l_y^2}+\frac{z^2}{l_z^2}\right)\right] dx\,dy\,dz \tag{10.2.10}$$

Denoting the sum $(x^2 + y^2 + z^2)$ as r^2 and recalling that $l_x^2 = l_y^2 = l_z^2 = l^2/3$,

$$p(r)\,dr = \left(\frac{3}{2n\pi l^2}\right)^{3/2} e^{-3r^2/2nl^2}\,dx\,dy\,dz \tag{10.2.11}$$

To obtain the probability that the free end of the chain lies not in the parallelepiped shown in Figure 10.1 but anywhere in a spherical shell of radius r and thickness dr, we appeal to the law of addition of probabilities. According to this law, the probability that any one of several events may occur is simply the sum of the probabilities of each of the events. Thus, the probability that the chain end may lie anywhere within the spherical shell is the sum of the probabilities of finding the chain end in each of the parallelepipeds constituting the spherical shell. Using Eq. (10.2.11) to carry out this summation, we see that the result is again Eq. (10.2.11), but with the right-hand side modified by replacing $dx\,dy\,dz$ with $4\pi r^2\,dr$, the volume of the spherical shell. Finally, then, we have

$$p(r)\,dr = \left(\frac{3}{2n\pi l^2}\right)^{3/2} e^{-3r^2/2nl^2} 4\pi r^2\,dr \tag{10.2.12}$$

which represents the probability that the free end of the chain is located at a distance r from the origin and contained in a spherical shell of thickness dr. This is shown graphically in Figure 10.2. Note that the presence of r^2 in Eq. (10.2.12) causes $p(r)$ to be zero at the origin, whereas the negative exponential drives $p(r)$ to zero at large values of r. As seen in Figure 10.2, $p(r)$ is maximum at an intermediate value of r^2. Also, because the sum of all the probabilities must equal unity, $\int_0^\infty p(r)\,dr = 1$.

At this point, it is useful to make the transition from the behavior of a single chain to that of a large collection of identical chains. It is logical to expect that the end-to-end distances traced out by a single chain as a function of time would be the same as the various end-to-end distances assumed by the collection of chains at a single time instant. Thus, time averages for the isolated chain ought to equal ensemble averages for the collection of chains. Using Eq. (10.2.12), then, the average values of the chain's end-to-end distance and square of the chain's end-to-end distance are as follows:

$$\langle r \rangle = \int_0^\infty rp(r)\,dr = 2l\left(\frac{2n}{3\pi}\right)^{1/2} \tag{10.2.13}$$

$$\langle r^2 \rangle = \int_0^\infty r^2\,p(r)\,dr = nl^2 \tag{10.2.14}$$

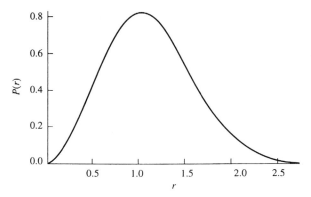

FIGURE 10.2 Distribution function $p(r)$ given by Eq. (10.2.12). (Reprinted from Treloar, L. R. G.: The Physics of Rubber Elasticity, 3rd ed., Clarendon, Oxford, U.K., 1975, by permission of Oxford University Press.)

where the angular brackets denote ensemble averages. Because the fully extended length of the chain (also called the *contour length*) is nl, Eq. (10.2.14) demonstrates that the mean square end-to-end distance is very considerably less than the square of the chain length. Therefore, the freely jointed chain behaves like a random coil and this explains the enormous extensibility of rubber molecules.

Having obtained the average value of the square of the chain end-to-end distance and the distribution of end-to-end values about this mean, it is worth pausing and again asking if there is any relation between these results and results for real polymer molecules. In other words, how closely do freely jointed chains approximate actual macromolecules? If the answer is "not very closely," then how do we modify the freely jointed chain results to make them apply to polymers?

The first response is that most polymer molecules do, indeed, resemble long flexible strings. This is because linear (unbranched) polymers with a large degree of polymerization have aspect ratios that may be as high as 10^4. They are thus fairly elongated molecules. Furthermore, despite the restriction to fixed bond angles and bond lengths, the possibility of rotation about chemical bonds means that there is little correlation between the position of one bond and another one that is five or six bond lengths removed. However, two consequences of these restrictions are that the contour length becomes less than the product of the bond length and the number of bonds and that the mean square end-to-end distance becomes larger than that previously calculated.

If bond angles are restricted to a fixed value θ, the following can be shown [4]:

$$\langle r^2 \rangle = nl^2 \frac{(1 - \cos \theta)}{(1 + \cos \theta)} \qquad (10.2.15)$$

If, in addition, there is hindered rotation about the backbone due to, say, steric effects, then we have

$$\langle r^2 \rangle = nl^2 \frac{(1 - \cos \theta)(1 + \cos\langle \phi \rangle)}{(1 + \cos \theta)(1 - \cos\langle \phi \rangle)} \qquad (10.2.16)$$

where $\langle \phi^2 \rangle$ is the average value of the torsion angle. Small-angle neutron scattering data have supported this predicted proportionality between $\langle r^2 \rangle$ and nl^2.

Because $\langle r^2 \rangle$ increases with each additional restriction but remains proportional to $\langle r^2 \rangle$ for a freely jointed chain, we can consider a polymer molecule a freely jointed chain having n' links, where n' is less than the number of bonds, but the length of each link l' is greater than the bond length, so that $\langle r^2 \rangle$ is again $n'l'^2$ and the contour length is $n'l'$.

Example 10.1: Polyethylene has the planar zigzag structure shown in Figure 10.3. If the bond length is l and the valence angle θ is 109.5°, what are the

FIGURE 10.3   The planar zigzag structure of polyethylene.

contour length R and the mean square end-to-end distance? Let the chain have n bonds and let there be free rotation about the bonds.

Solution: From Figure 10.3, it is clear that the projected length of each link is $l \sin(\theta/2)$. Using the given value of θ and noting that there are n links, the fully extended chain length is given by

$$R = nl \sin(54.75°) = \sqrt{\frac{2}{3}}\, nl$$

The mean square end-to-end distance is obtained from Eq. (10.2.15) as follows:

$$\langle r^2 \rangle = 2nl^2$$

When the mean square end-to-end distance of a polymer is given by Eq. (10.2.16), the polymer is said to be in its "unperturbed" state. What causes the polymer to be "perturbed" is the fact that in the derivation of Eq. (10.2.16), we have allowed for the possibility of widely separated atoms that make up different portions of the same polymer molecule to occupy the same space. In reality, those arrangements that result in overlap of atoms are excluded. This is known as the excluded-volume effect, and it results in dimensions of real polymer molecules becoming larger than the unperturbed value. It is customary to quantify this phenomenon by defining a coil expansion factor that is the ratio of the root mean square end-to-end distance of the real chain to the corresponding quantity for the unperturbed chain. In a very good solvent, there is a further increase in size, as determined by intrinsic viscosity measurements, and the coil-expansion factor can become as large as 2. In a poor solvent, on the other hand, the molecule shrinks, and if the solvent quality is poor enough, the coil expansion factor can become unity. In such a case, the solvent is called a theta solvent, and we have the theta condition encountered earlier in Chapter 9. It is, therefore, seen that the theta condition can be reached either by changing temperature without changing the solvent or by changing the solvent under isothermal conditions.

In closing this section, we re-emphasize that the size of a polymer molecule measured using the light-scattering technique discussed in Chapter 8 is the mean square radius of gyration $\langle s^2 \rangle$. For a freely jointed chain this quantity, defined as the square distance of a chain element from the center of gravity, is given by

$$\langle s^2 \rangle = \frac{1}{6} \langle r^2 \rangle \tag{10.2.17}$$

The radius of gyration is especially useful in characterizing branched molecules having multiple ends where the concept of a single end-to-end distance is not particularly meaningful.

10.3 ELASTIC FORCE BETWEEN CHAIN ENDS

If we return to the unconstrained chain depicted in Figure 10.1 and measure the time-dependent force needed to hold one of the chain ends at the origin of the coordinate system, we find that the force varies in both magnitude and direction, but its time average is zero due to symmetry. If, however, the other chain end is also held fixed so that a specified value of the end-to-end distance is imposed on the chain, the force between the chain ends will no longer average out to zero. Due to axial symmetry, though, the line of action of the force will coincide with r, the line joining the two ends. For simplicity of analysis, let this line be the x axis.

In order to determine the magnitude of the force between the chain ends, let us still keep one end at the origin but apply an equal and opposite (external) force f on the other end so that the distance between the two ends increases from x to $x + dx$. The work done on the chain in this process is

$$dW = -f \, dx \tag{10.3.1}$$

where the sign convention employed is that work done by the system and heat added to the system are positive.

If chain stretching is done in a reversible manner, a combination of the first and second laws of thermodynamics yields

$$dW = T \, dS - dU \tag{10.3.2}$$

where S is entropy and U is internal energy. Equating the right-hand sides of Eqs. (10.3.1) and (10.3.2) and dividing throughout by dx gives

$$f = -T \, \frac{dS}{dx} + \frac{dU}{dx} \tag{10.3.3}$$

From statistical thermodynamics, the entropy of a system is related to the probability distribution through the following equation:

$$S = k \, \ln p(x) \tag{10.3.4}$$

where k is Boltzmann's constant. In the present case, $p(x)$ is given by Eq. (10.2.9) so that

$$f = -T \frac{d}{dx}\left(-\frac{k}{2}\ln(2n\pi l_x^2) - \frac{kx^2}{2nl_x^2}\right) + \frac{dU}{dx} \qquad (10.3.5)$$

and carrying out the differentiation,

$$f = \frac{kTx}{nl_x^2} + \frac{dU}{dx} = \frac{3kTx}{nl^2} + \frac{dU}{dx} \qquad (10.3.6)$$

The internal energy term in this equation is related to changes in the internal potential energy arising from the making and breaking of van der Waals bonds. Because rubbers elongate very easily, we find that the second term on the right-hand side of Eq. (10.3.6) is small compared to the first term. Consequently,

$$f = \frac{3kT}{nl^2}x \qquad (10.3.7)$$

which is a linear relationship between the force and the distance between chain ends and is similar to the behavior of a linear spring. The constant of proportionality, $3kT/nl^2$, is the modulus of the material and its value increases as temperature increases. This explains why a stretched rubber band contracts on heating when it is above the polymer glass transition temperature.

   The positive force f in Eq. (10.3.7) is externally applied and is balanced by an inward-acting internal force, which, in the absence of the external force, tends to make the end-to-end distance go to zero. This, however, does not happen in practice because the spring force is not the only one acting on the chain; the equilibrium end-to-end distance is given by a balance of all the forces acting on the polymer molecule. This aspect of the behavior of isolated polymer molecules will be covered in greater detail in the discussion of constitutive equations for dilute polymer solutions in Chapter 14.

   If we were not aware of the assumptions that have gone into the derivation of Eq. (10.3.7), we might conclude that the force between the chain ends increases linearly and without bound as x increases. Actually, Eq. (10.3.7) is valid only for values of x that are small compared to the contour length of the chain. For larger extensions exceeding one-third the contour length, f increases nonlinearly with x, and we know that for values of x approaching nl, chemical bonds begin to be stretched. It can be shown that the right-hand side of Eq. (10.3.7) is merely the first term in a series expansion for f [1]:

$$f = \frac{kT}{l}\left[\frac{3x}{nl} + \frac{9}{5}\left(\frac{x}{nl}\right)^3 + \frac{297}{175}\left(\frac{x}{nl}\right)^5 + \cdots\right] = \frac{kT}{l}L^{-1}\left(\frac{x}{nl}\right) \qquad (10.3.8)$$

where L^{-1} is called the *inverse Langevin function*. The Langevin function itself is defined as

$$L(x) = \coth x - \frac{1}{x} \qquad (10.3.9)$$

Equations (10.3.7) and (10.3.8) are plotted in Figure 10.4 to show the region in which it is permissible to use the simpler expression, Eq. (10.3.7).

Example 10.2: What is the percentage error involved in using Eq. (10.3.7) when x/nl equals (a) 0.25? (b) 0.5?

Solution:

  (a) According to Eq. (10.3.8), $f = 0.78(kT/l)$, whereas according to Eq. (10.3.7), $f = 0.75(kT/l)$. Thus, the percentage error is 3.85%.
  (b) The corresponding values for f now are $1.78(kT/l)$ and $1.5(kT/l)$. The percentage error, therefore, increases to 15.7%.

A closer examination of Eq. (10.3.3) reveals a significant difference between the nature of rubbers and the nature of crystalline solids. In general, f includes contributions due to changes in entropy as well as changes in internal energy. In crystalline solids, the change in entropy on deformation is small and all

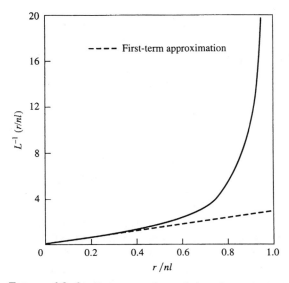

FIGURE 10.4 Force-extension relation for a freely jointed chain. (Reprinted from Treloar, L. R. G.: The Physics of Rubber Elasticity, 3rd ed., Clarendon, Oxford, UK, 1975, by permission of Oxford University Press.)

the work goes into increasing the internal (potential) energy. For rubbery polymers, on the other hand, the entropy change dominates and f depends entirely on changes in entropy. It is for this reason that polymer molecules are said to act as entropy springs. Note that the spring constant decreases (i.e., the spring becomes softer) as the polymer chain length increases. Also, because ΔU on deformation is zero, a consequence of entropic elasticity is that the work done on stretching a rubber must result in a release of heat if the process is isothermal. If the stretching is rapid, however, adiabatic conditions may result so that the temperature rises. The reverse situation occurs when the stretched rubber is released. For crystalline materials, on the other hand, stretching results in a storage of energy. On removal of load, no work is done against any external force and the recovered internal energy shows up as an increase in temperature.

10.4   STRESS–STRAIN BEHAVIOR

In this section, we are interested in determining how a block of rubber deforms under the influence of an externally applied force. The procedure for doing this is the same as the one employed for the isolated chain in the previous section. We assume that there are N chains per unit volume, and each behaves like an isolated chain in its unstrained, equilibrium state. When the block of rubber deforms, each chain making up the block of rubber deforms as well. It is assumed that the deformation is affine; that is, there is no slippage past chains and the macroscopic strain equals the microscopic strain. In other words, changes in the length of individual chains correspond exactly to changes in length of corresponding lines drawn on the exterior of the bulk rubber.

This assumption makes it possible to calculate the change in entropy on deformation of a single chain for a specified macroscopic strain. A summation over all chains gives the macroscopic change in entropy of the rubber block, and the subsequent application of Eq. (10.3.3) yields the desired force or stress corresponding to the imposed strain. Let us illustrate this process for some idealized situations. The more general case will be considered later.

Before proceeding further, we must define strain. In a tensile test, we find that materials such as metals extend only by 1% or less. We, therefore, define strain as the increase in length divided by either the original length or the final length. For rubbers, however, a doubling in length is easily accomplished, and the initial length l_0 and final length l are dramatically different. Consequently, the measure of infinitesimal strain that works for metals is inappropriate in this case; a measure of finite strain is needed instead. One popular measure is the Hencky strain $\ln l/l_0$ and another is the extension ratio $\lambda = l/l_0$. The latter quantity is more easily related to the force acting on one face of a block of rubber.

Consider, for example, a normal force F acting perpendicular to one face of an initially unstrained cube of rubber of edge l_0. Under the influence of this force, the cube transforms into a rectangular prism having dimensions l_1, l_2, and l_3, as shown in Figure 10.5. If we define λ_1 as l_1/l_0, λ_2 as l_2/l_0, and λ_3 as l_3/l_0, then the affine deformation assumption implies that the coordinates of the end-to-end vector of a typical polymer chain change from (x_0, y_0, z_0) to $(\lambda_1 x_0, \lambda_2 y_0, \lambda_3 z_0)$. Under this change in dimensions, the change in entropy of the chain is, from Eqs. (10.2.11) and (10.3.4), as follows:

$$\Delta S = -\frac{3k}{2nl^2}(\lambda_1^2 x_0^2 + \lambda_2^2 y_0^2 + \lambda_3^2 z_0^2 - x_0^2 - y_0^2 - z_0^2) \tag{10.4.1}$$

Because the chain is randomly oriented before it is stretched,

$$x_0^2 = y_0^2 = z_0^2 = \frac{r_0^2}{3} = \frac{nl^2}{3} \tag{10.4.2}$$

The change in entropy ΔS_t of all the chains in the cube of rubber is Nl_0^3 times the change in entropy of a single chain. In view of Eqs. (10.4.1) and (10.4.2), this quantity is

$$\Delta S_t = -\frac{k}{2}Nl_0^3(\lambda_1^2 + \lambda_2^2 + \lambda_3^2 - 3) \tag{10.4.3}$$

and the work done on the rubber is

$$W = -\frac{NkT}{2}l_0^3(\lambda_1^2 + \lambda_2^2 + \lambda_3^2 - 3) \tag{10.4.4}$$

which is also known as the strain-energy function. Note that, thus far, the treatment has been quite general, and the specific nature of the stress distribution has not been used.

Because rubber is incompressible, its volume does not change on deformation. Therefore, it must be true that

$$\lambda_1 \lambda_2 \lambda_3 = 1 \tag{10.4.5}$$

For the tensile deformation considered here, $\lambda_2 = \lambda_3$ from symmetry, so that a combination of Eqs. (10.4.4) and (10.4.5) yields

$$W = -\frac{NkT}{2}l_0^3\left(\lambda_1^2 + \frac{2}{\lambda_1} - 3\right) \tag{10.4.6}$$

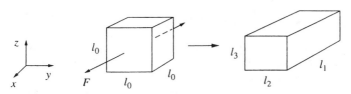

FIGURE 10.5 Uniaxial extension of a block of rubber.

From the definition of λ_1 it is obvious that $\Delta\lambda_1$ equals $\Delta x/l_0$. Because W must also equal $-\int F\,dx$, F is obtained by dividing the right-hand side of Eq. (10.4.6) by l_0 and differentiating the result with respect to λ_1. Thus,

$$\frac{F}{l_0^2} = NkT(\lambda_1 - \lambda_1^{-2}) \tag{10.4.7}$$

It can be shown that the form of this equation is unaffected by the presence of chains of unequal lengths; only the numerical value of the coefficient changes. The quantity NkT is called the *modulus G* of the rubber. The left-hand side of Eq. (10.4.7) is recognized to be the stress based on the undeformed area.

Example 10.3: When rubber is brought into contact with a good solvent, it swells in an isotropic manner. Consider a cube of rubber, initially of unit volume, containing N polymer chains. If in the swollen state the polymer volume fraction is ϕ_2 and the length of each edge is λ, how much work is done in the process of swelling?

Solution: Here, we use Eq. (10.4.4), with each extension ratio being equal to λ. Note that Eq. (10.4.5) does not apply because there is an obvious increase in volume. The total volume of the swollen rubber is equal to $1/\phi_2$, so $\lambda = \phi_2^{-1/3}$. Consequently,

$$W = \frac{3NkT}{2}(\phi_2^{-2/3} - 1)$$

This problem considers a particular kind of deformation—uniaxial extension. The same procedure can be applied to other kinds of deformation, and the result is a "material function" or, in the case of rubber, a material constant that relates a component of the stress to a component of the strain imposed on the material. More generally, though, we can determine the relationship between an arbitrary, three-dimensional deformation and the resulting three-dimensional stress. Such a relationship is called the *stress constitutive equation*. We will develop such a relationship for rubbers after we review the definitions of stress and the strain in three dimensions.

10.5 THE STRESS TENSOR (MATRIX)

If we isolate a rectangular parallelepiped of material having infinitesimal dimensions, as shown in Figure 10.6, we find that two kinds of forces act on the material element. These are body forces and surface forces. Body forces result from the action of an external field such as gravity upon the entire mass of material. Thus, the force of gravity in the z direction is $g_z\rho\,dx\,dy\,dz$, where g_z is the component

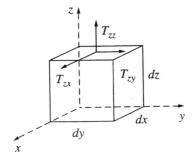

FIGURE 10.6 The stress matrix (tensor).

of the acceleration due to gravity in the positive z direction. Surface forces, on the other hand, express the influence of material outside the parallelepiped but adjacent to a given surface. Dividing the surface force by the area on which it acts yields the stress vector. Because the parallelepiped has six surfaces, there are six stress vectors. Because each of the 6 vectors can be resolved into 3 components parallel to each of the 3 coordinate axes, we have a total of 18 components. These are labeled T_{ij}, where the two subscripts help to identify a specific component. The first subscript, i, identifies the surface on which the force acts; the surface, in turn, is identified by the direction of the outward drawn normal. If the normal points in the positive coordinate direction, the surface is a positive surface; otherwise it is negative. The second subscript, j, identifies the direction in which the stress component acts. According to convention, a stress component is positive when directed in the positive coordinate direction on a positive face. It is also positive when directed in the negative direction on a negative face. Nine of the 18 components can be represented using a 3×3 matrix, called the *stress tensor*:

$$\begin{pmatrix} T_{xx} & T_{xy} & T_{xz} \\ T_{yx} & T_{yy} & T_{yz} \\ T_{zx} & T_{zy} & T_{zz} \end{pmatrix}$$

T_{zz}, for example, is the z component of the stress vector acting on the face whose outward drawn normal points in the positive z direction; T_{zy} is the corresponding component acting in the y direction. These are shown in Figure 10.6. The other nine components are the same as these, but they act on opposite faces.

By means of a moment balance on a cubic element, it can be shown (as in any elementary textbook of fluid mechanics) that T_{ij} equals T_{ji}. Thus, only six of the nine components are independent components. The utility of the stress tensor is revealed by examining the equilibrium of the tetrahedron shown in Figure 10.7.

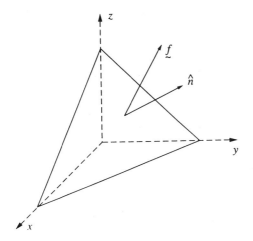

FIGURE 10.7 Equilibrium of a tetrahedron.

It can be demonstrated that if the normal to the inclined surface is $\hat{n}$, then the components of the surface stress **f** acting on that surface in a rectangular Cartesian coordinate system are as follows [8]:

$$
\begin{aligned}
f_x &= T_{xx}n_x + T_{xy}n_y + T_{xz}n_z \\
f_y &= T_{xy}n_x + T_{yy}n_y + T_{yz}n_z \\
f_z &= T_{xz}n_x + T_{yz}n_y + T_{zz}n_z
\end{aligned}
\tag{10.5.1}
$$

where

$$
\hat{\mathbf{n}} = n_x\hat{\mathbf{i}} + n_y\hat{\mathbf{j}} + n_z\hat{\mathbf{k}}
\tag{10.5.2}
$$

$$
\mathbf{f} = f_x\hat{\mathbf{i}} + f_y\hat{\mathbf{j}} + f_z\hat{\mathbf{k}}
\tag{10.5.3}
$$

and $\hat{\mathbf{i}}$, $\hat{\mathbf{j}}$, and $\hat{\mathbf{k}}$ are the three unit vectors.

If we represent x by 1, y by 2, and z by 3, Eq. (10.5.1) can be written in matrix notation as follows:

$$
\begin{pmatrix} f_1 \\ f_2 \\ f_3 \end{pmatrix} = \begin{pmatrix} T_{11} & T_{12} & T_{13} \\ T_{12} & T_{22} & T_{23} \\ T_{13} & T_{23} & T_{33} \end{pmatrix} \begin{pmatrix} n_1 \\ n_2 \\ n_3 \end{pmatrix}
\tag{10.5.4}
$$

or even more compactly as follows:

$$
\mathbf{f} = \underset{\sim}{\mathbf{T}} \cdot \hat{\mathbf{n}}
\tag{10.5.5}
$$

Knowing the six independent components of the stress tensor (matrix) $\underset{\sim}{\mathbf{T}}$, therefore, allows us to obtain the stress vector acting on any plane described by the unit normal $\hat{\mathbf{n}}$.

Throughout this chapter, we have discussed stress and strain, and in the minds of most people these two terms are intimately connected. Therefore, it is logical to ask if strain or deformation always results from the presence of a nonzero stress component. The answer is "not necessarily." Indeed, if a material is incompressible, no amount of pushing (i.e., the application of hydrostatic pressure) will cause it to compress or reduce in volume. It is only when pressures are unequal that a strain, which can be understood as a change in the distance between two neighboring particles, occurs. In essence, if we try to push the material in from one side, all it can do is squeeze out from another side. It is for this reason that it is usual to separate the stress tensor into two parts:

$$\begin{pmatrix} T_{11} & T_{12} & T_{13} \\ T_{12} & T_{22} & T_{23} \\ T_{13} & T_{23} & T_{33} \end{pmatrix} = \begin{pmatrix} -p & 0 & 0 \\ 0 & -p & 0 \\ 0 & 0 & -p \end{pmatrix} + \begin{pmatrix} \tau_{11} & \tau_{12} & \tau_{13} \\ \tau_{12} & \tau_{22} & \tau_{23} \\ \tau_{13} & \tau_{23} & \tau_{33} \end{pmatrix} \quad (10.5.6)$$

where p is the isotropic pressure whose presence causes no strain or deformation for incompressible materials, and the τ_{ij} terms are the components of the extra stress tensor whose presence causes strain to take place. Strain is therefore related to the extra stress $\underset{\sim}{\tau}$ rather than the total stress $\mathbf{T}$.

10.6 MEASURES OF FINITE STRAIN

When a material translates or rotates, it moves as a rigid body. In addition, it can deform (i.e., the distances between neighboring points can change). In general, we can relate the distance vector $d\mathbf{x}$ at time t between two neighboring points in a body to the distance vector $d\mathbf{x}'$ at time t' between the same two points after motion and deformation through an equation of the type

$$d\mathbf{x}' = \underset{\sim}{\mathbf{F}} \, d\mathbf{x} \quad (10.6.1)$$

where $\mathbf{F}$ is a 3×3 matrix called the *deformation gradient*. Let the components of $d\mathbf{x}$ be dx_1, dx_2, and dx_3 and those of $d\mathbf{x}'$ be dx'_1, dx'_2, and dx'_3. This situation is depicted in Figure 10.8. If the coordinates x'_i at time t' of a point located at position x_i at time t are represented as

$$\begin{aligned} x'_1 &= x_1 + X_1(x_1, x_2, x_3, t', t) \\ x'_2 &= x_2 + X_2(x_1, x_2, x_3, t', t) \\ x'_3 &= x_3 + X_3(x_1, x_2, x_3, t', t) \end{aligned} \quad (10.6.2)$$

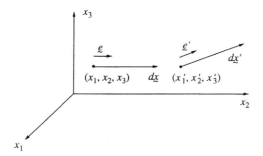

FIGURE 10.8 Deformation of a line element.

where X_i represents unknown functions, then, by similar reasoning, we have

$$x_1' + dx_1' = x_1 + dx_1 + X_1(x_1 + dx_1, \; x_2 + dx_2, \; x_3 + dx_3, \; t', \; t)$$
$$x_2' + dx_2' = x_2 + dx_2 + X_2(x_1 + dx_1, \; x_2 + dx_2, \; x_3 + dx_3, \; t', \; t) \qquad (10.6.3)$$
$$x_3' + dx_3' = x_3 + dx_3 + X_3(x_1 + dx_1, \; x_2 + dx_2, \; x_3 + dx_3, \; t', \; t)$$

Subtracting Eq. (10.6.2) from Eq. (10.6.3) and using a Taylor series expansion yields

$$dx_1' = dx_1 + \frac{\partial X_1}{\partial x_1} \, dx_1 + \frac{\partial X_1}{\partial x_2} \, dx_2 + \frac{\partial X_1}{\partial x_3} \, dx_3$$
$$dx_2' = dx_2 + \frac{\partial X_2}{\partial x_1} \, dx_1 + \frac{\partial X_2}{\partial x_2} \, dx_2 + \frac{\partial X_2}{\partial x_3} \, dx_3 \qquad (10.6.4)$$
$$dx_3' = dx_3 + \frac{\partial X_3}{\partial x_1} \, dx_1 + \frac{\partial X_3}{\partial x_2} \, dx_2 + \frac{\partial X_3}{\partial x_3} \, dx_3$$

Equation (10.6.2), however, gives the following:

$$\frac{\partial X_1}{\partial x_1} = \frac{\partial x_1'}{\partial x_1} - 1 \qquad (10.6.5)$$

$$\frac{\partial X_1}{\partial x_2} = \frac{\partial x_1'}{\partial x_2} \qquad (10.6.6)$$

and so on. Equation (10.6.4) therefore becomes

$$dx_i' = \sum_{j=1}^{3} \frac{\partial x_i'}{\partial x_j} \, dx_j, \quad i = 1, \, 2, \, 3 \qquad (10.6.7)$$

Comparing Eqs. (10.6.1) and (10.6.7) gives

$$F_{ij} = \frac{\partial x_i'}{\partial x_j} \tag{10.6.8}$$

In general, the deformation gradient depends on position. However, if it is independent of position, the displacement is said to be *homogeneous*. Note that a nonzero value of the deformation gradient does not, *ipso facto*, imply that deformation has taken place; for this to happen, distances between neighboring points must change. Let us pursue this point further.

If, as shown in Figure 10.8, $\mathbf{e}$ and $\mathbf{e}'$ are unit vectors along $d\mathbf{x}$ and $d\mathbf{x}'$, respectively, then the following relations hold:

$$d\mathbf{x} = dx\, \mathbf{e} \tag{10.6.9}$$
$$d\mathbf{x}' = dx'\, \mathbf{e}' \tag{10.6.10}$$

where dx is the magnitude of $d\mathbf{x}$ and dx' is that of $d\mathbf{x}'$. The terms dx and dx' are, however, also related through Eq. (10.6.1). Therefore,

$$dx'\, \mathbf{e}' = \underset{\sim}{\mathbf{F}} \cdot (dx\, \mathbf{e}) \tag{10.6.11}$$

or

$$\frac{dx'}{dx} \mathbf{e}' = \underset{\sim}{\mathbf{F}} \cdot \mathbf{e} \tag{10.6.12}$$

from which it follows that

$$\left(\frac{dx'}{dx} \mathbf{e}'\right)^{\mathrm{T}} \cdot \left(\frac{dx'}{dx} \mathbf{e}'\right) = (\underset{\sim}{\mathbf{F}} \cdot \mathbf{e})^{\mathrm{T}} \cdot (\underset{\sim}{\mathbf{F}} \cdot \mathbf{e}) \tag{10.6.13}$$

where the superscript T denotes transpose, or

$$\left(\frac{dx'}{dx}\right)^2 = \mathbf{e}^{\mathrm{T}} \cdot \underset{\sim}{\mathbf{F}}^{\mathrm{T}} \cdot \underset{\sim}{\mathbf{F}} \cdot \mathbf{e} \tag{10.6.14}$$

because the dot product of the transpose of a unit vector with itself is unity. Equation (10.6.14) can be rewritten as

$$\left(\frac{dx'}{dx}\right)^2 = \mathbf{e}^{\mathrm{T}} \cdot \underset{\sim}{\mathbf{C}} \cdot \mathbf{e} \tag{10.6.15}$$

where the product

$$\underset{\sim}{\mathbf{C}} = \underset{\sim}{\mathbf{F}}^{\mathrm{T}} \cdot \underset{\sim}{\mathbf{F}} \tag{10.6.16}$$

is called the *Cauchy tensor*. Clearly, if $\underset{\sim}{C} = \underset{\sim}{1}$, $dx' = dx$ and there is no deformation. Only if $\underset{\sim}{C} \neq \underset{\sim}{1}$ does deformation take place and interparticle distances change. In terms of components,

$$C_{ij} = \sum_{k=1}^{3} F_{ki}F_{kj} = \sum_{k=1}^{3} \frac{\partial x_k'}{\partial x_i} \frac{\partial x_k'}{\partial x_j} \tag{10.6.17}$$

which shows that $\underset{\sim}{C}$ is a symmetric matrix. It can be proven that for small strains, the Cauchy strain tensor defined as $\underset{\sim}{C} - 1$ reduces properly to the usual infinitesimal strain matrix encountered in mechanics. However, this is not the only possible measure of large strain. In fact, the matrix inverse of $\mathbf{C}$—called the *Finger tensor*, $\mathbf{B}$—is another valid measure of finite strain. Physically, $\mathbf{B}$ can be shown to be related to changes in distance between neighboring planes. In component form, the Finger tensor is given by

$$B_{ij} = \sum_{k=1}^{3} \frac{\partial x_i}{\partial x_k'} \frac{\partial x_j}{\partial x_k'} \tag{10.6.18}$$

Additional discussion of strain measures may be found in the literature [9].

Example 10.4: Obtain expressions for the deformation gradient and Cauchy tensors for the shear deformation illustrated in Figure 10.9. Here, the only nonzero velocity component is v_1, and it equals $\dot{\gamma}x_2$, where $\dot{\gamma}$ is the constant shear rate.

Solution: From the problem statement, it is clear that

$$x_1' = x_1 + \dot{\gamma}\Delta t x_2$$
$$x_2' = x_2$$
$$x_3' = x_3$$

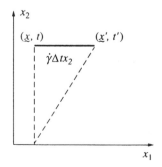

FIGURE 10.9 Shearing at constant shear rate.

where Δt equals $t' - t$. Using Eq. (10.6.8), then,

$$F_{ij} = \begin{pmatrix} 1 & \dot{\gamma}\Delta t & 0 \\ 0 & 1 & 0 \\ 0 & 0 & 1 \end{pmatrix}$$

and with the help of Eq. (10.6.16) or (10.6.17),

$$C_{ij} = \begin{pmatrix} 1 & 0 & 0 \\ \dot{\gamma}\Delta t & 1 & 0 \\ 0 & 0 & 1 \end{pmatrix} \begin{pmatrix} 1 & \dot{\gamma}\Delta t & 0 \\ 0 & 1 & 0 \\ 0 & 0 & 1 \end{pmatrix} = \begin{pmatrix} 1 & \dot{\gamma}\Delta t & 0 \\ \dot{\gamma}\Delta t & 1 + (\dot{\gamma}\Delta t)^2 & 0 \\ 0 & 0 & 1 \end{pmatrix}$$

Thus, it is clear that strain has taken place.

10.7   THE STRESS CONSTITUTIVE EQUATION

Let us revisit the stretching of the block of rubber pictured in Figure 10.5 and let us obtain expressions for the components of the Finger tensor. Even though it seems appropriate to denote distances in the stretched state by a prime and those in the equilibrium state without a prime, actual practice is just the reverse. This is the case because we want to use the same formalism for both liquids and solids; the absence of an equilibrium unstrained state for polymeric liquids forces us to use the current state at time t as the reference state. As a consequence, the deformed state is at a prior time t'. This somewhat confusing situation is clarified in Figure 10.10, which shows how the deformation is visualized. Clearly, then, we have the following:

$$\begin{aligned} x_1 &= \lambda_1 x_1' \\ x_2 &= \lambda_1^{-1/2} x_2' \\ x_3 &= \lambda_1^{-1/2} x_3' \end{aligned} \tag{10.7.1}$$

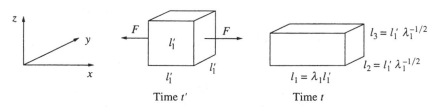

FIGURE 10.10   Uniaxial extension of rubber. Change in the reference state.

so that

$$C_{11}^{-1} = \lambda_1^2$$
$$C_{22}^{-1} = \lambda_1^{-1} \tag{10.7.2}$$
$$C_{33}^{-1} = \lambda_1^{-1}$$

with all other components being zero.

Because no shear stresses have been imposed, the nonzero components of the stress tensor are [in view of Eq. (10.5.6)] as follows:

$$T_{11} = -p + \tau_{11}$$
$$T_{22} = -p + \tau_{22} \tag{10.7.3}$$
$$T_{33} = -p + \tau_{33}$$

From equilibrium, at time t,

$$T_{11} = -p_a + \frac{F}{l_2^2}$$
$$T_{22} = -p_a \tag{10.7.4}$$
$$T_{33} = -p_a$$

where p_a is atmospheric pressure. Eliminating this quantity by taking differences between stress components gives the following:

$$T_{11} - T_{22} = \frac{F}{l_2^2} = \tau_{11} - \tau_{22} \tag{10.7.5}$$

From the discussion following Eq. (10.4.6) and from Problem 10.6, we have

$$\frac{F}{l_2^2} = G(\lambda_1^2 - \lambda_1^{-1}) \tag{10.7.6}$$

so that

$$\tau_{11} - \tau_{22} = G(\lambda_1^2 - \lambda_1^{-1}) \tag{10.7.7}$$

which can also come about if we let

$$\tau_{ij} = G C_{ij}^{-1} \tag{10.7.8}$$

Equation (10.7.8) is called a *stress constitutive equation*, and it relates a three-dimensional measure of strain to the three-dimensional stress. For rubbers, Eq. (10.7.8) obviously holds for the specialized case of uniaxial extension. By similar reasoning, it can be shown to hold for other idealized deformations such as biaxial extension and shear. Indeed, Eq. (10.7.8) is valid for all volume-preserving deformations [10]. The only material quantity appearing in this constitutive equation is the modulus G; its value is obtained by comparing predictions with experimental observations.

10.8   VULCANIZATION OF RUBBER AND SWELLING EQUILIBRIUM

Raw rubber behaves as an elastic solid only over a short time scale. At longer times, polymer chains in the stretched rubber begin to disentangle and slip past each other. This happens because natural rubber is a thermoplastic. For the same reason, it becomes soft and sticky in summer and, due to the onset of crystallization, hard and brittle in winter (recall Gough's cold water experiments) [11]. To overcome these problems, we cross-link rubber in its randomly coiled state using a material such as sulfur. This process is known as *vulcanization*, and the resulting cross-links prevent slippage of polymer chains. Indeed, as little as 1% of added sulfur is effective in ensuring that rubber retains its desired elasticity. Excessive cross-linking, though, makes the polymer hard and brittle, and this is similar to the influence of crystallization. Regarding the foregoing theory, we now assume that there are N chain segments per unit volume, where a chain segment is defined as the length of chain between cross-link points. Provided that a chain segment is long enough, it behaves like an isolated chain and all of the previous equations remain unchanged.

To determine the chain density N, we can compare the predictions of Eq. (10.4.7) with appropriate experimental data. This is done in Figure 10.11 using the uniaxial elongation data of Treloar on a sample of vulcanized natural rubber [1,12]. The theoretical line in the figure has been drawn using the best-fit value of the modulus, and there is fair agreement with data over a significant range of extension ratio values. Nonetheless, there is a slight mismatch at small values of λ_1 and a very large mismatch at values of λ_1 exceeding 6. The cause of the latter deviation is strain-induced crystallization in rubber; the crystallites that form at large extension ratios act as cross-links, resulting in an increase in the modulus. At moderate extensions, on the other hand, the affine deformation assumption begins to fail [7,13] and junction fluctuations cause a reduction in the modulus. In general, though, an increase in the cross-link density results in an increase in the best-fit modulus, but the cross-link density estimated from the extent of cross-linking is usually lower than the experimentally determined best-fit value. A part of the discrepancy is thought to be due to the presence of physical entanglements that act as cross-links over the time scale of the experiment.

A consequence of cross-linking is that the resulting gigantic molecule does not dissolve in any solvent; all it can do is swell when brought into contact with a good solvent. The equilibrium extent of swelling is determined by an interplay between the reduction in free energy due to polymer–solvent mixing and an

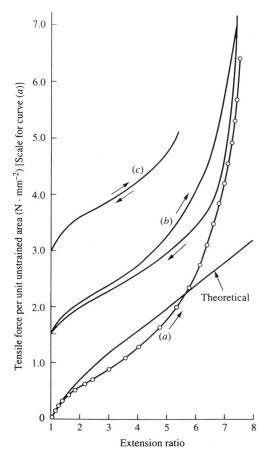

FIGURE 10.11 Simple extension. Comparison of experimental curve with theoretical form. (Reprinted from Treloar, L. R. G., The Physics of Rubber Elasticity, 3rd ed., Clarendon Oxford, UK, 1975, by permission of Oxford University Press.)

increase in free energy due to stretching of polymer chains. If we use the Flory–Huggins expression, Eq. (9.3.20), for the former free-energy change and the result of Example 10.3 for the latter free-energy change, then the total change in free energy on mixing unit volume of polymer containing n_2 moles of chain segments with n_1 moles of solvent is

$$\Delta G_M = RT(n_1 \ln \phi_1 + n_2 \ln \phi_2 + \chi_1 n_1 \phi_2) + \frac{3n_2 RT}{2}(\phi_2^{-2/3} - 1) \quad (10.8.1)$$

Noting that the total volume of the swollen rubber is the sum of the polymer and solvent volumes, we have

$$\frac{1}{\phi_2} = 1 + n_1 V_1 \tag{10.8.2}$$

where V_1 is the molar volume of the solvent.

Combining the two previous equations and recalling the definition of the free-energy change on mixing, we find that

$$G_{\text{mixture}} = n_1 g_1 + n_2 g_2 + RT(n_1 \ln \phi_1 + n_2 \ln \phi_2 + \chi_1 n_1 \phi_2)$$
$$+ \frac{3 n_2 RT}{2}[(1 + n_1 V_1)^{2/3} - 1] \tag{10.8.3}$$

where g_1 and g_2 are the molar free energies of the solvent and polymer, respectively.

Differentiating Eq. (10.8.3) with respect to n_1 and using the results of Eq. (9.3.30) gives the solvent chemical potential as

$$\mu_1 = \frac{\partial G_{\text{mixture}}}{\partial n_1} = g_1 + RT\left[\ln(1 - \phi_2) + \phi_2\left(1 - \frac{1}{m}\right) + \chi_1 \phi_2^2\right]$$
$$+ RTn_2 \phi_2^{1/3} V_1 \tag{10.8.4}$$

At phase equilibrium between the swollen rubber and the pure solvent, μ_1 must equal g_1, so that

$$\ln(1 - \phi_2^*) + \phi_2^* + \chi_1 \phi_2^{*2} + n_2 V_1 \phi_2^{*1/3} = 0 \tag{10.8.5}$$

in which ϕ_2^* is the polymer volume fraction at equilibrium. Also, $1/m$ has been neglected in comparison with unity. A measurement of the equilibrium amount of swelling, together with a knowledge of the polymer–solvent interaction parameter, then allows us to compute the chain density n_2. Indeed, Eq. (10.8.5) has proved to be a popular alternative to Eq. (10.4.7) for the determination of the number of chain segments per unit volume. We note, though, that as far as data representation is concerned, better agreement is obtained if, instead of Eq. (10.4.7), we use

$$\frac{F}{l_0^2} = 2(\lambda_1 - \lambda_1^{-2})(C_1 + C_2 \lambda_1^{-1}) \tag{10.8.6}$$

which can be derived in a phenomenological way by using the Mooney strain–energy function instead of Eq. (10.4.4) [14]. Here, C_1 and C_2 are constants. Note, again, that all comparisons with theory have to be made with data generated above the polymer glass transition temperature. Below T_g, polymer chains cannot move and rotate freely; they lose their elasticity and become glassy. The theory developed in this chapter is then inapplicable.

Example 10.5: Estimate a value for the chain density in units of moles per cubic centimeter for butyl rubber if the equilibrium swelling ratio, q, in cyclohexane is 8. Here, the swelling ratio is defined as the ratio of the equilibrium swollen volume to the original unswollen volume. In accordance with the data of Flory [15], let χ_1 be 0.3. The molar volume of cyclohexane is 105 cm^3.

Solution: From the definition of the swelling ratio, it is clear that $\phi_2^* = 1/q$, and the polymer volume fraction in the swollen network equals 0.125. Introducing this value into Eq. (10.8.5) along with the given values of χ_1 and V_1, we find that $n_2 = 7.32 \times 10^{-5}$ mol/cm^3.

10.9 CONCLUSION

We have seen that the use of a very simple model, the freely jointed chain, is adequate for explaining all of the qualitative observations made with elastomers. To obtain quantitative agreement, though, the theory needs to be modified, but without sacrificing the basic principles presented here; these modifications are explored elsewhere [5]. The utility of the theory, however, does not end with explaining the behavior of cross-linked rubber. A knowledge of the fundamentals of rubber elasticity allows us to synthesize other elastomers and to modify and optimize their properties. Indeed, the total production of synthetic rubbers such as styrene–butadiene rubber today exceeds that of natural rubber, and synthetic routes to polyisoprene have also been developed. Natural rubber, however, is not likely to disappear any time soon. Its superior heat-dissipation properties make it the preferred choice for the manufacture of heavy-duty truck tires. Another class of synthetic rubbers is that of thermoplastic elastomers, initially developed by the Shell Chemical Company in 1965 [6,16]. These are A-B-A-type block copolymers, where A is a thermoplastic such as polystyrene and B is an elastomer such as polybutadiene. These can be processed like thermoplastics because the hard segment, the A block, permits flow upon heating above its glass transition temperature. On cooling, the glassy domains of A act like cross-links within B and the copolymer has rubberlike properties. As a consequence, vulcanization is not required.

Rubberlike elasticity theory also has relevance beyond elastomers. On the practical side, networks that can imbibe large amounts of liquid form gels that act as superabsorbents. A common example is the use of such polymers in disposable diapers. A more "high-tech" application is the use of gels for concentrating dilute macromolecular solutions [17]. Such a "swellex process" can compete with membrane separation processes for purifying and separating biotechnology products such as proteins and enzymes [18]. On the theoretical side, rubberlike elasticity theory can be employed to derive equations of state for molten polymers. We assume that the cross-link density in Eq. (10.7.8) is not constant. This is done by identifying physical entanglements in polymer melts with cross-links in vulcanized rubber and by allowing the entanglements to be continually created and destroyed by flow. The resulting model is known as the rubberlike liquid model [19], and it has enjoyed great popularity among polymer rheologists. Polymer rheology is considered further in Chapter 14.

Before leaving the topic of elastomers, we mention that there is less than complete understanding of some rather important issues. One such issue is the mechanism by which added particulates influence the mechanical properties of rubbers. Carbon black, for example, is added [6,20,21] to natural rubber and silica is added [22] to silicone rubber to improve the tear strength and abrasion resistance of the elastomer. A second issue is the impact modification of polymers during the formation of microcomposites or macrocomposites by the addition of a rubbery phase. An example of this is high-impact polystyrene (HIPS); adding rubber to glassy polymers can raise their impact strength by an order of magnitude [23]. On a macrolevel, we use polyvinyl butyral as an interlayer in laminated safety glass to resist penetration from impacts; understanding the mechanism of window glazing continues to be a subject of current research.

REFERENCES

1. Treloar, L. R. G., *The Physics of Rubber Elasticity*, 3rd ed., Clarendon, Oxford, UK, 1975.
2. Kuhn, W., Beziehungen Zwischen Molekulgrosse, Statistischer Molekulgestalt und Elastischen Eigenschaften Hochpolymerer Stoffe, *Kolloid Z.*, 76, 258–271, 1936.
3. Kuhn, W., Molekulkonstellation und Kristallitorientierung als Ursachen Kautschu-kahnlicher Elastizitat, *Kolloid Z.*, 87, 3–12, 1939.
4. Flory, P. J., *Principles of Polymer Chemistry*, Cornell University Press, Ithaca, NY, 1953.
5. Smith, T. L., Molecular Aspects of Rubber Elasticity, *Treatise Mater. Sci. Technol.*, 10A, 369–451, 1977.
6. Eirich, F. R. (ed.), *Science and Technology of Rubber*, Academic Press, Orlando, FL, 1978.
7. Mark, J. E., The Rubber Elastic State, in *Physical Properties of Polymers*, J. E. Mark, A. Eisenberg, W. W. Graessley, L. Mandelkern, and J. L. Koenig (eds.), American Chemical Society, Washington, DC, 1984, pp. 1–54.
8. Shames, I. H., *Mechanics of Fluids*, 3rd ed., McGraw-Hill, New York, 1992.
9. Malvern, L. E., *Introduction to the Mechanics of a Continuous Medium*, Prentice-Hall, Englewood Cliffs, NJ, 1969.
10. Larson, R. G., *Constitutive Equations for Polymer Melts and Solutions*, Butterworths, Boston, MA, 1988.
11. Kauffman, G. B., and R. B. Seymour, Elastomers, *J. Chem. Ed.*, 67, 422–425, 1990.
12. Treloar, L. R. G., Stress–Strain Data for Vulcanised Rubber Under Various Types of Deformation, *Trans. Faraday Soc.*, 40, 59–70, 1944.

13. Queslel, J. P., and J. E. Mark, Advances in Rubber Elasticity and Characterization of Elastomer Networks, *J. Chem. Ed.*, 64, 491–494, 1987.
14. Mooney, M., A Theory of Large Elastic Deformation, *J. Appl. Phys.*, 11, 582–592, 1940.
15. Flory, P. J., Effects of Molecular Structure on Physical Properties of Butyl Rubber, *Ind. Eng. Chem.*, 38, 417–436, 1946.
16. Holden, G., E. T. Bishop, and N. R. Legge, Thermoplastic Elastomers, *J. Polym. Sci.*, C26, 37–57, 1969.
17. Cussler, E. L., M. R. Stokar, and J. E. Vaarberg, Gels as Size-Selective Extraction Solvents, *AIChE J.*, 30, 578–582, 1984.
18. Badiger, M. V., M. G. Kulkarni, and R. A. Mashelkar, Concentration of Macromolecules from Aqueous Solutions: A New Swellex Process, *Chem. Eng. Sci.*, 47, 3–9, 1992.
19. Lodge, A. S., *Elastic Liquids*, Academic Press, London, 1964.
20. Boonstra, B. B., Role of Particulate Fillers in Elastomer Reinforcement: A Review, *Polymer*, 20, 691–704, 1979.
21. Kraus, G., Reinforcement of Elastomers by Carbon Black, *Rubber Chem. Technol.*, 51, 297–321, 1978.
22. Polmanteer, K. E., and C. W. Lentz, Reinforcement Studies—Effect of Silica Structure on Properties and Crosslink Density, *Rubber Chem. Technol.*, 48, 795–809, 1975.
23. Bragaw, C. G., The Theory of Rubber Toughening of Brittle Polymers, *Adv. Chem. Ser.*, 99, 86–106, 1971.
24. Clough, S. B., Stretched Elastomers, *J. Chem. Ed.*, 64, 42–43, 1987.

PROBLEMS

10.1. Give the chemical structure and unique characteristics of each of the following synthetic rubbers: styrene–butadiene rubber, polybutadiene, neoprene, butyl rubber, nitrile rubber, and silicone rubber.

10.2. Two identical-looking, elastomeric balls are dropped from the same height onto a hard surface. One ball is made from neoprene ($T_g = -42°C$, maximum elongation of 500% at room temperature), whereas the other is made from polynorbornene, a linear polymer containing one rigid, five-membered ring and one double bond in each repeating unit ($T_g = -60°C$, maximum elongation of 400% at room temperature).

(a) Which ball will bounce to a higher height at room temperature? Why?

(b) What happens to the amount of bounce as the temperature is lowered? Why?

(c) If the two balls are cooled in ice water and then allowed to bounce, which one will bounce higher? Why?

10.3. Consider the polyethylene chain of Example 10.1. Determine the values n' and l' (in terms of n and l) of an equivalent freely jointed chain so that the two chains have the same contour length and the same mean square end-to-end distance.

10.4. If the polyethylene molecule can be considered a freely jointed chain, what will be the mean square end-to-end distance if the polymer molecular weight is 140,000? The length of a single bond is 1.54 Å. If the polymer sample contains Avogadro's number of molecules, how many molecules (at a given time) will actually have this end-to-end distance?

10.5. Seven identical spheres are located at equal intervals along a straight line. If each sphere has unit mass and if the distance between the centers of neighboring spheres is unity, what is the radius of gyration?

10.6. Obtain the equivalent form of Eq. (10.4.7) if stress is defined based on the deformed (actual) area.

10.7. Instead of the uniaxial deformation shown in Figure 10.5, consider equal biaxial extension: a force F acting parallel to the x axis and an identical force F acting parallel to the y axis. Relate F/l_0^2 to an appropriately defined extension ratio.

10.8. Repeat Problem 10.7 for the case in which the two forces are not the same and equal F_1 and F_2, respectively. Relate F_1/l_0^2 to λ_1 and λ_2 and F_2/l_0^2 to λ_1 and λ_2.

10.9. A catapult is made using a strip of the butyl rubber of Example 10.5. If, at 25°C, the strip is extended to twice its original length and used to hurl a 10-g projectile, what will be the maximum possible speed of the projectile? Let the volume of the rubber band be 1 cm^3.

10.10. What is the Young's modulus of the rubber sample used in Figure 10.11? How does it compare with the corresponding value for steel?

10.11. Does a block of rubber obey Hooke's law in (a) extension? (b) shear? Justify your answer.

10.12. A weight is attached to a 6-cm-long rubber band and the stretched length is measured as a function of temperature. Are the results shown here [24] quantitatively consistent with the theory of rubber elasticity?

Temperature (°C)	Length (mm)
20	163.0
35	158.5
45	155.5
48	154.0
57	151.0
61	149.0

10.13. Determine the Finger tensor for the shearing deformation considered in Example 10.4. Show that the same result is obtained by inverting the $\underset{\sim}{C}$ matrix calculated in that example.

10.14. A sample of the rubber used in Example 10.5 is stretched rapidly to five times its original length. If the temperature increases from 25°C to 35°C, what is the polymer specific heat? The density of rubber is 0.97 g/cm^3.

10.15. Show that for large values of the swelling ratio q, Eq. (10.8.5) reduces to

$$q^{5/3} = \frac{(1/2 - \chi_1)}{n_2 V_1}$$

11

Polymer Crystallization

11.1 INTRODUCTION

Low-molecular-weight materials, such as metals, typically exist as crystals in the solid state. The driving force behind the formation of crystals, which are structures having a long-range periodic order, is the lowering in free energy that accompanies the process of crystallization. Thus, if we were to plot the Gibbs free energy per unit volume, G, of a material in both the solid crystalline and molten forms as a function of temperature, we would get a plot of the type shown in Figure 11.1; the decrease in free energy with increasing temperature for both phases is due to the relative increase in the temperature–entropy term. The point of intersection of the two curves is the equilibrium melting point T_M^0, whereas the vertical difference between them represents the free-energy change, ΔG_v, between the two states at any temperature. Note that many materials (such as iron) exhibit polymorphism; that is, they exist in more than one crystalline form. In such a case, each form has its own G versus temperature curve.

Long-chain molecules can also crystallize, and they do so for the same energetic reasons as short-chain molecules. However, for a polymer to be crystallizable, its chemical structure should be regular enough that the polymer molecule can arrange itself into a crystal lattice. Thus, isotactic and syndiotactic polypropylenes crystallize easily, but atactic polypropylene does not. For the same reason, the presence of bulky side groups (as in polystyrene) hinders crystallization, but the possibility of hydrogen-bonding (as in polyamides)

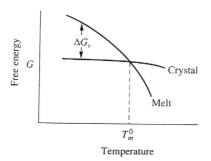

FIGURE 11.1 Variation with temperature of the Gibbs free energy per unit volume.

promotes the process. Figure 11.2 shows a schematic representation of nylon 66 [1]. A unit cell contains a total of one chemical repeat unit, and the molecules are in the fully extended zigzag conformation. The polymer chain direction is generally labeled the c axis of the cell. The cell itself is triclinic, which means that magnitudes of the three axes and the three interaxial angles are all different. A sketch of the unit cell is shown in Figure 11.3, in which arrows are used to designate hydrogen bonds. Whether a crystallizable polymer actually crystallizes or not, though, depends on the thermal history of the sample. For macromolecules, polymer mobility exists only above the glass transition temperature, and because the energetics are favorable only below the melting point, crystallization can take place only in a temperature range between T_m^0 and T_g. However, crystallization is not an instantaneous process; it takes place by nucleation and growth, and these steps take time. If the rate of cooling from the melt is rapid enough, a completely amorphous polymer can result. This is shown schematically in Figure 11.4, which is a continuous-cooling transformation curve [2]. If the cooling rate is such that we can go from T_m^0 to T_g without intersecting the curve labeled "crystallization begins in quiescent melts," no crystallization takes place. Therefore, it is possible to obtain completely amorphous samples of a slowly crystallizing polymer such as polyethylene terephthalate, but it is not possible for a rapidly crystallizing polymer such as polyethylene. In addition to temperature, the extent of crystallization also depends on factors such as the applied stress during processing, which tends to align polymer chains in the stress direction. This can alter the energetics of phase change and can lead to a very significant enhancement of the rate of crystallization. The phenomenon can be understood as a shift to the left in Figure 11.4, from the curve indicating the onset of quiescent crystallization to the curve labeled "crystallization begins in stretched melts." Because polymers are rarely completely crystalline, they are called *semicrystalline*.

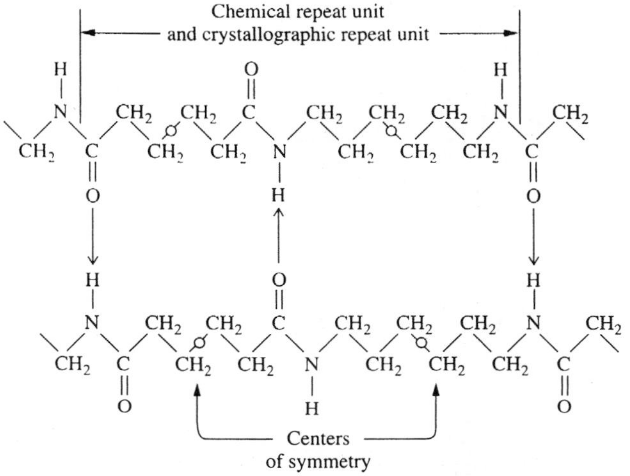

Chemical repeat unit
and crystallographic repeat unit

H | O   H |
  N   CH₂ CH₂   C   CH₂ CH₂ CH₂ N | CH₂
CH₂ C   CH₂ CH₂ N   CH₂ CH₂ CH₂ C
     ‖ | ‖
     O   H   O

H   O   H
N   CH₂ CH₂   C   CH₂ CH₂ CH₂ N   CH₂
CH₂ C   CH₂ CH₂ N   CH₂ CH₂ CH₂ C
     ‖ | ‖
     O   H   O

Centers
of symmetry

FIGURE 11.2   Schematic representation of nylon 66. (From Ref. 1.)

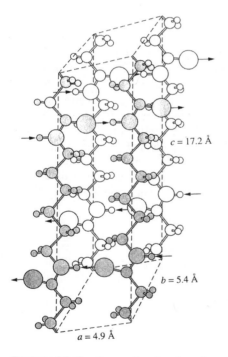

$c = 17.2$ Å

$b = 5.4$ Å

$a = 4.9$ Å

FIGURE 11.3   Perspective drawing of a unit cell of nylon 66. The viewpoint is 11 Å up, 10 Å to the right, and 40 Å back from the lower left corner of the cell. (From Ref. 1.)

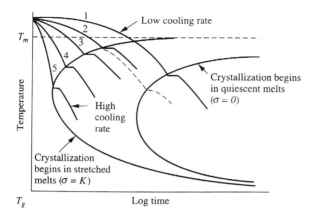

FIGURE 11.4   Schematic illustration of the concept of a "continuous-cooling trans-formation curve" showing the anticipated effect of stress in shifting such curves. (From Ref. 2.)

Crystallizable polymers that dissolve in a solvent can also be made to crystallize from solution. When this is done using dilute solutions, single crystals can be obtained [3]. The crystals can have a large degree of perfection and are usually in the form of lamellae or platelets having thickness of the order of 100 Å and lateral dimensions of the order of microns. The observed thickness depends on the temperature of crystallization. As shown in Figure 11.5, a single lamella is composed of chain-folded polymer molecules. Lamellae of different polymers have been observed in the form of hollow pyramids and hexagonal structures. When crystallized from quiescent melts, though, spherical structures called *spherulites* are observed. These spheres, which can grow to a few hundred microns in diameter, are made up of lamellae that arrange themselves along the

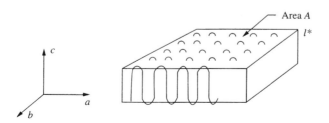

FIGURE 11.5   A chain-folded crystal lamella.

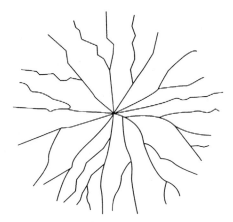

FIGURE 11.6   Schematic diagram of a spherulite. Each ray is a lamella.

radial direction in the sphere, as shown in Figure 11.6. The interlamellar regions as well as the region between spherulites are composed of amorphous or noncrystallizable fractions of the polymer. When crystallization takes place from a strained (deforming) solution or strained melt, the crystal shape can change to that of a shish kebab in the former case and to a row-nucleated crystal in the latter case; these are shown in Figure 11.7 [4]. The chain orientation in solution initially results in the formation of extended chain crystals, which give rise to the central core or "shish"; the "kebabs," which are lamellar, then grow radially outward from the shish. If the polymer sample is polydisperse, the higher-molecular-weight fraction crystallizes first, and this results in fractionation.

From this discussion, it should be clear that a solid semicrystalline polymer is a two-phase structure consisting essentially of an amorphous phase with a

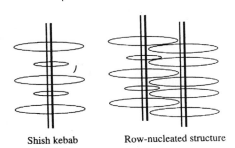

Shish kebab   Row-nucleated structure

FIGURE 11.7   Oriented morphologies appearing in polyethylene. (Reprinted from Gedde, U. W., Polymer Physics, Figure 7.38, copyright 1995, Chapman and Hall. With kind permission of Kluwer Academic.)

dispersed crystalline phase. Each phase is characterized by different values of a given physical property. The density of the crystals, for instance, is always greater than that of the amorphous polymer. Furthermore, process conditions determine the volume fraction of crystals, their shape, size, and size distribution, the orientation of polymer chains within the two phases, and how the crystalline regions are connected to the amorphous regions. Thus, whereas the properties of an amorphous polymer can be described as glassy or rubbery (depending on whether the temperature of measurement is below or above the glass transition temperature), the behavior of a semicrystalline polymer is much more complicated and is often anisotropic: If polymer chains are aligned in a particular direction, the material will be very strong in that direction but weak in a direction perpendicular to it. However, one of the two phases may be dominant in terms of influencing a particular overall property of the polymer. Thus, regarding mechanical properties, we find that increasing the spherulite size results in a decrease in the impact strength, an increase in the yield stress, and a reduction in the elongation to break in a tensile experiment while the Young's modulus goes through a maximum [5]. The solubility of a molecule in a polymer and also its diffusivity, though, are determined by the amorphous phase. As a consequence, the permeability, which is a product of these two quantities, decreases as the extent of crystallinity increases. The breakdown of electrical insulation, on the other hand, depends on the properties of the interspherulitic region in the polymer [6]. Other factors that are influenced by the structure include brittleness, environmental degradation, thermal properties, melting point, and glass transition temperature. If the solid structure that is formed is a nonequilibrium one, it can change later if conditions (especially temperature) are such that equilibrium can be approached. Thus, a polymer sample whose chains have been frozen in an extended position can shrink when chain alignment is lost on heating to a temperature above the glass transition temperature.

Although some of the influence of structure on properties can be rationalized by thinking of crystallites either as filler particles in an amorphous matrix or as permanent cross-links (as in vulcanized rubber), a proper understanding of structure development during processing is necessary to satisfy intellectual curiosity and to utilize crystallization knowledge for economic gain. As a result, we need to know what structure arises from a given set of processing conditions, how we can characterize (or measure) this structure, and how it affects a property of interest. It is, of course, much more difficult to reverse the process of thinking and inquire how we might obtain a particular structure in order to get specified values of properties of interest. This is the realm of engineered material properties and the subject of research of many industrial research laboratories. Before tackling greater problems, though, we must first get acquainted with some rather fundamental concepts.

11.2 ENERGETICS OF PHASE CHANGE

If the temperature of a liquid is lowered to below the melting point, material tends to solidify. As mentioned previously, the process is neither sudden nor instantaneous. Indeed, it proceeds relatively slowly and on a small scale if the temperature is only slightly below the melting point, and it involves two distinct steps. Initially, nuclei of the new phase must be formed, and the ease with which this happens depends on the extent of supercooling. This step is followed by growth of the nuclei, a procedure that involves diffusion of material to the phase boundary. The combined process of nucleation and growth is the same regardless of whether the crystallization behavior being observed is that of molten metals or molten polymers.

11.2.1 Homogeneous Nucleation

To understand the thermodynamics of nucleation, let us first consider homogeneous nucleation, also called *sporadic nucleation*, from an isothermal, quiescent melt whose temperature T is kept below the melting point T_m^0. Here, "homogeneous" refers to the appearance of the new solid phase in the middle of the old liquid phase.

Based on Figure 11.1, we would expect nucleation to be accompanied by a reduction in the Gibbs free energy equal to ΔG_v per unit volume. However, the system free energy is not reduced by the full amount of ΔG_v. This is because surface energy equal to γ per unit area has to be expended in creating the surface that bounds the nuclei. Thus, if the typical nucleus is a sphere of radius r, the net change in the free energy due to the formation of this particle is

$$\Delta G = 4\pi r^2 \gamma + \frac{4}{3}\pi r^3 \Delta G_v \qquad (11.2.1)$$

where ΔG_v is a negative number. For a small sphere, the surface area-to-volume ratio can be fairly large; therefore, ΔG initially increases with increasing r and goes through a maximum before becoming negative. This maximum (positive) value ΔG^* represents an energy barrier and must be overcome by the thermal motion of the molecules before a stable nucleus can be formed.

If we set the derivative of ΔG with respect to r equal to zero, then using Eq. (11.2.1) we find that r^*, the value of r corresponding to ΔG^*, is

$$r^* = -\frac{2\gamma}{\Delta G_v} \qquad (11.2.2)$$

with the following result:

$$\Delta G^* = \frac{16\pi\gamma^3}{3(\Delta G_v)^2} \qquad (11.2.3)$$

Because the magnitude of ΔG_v increases as the temperature is lowered, both r^* and ΔG^* decrease with decreasing temperature. This variation can be made explicit by noting that, at the melting point,

$$\Delta G_v = \Delta H_v - T_m^0 \Delta S_v = 0 \tag{11.2.4}$$

where H and S are respectively the enthalpy and entropy per unit volume. Consequently, we have the following:

$$\Delta S_v = \frac{\Delta H_v}{T_m^0} \tag{11.2.5}$$

Now, ΔS_v and ΔH_v depend only weakly on temperature, so that

$$\Delta G_v = \Delta H_v \left(1 - \frac{T}{T_m^0} \right) \tag{11.2.6}$$

which, when introduced into Eqs. (11.2.2) and (11.2.3), gives

$$r^* = -\frac{2\gamma T_m^0}{\Delta H_v \Delta T} \tag{11.2.7}$$

$$\Delta G^* = \frac{16\pi\gamma^3 (T_m^0)^2}{3\Delta H_v^2 \Delta T^2} \tag{11.2.8}$$

where ΔT equals the amount of subcooling $(T_m^0 - T)$ and ΔH_v is physically the latent heat of crystallization per unit volume and is a negative quantity. Clearly, lower temperatures favor the process of nucleation, as ΔG^* decreases rapidly with decreasing temperature.

Example 11.1: For a polyolefin it is found that $\Delta H_v = -3 \times 10^9$ ergs/cm^3 and $\gamma = 90$ ergs/cm^2. If the equilibrium melting point is 145°C, how do the radius r^* of a critical-sized nucleus and the associated energy change ΔG^* depend on the extent of subcooling, ΔT?

Solution: Using Eq. (11.2.7), we find that

$$r^*\Delta T = 2508$$

where r^* is measured in angstroms. Also, with the help of Eq. (11.2.8), we have

$$\Delta G^*(\Delta T)^2 = 2.37 \times 10^{-7} \text{ ergs K}^2$$

11.2.2 Heterogeneous Nucleation

Although the treatment of the previous section can be extended to nonspherical nuclei, we find that this is not needed in practice because the contribution of homogeneous nucleation to overall crystal growth is small compared to that of heterogeneous or predetermined nucleation, unless the temperature is significantly below the melting point. In the case of heterogeneous nucleation, crystal growth takes place on a pre-existing surface, which might be a dust particle, an impurity, part of the surface of the container, or an incompletely melted crystal. If we consider heterogeneous nucleation to take place on the surface of a pre-existing lamella, as shown in Figure 11.8 [7], we discover that if the crystal volume increases by an amount $nabl$, where abl is the volume of a single strand, the surface area increases by only $2b(l + na)$. Had we considered primary nucleation, the surface area would have gone up by $2b(l + na) + 2nal$. Note that due to chain folding, γ_e, the surface energy associated with the chain ends can be expected to be large compared to γ, the surface energy of the lateral surface.

In view of the foregoing, the free-energy change due to the deposition of n polymer strands is

$$\Delta G = 2bl\gamma + 2bna\gamma_e + nabl\Delta G_v \qquad (11.2.9)$$

and the free-energy change involved in laying down the $(n + 1)$st strand is obtained from Eq. (11.2.9) as

$$\Delta G(n + 1) - \Delta G(n) = 2ab\gamma_e + abl\Delta G_v \qquad (11.2.10)$$

Clearly, for this process to be energetically favorable, the right-hand side of Eq. (11.2.10) has to be negative. This condition requires that [8]

$$l > -\frac{2\gamma_e}{\Delta G_v} \qquad (11.2.11)$$

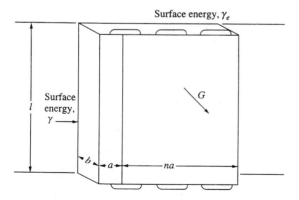

FIGURE 11.8 Crystal growth on a pre-existing surface.

which, in view of Eq. (11.2.6), implies that

$$l > -\frac{2\gamma_e T_m^0}{(\Delta H_v \Delta T)} \tag{11.2.12}$$

Because the argument leading up to Eq. (11.2.12) is valid for any value of n, it must also hold for the beginning of the process when n equals unity. Thus, for heterogeneous nucleation, the critical-sized nucleus occurs at $n = 1$, with the result that

$$\Delta G^* = 2bl^*\gamma + abl^*\Delta G_v + 2ab\gamma_e \tag{11.2.13}$$

with

$$l^* = -\frac{2\gamma_e T_m^0}{\Delta H_v \Delta T} \tag{11.2.14}$$

and we find that, just as with r^*, l^* depends inversely on ΔT. This is found to be true experimentally. Also, to a good approximation,

$$\Delta G^* \propto \frac{1}{\Delta T} \tag{11.2.15}$$

Example 11.2: Use the data given in Example 11.1 to confirm the validity of Eq. (11.2.15). Determine ΔG^* for $\Delta T = 10\,\text{K}$ and compare this value with that in Example 11.1. Assume that $\gamma_e \approx \gamma$.

Solution: As $\gamma_e \approx \gamma$, $l^* = r^* = 250.8\,\text{Å}$. Because a cannot be more than a few angstroms, $a \ll l^*$ and we can neglect the last term on the right-hand side of Eq. (11.2.13) in comparison with the first term. The second term in Eq. (11.2.13) can also be neglected, provided that the following holds:

$$a\Delta G_v \ll 2\gamma$$

or

$$\frac{a\Delta H_v \Delta T}{2\gamma T_m^0} \ll 1$$

Introducing numbers, $a\Delta H_v \Delta T / 2\gamma T_m^0 = a(3 \times 10^9 \times 10)/(2 \times 90 \times 418) = 3.98 \times 10^5 a$, which is much less than unity because $a < 10^{-7}\,\text{cm}$. Consequently, $\Delta G^* = 2bl^*\gamma = 4b\gamma\gamma_e T_m^0 / \Delta H_v \Delta T \propto 1/\Delta T$. For homogeneous nucleation, ΔG^* is equal to $2.37 \times 10^{-9}\,\text{ergs}$. For heterogeneous nucleation, $\Delta G^* = 4.51b \times 10^{-4}\,\text{ergs}$. Because $b < 10^{-7}\,\text{cm}$, the energy barrier for heterogeneous nucleation is much smaller than that for homogeneous nucleation when the temperature is close to the equilibrium melting point.

11.3   OVERALL CRYSTALLIZATION RATE

Even when a phase transformation is thermodynamically possible, the rate at which it happens is controlled by the existence of any barriers retarding an approach to equilibrium; the activation energy ΔG^* calculated in the previous section represents just such a barrier for crystallization. The probability that a group of molecules has an energy ΔG^* greater than the average energy at a specified temperature T is given by the Boltzmann relation

$$\text{Probability} \propto \exp\left(-\frac{\Delta G^*}{kT}\right) \tag{11.3.1}$$

which, when applied to the process of nucleation, means that if the total number of solid particles at any instant is N_0, the number N^* that actually possess the excess energy ΔG^* is given by

$$N^* = N_0 \exp\left(-\frac{\Delta G^*}{kT}\right) \tag{11.3.2}$$

Nuclei form by the addition of molecules; this is a process of diffusion for which the following relation holds:

$$\text{Rate} \propto \exp\left(-\frac{E_D}{kT}\right) \tag{11.3.3}$$

where E_D is the activation energy for diffusion. A combination of Eqs. (11.3.2) and (11.3.3) then implies that the rate of nucleation $\dot{N}$ in units of nuclei per unit time is given by [9]

$$\dot{N} = \dot{N}_0 \exp\left(-\frac{E_D}{kT}\right) \exp\left(-\frac{\Delta G^*}{kT}\right) \tag{11.3.4}$$

Because the growth of nuclei proceeds by the process of heterogeneous nucleation, the growth rate v can also be represented by an equation of the form

$$v = v_0 \exp\left(-\frac{E_D}{kT}\right) \exp\left(-\frac{\Delta G^*}{kT}\right) \tag{11.3.5}$$

and the overall linear growth rate G (see Fig. 11.8) having units of velocity will be proportional to some product of $\dot{N}$ and v. Thus,

$$G = G_0 \exp\left(-\frac{E_D}{kT}\right) \exp\left(-\frac{\Delta G^*}{kT}\right) \tag{11.3.6}$$

Around the glass transition temperature, the following holds [10]:

$$\frac{E_D}{kT} = \frac{c_1}{R(c_2 + T - T_g)} \tag{11.3.7}$$

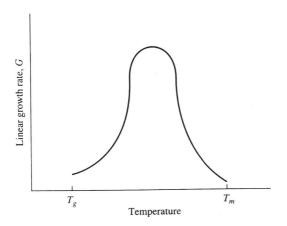

FIGURE 11.9 Qualitative temperature dependence of the linear growth rate G.

where c_1 is a constant with units of energy per mole, and c_2 is a constant with units of temperature. Also, if Eq. (11.2.15) is used for ΔG^*, we have

$$\frac{\Delta G^*}{kT} = \frac{A}{T\Delta T} \tag{11.3.8}$$

in which A is a constant. Finally, the overall growth rate is given by

$$G = G_0 \exp\left(-\frac{c_1}{R(T - T_\infty)}\right) \exp\left(-\frac{A}{T\Delta T}\right) \tag{11.3.9}$$

where T_∞ equals $T_g - c_2$ and A equals $-4b\gamma\gamma_e T_m^0/k\Delta H_v$.

An examination of Eq. (11.3.9) shows that the overall growth rate is independent of time under isothermal conditions. Also, the plot of G as a function of temperature is bell shaped (see Fig. 11.9). The rate is zero in the vicinity of the glass transition temperature because the rate of diffusion is small and the first exponential in Eq. (11.3.9) tends to zero, and it is also zero close to the melting point because the second exponential is driven to zero due to ΔT tending to zero. The rate is a maximum approximately midway between these two limits.

Example 11.3: Kennedy and co-workers used photomicroscopy to measure the radial growth rate of spherulites of isotactic polystyrene at a variety of constant temperatures [11]. These data are shown in Figure 11.10. Show that the results are consistent with Eq. (11.3.9).

Solution: Taking the natural logarithm of both sides of Eq. (11.3.9) and rearranging gives

$$\ln G + \frac{c_1}{R(T - T_\infty)} = \ln G_0 - \frac{A}{T\Delta T}$$

and a plot of the left-hand side of this equation versus $1/T\Delta T$ should be a straight line. This is, indeed, found to be the case when known values of 365 K and 503.8 K are used for T_g and T_m^0, respectively, and c_1 and c_2 are considered to be adjustable parameters. The final result is

$$\ln G = 15.11 - \frac{2073}{T - 290} - \frac{9.25 \times 10^4}{T(503.8 - T)}$$

This equation is plotted (on linear coordinates) in Figure 11.10 and gives excellent agreement with experimental growth rates.

In closing this section, we note that the size of spherulites is large when crystallization takes place near the melting point. This is because few nuclei are formed due to the large value of ΔG^*. However, once formed, nuclei grow easily. By contrast, when crystallization occurs near the glass transition temperature, a large number of small spherulites are observed; here, nuclei are formed readily, but they do not grow because the rate of diffusion is low. A common technique of

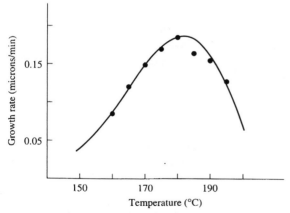

FIGURE 11.10 Comparison of experimentally determined growth rates of spherulites of i-polystyrene with (●) experimental (—) theoretical values. (From J. Polym. Sci. Polym. Phys. Ed., 21, Kennedy, M. A., G. Turturro, G. R. Brown, and L. E. St.-Pierre: Retardation of spherulitic growth rate in the crystallization of isotatic polystyrene due to the presence of a nucleant, Copyright ©1983 by John Wiley & Sons, Inc. Reprinted by permission of John Wiley & Sons, Inc.)

increasing the rate of crystallization is to intentionally add a fine powder such as silica or titanium dioxide to act as a nucleating agent.

11.4   EMPIRICAL RATE EXPRESSIONS: THE AVRAMI EQUATION

Equation (11.3.9) and its derivation are very useful in understanding how crystals nucleate and grow in a stagnant melt. However, the knowledge gained does not permit us to predict, *a priori*, the time dependence of the extent of crystallinity and the size distribution of spherulites in a polymer sample kept at a specified temperature between T_g and T_m^0. This information about the microstructure is crucial if we want to know in advance the mechanical properties likely to be observed in plastic parts produced by methods such as injection molding (see Chapter 15 for a description of the process). Because some of the very large-volume polymers such as polyethylene, polypropylene, and various nylons are injection molded, the question of the microstructure of semicrystalline polymers has received a considerable amount of attention [12]. This has led to the formulation of empirical expressions based on the theory originally developed by Avrami [13–15], Johnson and Mehl [16] and Evans [17] to explain the solidification behavior of crystallizable metals.

11.4.1   Isothermal Quiescent Crystallization

If we consider the isothermal crystallization of a quiescent polymer melt—whether by homogeneous or heterogeneous nucleation—then at time t, the volume of a spherulite that was nucleated at time τ ($\tau < t$) will be $V(t, \tau)$. As a consequence, the weight fraction, X', of material transformed will be as follows [18]:

$$X'(t) = \int_0^t \rho_s V(t, \tau) \, dn \tag{11.4.1}$$

where ρ_s is the density of the crystalline phase and dn is the number of nuclei generated per unit mass in the time interval τ and $\tau + d\tau$.

If the nucleation frequency per unit volume is $\dot{N}(\tau)$ and the liquid phase density is ρ_L, then

$$dn = \frac{\dot{N}(\tau) \, d\tau}{\rho_L} \tag{11.4.2}$$

Equation (11.4.1) therefore takes the following form:

$$X'(t) = \frac{\rho_s}{\rho_L} \int_0^t V(t, \tau) \dot{N}(\tau) \, d\tau \tag{11.4.3}$$

In order to make further progress, it is necessary to specify the time dependence of V and $\dot{N}$. This is done by appealing to *observed* crystallization behavior. It is now well established that, under isothermal conditions, the radius of a spherulite increases linearly with time [see also Eq. (11.3.9)], with the following result:

$$V(t, \tau) = \frac{4}{3}\pi G^3 (t - \tau)^3 \tag{11.4.4}$$

where G is the constant time rate of change of the spherulite radius.

For sporadic or homogeneous nucleation, $\dot{N}(\tau)$ is usually a constant, and a combination of Eqs. (11.4.3) and (11.4.4) yields

$$X'(t) = \frac{\pi}{3}\frac{\rho_s}{\rho_L} G^3 \dot{N} t^4 \tag{11.4.5}$$

For heterogeneous nucleation, on the other hand, the total number of nuclei per unit volume N_0 is independent of time, and an integration over time is not necessary.

The corresponding result for the mass fraction crystallinity is given by

$$X'(t) = \frac{\rho_s}{\rho_L}\frac{4}{3}\pi G^3 N_0 t^3 \tag{11.4.6}$$

These equations cannot be valid at long times, because they predict a physically meaningless unbounded increase in X'. In real interactions, the spherulites impinge on each other; growth slows and ultimately stops. The situation is easily remedied by assuming the following [19]:

$$\frac{dX}{dX'} = 1 - X \tag{11.4.7}$$

where dX is the actual amount of material transformed in time $d\tau$ and dX' is the amount of material that would be transformed in the same time interval in the absence of impingement. This equation simply expresses the fact that the effect of impingement is small when the amount of crystallization is small, and the rate of crystallization must decrease to zero as X tends to unity.

Equation (11.4.7) is easily integrated because X' is given either by Eq. (11.4.5) or Eq. (11.4.6). For homogeneous nucleation the result is

$$X = 1 - \exp\left(-\frac{\pi}{3}\frac{\rho_s}{\rho_L} G^3 \dot{N} t^4\right) \tag{11.4.8}$$

For heterogeneous nucleation, the corresponding result is

$$X = 1 - \exp\left(-\frac{\rho_s}{\rho_L}\frac{4}{3}\pi G^3 N_0 t^3\right) \tag{11.4.9}$$

or, more generally,

$$X = 1 - \exp(-kt^n) \tag{11.4.10}$$

which is known as the *Avrami equation*. All of the temperature dependence is embodied in the rate constant k, whereas the Avrami exponent, n, is usually considered to be the sum $p + q$, with p being 0 or 1 depending on predetermined or sporadic nucleation and q being 1, 2, or 3 depending on the dimensionality of crystal growth. Thus, n would equal 3 for the growth of disklike crystals by homogeneous nucleation but would only be 2 if the nucleation were heterogeneous.

For most polymers, crystallinity is never complete and Eq. (11.4.10) is modified by defining an effective fraction of transformed material X/X_∞, where X_∞ is the mass fraction crystallized at the end of the transformation. The result of this modification is given as follows [20]:

$$1 - \frac{X}{X_\infty} = \exp\left(-\frac{k}{X_\infty} t^n\right) \tag{11.4.11}$$

or

$$\ln\left(1 - \frac{X}{X_\infty}\right) = -kt^n \tag{11.4.12}$$

where X_∞ on the right-hand side of Eq. (11.4.12) has been absorbed into the constant k.

Figure 11.11 shows crystallization data for the degree of crystallinity X as a function of time, at several constant temperatures, for poly(ether-ether-ketone) (PEEK). This polymer has a glass transition temperature of 145°C and a melting point of 340°C, properties that make it a candidate for high-performance thermoplastic composite matrix applications. In Figure 11.11, the degree of crystallinity is determined as the ratio between the heat evolved during isothermal crystallization in a differential scanning calorimeter and the latent heat of crystallization of a perfect crystal. An examination of Figure 11.11 reveals that a certain induction time is needed before crystallization commences and that X_∞ the ultimate crystallinity at very long times depends on the temperature of crystallization.

When the data of Figure 11.11 are plotted on logarithmic coordinates, as suggested by Eq. (11.4.12), a set of parallel lines is obtained, shown in Figure 11.12. The slope of each of the lines is approximately 3, suggesting heterogeneous nucleation and three-dimensional spherulitic growth. Note, though, that in the latter stages of crystallization, growth slows and the Avrami expression is not obeyed. This phase of crystallization is called *secondary crystallization*, and it is characterized by the thickening of crystal lamellae and an increase in crystal perfection rather than an increase in spherulite radius.

Example 11.4: It is often found (see Fig. 11.13) that primary isothermal crystallization data at various temperatures superpose when X/X_∞ is plotted as

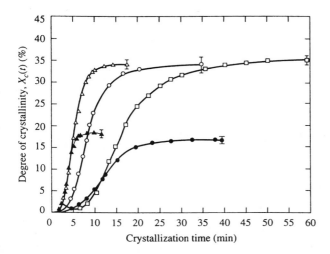

FIGURE 11.11 Crystallization data for PEEK. Development of absolute crystallinity with time for isothermal crystallization at 315°C (□), 312°C (○), 308°C (△), 164°C (▲), and 160°C (●). (From Ref. 21.)

Reprinted from Polymer, vol. 27, Cebe, P., and S. D. Hong: "Crystallization Behaviour of Poly(ether-ether-ketone)," pp. 1183–1192, Copyright 1986, with kind permission from Elsevier Science Ltd., The Boulevard, Langford Lane, Kidlington OX5 1GB, UK.

a function of $t/t_{1/2}$, where $t_{1/2}$ is the time for the percent crystallinity to reach 50% of the final value [22]. If the Avrami theory is valid, how is $k(T)$ related to $t_{1/2}$?

Solution: Allowing X/X_∞ to equal 0.5 in Eq. (11.4.12) gives

$$\ln 0.5 = -kt_{1/2}^n$$

or

$$k(T) = \frac{\ln 2}{t_{1/2}^n}$$

Although the Avrami equation is obeyed exactly by a large number of polymers, noninteger values of the exponent n are often observed [23]. Also note that, for heterogeneous crystallization, the constant k in Eq. (11.4.12) is related to the linear growth rate G of Eq. (11.3.6) as

$$G \sim k^{1/n} \tag{11.4.13}$$

in which n is the Avrami exponent and where it is obvious that k depends on temperature. An extensive treatment of the temperature dependence of G

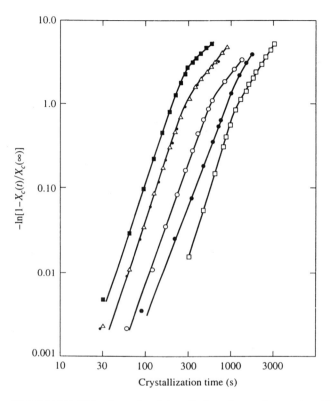

FIGURE 11.12 Avrami plot of the data shown in Figure 11.11. Plot of $\log\{-\ln[1 - X_c(t)/X_c(\infty)]\}$ versus time for isothermal crystallization at 315°C (□), 312°C (○), 308°C (△), 164°C (■), and 160°C (●). (From Ref. 21.)
Reprinted from Polymer, vol. 27, Cebe, P., and S. D. Hong: "Crystallization Behaviour of Poly(ether-ether-ketone)," pp. 1183–1192, Copyright 1986, with kind permission from Elsevier Science Ltd., The Boulevard, Langford Lane, Kidlington OX5 1GB, UK.

according to the original theory of Hoffman and Lauritzen [7], along with supporting experimental data, may be found in the book by Gedde [4], who also discusses some of the more recent theoretical developments.

11.4.2 Nonisothermal Quiescent Crystallization

During the processing of any crystallizable polymer, crystallization never occurs at a fixed, constant temperature. Instead, the polymer cools from the molten state to the solid state at some rate that is determined by the processing conditions, and

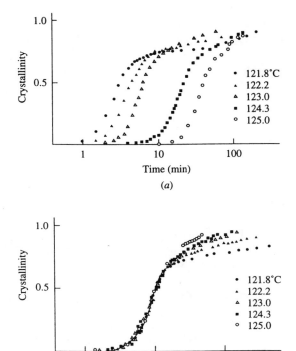

FIGURE 11.13 Isothermal crystallization data for high-density polyethylene. (*a*) Crystallinity–time curves in isothermal crystallization at various temperatures. (*b*) Plots of crystallinity versus relative time t/t^*, where t^* is time when the crystallization proceeds to 50%. (From Ref. 22.) From J. Appl. Polym. Sci., vol. 17, Nakamura, K., K. Katayama, and T. Amano: Some aspects of nonisothermal crystallization of polymers: II. Consideration of the isokinetic condition, Copyright © 1973 by John Wiley & Sons, Inc. Reprinted by permission of John Wiley & Sons, Inc.

the structure that results is due to crystallization that takes place over the entire temperature range between the melting point and the glass transition temperature. In principle, the final crystallinity at the end of the cooling process can again be calculated by combining Eqs. (11.4.3) and (11.4.7), with the result that

$$ X = 1 - \exp\left(-\frac{\rho_s}{\rho_L} \int_0^t V(t, \tau)\dot{N}(\tau)\, d\tau\right) \tag{11.4.14} $$

Now, G and $\dot{N}$ are no longer constants. They may, however, be evaluated from a knowledge of the temperature dependence of the isothermal rate constants [24].

$$\dot{N}(\tau) = \int_0^t \left(\frac{d\dot{N}}{dT}\right)\left(\frac{dT}{ds}\right) ds \tag{11.4.15}$$

$$V(t, \tau) = \frac{4}{3}\pi[r(t, \tau)]^3 \tag{11.4.16}$$

with

$$r(t, \tau) = \int_\tau^t \left(\frac{dr}{ds}\right) ds \tag{11.4.17}$$

$$\frac{dr}{ds} = G(s) = G(0) + \int_0^S \left(\frac{dG}{dT}\right)\left(\frac{dT}{d\tau}\right) d\tau \tag{11.4.18}$$

Insertion of these expressions into Eq. (11.4.14) then makes it possible to compute $X(t)$. However, it is obvious that an analytical result cannot be obtained unless gross assumptions are made. One such assumption involves allowing the ratio $\dot{N}/G$ to be constant, independent of temperature, resulting in an isokinetic process [25]. This is justified based on the fact that the shapes of both functions are similar when they are plotted in terms of temperature [see Eqs. (11.3.4) and (11.3.6)]. A consequence of this assumption, as demonstrated by Nakamura and co-workers [22, 26] is that Eq. (11.4.12), the Avrami equation, is modified to

$$\ln\left[1 - \frac{X}{X_\infty}\right] = -\left\{\int_0^t K[T(\tau)] \, d\tau\right\}^n \tag{11.4.19}$$

where

$$K(T) = k(T)^{1/n} \tag{11.4.20}$$

and n is the Avrami exponent determined from data on isothermal crystallization. It has been shown that the modified Avrami equation represents the nonisothermal crystallization behavior of high-density polyethylene very well [22].

11.5 POLYMER CRYSTALLIZATION IN BLENDS AND COMPOSITES

As discussed in Section 9.6 of Chapter 9, polymers are blended together with the expectation of obtaining a material having enhanced thermal, mechanical, or solvent-resistance properties relative to the blend constituents. If one of the components is crystallizable, the presence of the other component can influence the nature, rate, extent, and temperature range of crystallization. If the two polymers are immiscible, crystallization may occur in the domain of one polymer unaffected by the presence of the other polymer, or the other polymer may act as a nucleating agent, depending on which component solidifies first. However, the more interesting situation, from a theoretical viewpoint, is one in which the polymers are thermodynamically compatible and only one component is crystallizable. An example is the blending of PEEK with poly(etherimide) (PEI), an amorphous polymer with a high thermal resistance ($T_g \approx 215°C$). The amorphous polymer acts as a diluent for the crystallizable polymer, and the result is a lowering of the equilibrium melting point; this effect is similar to the depression in freezing point (see Sect. 8.3 of Chap. 8) of a liquid upon addition of a nonvolatile solute. If we use the Flory–Huggins theory developed in Section 9.3 of Chapter 9, it is possible to show the following [27]:

$$\frac{1}{T_m} - \frac{1}{T_m^0} = -\frac{Rv_2}{\Delta H_2 v_1} \chi_{12}(1 - \phi_2)^2 \tag{11.5.1}$$

where T_m and T_m^0 are the melting points of the crystals in the blend and pure states, ΔH_2 is the (positive) heat of fusion per mole of repeating unit, v_1 and v_2 are the molar volumes of the repeating units, χ_{12} is the polymer–polymer interaction parameter, and ϕ_2 is the volume fraction. (In a volume fraction, the amorphous polymer is denoted by subscript 1 and the crystalline polymer by subscript 2.) Note that for T_m to be smaller than T_m^0, χ_{12} has to be negative. This is consistent with the remark following Eq. (9.6.3) of Chapter 9 that polymer–polymer miscibility depends entirely on energetic effects.

A further effect of polymer blending is that the blend's glass transition temperature is bounded by the glass transition temperatures of the blend constituents; this intermediate value is given by Eq. (9.6.1) of Chapter 9. Thus, because crystallization takes place only in a temperature range between T_m and T_g, the addition of high-T_g amorphous diluent can serve to significantly contract the range of available crystallization temperatures.

Example 11.5: Estimate the lowering in melting point and the elevation in the glass transition temperature of PVF$_2$ ($T_m^0 = 170.6°C$, $T_g = -50°C$) when it is mixed with PMMA ($T_g = 90°C$) such that the blend contains 60% by weight of PVF$_2$. According to Nishi and Wang [27], $v_1 = 84.9\,\text{cm}^3/\text{mol}$, $v_2 = 36.4\,\text{cm}^3/\text{mol}$, $\Delta H_2 = 1.6\,\text{kcal/mol}$, and $\chi_{12} = -0.295$.

Solution: Using Eq. (11.5.1) gives

$$\frac{1}{T_m} = \frac{1}{443.6} + \frac{1.987 \times 36.4}{1600 \times 84.9} \times 0.295(1 - 0.6)^2$$

or

$$T_m = 438.7 \text{ K}$$
$$(T_m^0 - T_m) = 4.89 \text{ K}$$

Using Eq. (9.6.1) again gives

$$\frac{1}{T_g} = \frac{0.4}{363} + \frac{0.6}{223}$$

or $T_g = 263.7 \text{ K}$, and T_g increases by 40.7 K.

When a polymer crystallizes in the presence of a noncrystallizable component, the crystal growth rates can be significantly affected due to the inevitable rejection of the noncrystallizable material from the growing crystal. This effect becomes increasingly important at large undercoolings, where growth rates are dominated by diffusion. Thus, the maximum isothermal radial growth rate of isotactic polystyrene falls by almost a third on adding a 15% atactic polymer [28]. In this situation, the spherulite radius still increases linearly with time because the mixture composition at the crystal growth front remains unchanged as crystallization proceeds. However, if the rejected polymer diffuses more rapidly than the rate at which spherulites grow, the concentration of diluent at the growth front can increase sufficiently, to further reduce the rate of crystallization and make the spherulite radius change nonlinearly with time. An additional significant reduction in the growth rate can occur if the change in melt composition is accompanied by an increase in melt viscosity. This is especially important when the two components have widely separated glass transition temperatures. Furthermore, these kinetic effects can be accompanied by changes in crystal morphology, depending on where the amorphous polymer segregates itself [29]. Additionally, when crystallization takes place in the presence of cooling, the solidified blend may have very low levels of crystallinity due to all of the effects mentioned. Nadkarni and Jog [30] have summarized the crystallization behavior of commonly encountered crystalline/amorphous as well as crystalline/crystalline polymer blend systems.

An important effect that is observed during the melt processing of blends of condensation polymers, such as two polyamides or two polyesters, is the occurrence of interchange reactions [31]. The result of these transamidation or transesterification reactions is the formation of a block copolymer. Initially, a diblock copolymer is produced, but, with increasing processing time, this gives way to blocks of progressively smaller size; ultimately, a random copolymer results. This "processing" route to the synthesis of copolymers is often simpler and more economical than making the copolymers in a chemical reactor, and "reactive extrusion" is a major industry today [32]. If one homopolymer is semicrystalline and the other amorphous, noncrystallizable sequences will be built in between the crystallizable sequences of the semicrystalline polymer. This has a profound effect on the crystallization behavior of the semicrystalline polymer [33]. In particular, there is a reduction in the melting point of the crystals and a change in the glass transition temperature; the T_g of the resulting random copolymer can be estimated using Eq. (9.6.1) of Chapter 9 that was earlier shown to be valid for miscible polymer blends. There is also a decrease both in the crystallization rate and the total crystallinity of the blend, as compared with the crystallizable homopolymer.

In order to modify polymer properties, we commonly add fillers and reinforcements to plastics. The dimensions of these additives are usually no smaller than a few microns, and the fillers influence the behavior of crystallizable polymers only to the extent that they provide sites for nucleation; polymer morphology is generally not affected. In the recent past, though, it has become possible to add solids whose smallest dimension is of the order of 1 nm (10 Å). These materials are called nanomers, and the mixture is known as a nanocomposite [34]. The most extensively researched filler is montmorillonite (MMT), a clay that is a layered silicate made up of platelets or sheets that are each about 1 nm thick and which have an aspect ratio ranging from 100 to 300. MMT has a large surface area of about $750 \, m^2/g$, and the addition of just 1 wt% of well-dispersed clay to a polymer such as nylon 6 results in very significant property improvements: The Young's modulus, dimensional stability, heat-distortion temperature, solvent resistance, and flame resistance all increase. The extent of increase is what might normally be expected on adding more than 10 wt% glass fibers, say. Also, there is a reduction in gas and moisture permeability, and all this happens without loss of any other property of interest. Not surprisingly, nanocomposites are being researched for a wide variety of applications, including the original automotive applications. In terms of the crystallization behavior of polymers containing nanofillers, it has been found that the presence of silicate layers enhances the rate of isothermal crystallization [35]; this is not surprising because the clay platelets act as nucleating agents. What is surprising, though, is that for polymers such as polypropylene and nylon 6, the polymer morphology changes from spherulitic, in the absence of MMT, to fibrillar, in the presence of MMT— the fibrous structures grow in both length and diameter as crystallization proceeds. Also, crystallization can occur at high temperatures where the neat polymers do not crystallize [35].

11.6 MELTING OF CRYSTALS

When the temperature of a polymeric crystal is raised above the glass transition temperature, it can begin to melt. This process is the reverse of crystallization, but, surprisingly (and in contrast to the behavior of low-molecular-weight substances), melting takes place over a range of temperatures even for crystals of a monodisperse polymer. Furthermore, the melting point changes if the rate of heating is changed. These phenomena, although ostensibly unusual, can be explained in a straightforward manner using the theory developed in this chapter.

A polymer single crystal of the type shown in Figure 11.5 will actually melt at a temperature T_m, where T_m is less than T_m^0, when the net change in the Gibbs free energy is zero. Thus,

$$\Delta G = Al\Delta G_v - 2A\gamma_e = 0 \tag{11.6.1}$$

where l is the thickness of the crystal, A is the surface area of the fold surface as shown in Figure 11.5, and γ_e is the free energy of the fold surface. Here, ΔG_v and γ_e are both positive quantities, and the total surface area of the crystal has been approximated by $2A$.

Using Eq. (11.2.6) for ΔG_v gives

$$\Delta G_v = \Delta H_v \left(1 - \frac{T_m}{T_m^0} \right) \tag{11.6.2}$$

which, when introduced into Eq. (11.6.1), allows us to solve for the melting point as follows:

$$T_m = T_m^0 \left[1 - \left(\frac{2\gamma_e}{\Delta H_v l} \right) \right] \tag{11.6.3}$$

It is clear why the actual melting point must be less than the equilibrium melting point. The two melting points become equal only for infinitely thick crystals for which l tends to infinity. This is a situation that can prevail only for high-molecular-weight polymers and only for extended-chain crystals. Conversely, low-molecular-weight polymers must necessarily have a melting point that is less than T_m^0. Even when the molecular weight is kept fixed, Eq. (11.6.3) teaches us that the melting point varies as l varies. Because l depends on the temperature at which crystallization originally took place and increases with increasing temperature [see Eq. (11.2.14)], the melting point of a crystal formed at a given temperature is higher than the melting point of a similar crystal formed at a lower temperature. Thus, nonisothermal crystallization gives rise to crystals that do not have a single, sharply defined melting point. Indeed, it is even possible to heat a semicrystalline polymer to a temperature above T_g and to melt some crystals while allowing other crystals to form! Furthermore, because crystals tend to thicken on *annealing* (being held at a temperature above the glass transition temperature), slow heating of a crystal gives rise to a higher melting point than does fast heating. Finally, a conceptually easy way to measure the equilibrium melting point is to plot T_m against $1/l$ and extrapolate to a zero value of the abscissa. A more practical way is to plot T_m as a function of T_c, the temperature of crystallization, and extend the plot until it intersects the graph of $T_m = T_c$. The point of intersection yields T_m^0. This procedure is known as a Hoffman–Weeks plot, and it is illustrated in Figure 11.14 for a 90/10 blend of nylon 66 with an amorphous nylon [33]. Also shown in this figure are data on block copolymers of these two plastics. Copolymerization takes place simply on holding the blend in the melt state for an extended period of time, and as time in the melt increases, it results in the formation of progressively smaller blocks. The progressively smaller blocks lead to progressively less perfect crystals that have a progressively lower melting point.

Information generated about the melting point and the heat of fusion of a semicrystalline polymer by melting tiny samples in a differential scanning calorimeter can generally be applied to predict the melting behavior of large amounts of the same polymer in processing equipment. Such a heat transfer model for polymer melting in a single screw extruder is presented in Chapter 15; the rate of melting is determined by the sum of the heat generated in unit time by viscous dissipation and that which is provided by band heaters attached to the extruder barrel. When one goes to progressively larger extruders, though, the ratio of the surface area available for heat transfer to the volume of polymer in the

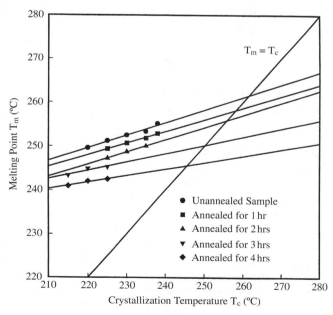

FIGURE 11.14 Hoffman–Weeks plots for a 90/10 nylon 66/amorphous nylon blend annealed in the melt state for different periods of time. (From Ref. 33.)

extruder decreases to such an extent that heat transfer from the outside becomes of secondary importance. In this case, the sequence of events that leads to the melting of a crystalline polymer has been elucidated by Shih and co-workers [36] by carrying out melting experiments on polyethylene in a heated internal (batch) mixer. The polymer was charged to the mixer, and a low but constant heating rate was imposed; the torque and temperature were measured, and the pellets were observed through a glass window. Initially, the polymer existed in the form of free-flowing pellets whose temperature increased as expected. As the temperature approached T_g, the pellets softened and were compacted. The nonelastic deformation of these compacted pellets resulted in large energy dissipation, especially in the presence of a small amount of clay that increased the coefficient of friction. The result was a sharp increase in both the mixing torque and the mixture temperature. This led to the formation of molten polymer that tended to lubricate the pellets, and there was a reduction in the torque. Ultimately, there was enough melt generated to subsume all of the solid particles in the form of a slurry. In large extruders, melting of pellets arises not due to heat transfer from the barrel but due to heat generated by the periodic deformation of the softened pellets. This can lead to rather rapid melting in a narrow region instead of gradual melting over a large region [37]. Indeed, it is this very rapid generation of energy coupled with the low thermal conductivity of typical polymers that is responsible for the poor "melt quality" that is often observed: The polymer that leaves the extruder can consist of islands of relatively cold, unmelted polymer floating in very hot molten liquid.

11.7 INFLUENCE OF POLYMER CHAIN EXTENSION AND ORIENTATION

Early work on polymer crystallization dealt exclusively with isothermal crystallization in stress-free, unoriented polymers and was useful for elucidating mechanisms and building theories. However, in practical polymer processing operations (such as fiber spinning and film blowing), crystallization takes place from oriented polymers under stress, and events arise that cannot be explained in a quantitative manner. Southern and Porter have found that when high-density polyethylene is extruded using a capillary viscometer, at a temperature close to the polymer melting point, crystallization can be induced in the polymer in the entry region of the capillary [38]. In this study, crystallization was so massive that the capillary was essentially blocked, resulting in a cessation of extrudate flow and a rapid increase in the extrusion pressure. Furthermore, the crystals that were formed had an extended chain structure and a higher melting point than normal. That polymer chain extension and orientation are responsible for the enhancement in the crystallization rate has been demonstrated most strikingly by straining polyethylene terephthalate in the glassy state and then measuring the rate of crystallization by annealing the samples above the glass transition temperature [39, 40]. Indeed, while under isothermal, quiescent conditions, the half-time of polyethylene terephthalate crystallization is about 1 min at the temperature of maximum crystallization rate [23]. During commercial fiber spinning, the same polymer crystallizes in less than one 1 s due to the influence of mechanical stretching [41]. These studies, however, have been largely qualitative in character, serving to prove that grossly enhanced crystallization rates are obtained when orientation occurs prior to crystallization.

More quantitative results have been reported by Katayama *et al.* [42], Dees and Spruiell [43], and many others; these authors have presented details of orientation and structure development along a monofilament spinline for fibers made from polymers such as polyethylene, polypropylene, and polybutene-1. These results, however, are given in terms of process variables such as spinning speed, and because the process is nonisothermal, it is impossible to reanalyze them in terms of fundamental variables such as residence time, orientation, and temperature. In general, though, we can say that, during polymer processing, polymer chain alignment and extension occur due to flow. We also find that elongational flow (see Chapter 14)—such as occurs on the spinline or in the converging, entry region of a capillary—is much more effective than shear flow in causing chain extension and orientation. A major consequence of chain extension for flexible macromolecules is a decrease in conformational entropy with a consequent increase in the free energy, a result similar to what happens in rubber elasticity [44]. In Figure 11.1, for example, this means that the free-energy curve for the melt is shifted vertically upward, resulting in an elevation of the equilibrium melting point and an increase in the driving force for crystallization at a fixed processing temperature. Further, because polymer molecules are uncoiled and stretched by the flow field, it is natural that extended-chain crystals be formed.

Although this reasoning provides a qualitative framework for all the observations, it does not allow us to obtain an explicit expression for the rate of crystallization from a deforming melt; such an expression is needed if we want to quantitatively simulate the processing of a semicrystalline polymer. The process of chain extension and orientation in a crystallizing melt has been examined theoretically by Ziabicki [45], who defined a scalar-valued orientation factor f in terms of the invariants of a suitable deformation tensor, and then allowed the nucleation and growth rates to depend explicitly on f under isothermal conditions. Later, Gupta and Auyeung actually measured isothermal crystallization rates for polyethylene terephthalate using a spinline and showed that these did, indeed, correlate with the instantaneous value of the polymer chain orientation in the surrounding (amorphous) melt [46]. However, simple expressions, analogous to the Avrami equation, are yet to be developed for oriented crystallization.

Despite the absence of theoretical expressions, the practical applications of oriented crystallization have been developed and commercialized. The most important application is in the synthesis of high-modulus fibers from conventional, flexible-chain, random-coil polymers such as polyethylene. In the process of solid-state extrusion (similar to the experiments of Southern and Porter), almost perfectly oriented, extended-chain structures are obtained by forcing a polymer billet through a tapered die. A major use of such fibers is in the reinforcement of composites.

11.8 POLYMERS WITH LIQUID-CRYSTALLINE ORDER

Although it is possible to spin high-strength, high-modulus fibers from flexible-chain polymers such as polyethylene, the procedure requires that polymer chains be extended and packed into a crystal lattice as tightly as possible. To achieve this, not only do polymer chains have to be extended by some means, they also have to be prevented from relaxing both before and during crystallization. An alternate approach is to employ crystallizable, rigid-chain polymers that appear to be rodlike or disklike in solution. It is found that many such polymers exist, and both kinds of rigid polymer organize themselves into an ordered liquid phase, called a *mesophase*, either in appropriate solvents or in the melt itself. Such liquid-crystalline solutions or melts can be processed into fibers so that the preordered domains are not only preserved but also enhanced by actual crystallization. Molecular alignment results in high strength, and the materials so formed are also chemically inert and dimensionally stable because crystalline melting points are typically in the 275–420°C range.

Rigidity in the chemical structure of most polymers that show liquid-crystalline order generally comes from para-linked, aromatic rings such as those found in aromatic polyamides, polyesters, and polyazomethines [47]. Often, the melting point is so high and so close to the degradation temperature that these materials are difficult to process in the molten state. In such a case, it is common to introduce flexible, aliphatic spacer units into the backbone to lower the melting point [48]. Among soluble polymers, synthetic polypeptides (e.g., poly-benzyl-L-glutamate) that form a helix in appropriate solvents have been extensively studied. The history of the development of polymeric liquid crystals may be found in the review by White [49].

Polymers that form a liquid-crystalline phase in solution are known as *lyotropic*. Three different physical structures are found to occur with rodlike molecules; these are shown in Figure 11.15. In the nematic phase, there is no

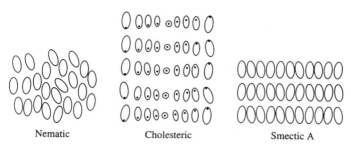

FIGURE 11.15 Schematic representation of mesophase types. (From Ref. 47.) Reprinted with permission from Wissbrun, K. F., "Rheology of Rod-Like Polymers in the Liquid Crystalline State," J. Rheol., 25, 619–662, 1981.

long-range order of positions, but there is a preferred direction called the director, although a distribution of angles with respect to the director is observed. Liquid crystals in the cholesteric phase show an increase in order over the nematic phase, with the direction of the director varying helically along an axis perpendicular to the plane of the director. Finally, the smectic phase shows the most order, albeit in only one dimension. The formation of lyotropic liquid crystals was theoretically predicted by Flory [50], who showed that a single ordered phase comes about when the concentration of a rodlike polymer in solution exceeds $12.5/x$, where x is the aspect ratio of the rod. The ordering itself can be promoted by electrical, magnetic, or mechanical forces and is accompanied by a sharp decrease in the solution viscosity, implying ease of processing. The viscosity of a 50/50 copolymer of n-hexyl and n-propylisocyanate in toluene is shown in Figure 11.16, which illustrates that it is easier for rods to slide past each other when they are oriented parallel to each other [51]. The best known example of a lyotropic liquid crystal is the polyaramid fiber Kevlar, manufactured by DuPont [52]. This is an extended-chain, para-oriented polyamide made by reacting p-phenylenediamine and terephthaloyol chloride. Products made from Kevlar fiber include composites for marine, aircraft, and aerospace applications, tire cord, ropes, belts, and bullet-proof vests. Note, though, that 100% sulfuric acid is the usual solvent for Kevlar, and the use of this solvent necessitates a certain amount of safety precautions.

Because lyotropic liquid-crystalline polymers cannot be extruded, injection molded, or blown into films, other polymers that can be melt processed have been developed. These thermotropic liquid-crystalline polymers convert to a mesophase when the solid polymer is heated to a temperature above the crystalline melting point. Thus, these polymers show three thermal transitions. In increasing order of temperature, these are glass transition temperature, crystalline melting

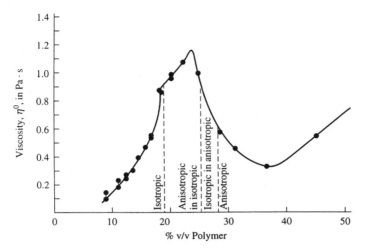

FIGURE 11.16 Viscosity as a function of polymer concentration for the system poly(50% n-hexyl + 50% n-propyl) isocyanate of $M_w = 41,000$ in toluene at 25°C. (From Ref. 51.)

Reprinted from Polymer, vol. 21, Aharoni, S. M.: "Rigid Backbone Polymers: XVII. Solution Viscosity of Polydisperse Systems," pp. 1413–1422, Copyright 1980, with kind permission from Elsevier Science Ltd., The Boulevard, Langford Lane, Kidlington OX5 1GB, UK.

point, and nematic-to-isotropic transition temperature. The most widely studied class of thermotropic polymers are aromatic polyesters, such as the copolyester of p-hydroxybenzoic acid (HBA) and polyethylene terephthalate (PET). More recently, the Hoechst–Celanese Company has commercialized a 73/27 copolymer of HBA and 6-hydroxy-2-napthoic acid (HNA) called Vectra A900, which has a nematic-to-isotropic transition temperature of about 370°C. Shown in Figure 11.17 is the melt viscosity as a function of HBA content of an HBA–PET copolymer at different shear rates [53]. It is seen that the viscosity behavior is similar to that shown in Figure 11.16 insofar as the viscosity goes through a maximum at a particular HBA content, which is around 30% in the present case. As might be guessed, this is due to the formation of a mesophase.

Crystallization in a thermotropic liquid-crystalline polymer is again a process of nucleation and growth [54]. It has been shown that the process can be followed easily using dynamic mechanical analysis (see Chap. 12) [55], in which we measure the stress response of the material to an imposed small-amplitude sinusoidal shear strain. Differential scanning calorimeter (DSC) data on the kinetics of crystallization show that the process is describable by an Avrami equation [56].

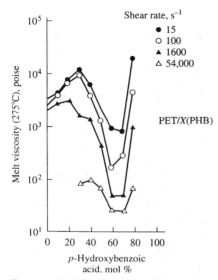

FIGURE 11.17   Melt viscosity of PET modified with *p*-hydroxybenzoic acid. (From Ref. 53.) From J. Polym. Sci. Polym. Chem. Ed., 14, Jackson, W. J., Jr., and H. Kuhfuss: Liquid crystal polymers: I. Preparation and properties of *p*-Hydroxybenzoic acid copolyesters, Copyright © 1976 by John Wiley & Sons, Inc. Reprinted by permission of John Wiley & Sons, Inc.

11.9   STRUCTURE DETERMINATION

As we have seen in the previous sections, a crystallizable polymer melt that has been solidified under quiescent conditions possesses a two-phase structure consisting of chain-folded crystals organized as spherulites in an amorphous matrix. Because the mechanical, optical, electrical, thermal, and transport properties of the two phases are generally quite different from each other, the observed behavior will be a weighted average of the properties of the two phases. We can expect the weighting function to be the fraction of the crystalline or amorphous phase, with the size and size distribution of the domains often playing a relatively minor role.

Even though polymer molecules are inherently anisotropic, with properties along the chain axis being vastly different from properties perpendicular to the chain axis, this difference does not show up in materials formed under quiescent conditions. This is because the polymer chains are typically randomly oriented. However, during processing operations such as fiber spinning and film blowing and, to a lesser extent, injection molding, the polymer chain axis naturally tends to align itself along the stretching direction, which makes properties of the solid polymer directional in nature. In other situations, we can intentionally create orientation by (1) simultaneously drawing and heat-treating a solid semicrystalline polymer or (2) forming oriented crystals either by employing an oriented melt or by stretching the glassy polymer before annealing. In such cases, we also need to know the average orientation of polymer molecules relative to some axis in each phase. This information is necessary for computing a particular average property of the polymer.

The orientation, relative to a specified direction, of the polymer chain axis in the amorphous region or of any of the three crystallographic axes (labeled a, b, and c in Fig. 11.5) in the crystalline region can be defined in a number of ways [57–59], and this is generally done in terms of the anisotropy of the polarizability tensor. Even though we expect that a solid part will contain a distribution of orientations, the inability to measure the complete distribution function forces us to use one or more moments of the distribution function. This, however, is not a severe limitation because most properties depend only on certain moments of the distribution [59]. Thus, for specifying polymer orientation in a fiber that has angular symmetry about the fiber axis, we generally use the second moment, which is commonly known as *Herman's orientation factor, f*, defined as follows (see Fig. 11.18):

$$f = \tfrac{1}{2}(3\langle\cos^2\ \phi\rangle - 1) \qquad\qquad (11.9.1)$$

where ϕ is the angle between the fiber axis and either the polymer chain axis or the c axis, depending on whether one is considering the noncrystalline region or the crystalline region, respectively. The angular brackets denote a spatial average.

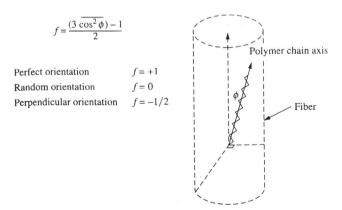

$$f = \frac{(3\ \overline{\cos^2\phi}) - 1}{2}$$

Perfect orientation	$f = +1$
Random orientation	$f = 0$
Perpendicular orientation	$f = -1/2$

FIGURE 11.18 Schematic representation of the significance of the Hermans' orientation function. (From Ref. 58.) From Samuels, R. J.: Structured Polymer Properties. Copyright © 1974 by John Wiley & Sons, Inc. This material is used by permission of John Wiley & Sons, Inc.

For perfect alignment, ϕ is zero and f equals unity. For perpendicular orientation, ϕ is a right angle and f equals $-\frac{1}{2}$. Further, a zero value of the orientation factor implies random orientation, which occurs at $\phi = 54.7°$.

In summary, then, it is necessary to measure the fraction of crystals, the crystalline orientation factor f_c, the amorphous orientation factor f_a, and possibly the size and size distribution of crystals in order to relate polymer structure to polymer properties. Although the extent of crystallinity is generally measured using density or heat-of-fusion methods, orientation is determined with the help of optical birefringence, dichroism, sonic modulus, or x-ray diffraction [60]. The size of crystals is observed with an optical or electron microscope.

11.9.1 Mass Fraction Crystallinity

The simplest method of determining the mass fraction crystallinity X of an unfilled, semicrystalline homopolymer is to measure the density ρ of a representative sample. If the material is free of voids and impurities, the total volume V of unit mass of polymer is given by

$$V = \frac{X}{\rho_c} + \frac{1-X}{\rho_a} \tag{11.9.2}$$

where ρ_c and ρ_a are the known densities of the crystalline and amorphous phases, respectively. Because the sample density ρ must equal $1/V$, we have

$$\frac{1}{\rho} = \frac{X}{\rho_c} + \frac{1-X}{\rho_a} \tag{11.9.3}$$

which can easily be solved for the desired mass fraction of crystals to give

$$X = \frac{1 - \rho_a/\rho}{1 - \rho_a/\rho_c} \tag{11.9.4}$$

The actual measurement of the density is carried out with the help of a density gradient column. This is a graduated glass cylinder filled with a mixture of two miscible liquids with appropriate densities such that a gradient of density exists along the column. Glass floats of different but known density are suspended at various locations along the length of the column, and the liquid density at any other position is obtained by interpolating between these values. The sample of unknown density is gently dropped into the column and allowed to settle slowly over a period of hours until it comes to rest at some vertical position where the sample density equals the local liquid density. In the case of polyethylene terephthalate, for example, where the density can vary between 1.335 and 1.455 g/cm^3, mixtures of carbon tetrachloride (1.594 g/cm^3) and toluene (0.864 g/cm^3) are used.

The column itself can be prepared using the scheme illustrated in Figure 11.19 [61]: The denser liquid is put in flask A and the other liquid, in flask B. The

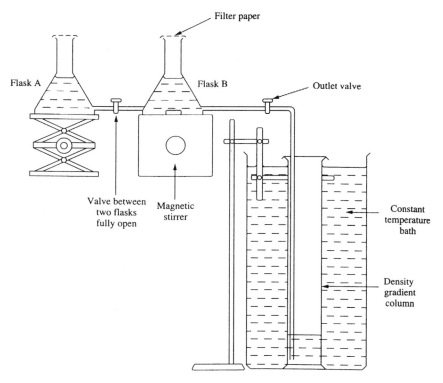

FIGURE 11.19   Setting up a density gradient column. (From Ref. 61.)

valve between the two flasks is opened, and the outlet valve is adjusted to generate a slow flow of liquid into the column; a 1-L column may take 2–3 hr to fill. If the column is kept covered and undisturbed, it can be stable for at least 1 month. To use the column, we hold the sample with a pair of tweezers, wet it with the lighter liquid, and drop it slowly into the column. After experimentation, old samples are removed using a wire-mesh basket that is normally kept sitting at the bottom of the column. This technique allows us to measure the density with an accuracy of at least 0.05%.

Example 11.6:   When a fiber made from PET is dropped in a density gradient column made from toluene and carbon tetrachloride, it comes to rest 70% of the way down the column. What is the percent crystallinity? For PET, ρ_a is equal to 1.335 g/cm^3 and ρ_c is equal to 1.455 g/cm^3.

Solution: By interpolating between the densities of the liquids making up the column, the sample density is

$$\rho = 0.864 + 0.7(1.594 - 0.864) = 1.375 \text{ g/cm}^3$$

and using Eq. (11.9.4) gives

$$X = \frac{1 - (1.335/1.375)}{1 - (1.335/1.455)} = 0.353$$

Thus, the mass percent crystallinity is 35.3%.

Other techniques that can be used to measure the degree of crystallinity include the measurement of properties such as specific heat, electrical resistivity, and the heat of fusion. If we can say that a property p of the polymer can be written as

$$p = Xp_c + (1 - X)p_a \qquad (11.9.5)$$

then the measurement of any such property, together with a knowledge of p_c and p_a (the properties of the individual phases), allows us to determine X. Even though we want to predict the properties from the structure, here we reverse the process to determine the structure from the knowledge of one such property. This, then, allows us to predict other properties of interest. Note that in some cases, such as heat of fusion, the amorphous contribution in Eq. (11.9.5) is zero, whereas in other cases, p_c may differ from the corresponding property of a perfect crystal.

Other techniques of determining crystallinity are wide-angle x-ray diffraction and small-angle x-ray scattering. These are described in standard texts [8].

11.9.2 Spherulite Size

The size and size distribution of spherulites in a semicrystalline polymer sample can be determined easily with a polarized light microscope. Every light microscope has a light source and a set of lenses that focus the light onto the sample and then produce a magnified image. If the sample is transparent and thin, we can use transmitted light microscopy, in which case the light beam passes directly through the sample and reveals details of the internal structure of the sample. Alternately, for opaque or thick samples, we use reflected light microscopy, wherein light is reflected back from the specimen revealing the surface topography. The major limitation of optical microscopy is the shallow depth of field, which limits its use to flat specimens. However, with the magnifications achievable (these depend on the focal lengths of the objective and the eyepiece), we can typically observe features that are about 1 μm in size and separated by distances of about 0.5 μm. All of this, though, requires that there be adequate contrast between different features. Contrast can be enhanced with the use of polarized light; a typical polarized light microscope is shown in Figure 11.20 [62]. The regular elements of the microscope are mirror M and convex lenses C and N for focusing light upon the specimen, plus the objective lenses O and eyepiece E. The elements that make the microscope a polarizing microscope are the polarizers P and A, kept in what is known as the crossed position. Polarizers are explained as follows.

As mentioned in Section 8.4 of Chapter 8, a light beam is a transverse wave made up of sinusoidally varying electric and magnetic field vectors, which are perpendicular to each other and also to the direction of propagation of the wave. When such a beam passes through a polarizer (which is a sheet of material having a characteristic direction), only those electric vectors that vibrate parallel to this direction are transmitted, and the emerging light is *plane-polarized* and has an amplitude A_m (see Fig. 11.21). Now, if a second polarizer, called an analyzer, is placed in the path of plane-polarized light, the light that emerges is still plane-polarized, but the electric vectors now vibrate parallel to the characteristic

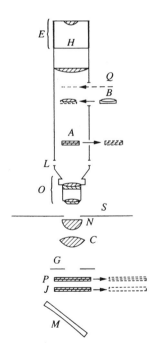

FIGURE 11.20 Arrangement of components in a typical polarizing microscope (diagrammatic).

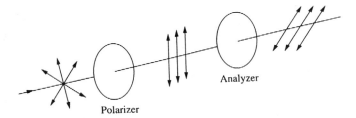

FIGURE 11.21 Cross-polarization of light.

direction of the analyzer. The amplitude of vibration, however, is now reduced to $A_m \cos \alpha$, where α is the angle between the characteristic directions of the polarizer and the analyzer. If α is a right angle, the polarizer and analyzer are said to be in the crossed position, and no light is transmitted through them, provided the intervening medium is isotropic. Spherulites, however, are birefringent entities; in other words, they are not isotropic. In particular, the refractive index n_r along the radial direction differs from n_t along the tangential direction. When $n_r > n_t$, the spherulite is positively birefringent. Conversely, when $n_t > n_r$, the spherulite is negatively birefringent. Because the largest refractive index is usually along the chain axis, most chain-folded polymer spherulites are negatively birefringent. Note that the velocity of light becomes less and less as the refractive index of a medium increases.

As explained by Marentette and Brown [63], if a spherulite is examined under a polarizing microscope, the anisotropic nature of the polymer causes the plane-polarized light of amplitude A to split up (double refraction) into two components having amplitudes $A \cos \theta$ and $A \sin \theta$ and vibrating in mutually perpendicular directions aligned with the principal refractive indices n_t and n_r. These directions are labeled n_1 and n_2 in Figure 11.22. On exiting the sample, the two components pass through the analyzer, which is kept in the crossed position. The result is a single component that vibrates along the characteristic direction of the analyzer and has an amplitude equal to $A \sin \theta \cos \theta + A \cos \theta \sin \theta$. Because of differences in the magnitude of the two refractive indices, though, the two components travel at different speeds through the sample, resulting in both constructive and destructive interference of specific wavelengths of white light. The result is a magnified image of the spherulite, but one that contains a Maltese cross pattern as shown in Figure 11.23. If the microscope employed is fitted with a hot stage, we can actually observe the process of nucleation and growth of spherulites in real time. Spherulite sizes can be obtained from photomicrographs in a trivial manner.

The foregoing is only a brief summary of optical microscopy as applied to polymers. For further details of issues such as sample preparation, the reader

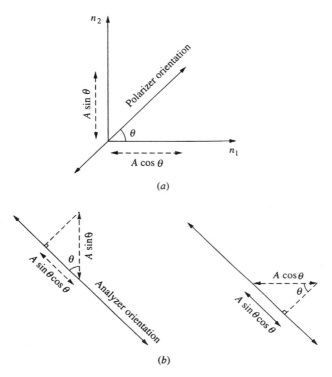

FIGURE 11.22 (*a*) Resolution of plane-polarized light of amplitude *A* into two components along the principal refractive indices of a sample, n_1 and n_2; (*b*), resolution of the light transmitted by the sample by an analyzer that is positioned at right angles to the polarizer in (*a*). (From Ref. 63.)
Reprinted with permission from *J. Chem. Education*, vol. 70, pp. 435–439; Copyright 1993, Division of Chemical Education, Inc.

should consult relevant review papers [64] or texts [62, 65]. Although optical microscopy makes it possible to observe spherulites, the resolution is not such that we can examine individual crystal lamellae. To accomplish this, we must use electron microscopes, whether of the scanning or transmission variety. With electron microscopes, it is possible to distinguish feature sizes that are of the order of nanometers [65].

11.9.3 Polymer Chain Orientation

Polymer chain orientation factors are most conveniently obtained by a combination of wide-angle x-ray diffraction and optical birefringence measurements on

FIGURE 11.23 Sketch of Maltese cross pattern exhibited by spherulites under polarized light.

semicrystalline polymer samples. Although theory and experimental methods for the general case are available in the literature [58, 66], here we illustrate the process for uniaxially oriented materials such as fibers.

Morphologists use x-ray techniques extensively for the determination of crystal structure. The basic principle is based on the interaction of electromagnetic radiation with the dimensional periodicity inherent in the crystal structure, which serves as a diffraction grating. The choice of the appropriate wavelength is critical for the proper resolution of the structure. Bragg's law relates the incidence angle θ, wavelength λ, and interatomic spacing d as follows (see Fig. 11.24):

$$n\lambda = 2d \sin \theta \tag{11.9.6}$$

Here, "wide angle" means that 2θ is allowed to take values all the way to $180°$.

X-rays can be generated by means of electron emission in a hot filament enclosed within an evacuated glass tube. The electrons are then accelerated by means of an applied voltage to a metal target—typically copper, iron, or molybdenum. A small fraction of the energy is converted to x-rays upon collision.

For determining the structure of semicrystalline polymers, we use the Debye–Scherrer method, in which a narrow beam of x-rays enters a cylindrical film cassette through a collimator and hits the sample situated at the center of the cassette. The resulting diffraction pattern is recorded on the photographic film, which is analyzed with the help of a microdensitometer. Typically, for unoriented, crystalline samples, the diffraction pattern is a series of concentric circles. If, however, the crystal axes have a preferred orientation, the rings change to arcs or

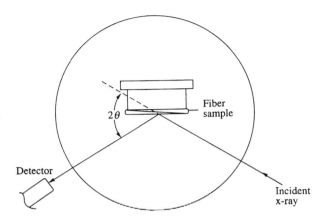

FIGURE 11.24 Top view of the sample holder showing the path taken by x-rays. (From Ref. 61.)

dots. These patterns allow us to determine the crystal planes responsible for the diffraction. Note that for quantitative work, the photographic film is replaced by an ionization counter or detector, which is mounted on a platform. The motion of this assembly, called a *diffractometer*, may be coupled with that of the sample when rotation of the sample is desired.

For determining the crystalline orientation factor of a fiber, we wrap several layers of the fiber around a glass slide and mount the slide on a device that permits rotation of the sample, as shown in Figure 11.25. The value of 2θ is set equal to that which corresponds to a crystal plane known to give rise to diffraction. The intensity $I(\phi)$ of the diffracted beam is measured at values of ϕ (the angle between the fiber axis and the axis of rotation of the sample holder) ranging from $0°$ to $90°$. The average value of the angle between the fiber axis and the c axis of the crystals is then given by

$$\langle\cos^2\phi\rangle = \frac{\int_0^{\pi/2} I(\phi)\sin\phi\cos^2\phi\,d\phi}{\int_0^{\pi/2} I(\phi)\sin\phi\,d\phi} \tag{11.9.7}$$

which, when combined with Eq. (11.9.1) yields f_c. Practical details regarding x-ray techniques can be found elsewhere [58, 67–69].

Unlike the crystalline orientation factor, the amorphous orientation factor cannot be measured directly. It is obtained by measuring the total birefringence Δ_T of the fiber and subtracting from it the crystalline contribution. Here, Δ_T is

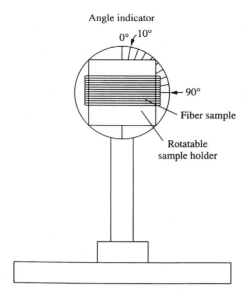

FIGURE 11.25 Front view of the sample holder used for x-ray diffraction experiments. (From Ref. 61.)

defined as the difference between the principal refractive indices perpendicular ($n_\perp$) and parallel $n_\parallel$ to the fiber axis. Thus,

$$\Delta_T = n_\parallel - n_\perp \tag{11.9.8}$$

Recall that the refractive index is a measure of the velocity of light in the medium and is related to the polarizability of the molecular chains in the sample. One way to determine the refractive indices parallel and perpendicular to the fiber axis is to immerse the fiber in oil of known refractive index and to observe the combination using a polarizing microscope with the plane of the polarized light first parallel and then perpendicular to the fiber, as sketched in Figure 11.26. When both the sample and the immersion oil have the same refractive index, the fiber is no longer visible.

Once the total birefringence has been determined, it is expressed as a sum of contributions from the amorphous and crystalline regions,

$$\Delta_T = \Delta_c X + \Delta_a (1 - X) \tag{11.9.9}$$

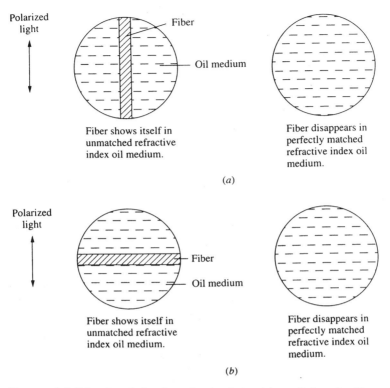

FIGURE 11.26 Search for the refractive index (a) parallel to the fiber axis and (b) perpendicular to the fiber axis. (From Ref. 61.)

in which the degree of crystallinity X is obtained from density measurements, and Δ_c and Δ_a are respectively the birefringences of the crystalline and amorphous phases. These can be separately written as

$$\Delta_c = f_c \Delta_c^0 \tag{11.9.10}$$

$$\Delta_a = f_a \Delta_a^0 \tag{11.9.11}$$

where Δ_c^0 and Δ_a^0 are the intrinsic birefringences of the fully oriented crystalline and fully oriented amorphous phases, respectively, and are known quantities. A knowledge of X and Δ_c allows us to calculate Δ_a from Eq. (11.9.9), and this, using Eq. (11.9.11), gives f_a.

Example 11.7: For a PET fiber, the total birefringence has been measured to be 0.034. If the crystalline orientation factor is 0.12 and the crystallinity value is

0.206, what is the amorphous orientation factor? For PET, Δ_c^0 is equal to 0.22 and Δ_a^0 is equal to 0.19.

Solution: Using Eq. (11.9.10) gives

$$\Delta_c = 0.12 \times 0.22 = 0.0264$$

which allows us to solve for Δ_a from Eq. (11.9.9):

$$\Delta_a = \frac{\Delta_T - \Delta_c X}{1 - X} = \frac{0.034 - 0.0264 \times 0.206}{1 - 0.206} = 0.036$$

Now, with the help of Eq. (11.9.11), we can calculate the amorphous orientation factor:

$$f_a = \frac{0.036}{0.19} = 0.19$$

Like the velocity of light, the velocity of sound differs in magnitude when measured along the polymer chain axis as compared to a direction transverse to the chain axis. A measurement of the velocity of sound can, therefore, also be used in place of the birefringence. We again assume that the crystalline and amorphous phases contribute in proportion to their relative amounts. Experimental details of this method, which involves a measurement of the "sonic modulus," can be found in the literature, which also describes other methods (such as infrared dichroism) that can be used to determine the amorphous orientation factor [58, 66].

11.10 WORKING WITH SEMICRYSTALLINE POLYMERS

As we have seen in this chapter, polymer mechanical properties are directional, depend strongly on temperature and molecular weight, vary with time of loading, and change as the processing history is changed. The net result is that chemically similar materials are found to have drastically different properties. This fact, which cannot be wished away, gives nightmares to the traditional design engineer. Indeed, as underscored by Samuels [58], it is possible for fibers produced from the same semicrystalline polymer to exhibit any of the four extremes of mechanical behavior available to a solid material. The fibers could be (1) brittle, (2) simultaneously tough and brittle, (3) ductile, or (4) elastic over fairly large strain values. The problem of how to work with such apparently unpredictable materials can be handled in one of two ways [70, 71]. The first option is take a macromechanical approach, wherein the microstructure of the polymer is ignored and the material is treated as a homogeneous material, but with different properties in different directions. The three-dimensional stress can then be related to the three-dimensional strain by means of a generalized Hooke's law that can, however, involve as many as 21 independent constants [71]. As a consequence, if we are interested in designing a polymeric load-bearing structure, we must first resort to exhaustive characterization so that the 21 constants can be determined. This is obviously an expensive and time-consuming process, and it must be repeated each time the processing history of the polymer is altered.

The alternative to the preceding unsatisfactory approach is to recognize the nonhomogeneous nature of the polymer and, instead, consider each component as a homogeneous but possibly anisotropic continuum. With this micromechanical approach, we can use the known properties of each phase and the mechanism of phase coupling to predict the 21 constants needed in the macromechanical approach, much in the same way as is done with composite materials [70]. Success in this endeavor often requires the use of models based on mechanical analogs such as springs and dashpots. The input to these models are structural parameters such as the percentage of crystals, the shape, size, and relative orientation of the two phases, the packing geometry, and the degree of adhesion between the phases. The practicality of the procedure has been demonstrated by McCullough, who used it to successfully predict the anisotropic elastic behavior of polyethylene [71]. The description of this micromechanical approach, which is the preferred approach, is beyond the scope of this book. However, this approach offers the potential for tailoring polymeric materials to specific applications, as attested to in the literature [72].

11.11 CONCLUSION

In this chapter we have examined how the morphology of semicrystalline polymers depends on the processing conditions. We have attempted to explain how the amount, shape, size, and size distribution of crystals and the orientation of polymer chains in both the crystal and amorphous regions is a consequence of the thermal and deformational histories witnessed by the material during conversion from the melt to the solid state. We have also discussed techniques of measuring these microscopic variables and pointed to methods whereby these material descriptors can be used to predict properties convenient from the viewpoint of an engineer who wants to design load-bearing structures. At first glance, this is a very formidable task, but it has been handled in a logical manner by polymer materials scientists, who have made very considerable headway as the literature cited in this chapter indicates. Although the subject matter of this chapter may be difficult to assimilate on first reading, a thorough familiarity with this material is essential if we want to relate polymer processing to polymer structure, and the structure, in turn, to polymer properties. Parallel developments have taken place in solution processing of polymers. Although space limitations prevent this discussion here, the theory and logic are the same, and good reviews are available [73].

REFERENCES

1. Clark, E. S., and F. C. Wilson, *Physical Structure of Nylons*, in Nylon Plastics, M. Kohan (ed.), Wiley, New York, 1973, pp. 271–305.
2. Spruiell, J. E., and J. L. White, Structure Development During Polymer Processing: Studies of the Melt Spinning of Polyethylene and Polypropylene Fibers, *Polym. Eng. Sci.*, 15, 660–667, 1975.
3. Geil, P. H., *Polymer Single Crystals*, R. E. Krieger, Huntington, NY, 1973.
4. Gedde, U. W., *Polymer Physics*, Chapman & Hall, London, 1995.
5. Magill, J. H., Morphogenesis of Solid Polymer Microstructures, *Treatise Mater. Sci. Technol.*, 10A, 1–368, 1977.
6. Krishnakumar, B., R. K. Gupta, E. O. Forster, and J. R. Laghari, AC Breakdown of Melt Crystallized Isotactic Polypropylene, *J. Appl. Polym. Sci.*, 35, 1459–1472, 1988.
7. Hoffman, J. D., and J. I. Lauritzen, Jr., Crystallization of Bulk Polymers with Chain Folding: Theory of Growth of Lamellar Spherulites, *J. Res. Natl. Bur. Std.*, 65A, 297–336, 1961.
8. Schultz, J. M., *Polymer Materials Science*, Prentice-Hall, Englewood Cliffs, NJ, 1974.
9. Turnbull, D., and J. C. Fisher, Rate of Nucleation in Condensed Systems, *J. Chem. Phys.*, 17, 71–73, 1949.
10. Van Krevelen, D. W., Crystallinity of Polymers and the Means to Influence the Crystallization Process, *Chimia*, 32, 279–294, 1978.
11. Kennedy, M. A., G. Turturro, G. R. Brown, and L. E. St.-Pierre, Retardation of Spherulitic Growth Rate in the Crystallization of Isotactic Polystyrene Due to the Presence of a Nucleant, *J. Polym. Sci. Polym. Phys. Ed.*, 21, 1403–1413, 1983.
12. Katti, S. S., and J. M. Schultz, The Microstructure of Injection-Molded Semicrystalline Polymers: A Review, *Polym. Eng. Sci.*, 22, 1001–1017, 1982.
13. Avrami, M., Kinetics of Phase Change: I. General Theory, *J. Chem. Phys.*, 7, 1103–1112, 1939.
14. Avrami, M., Kinetics of Phase Change: II. Transformation–Time Relations for Random Distribution of Nuclei, *J. Chem. Phys.*, 8, 212–224, 1940.
15. Avrami, M., Kinetics of Phase Change: III. Granulation, Phase Change, and Microstructure, *J. Chem. Phys.*, 9, 177–184, 1941.
16. Johnson, W. A., and R. F. Mehl, Reaction Kinetics in Processes of Nucleation and Growth, *Trans. Am. Inst. Mining Met. Eng.*, 135, 416–442, 1939.
17. Evans, U. R., The Laws of Expanding Circles and Spheres in Relation to the Lateral Growth of Surface Films and the Grain-Size of Metals, *Trans. Faraday Soc.*, 41, 365–374, 1945.
18. Mandelkern, L., *Crystallization of Polymers*, McGraw-Hill, New York, 1964.
19. Sharples, A., *Introduction to Polymer Crystallization*, St. Martin's Press, New York, 1966.
20. Mandelkern, L., F. A. Quinn, Jr., and P. J. Flory, Crystallization Kinetics in High Polymers: I. Bulk Polymers, *J. Appl. Phys.*, 25, 830–839, 1954.

21. Cebe, P., and S. D. Hong, Crystallization Behaviour of Poly(ether-ether-ketone), *Polymer*, 27, 1183–1192, 1986.
22. Nakamura, K., K. Katayama, and T. Amano, Some Aspects of Nonisothermal Crystallization of Polymers: II. Consideration of the Isokinetic Condition, *J. Appl. Polym. Sci.*, 17, 1031–1041, 1973.
23. Cobbs, W. H., Jr., and R. L. Burton, Crystallization of Polyethylene Terephthalate, *J. Polym. Sci.*, 10, 275–290, 1953.
24. Hay, J. N., Crystallization from the Melt, in *Flow Induced Crystallization in Polymer Systems*, R. L. Miller (ed.), Gordon and Breach Science, New York, 1979, pp. 69–98.
25. Ziabicki, A., *Fundamentals of Fibre Formation*, Wiley, London, 1976.
26. Nakamura, K., T. Watanabe, K. Katayama, and T. Amano, Some Aspects of Nonisothermal Crystallization of Polymers: I. Relation Between Crystallization Temperature, Crystallinity, and Cooling Conditions, *J. Appl. Polym. Sci.*, 16, 1077–1091, 1972.
27. Nishi, T., and T. T. Wang, Melting Point Depression and Kinetic Effects of Cooling on Crystallization in Poly(vinylidene fluoride)-Poly(methyl methacrylate) Mixtures, *Macromolecules*, 8, 909–915, 1975.
28. Runt, J. P., and L. M. Martynowicz, Crystallization and Melting in Compatible Polymer Blends, in *Multicomponent Polymer Materials*, D. R. Paul and L. H. Sperling (eds.), Advances in Chemistry Series 211, American Chemical Society, Washington, DC, 1986, pp. 111–123.
29. Crevecoeur, G., and G. Groeninckx, Binary Blends of Poly(ether ether ketone) and Poly(ether imide): Miscibility, Crystallization Behavior, and Semicrystalline Morphology, *Macromolecules*, 24, 1190–1195, 1991.
30. Nadkarni, V. M., and J. P. Jog, Crystallization Behavior in Polymer Blends, in *Two-Phase Polymer Systems*, L. A. Utracki (ed.), Hanser, Munich, 1991, pp. 213–239.
31. Kotliar, A. M., Interchange Reactions Involving Condensation Polymers, *J. Polym. Sci. Macromol. Rev.*, 16, 367–395, 1981.
32. Xanthos, M. (ed.), *Reactive Extrusion*, Hanser, New York, 1991.
33. Walia, P. S., R. K. Gupta, and C. T. Kiang, Influence of Interchange Reactions on the Crystallization and Melting Behavior of Nylon 6,6 Blended with Other Nylons, *Polym. Eng. Sci.*, 39, 2431–2444, 1999.
34. Dennis, H. R., D. L. Hunter, D. Chang, S. Kim, J. L. White, J. W. Cho, and D. R. Paul, Nanocomposites: The Importance of Processing, *Plast. Eng.*, 57, 56–60, Jan 2001.
35. Hambir, S., N. Bulakh, P. Kodgire, R. Kalgaonkar, and J. P. Jog, PP/Clay Nanocomposites: A Study of Crystallization and Dynamic Mechanical Behavior, *J. Polym. Sci. B, Polym. Phys.*, 39, 446–450, 2001.
36. Shih, C.-K., D. G. Tynan, and D.A. Denelsbeck, Rheological Properties of Multicomponent Polymer Systems Undergoing Melting or Softening during Compounding, *Polym. Eng. Sci.*, 31, 1670–1673, 1991.
37. Gogos, C. C., Z. Tadmor, and M. H. Kim, Melting Phenomena and Mechanisms in Polymer Processing Equipment, *Adv. Polym. Technol.*, 17, 285–305, 1998.
38. Southern, J. H., and R. S. Porter, Polyethylene Crystallized Under the Orientation and Pressure of a Pressure Capillary Viscometer—Part I, *J. Macromol. Sci. Phys.*, B4, 541–556, 1970.

39. Spruiell, J. E., D. E. McCord, and R. A. Beuerlein, The Effect of Strain History in the Crystallization Behavior of Bulk Poly(ethylene terephthalate), *Trans. Soc. Rheol.*, 16, 535–555, 1972.

40. Smith, F. S., and R. D. Steward, The Crystallization of Oriented Poly(ethylene terephthalate), *Polymer*, 15, 283–286, 1974.

41. Heuvel, H. M., and R. Huisman, Effect of Winding Speed on the Physical Structure of As-Spun Poly(ethylene terephthalate) Fibers, Including Orientational-Induced Crystallization, *J. Appl. Polym. Sci.*, 22, 2229–2243, 1978.

42. Katayama, K., T. Amano, and K. Nakamura, Structural Formation During Melt Spinning Process, *Kolloid Z. Z. Polymer*, 226, 125–134, 1968.

43. Dees, J. R., and J. E. Spruiell, Structure Development During Melt Spinning of Linear Polyethylene Fibers, *J. Appl. Polym. Sci.*, 18, 1053–1078, 1974.

44. Hay, I. L., M. Jaffe, and K. F. Wissbrun, A Phenomenological Model for Row Nucleation in Polymers, *J. Macromol. Sci. Phys.*, B12, 423–428, 1976.

45. Ziabicki, A., Theoretical Analysis of Oriented and Nonisothermal Crystallization, *Colloid Polym. Sci.*, 252, 207–221, 1974.

46. Gupta, R. K., and K. F. Auyeung, Crystallization Kinetics of Oriented Polymers, *Polym. Eng. Sci.*, 29, 1147–1156, 1989.

47. Wissbrun, K. F., Rheology of Rod-Like Polymers in the Liquid Crystalline State, *J. Rheol.*, 25, 619–662, 1981.

48. Jackson, W. J., Jr., Liquid Crystalline Aromatic Polyesters: An Overview, *J. Appl. Polym. Sci., Appl. Polym. Symp.*, 41, 25–33, 1985.

49. White, J. L., Historical Survey of Polymer Liquid Crystals, *J. Appl. Polym. Sci., Appl. Polym. Symp.* 41, 3–24, 1985.

50. Flory, P. J., Phase Equilibrium in Solutions of Rod-Like Particles, *Proc. R. Soc. London*, A234, 73–89, 1956.

51. Aharoni, S. M., Rigid Backbone Polymers: XVII. Solution Viscosity of Polydisperse Systems, *Polymer*, 21, 1413–1422, 1980.

52. Magat, E. E., Fibres from Extended Chain Aromatic Polyamides, *Phil. Trans. R. Soc. London*, A294, 463–472, 1980.

53. Jackson, W. J., Jr., and H. Kuhfuss, Liquid Crystal Polymers: I. Preparation and Properties of *p*-Hydroxybenzoic Acid Copolyesters, *J. Polym. Sci. Polym. Chem. Ed.*, 14, 2043–2058, 1976.

54. Lin, Y. G., and H. H. Winter, Formation of a High Melting Crystal in a Thermotropic Aromatic Copolyester, *Macromolecules*, 21, 2439–2443, 1988.

55. Bafna, S. S., Rheological Studies on Nematic Thermotropic Liquid Crystalline Polymers, Ph.D. thesis, Chemical Engineering, University of Massachusetts, Amherst, 1989.

56. Danielli, D., and L. L. Chapoy, Morphology and Crystallization Kinetics of a Rigid Rod, Fully Aromatic, Liquid Crystalline Copolyester, *Macromolecules*, 26, 385–390, 1993.

57. Stein, R. S., The X-ray Diffraction, Birefringence, and Infrared Dichroism of Stretched Polyethylene: II. Generalized Uniaxial Crystal Orientation, *J. Polym. Sci.*, 31, 327–334, 1958.

58. Samuels, R. J., *Structured Polymer Properties*, Wiley, New York, 1974.

59. White, J. L., and J. E. Spruiell, Specification of Biaxial Orientation in Amorphous and Crystalline Polymers, *Polym. Eng. Sci.*, 21, 859–868, 1981.
60. Stein, R. S., Studies of Solid Polymers with Light, *J. Chem. Ed.*, 50, 748–753, 1973.
61. Auyeung, K. F., Polymer Crystallization in Melt Spinning, Ph.D. thesis, Chemical Engineering, State University of New York at Buffalo, 1986.
62. Hartshorne, N. H., and A. Stuart, *Crystals and the Polarising Microscope*, 4th ed., Elsevier, New York, 1970.
63. Marentette, J. M., and G. R. Brown, Polymer Spherulites, *J. Chem. Ed.*, 70, 435–439, 1993.
64. Hobbs, S. Y., Polymer Microscopy, *J. Macromol. Sci. Rev., Macromol. Chem.*, C19, 221–265, 1980.
65. Sawyer, L. C., and D. T. Grubb, *Polymer Microscopy*, 2nd ed., Chapman & Hall, London, 1996.
66. Wilkes, G. L., The Measurement of Molecular Orientation in Polymeric Solids, *Adv. Polym. Sci.*, 8, 91–136, 1971.
67. Klug, H. P., and L. E. Alexander, *X-ray Diffraction Procedures*, 2nd ed., Wiley, New York, 1974.
68. Alexander, L. E., *X-ray Diffraction Methods in Polymer Science*, Wiley–Interscience, New York, 1969.
69. Kakudo, M., and N. Kasai, *X-ray Diffraction by Polymers*, Elsevier, Amsterdam, 1972.
70. McCullough, R. L., *Concepts of Fiber–Resin Composites*, Marcel Dekker, New York, 1971.
71. McCullough, R. L., Anisotropic Elastic Behavior of Crystalline Polymers, *Treatise Mater. Sci. Technol.*, 10B, 453–540, 1977.
72. Kamal, M. R., and F. H. Moy, Microstructure in Polymer Processing. A Case Study—Injection Molding, *Polym. Eng. Rev.*, 2, 381–416, 1983.
73. McHugh, A. J., Mechanism of Flow Induced Crystallization, *Polym. Eng. Sci.*, 22, 15–26, 1982.
74. Barnes, W. J., W. G. Luetzel, and F. P. Price, Crystallization of Poly(ethylene oxide) in Bulk, *J. Phys. Chem.*, 65, 1742–1748, 1961.
75. Garg, S. N., and A. Misra, Crystallization Behavior of a Polyester–Polyamide Block Copolymer, *J. Polym. Sci. Polym. Lett. Ed.*, 23, 27–31, 1985.
76. Wu, S. S., D. S. Kalika, R. R. Lamonte, and S. Makhija, Crystallization, Melting, and Relaxation of Modified Poly(phenylene sulfide): I. Calorimetric Studies, *J. Macromol. Sci. Phys. Ed.*, B35, 157–178, 1996.

PROBLEMS

11.1. If, instead of a sphere, the nucleus in Section 11.2.1 is a cylinder of radius r and length l, derive expressions for ΔG^*, r^*, and l^*. Let the surface energy be γ_e for the flat surfaces and γ_s for the curved surface of the cylinder.

11.2. If ΔT in Example 11.2 were 100 K, would crystallization still proceed via heterogeneous nucleation? Justify your answer by doing some simple calculations.

11.3. The data of Barnes et al. for the radial growth rate of poly(ethylene oxide) spherulites are given as follows [74]:

T (°C)	G (mm/min)
47.6	0.1070
48.2	0.0910
50.5	0.0212
51.0	0.0253
53.8	0.00565
56.4	0.000096

Estimate a value of $\gamma_e\gamma$ if E_D/kT in Eq. (11.3.6) can be taken to be a constant over this narrow temperature range. For poly(ethylene oxide) (PEO), it is known that $T_m^0 = 330\,\text{K}$ and $\Delta H_v = 45\,\text{cal/cm}^3$. Let b equal 10 Å.

11.4. Everything else being equal, how would the growth rate of a spherulite change on increasing the polymer molecular weight? Justify your answer.

11.5. Use the data given in Figure 11.11 to obtain the rate constant k in Eq. (11.4.12). Plot your results as a function of temperature.

11.6. Will sample thickness have any influence on the nature of crystallization and the kinetics of crystallization as the thickness is reduced to a few tens of microns?

11.7. Derive Eq. (11.4.9) beginning with Eq. (11.4.1).

11.8. When Garg and Misra plotted $-\log(1 - X)$ versus time using isothermal crystallization data on a polyester–polyamide copolymer, they obtained two straight-line segments [75]. At short times, the slope was close to unity, whereas at later times the slope changed to a value close to 3. What can you speculate about the morphology of the copolymer crystals?

11.9. Use Eq. (11.4.19) and the data of Figure 11.13 to predict how the crystallinity of an initially amorphous high-density polyethylene sample changes with time when the temperature is lowered at a constant rate from 125°C to 121.8°C over a 30-min period. Compare the final crystallinity with the corresponding value for isothermal crystallization at (a) 125°C and (b) 121.8°C.

11.10. Show that Eq. (11.4.19) can be solved to give $K(T)$ explicitly as

$$K[T(t)] = \frac{1}{n}[-\ln(1 - \theta)]^{(1-n)/n} \frac{1}{(1 - \theta)} \frac{\theta}{dt}$$

where $\theta = \dfrac{x}{x_\infty}$

11.11. Wu et al. have measured the crystallization and corresponding melting temperatures of poly(phenylene sulfide) [76]. Their data are given as follows:

T_c (°C)	T_m (°C)
260	283
265	287
270	290
275	294
280	298

What is the equilibrium melting temperature T_m^0?

11.12. If 5 mg of the PET sample of Example 11.6 is melted in a differential scanning calorimeter, how much energy will be needed for the phase change? The latent heat of fusion for PET crystals is 140 J/g.

11.13. Wide-angle x-ray scattering data for a semicrystalline PET fiber sample are as follows [61]:

ϕ (deg)	$I(\phi)$
0	19.6
10	20.6
20	18.1
30	15.0
40	13.5
50	13.0
60	10.45
70	8.9
80	8.7
90	8.2

What is the crystalline orientation factor?

12

Mechanical Properties

12.1 INTRODUCTION

It is usually material properties, in addition to cost and availability, that determine which class of materials–polymers, metals, or ceramics—and which particular member within that class are used for a given application. Many commodity thermoplastics, for example, begin to soften around 100°C, and this essentially limits their use to temperatures that are a few tens of degrees Celsius below this value. A major factor in favor of polymers, though, is their low density (by a factor of 4 or 5) relative to metals; the possibility of a large weight savings, coupled with high strength, makes plastics very attractive for automotive, marine, and aerospace applications. In terms of choosing a specific polymer, however, it is necessary to consider whether the application of interest is structural or nonstructural. In the former case, mechanical properties such as tensile strength, stiffness, impact strength, and chemical resistance might be relevant, whereas important considerations in the latter case might include surface finish, ease of painting, and the influence of humidity and ultraviolet radiation on the tendency of the material to crack. In this chapter, we will consider mechanical properties of polymers at small strains as well as large strains. In general, the mode of deformation could be tension, compression, shear, flexure, torsion, or a combination of these. To keep the discussion manageable, we will restrict ourselves to tension and shear. Note, however, that we can use viscoelasticity theory [1], especially at small strains, to predict the behavior in one mode of deformation from measurements made in another mode of deformation. As with metals, we expect that the measured properties depend on the chemical nature of the polymer and the temperature of measurement. However, what makes data analysis and interpretation both fascinating and challenging are the facts that results also depend on time of loading or the rate of deformation, polymer molecular weight, molecular-weight distribution, chain branching, degree of cross-linking, chain orientation, extent of crystallization, crystal structure, size and shape of crystals, and whether the polymer was solution cast or melt processed. These variables are not all independent; molecular weight, for example, can determine chain orientation and crystallinity in a particular processing situation. To explain the separate influence of some of these variables, we present data on polystyrene, a polymer that can be synthesized in narrow molecular-weight fractions using anionic polymerization. Methods of improving polymer mechanical properties are again illustrated using polystyrene. This chapter therefore focuses on the (glassy) behavior of polymers below their glass transition temperature.

12.2 STRESS–STRAIN BEHAVIOR

When discussing the theory of rubber elasticity in Chapter 10, we were concerned with fairly large extensions or strains. These arose because polymer molecules could uncoil at temperatures above T_g. For materials used as structural elements (such as glassy polymers), we usually cannot tolerate strains of more than a fraction of 1%. Therefore, it is customary to employ measures of infinitesimal strain. In a tensile test, we usually take a specimen with tabs at the ends and stretch it, as shown in Figure 12.1. One end of the sample is typically fixed, whereas the other is moved outward at a constant velocity. The force F necessary to carry out the stretching deformation is monitored as a function of time along with the instantaneous sample length, L. From the measured load versus extension behavior, we can calculate the stress and strain as follows:

$$\text{Stress } (\sigma) = \frac{\text{Force (F)}}{\text{Cross-sectional area}} \qquad (12.2.1)$$

If the cross-sectional area is the undeformed, original cross-sectional area, the stress is called *engineering stress*, and if the actual, instantaneous area is used, the *true stress* is measured.

$$\text{Strain } (\epsilon) = \frac{L - L_0}{L_0} \qquad (12.2.2)$$

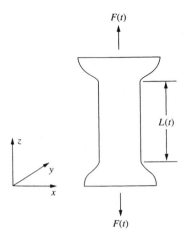

FIGURE 12.1 Typical specimen for a tensile test.

where L_0 is the initial sample length, and the strain, so defined, is known as the *engineering strain*. Note that this strain is related to the Hencky strain, also called the *true strain*, as follows:

$$\epsilon_{true} = \ln(1 + \epsilon_{eng}) \tag{12.2.3}$$

and the two strain measures are identical for small strains.

As the sample is stretched in the z direction, its cross-sectional area decreases, and this implies that the material suffers a negative strain in the x direction, which is perpendicular to the stretching direction. This is quantified using the Poisson ratio v, defined as

$$\epsilon_x = -v\epsilon_z \tag{12.2.4}$$

For incompressible materials such as rubber, it is easy to show that Poisson's ratio equals 0.5. For glassy polymers the sample volume increases somewhat on stretching, and Poisson's ratio ranges from 0.3 to 0.4.

Typical stress–strain data for glassy polystyrene are shown in Figure 12.2 in both tension and compression [2]. The slope of the stress–strain curve evaluated at the origin is termed the *elastic modulus*, E, and is taken to be a measure of the stiffness of the material. It is seen in this particular case that the modulus in tension differs from that in compression. The two curves end when the sample fractures. The stress at fracture is called the *strength* of the material. Because materials fracture due to the propagation of cracks, the strength in tension is usually less than that in compression because a compressive deformation tends to heal any cracks that form (provided the sample does not buckle). The strain at

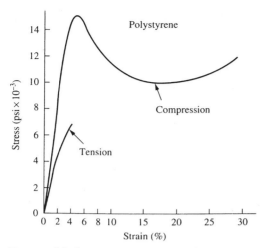

FIGURE 12.2   Stress–strain behavior of a normally brittle polymer such as polystyrene under tension and compression.
(Reprinted from Nielsen, L. E., and R. F. Landel: Mechanical Properties of Polymers and Composites, 2nd ed., Marcel Dekker, Inc., New York, 1994, p. 250, by courtesy of Marcel Dekker, Inc.)

fracture is known as the *elongation-to-break*; the larger the value of this quantity, the more *ductile* is the material being tested. Glassy polystyrene is not ductile in tension; indeed, it is quite brittle. Finally, the area under the stress–strain curve is called the *toughness* and has units of energy per unit volume. For design purposes, the materials generally sought are stiff, strong, ductile, and tough.

    For materials that are liquidlike, such as polymers above their softening point, it is easier to conduct shear testing than tensile testing. This conceptually involves deforming a block of material, as shown in Figure 12.3. The force F is

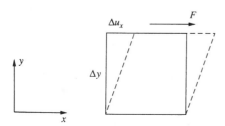

FIGURE 12.3   Shear deformation.

again monitored, but now as a function of the displacement Δu_x. Stress and strain are now defined as follows:

$$\text{Shear stress } (\tau) = \frac{\text{Force (F)}}{\text{Surface area}} \qquad (12.2.5)$$

$$\text{Shear strain } (\gamma) = \frac{\Delta u_x}{\Delta y} \qquad (12.2.6)$$

At temperatures above the polymer glass transition temperature, shear testing is done using a variety of viscometers (see Chap. 14). We might, for example, keep the sample in the annular region between two concentric cylinders and measure the torque while rotating one cylinder relative to the other. Stress–strain data in shear look qualitatively similar to the tensile data shown earlier in Figure 12.2. The initial slope is called the *shear modulus*, G. For elastic materials the moduli in shear and tension are related by the following expression:

$$E = 2G(1 + v) \qquad (12.2.7)$$

so that E equals 3 G for incompressible, elastic polymers. Note that when material properties are time dependent (i.e., viscoelastic), the modulus and strength increase with increasing rate of deformation [3], whereas the elongation-to-break generally reduces. Viscoelastic data are often represented with the help of mechanical analogs.

Example 12.1: A polymer sample is subjected to a constant tensile stress σ_0. How does the strain change with time? Assume that the mechanical behavior of the polymer can be represented by a spring and dashpot in series, as shown in Figure 12.4.

Solution: The stress-versus-strain behavior of a Hookean spring is given by

$$\sigma = E\epsilon$$

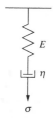

FIGURE 12.4   A Maxwell element.

For a Newtonian dashpot, the relation is

$$\sigma = \eta \, \frac{d\epsilon}{dt}$$

The terms E and η are the spring modulus and dashpot viscosity, respectively.

For the spring and dashpot combination, often called a *Maxwell element*, the total elongation or strain is the sum of the individual strains. The stress for the spring and for the dashpot is the same,

$$\text{Total strain} = \frac{\sigma_0}{E} + \frac{\sigma_0}{\eta} t$$

and it is seen that the strain increases linearly with time. This behavior is known as *creep*. Although a simple mechanical analog such as a Maxwell element cannot be expected to portray true polymer behavior, it does illustrate the usually undesirable phenomenon of creep. A better model for the quantitative representation of creep is a four-element model which is a linear combination of a Maxwell element and a *Voigt element*; the latter is composed of a spring and a dashpot in parallel.

A polymer sample creeps because polymer molecules are held in place by secondary bonds only, and they can rearrange themselves under the influence of an applied load. This is especially easy above the polymer glass transition temperature, but it also happens below T_g and strain gauges have to be employed for accurate measurements. To illustrate the latter point, we show long-term creep data, in the form of circles in Figure 12.5, on samples of polyvinyl chloride (PVC) at constant values of tensile stress, temperature, and relative humidity [4]. Note that data for the first 1000 h are shown separately, followed by all of the data using a compressed time scale. It is seen that the total creep can be several percent, and a steady state is not reached even after 26 years! These and similar data can be represented by the following simple equation shown by solid lines in Figure 12.5:

$$\epsilon(t) = \epsilon^0 + \epsilon^+ t^n \tag{12.2.8}$$

in which ϵ^0, ϵ^+, and n are constants. Although n is often independent of temperature and imposed stress, the other two constants are stress and temperature dependent. If creep is not arrested, it can lead to failure, which may occur either by the process of crazing or by the formation of shear bands; these failure mechanisms are discussed later in the chapter. Equation (12.2.8) is an empirical equation that is known as the Findley model. It may sometimes contain a second time-dependent term if failure can occur by two different mechanisms. Creep can generally be reduced by lowering the test temperature, raising the polymer T_g, cross-linking the sample, or adding either particulates or short fibers. Conversely, anything that lowers the T_g, such as exposure to atmospheric moisture, promotes creep. Physical aging (described later) also affects the extent of creep.

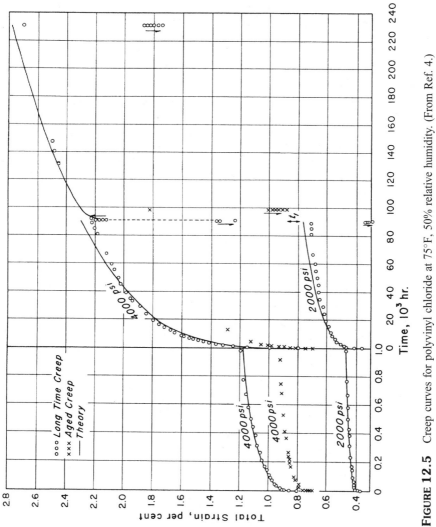

FIGURE 12.5 Creep curves for polyvinyl chloride at 75°F, 50% relative humidity. (From Ref. 4.)

12.2.1 Influence of Variables such as Molecular Weight and Temperature

The strength and stiffness of one glassy polymer can be expected to differ from that of another glassy polymer due to differences in intermolecular forces as a result of differences in chemical structure and the presence or absence of secondary bonds (e.g., hydrogen-bonding). Given these differences, the two variables that influence the mechanical properties of amorphous polymers the most are molecular weight and temperature. However, the elastic moduli and other small-strain properties of strain-free glassy polymers such as polystyrene (PS) are found not to depend on the molecular weight or molecular-weight distribution, except at very low molecular weights [5–7]. The tensile strength, σ_f, of polymers having a narrow molecular-weight distribution, however, is negligible at low molecular weight, increases with increasing molecular weight, and, ultimately reaches an asymptotic value [8]. This behavior can often be represented by the following equation [6, 9]:

$$\sigma_f = A - B/\overline{M}_n \tag{12.2.9}$$

where A and B are constants. Data for polystyrene, shown in Figure 12.6, support these conclusions [10]. From an examination of this figure, it is obvious that the addition of a low-molecular-weight fraction is bound to affect the tensile strength of any polymer. However, for polydisperse samples, data do not follow Eq. (12.2.9) exactly; results vary with the polydispersity index, even when the number-average molecular weight is held fixed.

The data just discussed are related to amorphous polymer samples for which the polymer chains were randomly oriented. One method of increasing both strength and stiffness is to use samples wherein polymer chains are oriented

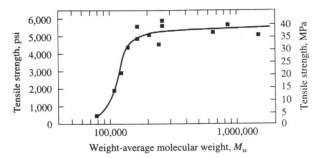

FIGURE 12.6 Tensile strength of monodisperse polystyrene as a function of molecular weight. From Hahnfeld, J. L., and B. D. Dalke: General purpose polystyrene, in Encyclopedia of Polymer Science and Engineering, 2nd ed., vol. 16, H. F. Mark, N. M. Bikales, C. G. Overberger, and G. Menges (eds.) Copyright © 1989 by John Wiley & Sons, Inc. This material is used by permission of John Wiley & Sons, Inc.

along the stretching direction. By using this technique, we can very significantly increase the modulus of polystyrene and hope to get strength that approaches the strength of primary chemical bonds [11]. Indeed, as discussed in Chapter 11, mechanical property enhancement using chain alignment is the reason for the popularity of polymers that possess liquid-crystalline order. Properties in a direction perpendicular to the chain axis, however, are likely to be inferior to those along the chain axis.

When the Young's modulus of any polymer is plotted as a function of temperature, we find that this quantity is of the order of 10^5–10^6 psi at low temperatures and decreases slowly with increasing temperature. This region is known as the *glassy region*. At the glass transition temperature T_g (see also Chap. 2), which varies for different polymers, the modulus drops suddenly by at least three orders of magnitude and can reach extremely low values for low-molecular-weight polymers. Figure 12.7 shows the Young's modulus of polystyrene in a temperature range of $-200°C$ to $25°C$ [12]. Figure 12.8 shows shear stress versus shear strain data for an entangled polystyrene in a temperature range of $160°C$–$210°C$ [13]. If we disregard the numerical difference between the Young's modulus and the shear modulus and note that 1 MPa equals 145 psi, we find that the modulus calculated from data in Figure 12.8 is several orders of magnitude smaller than the number expected on the basis of extrapolating the curve in Figure 12.7. This happens because the T_g of polystyrene is 100°C. The behavior of the Young's modulus, in qualitative terms, is sketched in Figure 12.9 over a temperature range that includes T_g. If the polymer molecular weight is above that needed for entanglement formation (for polystyrene, this is approximately 35,000), the presence of these entanglements temporarily arrests the fall in modulus on crossing T_g. This region of almost constant modulus is called the *rubbery plateau*, and the result is a rubbery polymer. Because crystals act in a manner similar to entanglements, the modulus of a semicrystalline polymer does

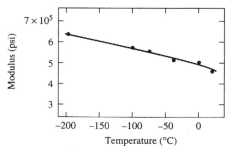

FIGURE 12.7 Effect of temperature on Young's modulus of polystyrene. (From Ref. 12.) Reprinted with permission from J. Appl. Phys., vol. 28, Rudd, J. F., and E. F. Gurnee: Photoelastic properties of polystyrene in the glassy state: II. Effect of temperature, 1096–1100, 1957. Copyright 1957 American Institute of Physics

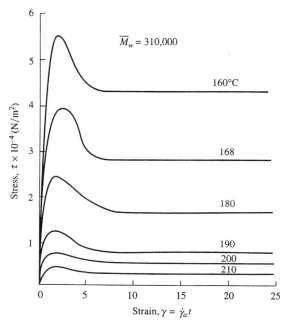

FIGURE 12.8   Effect of temperature on the stress–strain curves of polystyrene melts. (From Ref. 13.)

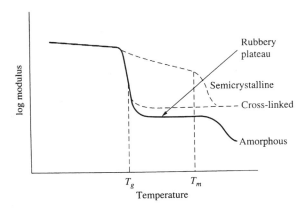

FIGURE 12.9   Qualitative effect of temperature on the elastic modulus of polymers.

not fall as precipitously as that of amorphous polymers for temperatures between the T_g and the melting point of the crystals. Of course, if chemical cross-links are present, the polymer cannot flow and the temperature variation of the modulus above T_g is given by the theory of rubber elasticity. Understanding and relating mechanical properties of a semicrystalline polymer to the different variables that characterize its structure has been discussed in Chapter 11 and is treated in detail by Samuels [14].

12.3   THE GLASS TRANSITION TEMPERATURE

As discussed in Chapter 2, the glass transition temperature separates regions of dramatically different polymer properties. In particular, a polymer behaves like a hard, brittle, elastic solid below T_g. In this glassy region, the motion of polymer chains is frozen and strain occurs by the stretching of bonds. The elastic modulus decreases with increasing temperature. On heating above T_g, an entangled, amorphous polymer displays a rubbery region in which it is soft and pliable due to the ability of polymer chain segments and entire polymer chains to move past each other in a reversible manner. In this region, the elastic modulus can increase with an increase in temperature; this property has been explained theoretically in Chapter 10. Structural applications clearly require a polymer T_g above room temperature, whereas applications where material flexibility is important, such as in films used for packaging, require that the T_g be below room temperature.

  Although we can use observations of the change in mechanical properties as a means of measuring T_g, we also find that thermodynamic properties change slope on going through the glass transition. Thus, if we plot the volume of a sample or its enthalpy as a function of temperature, behavior depicted qualitatively in Figure 12.10 is observed: The slope in the liquid phase is larger than the slope in the solid phase. By contrast, for a crystalline solid, there would be a discontinuity or jump in the value of these thermodynamic variables at the crystalline melting point. Note that all polymers exhibit a T_g, but only crystallizable ones show a T_m (melting temperature); the latter phenomenon is called a *first-order transition*, whereas the former is called a *second-order transition*. Clearly, the specific heat of the rubbery phase exceeds that of the glassy phase. The exact temperature where the change in slope occurs, though, depends on the cooling rate, and we obtain a range, albeit a narrow one, for the transition temperature. This happens because the rearrangement of polymer molecules into a glassy structure is a kinetic process. The greater the time available for the transition is, the more orderly the packing and the lower the observed T_g. This effect, however, is reversed on rapid heating, and the slowly cooled material overshoots the original T_g. This change in T_g can be related to the free volume

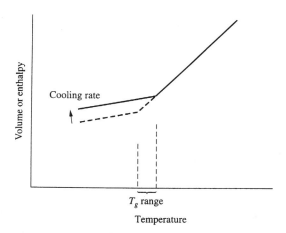

FIGURE 12.10 Variation of volume or enthalpy of polymers with temperature.

mentioned in Chapter 2. To recapitulate, the polymer free volume is the difference in the sample volume and the actual volume occupied by the atoms and molecules. The free volume is zero at absolute zero temperature and it increases as the temperature increases. Slow cooling allows for a closer approach to equilibrium and a lower free volume relative to material subjected to rapid cooling. Thus, the slowly cooled sample has to be heated to a higher temperature in order that there be enough free volume for the molecules to move around, and this implies a higher T_g. In addition to changes in T_g with cooling rate, we also observe volume relaxation when a polymer sample that was rapidly cooled is subsequently heated to a temperature close to T_g and held there for some time. Material shrinkage also occurs, accompanied by changes in the mechanical properties of the solid polymer. The phenomenon is known as *physical aging* [15] and is the subject of considerable research because of its influence on properties such as creep [16].

The glass transition temperature of a polymer depends on a number of factors, including the polymer molecular weight. The molecular-weight dependence can be seen in Figure 12.11, where the T_g of polystyrene is plotted as a function of the number-average molecular weight [3,17]. These data can be represented mathematically by the following equation [18]:

$$T_g = T_{g\infty} - \frac{K}{\bar{M}_n} \qquad (12.3.1)$$

This variation of T_g with molecular weight can again be related to the free volume [19]. As the molecular weight decreases, the number density of chain ends increases. Because each chain end is assumed to contribute a fixed amount of free

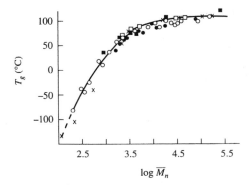

FIGURE 12.11 Glass transition temperature of polystyrene as a function of M_n as determined by various methods: (○) and (●) dilatometry, (■) Differential thermal analysis (DTA), (□) differential scanning calorimetry, (×) electron spin resonance. (From Ref. 3.)

volume, the total free volume increases on lowering the molecular weight, which explains the data of Figure 12.11. On increasing the chain length beyond a certain value, the contribution of chain ends becomes negligible and T_g becomes constant.

If it is assumed that the ratio of the volume of the polymer chain segment that moves to the free volume associated with that segment is the same for all polymers at the glass transition temperature, the variation of T_g with chemical structure becomes easy to understand [19]. Any structural change that increases the segmental volume requires a larger free volume per segment and results in a larger T_g because, as previously explained, the free volume increases on increasing temperature. Thus, T_g increases as a result of increasing chain stiffness, adding stiff or bulky side groups, and introducing steric hindrances. Similarly, hydrogen-bonding raises the T_g because such a polymer expands less than a non-hydrogen-bonded polymer on increasing temperature. Consequently, a higher temperature is necessary to get the same free-volume level. Finally, the presence of plasticizers or low-molecular-weight additives increases the free volume and lowers the T_g; plasticizers such as dioctyl phthalate are routinely added to PVC to convert it from a rigid to a more flexible material. The glass transition temperatures of common polymers are listed in the *Polymer Handbook* [20], and selected values are given in Table 12.1.

One of the most convenient methods of measuring T_g is through the use of a differential scanning calorimeter (DSC) [21]. The principle of operation of this instrument is shown schematically in Figure 12.12. A DSC contains two sample holders, each provided with its own heater. The actual sample is placed in one of the sample holders in an aluminum pan and the other sample holder contains an

TABLE 12.1 Glass Transition Temperature of
Common Polymers

Polymer	T_g (°C)
Natural rubber (polyisoprene)	−70
Nylon 6 (dry)	100
Nylon 66	50
Polycarbonate of bisphenol A	157
Polyethylene	−38 to −33
Polyethylene oxide	−70
Polyethylene terephthalate	67
Polymethyl methacrylate	105
Polypropylene	−15 to −3
Polystyrene	80–100
Polyvinyl chloride	70–100
Styrene–butadiene rubber	−64 to −59

empty pan. The temperature of both the sample holders is increased at a constant
rate, such as 10°C/min, and we measure the difference in the energy H supplied
to the two pans to keep them at the same temperature at all times. From an energy
balance, it is obvious that the rate of differential heat flow must be as follows:

$$\frac{dH}{dt} = mc_p \frac{dT}{dt} \tag{12.3.2}$$

where m is the mass of the sample (typically a few milligrams), c_p is the specific
heat, and dT/dt is the programmed rate of temperature increase.

If the specific heat increases on heating the polymer sample through the
glass transition temperature, dH/dt must go from one constant value to a higher
constant value at T_g. Thus, T_g can be identified by plotting dH/dt as a function of
the instantaneous sample temperature. This is usually done using a thermogram
of the kind shown in Figure 12.13 for a sample of amorphous nylon. If the

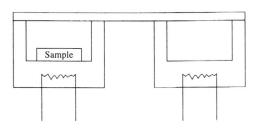

FIGURE 12.12 Schematic diagram of a differential scanning calorimeter.

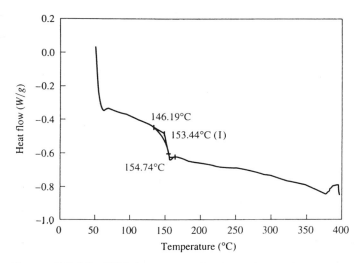

FIGURE 12.13 DSC thermogram of an amorphous nylon ($T_g = 153°C$).

polymer is semicrystalline, it must be quenched from the melt state rapidly to give a wholly amorphous structure; otherwise, the presence of crystals can impede the motion of polymer chains and result in a T_g value that is higher than the true value. For some very crystallizable polymers such as nylon 66, amorphous samples cannot be obtained and a DSC fails to even pick up a glass transition. In such a case, we turn to dynamic mechanical analysis, wherein a polymer sample, whether glassy or rubbery, is deformed in an oscillatory manner (in tension or shear, as appropriate) such that the maximum strain amplitude is infinitesimal in magnitude.

12.4 DYNAMIC MECHANICAL EXPERIMENTS

If a polymer is subjected to a sinusoidal strain γ of infinitesimal amplitude γ_0 and fixed frequency ω,

$$\gamma = \gamma_0 \sin \omega t \tag{12.4.1}$$

then the stress response τ will be linear (i.e., sinusoidal) but will, in general, be out of phase by an angle δ and have a different amplitude τ_0. Thus,

$$\tau = \tau_0 \sin(\omega t + \delta) \tag{12.4.2}$$

or

$$\tau = (\tau_0 \cos \delta) \sin \omega t + (\tau_0 \sin \delta) \cos \omega t \tag{12.4.3}$$

On dividing the stress by the strain amplitude, one obtains the modulus G as

$$G = G'(\omega) \sin \omega t + G''(\omega) \cos \omega t \qquad (12.4.4)$$

where $G' = \tau_0 \cos \delta / \gamma_0$ and $G'' = \tau_0 \sin \delta / \gamma_0$. The term G', called the *storage modulus*, is the in-phase component of the modulus and represents storage of energy, whereas G'', the *loss modulus*, is the out-of-phase component and is a measure of energy loss. The ratio of the loss to storage modulus, G''/G', is $\tan \delta$ and is an alternate measure of energy dissipation. One may conduct dynamic experiments in an isochronal manner by varying the temperature at a fixed frequency, or in an isothermal manner by varying the frequency at a fixed temperature. The former kinds of experiment are discussed in this section, whereas the latter are considered in the next section.

For a perfectly elastic material, stress and strain are always in phase and G' equals the elastic modulus and G'' is zero. For viscoelastic polymers, on the other hand, the work of deformation is partly stored as potential energy, and the remainder is converted to heat and shows up as mechanical damping. This is independent of the mode of deformation, which could be extension, shear, bending, or torsion. If a polymer is glassy, it will act essentially as an elastic solid and dynamic experiments will allow us to measure the modulus or stiffness. This value is typically of the order of 10^9 Pa. Similarly, in the rubbery region, the polymer is again elastic but with a much smaller modulus of the order of 10^6 Pa.

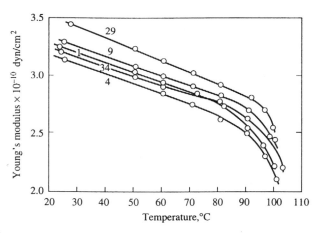

FIGURE 12.14 Polystyrene data: dynamic modulus versus temperature for fractions. Numbers on curves are fraction numbers. (Reprinted with permission from Merz, E. H., L. E. Nielsen, and R. Buchdahl: "Influence of Molecular Weight on the Properties of Polystyrene," Ind. Eng. Chem., vol. 43, pp. 1396–1401, 1951. Copyright 1951 American Chemical Society.)

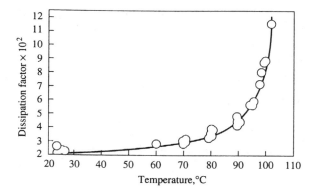

FIGURE 12.15   Polystyrene data: mechanical dissipation factor versus temperature for fractions. Fractions 1, 4, 9, 29, and 34 were tested. (Reprinted with permission from Merz, E. H., L. E. Nielsen, and R. Buchdahl: "Influence of Molecular Weight on the Properties of Polystyrene," Ind. Eng. Chem., vol. 43, pp. 1396–1401, 1951. Copyright 1951 American Chemical Society.)

Thus, a plot of storage modulus with temperature will mirror the plot of Young's modulus versus temperature and allow us to determine the glass transition temperature. Figures 12.14 and 12.15 show typical data for the storage modulus and $\tan \delta$ values of various polystyrene fractions as a function of temperature. The frequency range here is 20–30 Hz. As expected, the glass transition temperature is approximately 100°C. Note that both G'' and $\tan \delta$ go through a maximum at the T_g because the ability of a spring to store energy depends on its modulus [22]. On passing through the T_g, the polymer goes from a stiff spring to a soft one that cannot store as much energy. The difference in energy is dissipated in the transition from the glassy to the rubbery states. Note that T_g measured using dynamic mechanical analysis is usually slightly larger than that measured using a DSC. This discrepancy increases with increasing frequency of oscillation.

Figures 12.14 and 12.15 show data obtained in tension using cast films oscillated with the help of an electromagnetic reed vibrator operating at resonance. Commercial instruments available today use forced vibrations without resonance. These are desirable because they allow the user to vary temperature and frequency over wide intervals. For example, in the dynamic mechanical thermal analyzer (DMTA), an instrument made by the Rheometrics Company, a bar sample is clamped rigidly at both ends and its central point is vibrated sinusoidally by the drive clamp. The stress experienced by the sample is proportional to the current supplied to the vibrator. The strain in the sample is proportional to the sample displacement and is monitored by a nonloading eddy current transducer and a metal target on the drive shaft. In this instrument, the frequency can be varied from 0.033 to 90 Hz and the temperature changed from $-150°C$ to $300°C$. Descriptions of other instruments can be found in the book by Nielsen and Landel [2]. Note that liquidlike materials are often supported on glass braids [23].

Example 12.2: Determine the storage and loss moduli of a polymer whose mechanical behavior can be represented by the Maxwell element shown earlier in Figure 12.4.

Solution: Because the total strain γ is the sum of the individual strains, we have

$$\dot{\gamma} = \frac{\dot{\sigma}}{E} + \frac{\sigma}{\eta}$$

Substituting for the strain using Eq. (12.4.1) and rearranging gives

$$\dot{\sigma} + \frac{E}{\eta}\sigma = E\gamma_0\omega\cos\omega t$$

whose solution for $t \to \infty$ is

$$\sigma = \left(\frac{\eta\gamma_0\omega^2\theta}{1+\theta^2\omega^2}\right)\sin\omega t + \left(\frac{\eta\gamma_0\omega}{1+\theta^2\omega^2}\right)\cos\omega t$$

where $\theta = \eta/E$. Thus, the storage and loss moduli are given by the following:

$$G'(\omega) = \frac{E\omega^2\theta^2}{1+\omega^2\theta^2}, \qquad G''(\omega) = \frac{E\omega\theta}{1+\omega^2\theta^2}$$

Dynamic mechanical analysis is an extremely powerful and widely used analytical tool, especially in research laboratories. In addition to measuring the temperature of the glass transition, it can be used to study the curing behavior of thermosetting polymers and to measure secondary transitions and damping peaks. These peaks can be related to phenomena such as the motion of side groups, effects related to crystal size, and different facets of multiphase systems such as miscibility of polymer blends and adhesion between components of a composite material [24]. Details of data interpretation are available in standard texts [1,2,25]. In the next section, we consider time–temperature superposition, which is another very useful application of dynamic mechanical data.

12.5 TIME–TEMPERATURE SUPERPOSITION

If we plot isothermal shear data for the storage modulus as a function of the circular frequency at a series of temperatures, we obtain results of the type shown in Figure 12.16 [26]. The polymer is a polystyrene melt of narrow molecular-

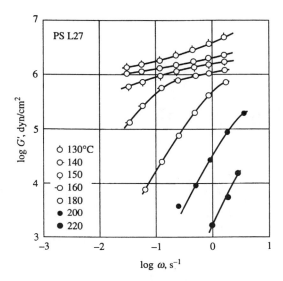

FIGURE 12.16 Frequency dependence of G' for narrow-distribution polystyrene L27 (molecular weight 167,000) at various temperatures. (Reprinted with permission from Onogi, S., T. Masuda, and K. Kitagawa: "Rheological Properties of Anionic Polystyrenes: I. Dynamic Viscoelasticity of Narrow-Distribution Polystyrenes," Macromolecules, vol. 3, pp. 109–116, 1970. Copyright 1970 American Chemical Society.)

weight distribution at temperatures from 130°C to 220°C. The remarkable feature of these data and similar data on other polymer molecular-weight fractions or other polymer melts is that all of the different curves can be made to collapse into a single curve by means of a horizontal shift. Thus, if we move the curve for 180°C to the left until it bumps into the 160°C curve, we find that it overlaps with it nicely and the composite curve extends to lower frequencies. The range of data at 160°C, taken to be the reference temperature, can be extended further toward lower frequencies by shifting the 200°C and 220°C curves to the left as well. To make the 130°C, 140°C, and 150°C curves line up with the 160°C data, though, these curves have to be moved to the right. The final result is a single master curve, as shown in Figure 12.17. Note that sometimes the different curves have to be moved slightly in the vertical direction as well to obtain perfect alignment. Figure 12.17 shows master curves for data on other molecular-weight fractions also; the molecular weights range from 8900 (curve L9) to 581,000 (curve L18). The reference temperature in each case is 160°C. Because changes in temperature appear to be equivalent to changes in frequency or time, the process of generating a master curve is called *time–temperature superposition.*

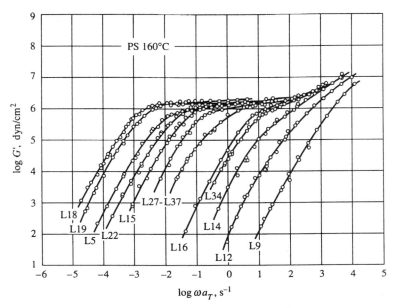

FIGURE 12.17 Master curves of G' for narrow-distribution polystyrenes having different molecular weights. The reference temperature is 160°C. (Reprinted with permission from Onogi, S., T. Masuda, and K. Kitagawa: "Rheological Properties of Anionic Polystyrenes: I. Dynamic Viscoelasticity of Narrow-Distribution Polystyrenes," Macromolecules, vol. 3, pp. 109–116, 1970. Copyright 1970 American Chemical Society.)

The fact that data at different temperatures superpose allows us to obtain low-frequency data, which would ordinarily require a significant amount of experimentation time at a given temperature, by simply making measurements at a higher temperature. Furthermore, because data at a given frequency at temperatures lower than the reference temperature correspond to high-frequency data at the reference temperature, we can think of the high-frequency end of the storage modulus master curve as modulus data characteristic of the glassy region. Similarly, data at low frequencies are representative of the viscous flow region, and the flat central plateau corresponds to the rubbery region. Figure 12.17 reveals, as expected, that the rubbery region becomes progressively larger on increasing polymer molecular weight, and it disappears entirely for samples having very low molecular weights, which are below the value needed for entanglement formation.

On a logarithmic plot, the storage modulus at reference temperature T_R and frequency ω_R equals the storage modulus at temperature T and frequency ω:

$$G'(\log \omega_R, T_R) = G'(\log \omega, T) \tag{12.5.1}$$

However, because the difference between $\log \omega_R$ and $\log \omega$ is a constant equal to $\log a_T$, where a_T is called the *temperature shift factor*, we have

$$G'(\omega a_T, T_R) = G'(\omega, T) \tag{12.5.2}$$

and we must plot $G'(\omega)$ versus ωa_T in order to get the superposition. Obviously, a_T equals unity at T_R and is less than unity if T exceeds T_R.

When the logarithm of the shift factor for each of the datasets shown in Figure 12.17 is plotted versus temperature, a single curve independent of molecular weight is obtained, provided that the molecular weight exceeds the value needed for entanglement formation. This result, shown in Figure 12.18, can be represented mathematically by the following equation:

$$\log a_T = \frac{-7.14(T - 160)}{112.1 + (T - 160)} \tag{12.5.3}$$

If instead of using $160°C$ as the reference temperature, we use T_g, the equivalent form of Eq. (12.5.3) is given by

$$\log a_T = \frac{-17.44(T - T_g)}{51.6 + (T - T_g)} \tag{12.5.4}$$

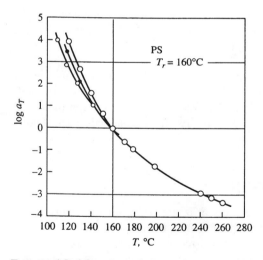

FIGURE 12.18 The logarithm of the shift factor a_T plotted against temperature for narrow-distribution polystyrenes. Large open circles indicate the results for $M \geq M_c$, closed circles for $M = 14,800$ (L12), and small open circles for $M = 8900$ (L9). (Reprinted with permission from Onogi, S., T. Masuda, and K. Kitagawa: "Rheological Properties of Anionic Polystyrenes: I. Dynamic Viscoelasticity of Narrow-Distribution Polystyrenes," Macromolecules, vol. 3, pp. 109–116, 1970. Copyright 1970 American Chemical Society.)

which is known as the *WLF equation*, after Williams, Landel, and Ferry, who first proposed it [27]. Very surprisingly, the WLF equation has been found to describe not only the temperature dependence of the storage modulus of other polymers but also the temperature dependence of the shift factors for other viscoelastic functions, provided that the temperature lies between T_g and $T_g + 100°C$ [28]. Thus, identical temperature-shift factors are calculated irrespective of whether one uses storage modulus, loss modulus, or stress relaxation data. The WLF equation can also be derived theoretically by appealing to observed shear viscosity behavior and employing the free-volume theory [26]. Above T_g and outside the range of validity of the WLF equation, we can represent the shift factors in an Arrhenius form:

$$a_T = \exp\left[\frac{E}{T}\left(\frac{1}{T} - \frac{1}{T_R}\right)\right] \qquad (12.5.5)$$

where E is an activation energy and R is the universal gas constant.

Because we can use linear viscoelastic theory (see Chapter 14 and Ref. 1) to relate one viscoelastic function to another, the use of a limited amount of data along with the time–temperature superposition principle makes it possible to obtain any small-strain property of a polymer.

Example 12.3: Use Figures 12.17 and 12.18 to determine the storage modulus at 200°C and 1 rad/sec of the polystyrene fraction labeled L27. Compare the result with that obtained with the use of Figure 12.16.

Solution: From Figure 12.18, the value of a_T at 200°C is 0.013. Thus, ωa_T is 0.013 rad/sec and $\log(\omega a_T)$ equals -1.89. The use of Figure 12.17 then reveals that the corresponding value of the storage modulus is approximately 2.5×10^4 dyn/cm^2. As expected, the same result is obtained when Figure 12.16 is used.

12.6 POLYMER FRACTURE

If a glassy polymer is stressed very rapidly or is stressed at a temperature that is significantly below its glass transition temperature, it tends to break or fracture in a brittle manner (i.e., without any plastic or irrecoverable deformation). Conversely, at temperatures above T_g, an amorphous polymer tends to draw down in a homogeneous manner and displays large strains before fracturing. At intermediate temperatures and low rates of deformation, the polymer can yield somewhat before fracturing or fracture in a ductile manner by neck formation. For anyone interested in structural applications, it is essential to know how and why polymers fracture in the glassy region. After all, stiffness and strength are two of the main criteria used in evaluating the potential utility of a material for load-bearing applications.

The theoretical strength of a polymer can be estimated based on known values of interatomic forces and surface energies, and the result for most materials is a number between 10^6 and 10^7 psi [29]. An examination of Figure 12.6, however, reveals that this theoretical strength is a couple of orders of magnitude larger than the experimentally observed tensile strength of polystyrene. A clue to the reason for this discrepancy between theory and practice is provided by the behavior of glass fibers, which can be manufactured to near-theoretical strengths provided that care is taken to ensure that the fiber surface is smooth and free of imperfections. Indeed, as explained by the classical Griffith theory [30], it is the presence of small surface cracks that is responsible for the reduction in tensile strength of glasses.

To understand why tensile strength is not a material property like the modulus but depends, rather, on sample preparation, consider the situation shown in Figure 12.19. A flat sheet of glassy polymer of width w and containing an elliptical crack is in plane strain under the influence of stress σ_{11}. The major axis of the ellipse has a length $2c$, whereas the minor axis has a length $2h$. Even though the stress distribution is not influenced by the presence of the crack at positions far from the crack, near the crack itself, and in particular at the crack tip, the stress can be significantly greater than the average imposed stress σ_{11}. Using elasticity theory [31], it is possible to show that the maximum stress occurs at the

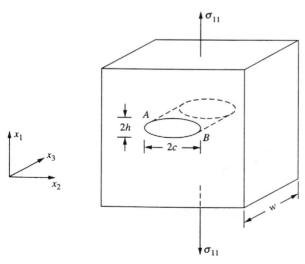

FIGURE 12.19 Crack propagation in a glassy polymer.

edge of the ellipse at points A and B and tends to cause the crack to open up. The value of this maximum stress is given by the following expression:

$$\sigma_m = \sigma_{11}\left(1 + \frac{2c}{h}\right) \tag{12.6.1}$$

Because c is typically much greater than h, σ_m/σ_{11} equals $2c/h$. Thus, the localized stress depends on the crack length and can be greater than the average stress by orders of magnitude; the larger the crack length is, the greater the stress concentration.

Because microscopic surface cracks are the inevitable result of any polymer processing operation, it is important to know whether or not a particular crack will grow and lead to specimen fracture under the influence of an applied stress. Returning to the situation depicted in Figure 12.19, we find that crack propagation results in elastic energy being released from regions of the sample above and below the broad surface of the crack, because these experience a decrease in strain [30]. However, new surfaces are simultaneously created, and there is an energy cost associated with this. Consequently, the crack grows only if there is a net release of energy. By equating the release of elastic energy to the loss of surface energy, we find that the critical stress for crack propagation is as follows: [30]:

$$(\sigma_{11})_c = \sqrt{\frac{2\gamma E}{\pi(1 - v^2)c}} \tag{12.6.2}$$

where γ is the specific surface energy, E is the Young's modulus, and v is Poisson's ratio.

Equation (12.6.2) clearly shows that the critical stress for crack growth depends on the largest crack and varies inversely as the square root of the crack length. Thus, fracture occurs at progressively smaller stress values as the crack length increases. Also, once crack propagation begins, the process results in catastrophic failure. Although Eq. (12.6.2) has been verified by Feltner by subjecting polymethyl methacrylate to cyclic deformation [32], critical stress values, in general, are underpredicted to a significant amount by this equation. The reason this happens is that energy is also taken up in plastic deformation of the region ahead of the crack tip, which is equivalent to having a larger value of the specific surface energy. This phenomenon is called crazing and occurs only under tensile loading. Quantitatively, we can account for crazing by using a larger value of γ in Eq. (12.6.2). Because plastic deformation requires a considerable absorption of energy, we find that the energy required to propagate a crack is almost totally used in promoting viscous flow.

Example 12.4: Estimate the critical crack length using Eq. (12.6.2) and physical property values characteristic of glassy polymers.

Solution: If we let $(\sigma_{11})_c$ be 4×10^8 dyn/cm^2, $\gamma = 100$ dyn/cm, $E = 3 \times 10^{10}$ dyn/cm^2, and $v = 0.3$, we find that

$$
\begin{aligned}
c &= \frac{2\gamma E}{\pi(1 - v^2)(\sigma_{11})_c^2} \\
&= \frac{2 \times 100 \times 3 \times 10^{10}}{3.142 \times 0.91 \times 16 \times 10^{16}} \\
&= 1.31 \times 10^{-5} \text{ cm} = 0.131 \ \mu\text{m}
\end{aligned}
$$

From everyday experience with plastic materials, we know that this calculated value is unrealistically low. Indeed, γ needs to be taken to be of the order of 10^5–10^6 ergs/cm^2 to obtain agreement with experimental data [2].

12.7 CRAZING AND SHEAR YIELDING

Nonlinearities in the stress–strain curve of a glassy polymer usually indicate the presence of irrecoverable deformation. Although the extent of yielding depends on the test temperature and the rate of strain, it is generally true that most glassy polymers do show some amount of plastic flow before fracture. This plastic flow contributes to the toughness of the polymer and, for this reason, is a desirable feature that needs to be examined and understood. The first point to note is that even in a sample that is loaded in tension, the yielding could be in shear. If the test piece shown in Figure 12.20 is sectioned along the dotted line, then, from equilibrium, the stress σ acting on the plane that makes an angle θ with the horizontal is given by

$$\sigma = \sigma_{zz} \cos \theta \tag{12.7.1}$$

and it can be resolved into a component σ_n normal to the surface and σ_s parallel to the surface,

$$\sigma_n = \sigma_{zz} \cos^2 \theta \tag{12.7.2}$$
$$\sigma_s = \sigma_{zz} \cos \theta \sin \theta \tag{12.7.3}$$

and, depending on the shear strength, deformation may take place in directions other than along the tensile axis. Although thin films of polystyrene show normal stress yielding or crazing when strained in tension in air, those made from polycarbonate rarely exhibit crazing under the same testing conditions [33]. Instead, they show shear yielding. Still other polymers such as polystyrene–acrylonitrile show both modes of deformation.

A craze, though not a crack, looks like a crack and is, in fact, a precursor to a crack. It runs perpendicular to the loading direction in a uniaxial tensile test, as seen in Figure 12.21 [34]. Unlike a crack, a craze can support a load because its two surfaces are bridged by a multitude of fine fibers ranging in diameter from 5

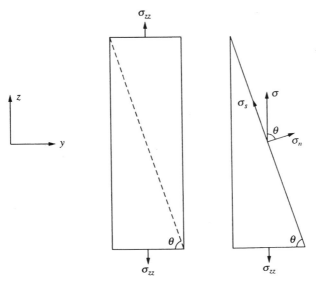

FIGURE 12.20   Presence of a shear component under the influence of a tensile stress.

to 30 nm; at the tip, a craze can be as thin as 100 Å. Although crazes form mainly at the surface, they can nucleate in the interior of the polymer as well. There is typically a sharp boundary between crazed and uncrazed material due partly to the fact that crazes propagate by the formation of voids with a consequent reduction in the density of the crazed material. Details of the process of crazing have been reviewed by Kambour [35], among others. As opposed to crazing, shear yielding appears as kink bands at an angle to the tensile axis [36]. From Eq. (12.7.3), we can figure out that the shear stress is a maximum when θ equals 45°, suggesting that shear bands should form at a 45° angle. This, to a large extent, is what is observed. A further difference between crazing and shear band formation is that shear yielding takes place without a change in density.

    Criteria that predict whether crazing or shear yielding will occur have been studied by Sternstein and co-workers [37]. For the loading situation shown in Figure 12.22, it is found that shear yielding takes place according to a modified von Mises criterion,

$$\tfrac{1}{3}[(\sigma_1 - \sigma_2)^2 + (\sigma_2 - \sigma_3)^2 + (\sigma_3 - \sigma_1)]^{1/2} \geq A - \tfrac{1}{3}B(\sigma_1 + \sigma_2 + \sigma_3)$$

$$(12.7.4)$$

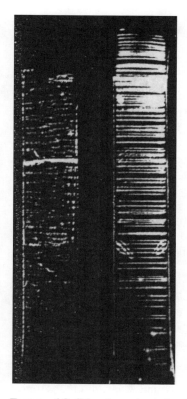

FIGURE 12.21 Craze formation in polystyrene. (From Ward, I. M., and D. W. Hadley: An Introduction to the Mechanical Properties of Solid Polymers, Wiley, Chichester, UK, 1993. Copyright John Wiley & Sons Limited. Reproduced with permission.)

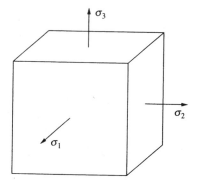

FIGURE 12.22 General tensile stress distribution.

where the left-hand side is called the *octahedral shear stress*, τ_{oct}. Furthermore, crazing (also called *normal stress yielding*) obeys a criterion of the following form:

$$|\sigma_1 - \sigma_3| \geq C + \frac{D}{\sigma_1 + \sigma_2 + \sigma_3} \tag{12.7.5}$$

which is similar to the Tresca yield condition for metals. Here, A, B, C, and D are constants that may depend on temperature, and the stress components are such that $\sigma_1 > \sigma_2 > \sigma_3$.

Although the von Mises criterion states that yielding arises whenever the stored energy exceeds a critical value, Eqs. (12.7.4) and (12.7.5) are essentially empirical equations. A general conclusion that can be drawn on examining them is that tension promotes yielding and compression hinders it. Although these equations have been quantitatively tested by Sternstein and Myers by conducting experiments on polymethyl methacrylate (PMMA) samples [38], some experimental evidence suggests that they may not work under all conditions [34].

Example 12.5: For the case of biaxial stress loading ($\sigma_1 \geq 0$, $\sigma_2 \geq 0$, $\sigma_3 = 0$), it is found that PMMA exhibits shear yielding at 70°C when τ_{oct} equals 2875 psi and the mean normal stress, $(\sigma_1 + \sigma_2 + \sigma_3)/3$, equals 2000 psi [38]. When the mean normal stress is increased to 3500 psi, τ_{oct} decreases to 2675 psi. At what stress value σ_1 will PMMA show the onset of shear bands if the sample is loaded in uniaxial tension?

Solution: Using the data provided, we find that A and B in Eq. (12.7.4) must have the values 3152 psi and 2/15, respectively. For uniaxial tension, $\sigma_2 = \sigma_3 = 0$ and Eq. (12.7.4) becomes

$$\frac{\sqrt{2}}{3}\sigma_1 \geq A - \frac{B}{3}\sigma_1$$

Using the equality sign and solving for σ_1 gives the following:

$$\sigma_1 = \frac{3A}{B + \sqrt{2}} = \frac{3 \times 3152}{0.133 + 1.414} = 6112 \text{ psi}$$

Once the criterion for craze initiation is satisfied, localized plastic flow produces microcavities whose rate of formation at a given temperature depends on the value of the applied stress and increases with increasing stress. This is a slow, time-dependent process that results in an interconnected void network. The subsequent growth of the craze is thought to occur by the repeated breakup of the concave air–polymer interface at the craze tip [39], as shown schematically in Figure 12.23 [40]. The process is similar to what happens when two flat plates

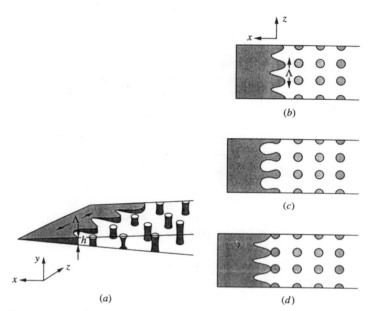

FIGURE 12.23 Schematic drawing of craze tip advance by the meniscus instability mechanism. (a) Drawing showing wedge of deformed polymer ahead of the void fingers and trailing fibrils; (b)–(d) xz sections through the craze showing the sequence of events as the craze tip advances by one fibril spacing. The void finger and fibril spacing in this drawing shows a much more regular structure than observed experimentally. (From Ref. 40.) Reprinted with permission from Kramer, E. J.: Microscopic and Molecular Fundamentals of Crazing, Adv. Polm. Sci., vol. 52/53, pp. 1–56, 1983. Copyright 1983 Springer-Verlag GmbH & Co. KG.

containing a layer of liquid between them are separated; an instability known as the Taylor meniscus instability arises and fibrils are left behind as flow fingers through in the outward direction. In the present case, as explained by Kramer [40], deformed polymer constitutes the fluid layer into which the craze tip meniscus propagates, whereas the undeformed polymer outside the craze acts as the rigid plates that constrain the fluid. As craze expansion continues, the craze also thickens in a direction perpendicular to the craze surface. This typically happens by surface drawing, wherein new polymer is drawn into the fibrils from the craze interfaces in such a way that the extension ratio of the fibrils remains constant with time, provided that the stress level is not changed [40].

Cracks propagate by the mechanism of craze fibril breakdown. Knowing how to prevent craze formation is therefore useful, because this might also prevent cracking. Studies have shown that crazing might be reduced by orienting polymer chains parallel to the stress direction [41] or by increasing the entanglement density (see Chap. 14) of the polymer chains [42]. Increases in entanglement density can entirely suppress crazing and change the mode of yielding to shear band formation.

The discussion of crazing has thus far been restricted to situations where the polymer is strained in air. It turns out that if the test is conducted in a liquid environment or in the presence of organic vapors, the stress needed to initiate crazing is considerably reduced. This common phenomenon is known as *environmental stress cracking* and it is important in situations where plastics necessarily come into contact with fluids, as in beverage containers or in tubes and hoses. It is found that variables that most influence crazing under these circumstances are the spreading coefficient, the diffusivity, and the solubility parameter of the crazing agent. Theories that have been advanced to explain the ease of cracking include (1) polymer plasticization, resulting in a lowered value of both the T_g and the viscosity of the glassy material, and (2) a lowering in the surface energy, which allows new cavities to form with ease [35].

12.8   FATIGUE FAILURE

We have seen that solid polymers in a tensile test fracture at stress values considerably lower than those that might be calculated theoretically. This diminution in strength is the result of local stress concentrations arising from the presence of small cracks and other flaws. The point of onset of failure can be predicted based on a modified Griffith criterion or, in the case of crazing, a modified Tresca criterion. It is on these observed values that design calculations have to be based. Even worse, we find that if the load is cycled (instead of being kept constant), failure takes place at stress amplitudes significantly lower than the breaking stress in a tensile test. This behavior is called fatigue, and typical fatigue data for glassy polystyrene having a tensile strength of about 6 ksi are shown in Figure 12.24 [43]. These data were collected under essentially isothermal conditions in uniaxial tension–compression at a constant frequency of 0.1 Hz. The fatigue life, measured in the number of stress cycles before failure, is very low in region I, where the strain amplitude is close to the tensile strength. Under these conditions, fracture occurs due to extensive crazing in a manner analogous to tensile failure. Reducing the stress amplitude yields region II, where a linear relationship develops between the stress applied and the logarithm of the life to failure. Here, again, fracture takes place by craze formation, craze growth, crack nucleation, and crack propagation. Finally, in region III, no failure is observed if the stress amplitude is kept below a critical value called the *endurance limit*. An observation of the fracture surfaces suggests that in region II, both craze growth and crack growth occur incrementally on each tension half-cycle, resulting in slow cracking. Once a critical crack size is reached, however, the crack moves

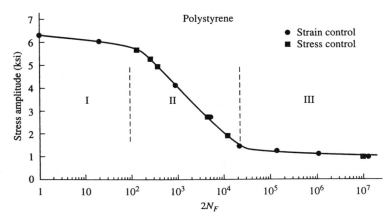

FIGURE 12.24 Polystyrene stress–fatigue life relation. (From Ref. 43.)

rapidly through previously uncrazed material, resulting in brittle fracture. We also find that, beyond a certain frequency level, an increase in frequency decreases fatigue life, whereas an increase in polymer molecular weight increases fatigue life up to a limiting molecular weight [3].

Typically, two kinds of fatigue failure are observed in glassy polymers: brittle failure of the sort just described and thermal softening. For polymers whose tensile strength is reduced on increasing temperature, a temperature rise in the sample resulting from high damping (a large value of tan δ) can lead to fatigue failure. To understand this, let us calculate E_c, the amount of energy dissipated per unit volume during each cycle. Using Eqs. (12.4.1) and (12.4.3) gives the following:

$$
\begin{aligned}
E_c &= \int \tau \, d\gamma = 2\tau_0 \gamma_0 \int_0^\pi (\cos \delta \sin \omega t + \sin \delta \cos \omega t) \cos \omega t \, d(\omega t) \\
&= \pi \tau_0 \gamma_0 \sin \delta
\end{aligned}
\tag{12.8.1}
$$

Because the strain amplitude is assumed to be small, we can estimate γ_0 using elasticity theory [44]. Thus,

$$
E_c = \frac{\pi \tau_0^2 \sin \delta}{E}
\tag{12.8.2}
$$

where E is Young's modulus. The energy dissipated per unit time is given by

$$
Q = \frac{\omega \tau_0^2 \sin \delta}{2E}
\tag{12.8.3}
$$

and it increases with increasing stress amplitude, frequency, and phase angle.

At steady state, all of this energy is lost to the atmosphere by convection from the surface of the sample. Thus, using Newton's law of cooling gives

$$Q = hA(T - T_0) \tag{12.8.4}$$

where h is the heat transfer coefficient, A is the surface area per unit volume, T is the average steady-state temperature, and T_0 is the temperature of the surroundings.

On equating the two preceding expressions for Q, we can obtain rather large values of $(T - T_0)$ if the ambient temperature is high [44]. Indeed, values of the average temperature rise as high as $60°C$ have been measured [44]. Because the stress at the crack tip is much larger than the average stress, the temperature rise there can be expected to be proportionately higher. As a consequence, failure can occur simply by melting of the polymer.

Regarding brittle fatigue fracture, one theory that has been extensively examined, especially in the Russian literature, is the *linear damage theory*. This is a phenomenological theory that assumes that cyclic stresses slowly damage the polymer (say, by chain scission) and this damage continues to accumulate. Once a critical damage level is reached, the material fails. According to this theory, the time for failure t is given by the following [45]:

$$t = t_0 \exp\left(\frac{U - \sigma V}{RT}\right) \tag{12.8.5}$$

where t_0 is a constant, U is an activation energy, σ the applied stress, and V is an activation volume. The use of Eq. (12.8.5) often leads to good agreement with experimental data. Note, though, that this equation can only be an approximation because it predicts a finite time to break when σ is zero [45].

12.9   IMPROVING MECHANICAL PROPERTIES

Over the last several decades, plastics have gone from being inexpensive substitutes for metals, concrete, and timber to becoming materials of choice. This has come about due to a combination of factors. As the scientific community has come to better understand the behavior of polymeric materials, their structure and properties have been improved to such an extent that they now compete with other materials for traditional applications on both quality and price. In addition, some of the unique properties of polymers have resulted in the creation of new markets in such diverse areas as transportation, housing, food packaging, health care, and information and communication [46]. Indeed, workers in both academic and industrial research laboratories are constantly striving to improve polymer properties and to create new polymers with enhanced properties in order to bring improved products to market and also find new applications for existing polymers.

Several strategies have been followed to achieve these objectives. For homopolymers, improving mechanical properties has largely been a process of relating the internal structure of the polymer to its properties. For amorphous polymers, we seek to align all of the polymer chains in the same direction; this anisotropy results in a higher glass transition temperature and an increase in both stiffness and strength in the direction of molecular chain orientation. Semicrystalline polymers can be annealed to induce crystallization. This, again, makes the polymer stiffer. Different structures can be formed by drawing the polymer and by varying the temperature and conditions of crystallization. These structures differ from each other in the size, shape, and amount of crystals and in the orientation of polymer chains in the amorphous and crystalline regions, resulting in different mechanical properties. Textile engineers have long studied the effect of these different variables using polymers such as nylons, polyethylene terephthalate, and polypropylene; a considerable body of knowledge now exists [47,48]. This has allowed for the production of not only high-quality textile yarns but also of high-strength, high-modulus fibers from conventional polymers such as high-density polyethylene of high molecular weight ($> 10^6$). By careful control of molecular orientation and packing density, it is now possible to manufacture high-density polyethylene fibers having a modulus of 100 GPa and a strength of 1.4 GPa. Of course, these results have been achieved only through the development of special polymer processing techniques such as solid-state extrusion and gel spinning [46].

If we incorporate aromatic structures into the polymer backbone, not only can we make the molecule rigid and rodlike, we can also raise its T_g, make it thermally stable at high temperatures, and impart high strength and stiffness. Thus, linear aromatic thermoplastics such as semicrystalline polyetherether ketone have a T_g of 143°C and a maximum continuous-use temperature of 250°C [49]. Ultrastiff, rodlike molecules can also be made to form liquid crystals from both the melt and solution [50]; molecules such as poly-p-phenyleneterephthalamide and thermotropic copolyesters can be spun into fibers in a highly oriented, extended-chain form to yield strength and stiffness values up to 3 GPa and 140 GPa, respectively.

In addition to manipulating the physical and chemical structures of homopolymers, we can chemically react two monomers whose homopolymers have two different desirable properties. We do so with a view toward obtaining both properties in the copolymer. Thus, a random copolymer of styrene with butadiene provides strength with flexibility, whereas reacting styrene with acrylonitrile gives toughness and solvent resistance to polystyrene. On the other hand, block copolymers such as polystyrene–polybutadiene–polystyrene form two-phase systems with the butadiene block constituting a continuous three-dimensional elastomeric network and the polystyrene phase serving as junction points [51]. This material behaves like vulcanized rubber at room temperature but flows like a thermoplastic above the T_g of polystyrene. Finally, grafting pendant groups such as sodium methacrylate to the polystyrene backbone can help raise the glass transition temperature of the polymer.

Most current-day research, though, is not concentrated in developing new polymers. Instead, it is focused on blending existing polymers to formulate new and improved materials. The classical example [52] of this approach is the blending of polystyrene with polyphenylene ether (PPE) to give a single-phase miscible blend. Not only does adding polystyrene to PPE allow the latter polymer to be melt processed at a lower temperature and over a wider temperature range, but mechanical properties, such as the modulus, show a significant synergistic behavior. Most commercial polymer blends, however, are thermodynamically incompatible, two-phase mixtures. They are still useful because they typically have improved impact strength and toughness, properties especially important in automotive applications. A characteristic example of this is high-impact polystyrene (HIPS), which is made by mixing or grafting polybutadiene onto polystyrene [53]. Another example of great commercial interest is acrylonitrile-butadiene-styrene (ABS), in which styrene–acrylonitrile is grafted onto poly(-butadiene) and then these graft polymers are mixed with styrene–acrylonitrile copolymers [54]. The dispersed rubber particles initiate crazes without crack formation, leading to a tough, self-reinforcing composite material.

The quintessential method of improving the strength and stiffness of polymers is to form reinforced composites by adding filler particles, whiskers, short fibers, or long fibers to polymer matrices such as epoxies, unsaturated polyesters, and vinyl esters [49,55]. Composite materials containing 50–70% by weight of fibers of glass, carbon, or polyaramid in thermoplastic or thermosetting polymer matrices can be lighter than aluminum and stronger than steel. Although most of the development in this area has been motivated by aircraft, aerospace, and automotive applications, future growth is likely to be in civil engineering construction [46].

REFERENCES

1. Ferry, J. D., *Viscoelastic Properties of Polymers*, 3rd ed., Wiley, New York, 1980.
2. Nielsen, L. E., and R. F. Landel, *Mechanical Properties of Polymers and Composites*, 2nd ed., Marcel Dekker, New York, 1994.
3. Nunes, R. W., J. R. Martin, and J. F. Johnson, Influence of Molecular Weight and Molecular Weight Distribution on Mechanical Properties of Polymers, *Polym. Eng. Sci.*, 22, 205–228, 1982.
4. Findley, W. N., 26-Year Creep and Recovery of Poly(Vinyl Chloride) and Polyethylene, *Polym. Eng. Sci., 27, 582–585, 1987.*
5. Martin, J. R., J. F. Johnson, and A. R. Cooper, Mechanical Properties of Polymers: The Influence of Molecular Weight and Molecular Weight Distribution, *J. Macromol. Sci. Macromol. Chem.*, C8, 57–199, 1972.

6. Merz, E. H., L. E. Nielsen, and R. Buchdahl, Influence of Molecular Weight on the Properties of Polystyrene, *Ind. Eng. Chem.*, 43, 1396–1401, 1951.
7. Fellers, J. F., and T. Chapman, Deformation Behavior of Polystyrene as a Function of Molecular Weight Parameters, *J. Appl. Polym. Sci.*, 22, 1029–1041, 1978.
8. Vlachopoulos, J., N. Hadjis, and A. E. Hamielec, Influence of Molecular Weight on the Tensile Properties of Nearly Monodisperse Polystyrenes, *Polymer*, 19, p. 115, 1978.
9. Flory, P. J., Tensile Strength in Relation to Molecular Weight of High Polymers, *J. Am. Chem. Soc.*, 67, 2048–2050, 1945.
10. Hahnfeld, J. L., and B. D. Dalke, General Purpose Polystyrene, in *Encyclopedia of Polymer Science and Engineering*, 2nd ed., Wiley, New York, Vol. 16, pp. 62–71, 1989.
11. Andrews, R. O., and J. F. Rudd, Photoelastic Properties of Polystyrene in the Glassy State: I. Effect of Molecular Orientation, *J. Appl. Phys.*, 28, 1091–1095, 1957.
12. Rudd, J. F., and E. F. Gurnee, Photoelastic Properties of Polystyrene in the Glassy State: II. Effect of Temperature, *J. Appl. Phys.*, 28, 1096–1100, 1957.
13. Maxwell, B., and M. Nguyen, Measurement of the Elastic Properties of Polymer Melts, *Polym. Eng. Sci.*, 19, 1140–1150, 1979.
14. Samuels, R. J., Structured Polymer Properties, *Chemtech*, 169–177, March 1974.
15. Struik, L. C. E., *Physical Aging of Amorphous Polymer and Other Materials*, Elsevier, Amsterdam, 1978.
16. Ogale, A. A., and R. L. McCullough, Physical Aging Characteristics of Polyether Ether Ketone, *Composite Sci. Technol.*, 30, 137–148, 1987.
17. Kumler, P. L., S. E. Keinath, and R. F. Boyer, ESR Studies of Polymer Transitions: III. Effect of Molecular Weight Distribution on T_g Values of Polystyrene as Determined by ESR Spin–Probe Studies, *J. Macromol. Sci. Phys.*, 13, 631–646, 1977.
18. Fox, T. G., and P. J. Flory, Second–Order Transition Temperatures and Related Properties of Polystyrene: I. Influence of Molecular Weight, *J. Appl. Phys.*, 21, 581–591, 1950.
19. Bueche, F., *Physical Properties of Polymers*, Krieger, Huntington, NY, 1979.
20. Brandrup, J., and E. H. Immergut, *Polymer Handbook*, 3rd ed., Wiley–Interscience, New York, 1989.
21. McNaughton, J. L., and C. T. Mortimer, Differential Scanning Calorimetry, *IRS Physi. Chem. Series 2*, 10, 1–44, 1975.
22. Nielsen, L. E., Dynamic Mechanical Properties of High Polymers, *Soc. Plast. Eng. J.*, 16, 525–533, 1960.
23. Gillham, J. K., and J. B. Enns, On the Cure and Properties of Thermosetting Polymers Using Torsional Braid Analysis, *Trends Polym. Sci.*, 2, 406–419, 1994.
24. Eisenberg, A., and B. C. Eu, Mechanical Spectroscopy, An Introductory Review, *Annu. Rev. Mater. Sci.*, 6, 335–359, 1976.
25. Murayama, T., *Dynamic Mechanical Analysis of Polymer Material*, Elsevier, Amsterdam, 1978.
26. Onogi, S., T. Masuda, and K. Kitagawa, Rheological Properties of Anionic Polystyrenes: I. Dynamic Viscoelasticity of Narrow–Distribution Polystyrenes, *Macromolecules*, 3, 109–116, 1970.

27. Williams, M. L., R. F. Landel, and J. D. Ferry, The Temperature Dependence of Relaxation Mechanisms in Amorphous Polymers and Other Glass–Forming Liquids, *J. Am. Chem. Soc.*, 77, 3701–3706, 1955.
28. Markovitz, H., Superposition in Rheology, *J. Polym. Sci.*, 50, 431–456, 1975.
29. Hayden, H. W., W. G. Moffatt, and J. Wulff, *The Structure and Properties of Materials*, Wiley, New York, 1965, Vol. 3.
30. Griffith, A. A., The Phenomena of Rupture and Flow in Solids, *Phil. Trans. Roy. Soc. London*, A221, 163–198, 1921.
31. Roark, R. J., *Formulas for Stress and Strain*, 3rd ed., McGraw–Hill, New York, 1954.
32. Feltner, C. E., Cycle–Dependent Fracture of Polymethyl Methacrylate, *J. Appl. Phys.*, 38, 3576–3584, 1967.
33. Donald, A. M., and E. J. Kramer, *The Competition Between Shear Deformation and Crazing in Glassy Polymers*, Cornell Materials Science Center Report 4568, 1981.
34. Ward, I. M., and D. W. Hadley, *An Introduction to the Mechanical Properties of Solid Polymers*, Wiley, Chichester, 1993.
35. Kambour, R. P., A Review of Crazing and Fracture in Thermoplastics, *J. Polym. Sci. Macromol. Rev.*, 7, 1–154, 1973.
36. Schultz, J. M., *Polymer Materials Science*, Prentice–Hall, Englewood Cliffs, NJ, 1974.
37. Sternstein, S., Mechanical Properties of Glassy Polymers, *Treatise Mater. Sci. Technol.*, 10B, 541–598, 1977.
38. Sternstein, S. S., and F. A. Myers, Yielding of Glassy Polymers in the Second Quadrant of Principal Stress Space, *J. Macromol. Sci. Phys.*, B8, 539–571, 1973.
39. Argon, A. S., J. G. Hanoosh, and M. M. Salama, Initiation and Growth of Crazes in Glassy Polymers, *Fracture*, 1, 445–470, 1977.
40. Kramer, E. J., Microscopic and Molecular Fundamentals of Crazing, *Adv. Polym. Sci.*, 52/53, 1–56, 1983.
41. Farrar, N. R., and E. J. Kramer, Microstructure and Mechanics of Crazes in Oriented Polystyrene, *Polymer*, 22, 691–698, 1981.
42. Kramer, E. J., *Craze Fibril Formation and Breakdown*, Cornell Materials Science Center Report 5038, 1983.
43. Rabinowitz, S., A. R. Krause, and P. Beardmore, Failure of Polystyrene in Tensile and Cyclic Deformation, *J. Mater. Sci.*, 8, 11–22, 1973.
44. Schultz, J. M., Fatigue Behavior of Engineering Polymers, *Treatise Mater. Sci. Technol.*, 10B, 599–636, 1977.
45. Sacher, E., Effect of Molecular Motions on Time–Dependent Fracture, *J. Macromol. Sci.*, B15, 171–181, 1978.
46. National Research Council, *Polymer Science and Engineering, The Shifting Research Frontiers*, National Academy Press, Washington, DC, 1994.
47. Samuels, R. J., *Structured Polymer Properties*, Wiley, New York, 1974.
48. Ziabicki, A., *Fundamentals of Fibre Formation*, Wiley, London, 1976.
49. Mallick, P. K., *Fiber–Reinforced Composites*, 2nd ed., Marcel Dekker, New York, 1993.
50. White, J. L., Historical Survey of Polymer Liquid Crystals, *J. Appl. Polym. Sci. Appl. Polym. Symp.*, 41, 3–24, 1985.
51. Holden, G., E. T. Bishop, and N. R. Legge, Thermoplastic Elastomers, *J. Polym. Sci.*, C26, 37–57, 1969.

52. Utracki, L. A., *Polymer Alloys and Blends*, Hanser, Munich, 1990.
53. Bucknall, C. B., *Toughened Plastics*, Applied Science Publishers, London, 1977.
54. Kranz, D., L. Morbitzer, K. H. Ott, and R. Casper, Struktur und Eigenschaften von ABS-Polymeren, *Angew. Makromol. Chem.*, 58/59, 213–226, 1977.
55. Agarwal, B. D., and L. J. Broutman, *Analysis and Performance of Fiber Composites*, 2nd ed., Wiley, New York, 1990.
56. Takaki, T., *The Extensional and Failure Properties of Polymer Melts*, M.S. thesis, University of Tennessee, Knoxville, 1973.

PROBLEMS

12.1. Show that Eq. (12.2.3) reduces to Eq. (12.2.2) for small strains.

12.2. By considering the stretching of a cylindrical sample, show that the Poisson ratio for an incompressible material is 0.5.

12.3. How does the stress change with time if at $t = 0$ the Maxwell element of Figure 12.4 is stretched to a total strain of γ_0 which is kept unchanged thereafter?

12.4. In an adhesion application, an epoxy is subjected to 1% shear strain. What is the corresponding shear stress? The tensile modulus of the epoxy is 4×10^9 Pa and the Poisson ratio is 0.3.

12.5. A cube of the epoxy of Problem 12.4 and having unit volume is put under a tensile stress of 10 MPa. What is the percent change in volume?

12.6. The $T_{g\infty}$ for a polymer made by step growth polymerization is 200°C. If, during synthesis by bulk polymerization, the reactor temperature is 170°C, would you expect to obtain a high-molecular-weight polymer? Justify your answer and suggest a remedy if your answer is no.

12.7. How does the storage modulus of the Maxwell element vary with frequency as (a) $\omega \to 0$ and (b) $\omega \to \infty$? By comparing your answers with the data given in Figure 12.17, determine if the model gives results that are at least qualitatively correct.

12.8. Shown in Figure P12.8 is a Voigt element: -a spring and a dashpot in parallel. Determine the creep response of this combination and compare the results with those obtained in Example 12.1.

12.9. Given below are data for the tensile strain as a function of time at three different temperatures obtained by hanging a constant weight on polystyrene samples [56]. Show that it is possible to obtain a master curve by means of time–temperature superposition. Calculate the shift factors and compare them with those obtained using shear data and represented by

$$\log a_T = 800.4\left(\frac{1}{T - 47.9} - \frac{1}{T_R - 47.9}\right)$$

in which the reference temperature T_R is 160°C.

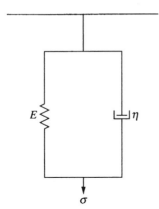

FIGURE P12.8 A Voigt element.

	Time (sec)		
Strain	180°C	190°C	200°C
0	0	0	0
0.2	14.0	5.4	1.8
0.4	29.6	12.1	4.6
0.6	43.9	16.4	6.4
0.8	58.6	20.4	8.6
1.0	72.5	23.9	10.0
1.2	83.6	25.7	10.7

12.10. Show that σ_s in Eq. (12.7.3) takes on its maximum value when $\theta = 45°$. Thus, determine $(\sigma_s)_{max}/\sigma_{zz}$.

12.11. For biaxial stress loading, Eq. (12.7.5) becomes

$$|\sigma_1 - \sigma_2| \geq C + \frac{D}{\sigma_1 + \sigma_2}$$

Under these conditions, the PMMA sample of Example 12.5 exhibits crazing at 70°C when $|(\sigma_1 - \sigma_2)|$ equals 1000 psi and $1/(\sigma_1 + \sigma_2)$ equals 2.15×10^4 (psi)$^{-1}$ [38]. On increasing the value of the former variable to 3000 psi, the needed value of the latter variable is 3.11×10^{-4} (psi)$^{-1}$. Compute the values of the constants C and D and plot the normal stress yielding criterion (the preceding equation) as σ_1 versus σ_2 for $\sigma_1 \geq 0$ and $\sigma_2 \geq 0$.

12.12. The PMMA sample of Example 12.5 and Problem 12.11 is stretched in uniaxial tension. Will shear yielding or crazing occur first? At what value of σ_1?

12.13. A polystyrene sample of unit volume is subjected to a fatigue test at a temperature of 50°C and frequency of 10 Hz. Use the data of Figures 12.7 and 12.15 to estimate the initial rate of temperature rise in the sample. Assume adiabatic conditions and let γ_0 be 0.01 and the polymer density be 1 g/cm³. The specific heat of polystyrene is 0.3 cal/g. How would the answer change if the test temperature was much closer to T_g?

12.14. Consider the continuous-fiber reinforced rod shown in Figure P12.14. If the fiber volume fraction is ϕ, determine Young's modulus, E, of the rod. Let the moduli of the fibers and matrix be E_1 and E_2, respectively. What can you say about the magnitude of E_1 relative to E_2 if the purpose of the reinforcement is to develop a material having a modulus that is significantly greater than that of the matrix polymer?

Force F

F

FIGURE P12.14 Unidirectional continuous-fiber reinforcement of plastics

13

Polymer Diffusion

13.1 INTRODUCTION

In engineering practice, we routinely encounter the diffusion of small molecules
through solid polymers, the diffusion of polymer molecules in dilute or concentrated solution, and the transport of macromolecules through polymer melts. We
came across diffusion in dilute solution when we developed the theory of the
ultracentrifuge in Chapter 8 as a method of determining polymer molecular
weight. Similarly, solution polymerization involves diffusion in a concentrated
solution. The reverse situation of mass transfer of small molecules through
polymers has great technological importance. Thus, anisotropic cellulose acetate
membranes can be used for desalination of water by reverse osmosis [1], and
ethyl cellulose membranes can be used to separate gas mixtures such as air to
yield oxygen [2]. Other common situations include the drying of polymeric
coatings [3] and the removal of the monomer and other unwanted volatiles from
finished polymer by the process of devolatilization [4]. In the field of medicine,
polymeric drug delivery systems have become a reality [5]. For example, there is
now a commercially available implant for glaucoma therapy consisting of a
membrane-controlled reservoir system made from an ethylene–vinyl acetate
copolymer [6]. This implant is placed in the lower eyelid's conjunctival cul-de-
sac, and it delivers the drug pilocarpine continuously over a 1-week period;
normally, patients would receive eyedrops of this drug four times each day. A few
other examples involving diffusion through polymers are biomedical devices such

526

as blood oxygenators and artificial kidneys. Finally, polymer diffusion in polymer melts is relevant to the self-adhesion of polymer layers. Polyimide layers, for instance, are used as insulators in electronic packaging, and the peel strength of such a bilayer is found to correlate with the interdiffusion distance [7].

A fundamental study of diffusion of and through polymers is clearly necessitated by all the applications just cited. For a rational design of devices employing polymer diffusion, it is necessary to know the mechanism of diffusion and how the rate of diffusion is affected by variables such as temperature, concentration, and molecular weight. Another extremely important variable is polymer physical structure. Because the transport of small molecules through a nonporous polymer is a solution-diffusion process [8], the flux depends on both the solubility and the diffusivity. Because of the small amount of free volume in a glassy polymer, diffusion coefficients are low and decrease rapidly with increasing molecular size of the diffusing species. This fact can be used to advantage in separating small molecules of air, for example, from larger organic vapors. Conversely, because solubility increases with increasing condensability, rubbery polymers permit easier transport of larger molecules because these are more condensable [9]. Polymer structure can affect properties in more subtle ways as well. Small changes in crystallinity and polymer chain orientation can alter the diffusion path and adversely affect the dyeability of knitted and woven fabrics and lead to color nonuniformities known as barre [10]. Such mechanistic information is also useful for testing molecular theories of polymer behavior, especially because transport properties such as diffusivity and viscosity are closely interrelated. In this chapter, therefore, we define the various diffusion coefficients, show their relevance, discuss methods of measuring the mutual diffusion coefficient, present typical data, and see how these might be explained by available theories.

13.2 FUNDAMENTALS OF MASS TRANSFER

When concentration gradients exist in a multicomponent system, there is a natural tendency for the concentration differences to be reduced and, ultimately, eliminated by mass transfer. This is the process of diffusion, and mass transfer occurs by molecular means. Thus, water evaporates from an open dish and increases the humidity of the air. However, the rate of mass transfer can be increased by blowing air past the dish. This is called *convective mass transfer* or *mass transfer due to flow*.

The basic equation governing the rate of mass transfer of component A in a binary mixture of A and B is (in one dimension) given by

$$J_{A,z} = -D_{AB}c \, \frac{dx_A}{dz} \qquad (13.2.1)$$

which is often known as Fick's first law. Here, $J_{A,z}$ is the flux of A in the z direction in units of moles per unit time per unit area, D_{AB} (assumed constant) is the mutual or interdiffusion coefficient, c is the molar concentration of the mixture, and x_A is the mole fraction of A. Clearly, dx_A/dz is the mole fraction gradient, and diffusion occurs due to the presence of this quantity.

Although Eq. (13.2.1) is similar in form to Fourier's law of heat conduction and Newton's law of viscosity, the similarity is somewhat superficial. This is because the flux given by Eq. (13.2.1) is not relative to a set of axes that are fixed in space, but are relative to the molar average velocity; that is,

$$J_{A,z} = cx_A(v_{A,z} - V_z) \tag{13.2.2}$$

where the molar average velocity is defined as

$$V_z = x_A v_{A,z} + x_B v_{B,z} \tag{13.2.3}$$

in which $v_{A,z}$ and $v_{B,z}$ are the velocities in the z direction of the two components relative to a fixed coordinate system.

Because $N_{A,z}$, the molar flux of A relative to the fixed axes, is $cx_A v_{A,z}$, this quantity must also equal the flux of A due to the mixture (molar) average velocity plus the flux of A relative to this average velocity; that is,

$$N_{A,z} = cx_A V_z - D_{AB}c \frac{dx_A}{dz} \tag{13.2.4}$$

and we can write similar equations for fluxes in the x and y directions as well.

If we choose to work in terms of mass units, the corresponding form of Eq. (13.2.1) is as follows [11]:

$$j_{A,z} = -D_{AB}\rho \frac{dw_A}{dz} \tag{13.2.5}$$

where ρ is the mixture density and w_A is the mass fraction of A. Now the mass flux $j_{A,z}$ is given relative to the mass average velocity,

$$v_z = w_A v_{A,z} + w_B v_{B,z} \tag{13.2.6}$$

which then leads to the following expression for the mass flux $n_{A,z}$ relative to fixed axes:

$$n_{A,z} = -D_{AB}\rho \frac{dw_A}{dz} + \rho_A v_z \tag{13.2.7}$$

where ρ_A is the mass concentration of A and equals ρw_A. Note that the diffusion coefficient D_{AB} appearing in Eqs. (13.2.4) and (13.2.7) is the same quantity and that Eq. (13.2.4) can be obtained by dividing both sides of Eq. (13.2.7) by M_A, the molecular weight of A.

The binary diffusion coefficient is not the only kind of diffusion coefficient that we can define. If we label some molecules of a pure material and follow their motion through the unlabeled molecules, the foregoing equations would still apply, but the diffusion coefficient would be called a *self-diffusion coefficient*. A similar experiment can be conducted by labeling some molecules of one component in a uniform mixture of two components. The motion of the labeled molecules would give the *intradiffusion coefficient* of this species in the mixture [12]. Still other diffusion coefficients can be defined for mass transfer in multicomponent systems [13].

Example 13.1: Show that $cx_A V_z$ in Eq. (13.2.4) can also be written as $x_A(N_{A,z} + N_{B,z})$, where $N_{B,z}$ is the flux of B relative to fixed axes.

Solution: Using the definition of the molar average velocity and the mole fraction gives the following:

$$cx_A V_z = cx_A(x_A v_{A,z} + x_B v_{B,z}) = cx_A\left(\frac{c_A}{c}v_{A,z} + \frac{c_B}{c}v_{B,z}\right)$$
$$= x_A(c_A v_{A,z} + c_B v_{B,z})$$
$$= x_A(N_{A,z} + N_{B,z})$$

An examination of the foregoing equations shows that we need the diffusion coefficient and the concentration profile before we can determine the flux of any species in a mixture. If, for the moment, we assume that we know the interdiffusion coefficient, we obtain the concentration profile by solving the differential mass balance for the component whose flux is desired. The general mass balance equation itself can be derived in a straightforward way, as follows.

If we consider the mass transport of species A through the rectangular parallelepiped shown in Figure 13.1, then it is obvious that the mass of component A inside the parallelepiped at time $t + \Delta t$ equals the mass of A that was present at time t plus the mass of A that entered during time interval Δt minus the mass that left during time interval Δt. In mathematical terms, therefore,

$$\rho_A \Delta x \Delta y \Delta z|_{t+\Delta t} = \rho_A \Delta x \Delta y \Delta z|_t + n_{A,x} \Delta y \Delta z \Delta t|_x$$
$$- n_{A,x} \Delta y \Delta z \Delta t|_{x+\Delta x} + n_{A,y} \Delta x \Delta z \Delta t|_y$$
$$- n_{A,y} \Delta x \Delta z \Delta t|_{y+\Delta y} + n_{A,z} \Delta x \Delta y \Delta t|_z$$
$$- n_{A,z} \Delta x \Delta y \Delta t|_{z+\Delta z}$$

(13.2.8)

Dividing the above equation by $\Delta x \Delta y \Delta z \Delta t$, rearranging, and taking limits yields

$$\frac{\partial \rho_A}{\partial t} + \frac{\partial}{\partial x}n_{A,x} + \frac{\partial}{\partial y}n_{A,y} + \frac{\partial}{\partial z}n_{A,z} = 0$$

(13.2.9)

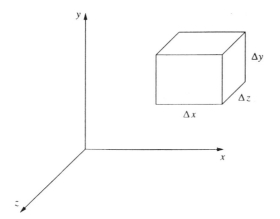

FIGURE 13.1 Coordinate system used for deriving the mass balance equation.

Replacing the flux components with expressions of the type given by Eq. (13.2.7) and noting that the overall mass balance is $\mathbf{V} \cdot \mathbf{v} = 0$ for incompressible materials [11] we have

$$\frac{\partial \rho_A}{\partial t} + v_x \frac{\partial \rho_A}{\partial x} + v_y \frac{\partial \rho_A}{\partial y} + v_z \frac{\partial \rho_A}{\partial z} = D_{AB}\left(\frac{\partial^2 \rho_A}{\partial x^2} + \frac{\partial^2 \rho_A}{\partial y^2} + \frac{\partial^2 \rho_A}{\partial z^2}\right)$$

(13.2.10)

where it has been assumed that both ρ and D_{AB} are constant. Also, ρw_A is equal to ρ_A.

   If there is no bulk fluid motion, the mass average velocity components are zero and

$$\frac{\partial \rho_A}{\partial t} = D_{AB}\nabla^2 \rho_A$$

(13.2.11)

which is often called Fick's second law. The term ∇^2 is Laplace's operator $(\partial^2/\partial x^2 + \partial^2/\partial y^2 + \partial^2/\partial z^2)$ in rectangular coordinates. At steady state,

$$\nabla^2 \rho_A = 0$$

(13.2.12)

and the equivalent forms of Eqs. (13.2.9)–(13.2.12) in mole units are obtained by dividing these equations by the molecular weight of A. Similar expressions in curvilinear coordinates are available in standard textbooks [11,13].

   In any given situation of practical interest, the concentration profile is obtained by solving the appropriate form of Eq. (13.2.9). For diffusion through solids and liquids (this is the situation of interest to us here), the flow terms are always small in comparison to the other terms. Consequently, we solve Eq.

(13.2.11) to determine the concentration profile if the diffusion coefficient is known. Alternately, the solution of Eq. (13.2.11) can be used in conjunction with experimental data to obtain the mutual diffusion coefficient. Because a new solution is generated each time a boundary condition is changed, a very large number of methods exist for the experimental determination of the diffusivity. Some of these methods are discussed in the next section. Note that if the diffusivity is not constant but depends on the mixture composition, the mass transfer situation is termed *non-Fickian*; this is examined later in the chapter in the discussion of theoretical predictions of the measured diffusion coefficients.

13.3 DIFFUSION COEFFICIENT MEASUREMENT

In this section, we examine some common experimental techniques for measuring the diffusion coefficient in liquids and solids under isothermal conditions. Both steady-state and transient conditions are encountered in these methods. Either Eq. (13.2.1) or (13.2.5) is employed for the former situation, whereas Eq. (13.2.11) is used in the latter case, because the flow terms are either identically zero or negligible.

13.3.1 Diffusion in the Liquid Phase

The simplest means of obtaining binary mutual diffusion coefficients, especially for liquid mixtures of low-molecular-weight materials, is through the use of a diaphragm cell, shown schematically in Figure 13.2. This method was introduced originally by Northrop and Anson and consists of two compartments separated by a porous diaphragm made of glass or stainless steel [14]. A concentrated solution is placed in the lower compartment of volume V_2 and a dilute solution is kept in the upper compartment of volume V_1. Both solutions are mechanically stirred to eliminate concentration gradients within the respective compartments, and diffusion is allowed to occur through the channels in the diaphragm. If the initial solute concentration in the upper chamber is $c_1(0)$ and that in the lower chamber is $c_2(0)$, these concentrations will change with time at a rate (assuming quasi-steady-state conditions) given by the following:

$$V_1 \frac{dc_1}{dt} = J(t)A \qquad (13.3.1)$$

$$V_2 \frac{dc_2}{dt} = -J(t)A \qquad (13.3.2)$$

in which A is the total area of the channels in the diaphragm and $J(t)$ is the solute flux across the diaphragm given by Eq. (13.2.1), and this flux varies with time

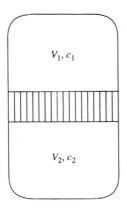

FIGURE 13.2   Schematic diagram of the diaphragm cell.

because the concentration gradient varies with time. This time dependence is given by

$$J(t) = -\frac{D_{12}(c_1 - c_2)}{L} \qquad (13.3.3)$$

where D_{12} is the mutual diffusion coefficient and L is the diaphragm thickness. Combining the three preceding equations gives

$$-\frac{d}{dt}\ln(c_2 - c_1) = D_{12}A\left(\frac{1}{V_1} + \frac{1}{V_2}\right)L^{-1} \qquad (13.3.4)$$

Integrating from $t = 0$ to t gives

$$\ln\left(\frac{c_2(0) - c_1(0)}{c_2 - c_1}\right) = D_{12}At\left(\frac{1}{V_1} + \frac{1}{V_2}\right)L^{-1} \qquad (13.3.5)$$

which allows for the determination of D_{12} from experimentally measurable quantities. If the diffusion coefficient varies with concentration, the procedure just illustrated will yield an average value. Details of specific cell designs and operating procedures are available in standard books on the topic [12].

Example 13.2: Northrop and Anson examined the diffusion of HCl in water by means of a diaphragm cell fitted with a porous aluminum membrane; this membrane separated pure water from 0.1 N HCl. Over a period of 30 min, the amount of acid that diffused through the membrane was equivalent to 0.26 cm^3 of 0.1 N HCl. What is the value of the membrane constant L/A? It is known that D_{12} equals 2.14×10^{-5} cm^2/sec. Assume that c_1 and c_2 remain unchanged over the course of the experiment.

Solution: Using Eq. (13.3.3) and noting that $c_1 = 0$, we find the following:

$$\text{Amount diffused} = \frac{D_{12}c_2 tA}{L}$$

or

$$\frac{L}{A} = \frac{D_{12}c_2 t}{\text{Amount diffused}} = \frac{2.14 \times 10^{-5} \times c_2 \times 30 \times 60}{c_2 \times 0.26} = 0.148 \text{ cm}^{-1}$$

Although the diaphragm cell has been used by some researchers to measure diffusion coefficients in polymer solutions, the results are likely to be influenced by the mechanical stirring of the solutions, which can cause extension and orientation of the polymer molecules. Polymer molecules can also adsorb on the membrane. As a consequence, it is preferable to use methods that do not require the fluid to be subjected to any shear stress. This can be achieved in *free-diffusion* experiments [15].

If, in the situation shown in Figure 13.2, the two compartments were infinitely long and the barrier separating them infinitely thin, then initially c would equal $c_1(0)$ for $z > 0$ and c would equal $c_2(0)$ for $z < 0$; here, z is measured from the plane separating the two solutions and taken to be positive in the upward direction. If the barrier were instantly removed at $t = 0$, there would be interdiffusion and the time dependence of the concentration would be given by the solution of Eq. (13.2.11). This situation is termed free diffusion because the solute concentration remains unchanged at the two ends of the cell. Thus, c equals $c_1(0)$ at $z = \infty$ and c equals $c_2(0)$ at $z = -\infty$ for all times. To obtain $c(z, t)$, we use a combination of variables as the new independent variable [16]:

$$\xi = \frac{z}{\sqrt{4D_{AB}t}} \tag{13.3.6}$$

so that Eq. (13.2.11) becomes the following, in molar units:

$$\frac{d^2c}{d\xi^2} + 2\xi \frac{dc}{d\xi} = 0 \tag{13.3.7}$$

subject to $c(\infty) = c_1(0)$ and $c(-\infty) = c_2(0)$.

The solution to Eq. (13.3.7) is given as follows [17,18]:

$$\frac{c - \bar{c}}{c_1(0) - \bar{c}} = \text{erf}\left(\frac{z}{\sqrt{4D_{AB}t}}\right) \tag{13.3.8}$$

in which $\bar{c} = [c_1(0) + c_2(0)]/2$ and erf is the error function,

$$\text{erf } x = \frac{2}{\sqrt{\pi}} \int_0^x e^{-u^2} \, du \tag{13.3.9}$$

and D_{AB} is obtained by comparing the measured concentration profile with Eq. (13.3.8).

Example 13.3: In a free-diffusion experiment, if $c_1(0) = 0$, how does the flux of the species diffusing across the plane that initially separated the two solutions vary with time?

Solution: Because $c_1(0) = 0$, Eq. (13.3.8) becomes

$$\frac{c}{c_2(0)} = \frac{1}{2}\left[1 - \text{erf}\left(\frac{z}{\sqrt{4D_{AB}t}}\right)\right]$$

The flux across the plane at $z = 0$ is given by Eq. (13.2.1) as follows:

$$J = -D_{AB}\frac{dc}{dz}\bigg|_{z=0}$$

Carrying out the differentiation gives

$$\frac{dc}{dz} = -\frac{c_2(0)}{\sqrt{\pi}}\frac{1}{\sqrt{4D_{AB}t}}\exp\left(-\frac{z^2}{4D_{AB}t}\right)$$

Evaluating this expression at $z = 0$ and introducing the result in the equation for the flux gives the following:

$$J = \frac{c_2(0)}{2}\sqrt{\frac{D_{AB}}{\pi t}}$$

A large number of optical methods are available for measuring the time-dependent concentration profiles for diffusion in solutions [19]; the accuracy of interferometric techniques is very good; data can be obtained with a precision of 0.1% or better [19]. For the interdiffusion of polymer melts, however, the number of techniques is very limited because diffusion coefficients can be as low as 10^{-15} cm^2/sec; this means that the depth of penetration measured from the interface is very small even if the experiment is allowed to run for several days. Analysis methods that have the necessary resolution capability include infrared microdensitometry, [20], forward recoil spectrometry, [21], and marker displacement used in conjunction with Rutherford backscattering spectrometry [22]. If the diffusion coefficient is large, we can also use radioactive labeling and nuclear magnetic resonance [20]. Clearly, making measurements of diffusion coefficients in polymer melts is a nontrivial exercise.

A third popular method of measuring diffusion coefficients in the liquid state (especially self-diffusion coefficients of ions in dilute solution) is the open-ended capillary of Anderson and Saddington [23]. Here, as shown in Figure 13.3, a capillary of length L, closed at the bottom, is filled with solution of a known and

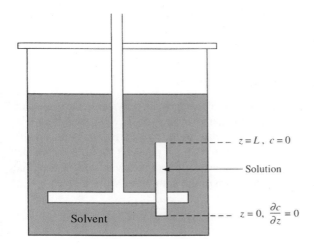

FIGURE 13.3 The open-ended capillary.

uniform concentration, c_0, and immersed in a large tank of pure solvent. Because of the concentration difference, solute diffuses out of the capillary and into the tank, but the tank concentration remains essentially unchanged at zero. After diffusion has taken place for a few hours to a few days, the capillary is withdrawn and the average solute concentration, $\bar{c}$, determined. This measured quantity is then related to the diffusion coefficient. An expression for $\bar{c}/c_0$ is derived by solving Eq. (13.2.11) by the method of separation of variables [19,24,25] subject to the initial and boundary conditions,

$$c = c_0 \quad \text{for } 0 < z < L \quad \text{when } t = 0$$
$$\frac{\partial c}{\partial z} = 0 \quad \text{for } z = 0$$
$$c = 0 \quad \text{for } z = L$$

The solution is as follows [24,25]:

$$c = \frac{4c_0}{\pi} \sum_{n=0}^{\infty} \frac{(-1)^n}{2n+1} \exp\left(-\frac{(2n+1)^2 \pi^2}{4L^2} D_{AB} t\right) \cos\left(\frac{(2n+1)\pi z}{2L}\right)$$

(13.3.10)

which can be integrated over the length of the capillary to yield

$$\frac{\bar{c}}{c_0} = \frac{8}{\pi^2} \sum_{n=0}^{\infty} \frac{1}{(2n+1)^2} \exp\left(-\pi^2 (2n+1)^2 \frac{D_{AB} t}{4L^2}\right)$$

(13.3.11)

The solute is usually radioactively labeled and the concentration determined using a scintillation counter. Other methods of measuring the diffusion coefficient can be found in standard books [12,18,19,25].

13.3.2 Diffusion in Solid Polymers

Diffusion of gases and vapors through solid, nonporous polymers is a three-step process. In the first step, the gas has to dissolve in the polymer at the high-partial-pressure side. Then, it has to diffuse as a solute to the low-partial-pressure side. In the third step, the solute evaporates back to the gas phase. Thus, if we consider steady-state diffusion through a membrane of thickness L exposed to a partial pressure difference Δp, the mass flux through the membrane will be given by Eq. (13.2.1) as follows:

$$J = DS\frac{-\Delta p}{L} \tag{13.3.12}$$

where S is the solubility of the gas in the polymer at pressure p such that the concentration of the gas in the polymer is given by

$$c = Sp \tag{13.3.13}$$

For simple gases above their critical temperature and dissolved in rubbery polymers, S is Henry's law constant, which is independent of p.

It is evident from Eq. (13.3.12) that a measurement of the steady-state flux alone does not yield D, but instead gives the product DS, called the *permeability*. To obtain the diffusivity, we have to know the value of the solubility or make one additional measurement; this is usually the time lag before a steady state is reached in the permeation experiment [26]. The time lag can be related to D by solving Eq. (13.2.11) for one-dimensional transient diffusion through the initially solute-free membrane, subject to the boundary conditions (see Fig. 13.4):

$$c(0, t) = c_1$$
$$c(L, t) = 0$$
$$c(0 < x < L, 0) = 0$$

The time-dependent concentration profile can be shown to be [25,26]:

$$c = c_1\left(1 - \frac{x}{L}\right) - \frac{2c_1}{\pi}\sum_{n=1}^{\infty}\frac{1}{n}\sin\left(\frac{n\pi x}{L}\right)\exp\left(-\frac{Dn^2\pi^2 t}{L^2}\right) \tag{13.3.14}$$

Note that as $t \to \infty$, the exponential terms vanish and we have the expected linear concentration profile. The flux of gas leaving the membrane is $-D(\partial c/\partial x)$

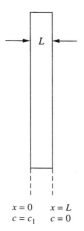

$x = 0$ $x = L$
$c = c_1$ $c = 0$

FIGURE 13.4 Gas diffusion through a membrane.

evaluated at $x = L$. When this term is integrated over time, we get $M(t)$, the amount of material that has actually passed through the membrane in time t:

$$\frac{M(t)}{Lc_1} = \frac{Dt}{L^2} - \frac{1}{6} - \frac{2}{\pi^2} \sum_1^\infty \frac{(-1)^n}{n^2} \exp\left(-\frac{Dn^2\pi^2 t}{L^2}\right) \qquad (13.3.15)$$

which reduces to the equation of a straight line as $t \to \infty$; that is,

$$M(t) = \frac{Dc_1}{L} t - \frac{Lc_1}{6} \qquad (13.3.16)$$

Clearly, this straight line intersects the t axis at the point $t = L^2/6D$, and this allows us to obtain the diffusivity D from a plot of $M(t)$ versus time.

Example 13.4: Guo et al. have used a permeation cell containing a 0.75-mm silicone rubber membrane to measure the cumulative mass of o-xylene vapor passing through the initially solute-free membrane at 303 K [27]. These data are shown in Figure 13.5. Determine the diffusion coefficient.

Solution: On extrapolating the line marked vapor back toward the origin, we find that this line intersects the time axis at a value of t equal to 40 minutes. Thus,

$$D = \frac{L^2}{6t} = \frac{(0.075)^2}{6 \times 40 \times 60} = 3.9 \times 10^{-7} \frac{cm^2}{sec}$$

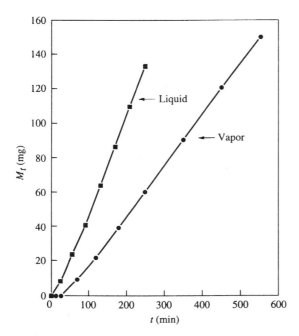

FIGURE 13.5 Permeation curves of o-xylene in silicone rubber at 303 K. (From Ref. 27.) From J. Appl. Polym. Sci., vol. 56, Guo, C. J., D. DeKee, and B. Harrison: Diffusion of organic solvents in rubber membranes measured via a new permeation cell, Copyright © 1995 by John Wiley & Sons, Inc. Reprinted by permission of John Wiley & Sons, Inc.

The measurement of water vapor transmission rate is very important for assessing the barrier properties of polymer films used in packaging applications. A very simple method of accomplishing this is to use the film of interest to seal the mouth of a dish containing a desiccant. The rate of moisture transport is determined simply by periodically weighing the dish. A more sophisticated technique used in commercial instruments is to generate the flow of air of controlled relative humidity on one side of the membrane and the flow of moisture-free nitrogen gas on the other side; the mass transfer rate is computed with the use of a sensor that detects the increase in moisture level of the nitrogen leaving the diffusion cell. Such instruments can operate over wide ranges of temperature, flow rate, and relative humidity, and they come equipped with multiple cells and data analysis software.

The permeation method works well for simple gases diffusing through polymers above the glass transition temperature because in this situation, the solubility and diffusivity are both constant, independent of concentration, at a given temperature. With increasing temperature, the diffusion coefficient increases, but the solubility decreases, with the result that the permeability can exhibit increases or decreases with temperature. This behavior can, however, be predicted from a limited amount of experimental data, because both the diffusivity and permeability obey the Arrhenius relation [28]. The process of predicting the permeability is also helped by the observation that, for any given pair of gases, the ratio of the permeabilities is constant, independent of the type of polymer used [29]. This is also found to be true for the ratio of the permeability activation energies [30].

In contrast to simple gases, organic vapors (at temperatures below the critical temperature) of the kind encountered in Example 13.4 tend to interact with polymers and often cause them to swell. A result of this interaction is that, even for the polymers above the glass transition temperature, the diffusion coefficient becomes concentration dependent and Henry's law is no longer obeyed. Consequently, the use of Eq. (13.3.16) gives the diffusivity in the low-concentration regime. Although analyses have been developed to account for this variation of D in a permeation experiment [31–33], the preferred method for this situation appears to be the sorption experiment. Here, a polymer film initially at equilibrium with a given partial pressure of vapor is suddenly exposed to a different constant partial pressure of the same vapor [34]. The diffusion coefficient is obtained from analyzing data on the gain or loss in mass of the polymer film with time.

If a membrane having an initially uniform solute concentration c_0 and thickness $2L$ is exposed to vapor at a different partial pressure such that the surfaces located at $x = \pm L$ instantly reach a constant concentration c_1, then the solution of Eq. (13.2.11) is given as follows [25]:

$$\frac{c - c_0}{c_1 - c_0} = 1 - \frac{4}{\pi}\sum_{n=0}^{\infty}\frac{(-1)^n}{2n+1}\exp\left(-\frac{D(2n+1)^2\pi^2 t}{4L^2}\right)\cos\left(\frac{(2n+1)\pi x}{2L}\right)$$

$$(13.3.17)$$

Note that this situation is mathematically equivalent to the open capillary considered earlier. Indeed, Eq. (13.3.10) is recovered by setting c_1 equal to zero in Eq. (13.3.17).

The expression in Eq. (13.3.17) can be used to calculate the ratio of mass of vapor, $M(t)$, that has diffused into the membrane at any time to the equilibrium mass of vapor, M_∞. At short times, this quantity is given by

$$\frac{M(t)}{M_\infty} = 2\left(\frac{Dt}{L^2}\right)^{1/2}\left[\pi^{-1/2} + 2\sum_{n=1}^{\infty}(-1)^n\,\text{ierfc}\left(\frac{nL}{\sqrt{Dt}}\right)\right]$$

$$(13.3.18)$$

FIGURE 13.6 Transmission electron micrograph of vinyl ester/clay nanocomposite containing 0.5 wt% organically treated montmorillonite. The magnification is 100,000×. (From Ref. 37.)

which says that $M(t)/M_\infty$ is a unique function of Dt/L^2. Here, ierfc(x) is given by $(1/\sqrt{\pi})e^{-x^2} - x(1 - \text{erf } x)$. On plotting Eq. (13.3.18), we find that [18,34]

$$\frac{M(t)}{M_\infty} = 2\left(\frac{Dt}{\pi L^2}\right)^{1/2} \tag{13.3.19}$$

until the point that $M(t)/M_\infty$ equals 0.5. This linear relationship on logarithmic coordinates facilitates the evaluation of D from experimental data. Details of how we might conduct such experiments above the polymer glass transition temperature have been provided by Duda et al. [35]. Duda and Vrentas have also shown how we might determine the concentration dependence of diffusion coefficients from a minimum amount of data from sorption experiments [36].

Equations (13.3.18) and (13.3.19) have been employed by Shah et al. to determine the diffusivity of water in vinyl ester nanocomposites [37]. As discussed in Chapter 11, nanocomposites are made by dispersing platelets of montmorrilonite (clay) in polymer, and Figure 13.6, taken from the work of Shah et al., shows individual platelets in the form of dark lines. Rectangular samples of the nanocomposite were immersed in 25°C water, and the increase in weight was noted with increasing time of immersion. Representative sorption results for nanocomposites containing 0.5 wt% clay are shown in Figure 13.7, and, as expected, data on samples of different thicknesses superpose when plotted as $M(t)/M_\infty$ versus $t^{1/2}/L$. The value of the diffusion coefficient computed using Eq. (13.3.19) can be inserted into Eq. (13.3.18) to give the complete theoretical curve, and this is also shown in Figure 13.7; the fit between Fickian theory and experiment is excellent. It is also remarkable that the addition of such a minute amount of clay is found to reduce the moisture diffusion coefficient by 50%.

In the case of a desorption experiment (during the final stages), the equivalent form of Eq. (13.3.18) is given by the following [18]:

$$\frac{d}{dt}\{\ln[M(t) - M_\infty]\} = -\frac{\pi^2 D}{4L^2} \tag{13.3.20}$$

and a plot of $\ln[M(t) - M_\infty]$ versus time should approach a straight line, whose slope is given by the right-hand side of Eq. (13.3.20). The use of this technique for diffusivity measurement is illustrated in the next example. Extensive data on the diffusion coefficient of gases and vapors in polymers can be found in the book edited by Crank and Park [18,34].

Example 13.5: Figure 13.8 shows the gravimetric data of Saleem et al. for the desorption of chloroform from a 0.15-mm-thick film of low-density polyethylene at 25°C [38]. Determine the diffusivity.

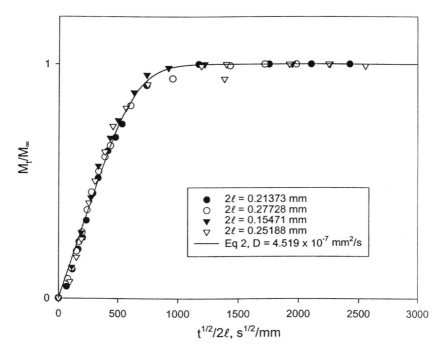

FIGURE 13.7   Water sorption curve at 25°C for the nanocomposite shown in Fig. 13.6. (From Ref. 37.)

Solution: The slope of the straight line in Figure 13.8 is

$$\frac{\ln 10^{-2} - \ln 10^{-1}}{(120 - 70)60} = -7.68 \times 10^{-4}$$

and this must equal $-\pi^2 D/(0.015)^2$. As a result, $D = 1.75 \times 10^{-8}$ cm^2/sec.

13.4   DIFFUSIVITY OF SPHERES AT INFINITE DILUTION

Theoretical prediction of the diffusion coefficient of spheres moving through a low-molecular-weight liquid is a problem that was examined by Einstein at the turn of this century [39]. This situation is of interest to the polymer scientist because isolated polymer molecules in solution act as random coils.

The analysis begins by showing that the osmotic pressure is the driving force for diffusion. This is done by carrying out the experiment illustrated in

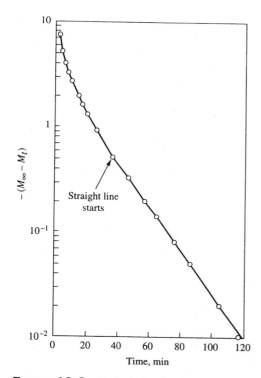

FIGURE 13.8 Typical plot for the desorption of chloroform at 25°C. (From Ref. 38.) From J. Appl. Polym. Sci., vol. 37, Saleem, M., A.-F. A. Asfour, D. DeKee, and B. Harrison: Diffusion of organic penetrants through low density polyethylene (LDPE) films: Effect of shape and size of the penetrant molecules, Copyright © 1989 by John Wiley & Sons, Inc. Reprinted by permission of John Wiley & Sons, Inc.

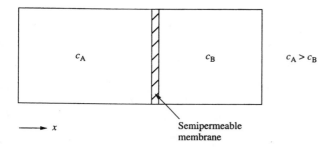

FIGURE 13.9 Diffusion through a semipermeable membrane of area A.

Figure 13.9. If a piston of area A and fabricated from a semipermeable membrane separates dilute solutions of two different molar compositions, there is a tendency for the piston to move toward one of the two sides. This happens because there is the transport of solvent from the dilute to the concentrated side as an attempt to eliminate the difference in concentration. Thus, if $c_A > c_B$, the piston tends to move to the right as the solvent diffuses into the compartment on the left. If we wanted to prevent the piston from moving, we would have to apply an external force equal to $(p_A - p_B)A$ to the piston in the negative x direction; here $p_A - p_B$ is the difference in osmotic pressure across the membrane. This clearly shows that osmotic pressure differences cause diffusion. Indeed, diffusion stops when the osmotic pressure difference is balanced by the application of an external force.

Now, consider the situation shown in Figure 13.10. The slice of material of thickness dx is taken from a dilute solution in which there is a concentration gradient in the x direction. Because there is a concentration gradient, there must necessarily be an osmotic pressure gradient as well. Because the osmotic pressure is RTc, where c is the molar concentration of spheres [see Eq. (8.3.18)], the difference in force across the slice is given by

$$-ART\,\frac{dc}{dx}\,dx$$

and it points in the negative x direction if c increases with increasing x. As the total number of spheres in the slice equals $A\,dx\,cN_A$, where N_A is Avogadro's number, the force acting on each sphere is

$$\frac{\text{Force}}{\text{Sphere}} = \left(-ART\,\frac{dc}{dx}\,dx\right)(A\,dx\,cN_A)^{-1} = -\frac{kT}{c}\frac{dc}{dx} \tag{13.4.1}$$

in which k is Boltzmann's constant.

The right-hand side of Eq. (13.4.1) must be the drag force exerted by the fluid on the sphere. This is given by Stokes' law and equals $6\pi\eta vR$, because the

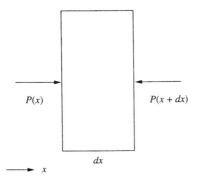

$P(x)$ $P(x + dx)$

dx

x

FIGURE 13.10   Force balance across a slice of fluid having a concentration gradient.

spheres move at a velocity v and do not accelerate [13]. Equating the two expressions for the drag force gives

$$vc = -\frac{kT}{6\pi\eta R}\frac{dc}{dx} \tag{13.4.2}$$

where η is the suspending liquid viscosity and R is the sphere radius. The left-hand side of Eq. (13.4.2.), however, is the molar flux of spheres, which is given in terms of the diffusion coefficient D by Eq. (13.2.1). Clearly, then,

$$D = \frac{kT}{6\pi\eta R} \tag{13.4.3}$$

which is known as the Stokes–Einstein equation. It is valid only for very dilute solutions, because only under these conditions is the osmotic pressure expression that is used correct. A further consequence of dilute conditions is that D is both the mutual diffusion coefficient and the self-diffusion coefficient of the spheres at infinite dilution. Also, the use of Stokes' law requires that the spheres be rigid and the flow regime be one of creeping flow. In addition, the size of the spheres has to be large in comparison to the size of the molecules of the suspending liquid. Under these conditions, D is typically of the order of 10^{-5} cm^2/sec. Equation (13.4.3) does a fair job of predicting not only the absolute value of D but also the temperature dependence of this quantity. More accurate predictions can be made, however, with the help of empirical relationships [40,41].

Example 13.6: Northrop and Anson measured the diffusion coefficient of hemoglobin in water to be 4.86×10^{-7} cm^2/sec at 5°C [14]. Estimate the radius of the hemoglobin molecule. The viscosity of water at 5°C is 0.0152 P.

Solution: Using Eq. (13.4.3) gives us the radius,

$$R = \frac{kT}{6\pi\eta D} = \frac{8.3 \times 10^7 \times 278}{6.02 \times 10^{23} \times 6 \times \pi \times 0.0152 \times 4.86 \times 10^{-7}}$$
$$= 2.75 \times 10^{-7}\text{cm}$$

If the diffusing species is not a sphere, we merely need to use an appropriately defined equivalent radius R_e instead of R in Eq. (13.4.3) [41]. Thus, we have the following:

$$D = \frac{kT}{6\pi\eta R_e} \tag{13.4.4}$$

Equation (13.4.4) gives the correct order of magnitude for D. However, the diffusivity of nonspherical particles such as ellipsoids may be different in different directions.

For infinitely dilute, linear, flexible-chain polymer solutions under theta conditions, we have the following [41,42]:

$$R_e = 0.676 \langle s^2 \rangle_0^{1/2} \tag{13.4.5}$$

in which $\langle s^2 \rangle_0$ is the mean square radius of gyration at theta conditions and can be measured using light-scattering techniques. Note that Eq. (13.4.5) also reveals the molecular-weight dependence of the diffusivity. In general, though, we can rewrite Eq. (13.4.3) as follows:

$$D = \frac{kT}{f} \tag{13.4.6}$$

where f is the friction coefficient. This is the force needed to drag the diffusant through the liquid at unit speed.

An example of the use of Eq. (13.4.6) is the prediction of the diffusion coefficient of spheres moving through a constant-viscosity elastic liquid, [43], (i.e., an extremely dilute polymer solution). Chhabra et al. have shown that sphere drag decreases with increasing fluid elasticity so that [44]

$$f = 6\pi \eta R X_e \tag{13.4.7}$$

where $X_e \approx 0.74$ for highly elastic liquids. We would expect this to lead to a 35% increase in the diffusivity compared to the situation for which X_e equals unity. Indeed, Wickramasinghe et al. have found this to be the case [15]. For more concentrated polymer solutions (where the solution viscosity differs significantly from the solvent viscosity), the bulk or macroscopic viscosity of the surrounding medium is not indicative of the actual flow resistance experienced by the solute. This is because a solute sees a local environment of a sea of solvent with polymer molecules serving merely to obstruct the motion of this particle in a minor way. With reference to Eq. (13.4.7), therefore, it is the local or microscopic viscosity that is relevant rather than the macroscopic viscosity [45].

In closing this section, we add that the Stokes–Einstein equation breaks down when the ratio of the solute-to-solvent radius becomes less than 5. Errors become quite large in high-viscosity solvents, and we find that the product $D\eta^{2/3}$ tends to become a constant [46].

13.5 DIFFUSION COEFFICIENT FOR NON-THETA SOLUTIONS

In a good solvent, polymer coils expand, increasing the effective molecular radius, which results in a lowering of the diffusion coefficient. Coil expansion is usually expressed as follows [see Eq. (8.6.14)]:

$$\langle s^2 \rangle = \alpha^2 \langle s^2 \rangle_0 \tag{13.5.1}$$

where $\langle s^2 \rangle$ is the mean square radius of gyration under conditions of infinite dilution. In view of Eqs. (13.4.4), (13.4.5), and (13.5.1),

$$D = \frac{D_\theta}{\alpha} \tag{13.5.2}$$

where D_θ is the infinite dilution diffusivity under theta conditions and α (the coil expansion factor) may be either measured experimentally or estimated theoretically. To carry out the latter exercise, we can use the expression of Yamakawa and Tanaka [47], in conjunction with intrinsic viscosity data represented in terms of the Mark–Houwink–Sakurada expression. Further details are available in Ref. 42.

Because dilute polymer solutions are encountered more frequently than infinitely dilute solutions, we need to be able to predict the diffusivity of dilute solutions as well. The diffusion coefficient in a solution having a mass concentration of polymer equal to ρ can be written in series form as follows [42]:

$$D(\rho) = D(1 + k_D\rho + \cdots) \tag{13.5.3}$$

and the diffusion coefficient may increase or decrease relative to its value at infinite dilution. In Eq. (13.5.3), k_D is given as follows:

$$k_D = 2A_2M - k_s - b_1 - 2\hat{v}_{20} \tag{13.5.4}$$

where A_2 is the second virial coefficient, M is the polymer molecular weight, $\hat{v}_{20}$ is the partial specific volume of the polymer at zero polymer concentration, and k_s is a coefficient in a series expansion for the concentration dependence of the friction coefficient:

$$f = f_0(1 + k_s\rho + \cdots) \tag{13.5.5}$$

Furthermore, b_1 is defined by

$$\hat{v}_1 = \hat{v}_1^0(1 + b_1\rho + \cdots) \tag{13.5.6}$$

in which $\hat{v}_1$ is the partial specific volume of the solvent and $\hat{v}_1^0$ is the specific volume of the pure solvent.

In the preceding discussion, b_1 and $\hat{v}_2^0$ can be obtained from density measurements, A_2 is available in the literature for many polymer–solvent pairs, and k_s can be estimated by the procedures suggested by Vrentas and Duda [42].

13.6 FREE-VOLUME THEORY OF DIFFUSION IN RUBBERY POLYMERS

As mentioned in Section 13.3.2, the diffusion coefficient of simple gases in polymers above the glass transition temperature is independent of concentration and follows the Arrhenius relation. In addition, Henry's law is obeyed, and solubilities appear to be correlated by the following [48]:

$$\ln S = -4.5 + \frac{10T'}{T} \tag{13.6.1}$$

where T' is the boiling point of the permeate and T is the temperature of interest. Furthermore, the diffusivities of a number of gases in a variety of polymers can be found in the literature as a function of temperature [49]. These factors make it easy to estimate, using limited experimental data, the diffusional properties of a gas in a given rubbery polymer [50]. For organic vapors, however, both the solubility and diffusivity are strongly concentration dependent, and simple correlations in terms of physical properties do not work [51]. Consequently, we need a model to predict the concentration, pressure, and temperature dependence of these quantities. The most successful model appears to be the free-volume theory [42,52,53], which is valid in a temperature range from T_g to 100°C above T_g, provided that the polymer weight fraction exceeds 0.2. Also, the polymer relaxation time taken to be the ratio of the viscosity to the modulus must be small in comparison to the characteristic time needed for diffusion. In this theory, the mutual diffusion coefficient is related to the self-diffusion coefficient, which, in turn, is expressed as a function of the free volume.

To understand free volume, consider Figure 13.11. At any temperature, the volume $\hat{V}$ occupied by unit mass of polymer equals the volume directly occupied by the molecules plus $\hat{V}_F$, the free volume. The latter quantity itself is made up of the interstitial free volume $\hat{V}_{F1}$ and the hole free volume $\hat{V}_{FH}$. At 0 K, there is no free volume, so that

$$\hat{V}(0) = \hat{V}_0 \tag{13.6.2}$$

where $\hat{V}_0$ is the specific occupied volume.

On increasing the temperature above 0 K, the volume increases by homogeneous expansion and by the formation of holes, which are distributed discontinuously throughout the material at any instant. The former kind of change is called the *interstitial free volume*, whereas the latter is called the *hole free volume*. Thus, we have

$$\hat{V} = \hat{V}_0 + \hat{V}_{FI} + \hat{V}_{FH} \tag{13.6.3}$$

so that

$$\hat{V}_{FH} = \hat{V} - (\hat{V}_0 + \hat{V}_{FI}) \tag{13.6.4}$$

In this theory, it is assumed that the hole free volume can be redistributed with no increase in energy and is, therefore, available for molecular transport. For a binary

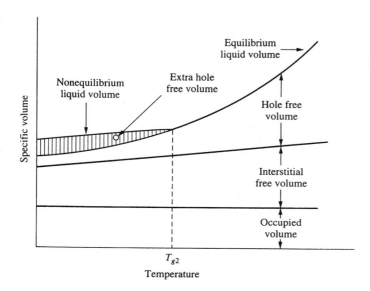

T_{g2}

Temperature

FIGURE 13.11 Illustration of the division of the specific volume of an amorphous polymer. (From Vrentas, J. S., and J. L. Duda: "Molecular Diffusion in Polymer Solutions," *AIChE J.*, vol. 25, 1–24. Reproduced with the permission of the American Institute of Chemical Engineers Copyright © 1979 AIChE. All rights reserved.)

mixture of solute and solvent, the hole free volume per unit mass of the mixture is taken to be

$$\hat{V}_{\mathrm{FH}} = w_1 \hat{V}_{\mathrm{FH1}} + w_2 \hat{V}_{\mathrm{FH2}} \tag{13.6.5}$$

where w_1 and w_2 are the solute and polymer mass fractions, respectively. By postulating separate coefficients of thermal expansion for $\hat{V}$ and $\hat{V}_0 + \hat{V}_{F1}$ and integrating the temperature rate of change of these quantities between T_g and T [54], we can show that

$$\hat{V}_{\mathrm{FH1}} = K_{11}(K_{21} + T - T_{g1}) \tag{13.6.6}$$
$$\hat{V}_{\mathrm{FH2}} = K_{12}(K_{22} + T - T_{g2}) \tag{13.6.7}$$

in which the K terms are constants and the T_g terms are the respective glass transition temperatures.

By appealing to the similarity in the behavior of the diffusion coefficient and the reciprocal of the viscosity [see Eq. (13.4.3)] and by assuming that the statistical redistribution of the free volume opens up voids large enough for diffusion to occur, Cohen and Turnbull obtained the following expression for the self-diffusion coefficient of low-molecular-weight materials [55]:

$$D_1 \propto \exp\left(-\frac{\gamma v_1^*}{V_{FH}^+}\right) \tag{13.6.8}$$

in which γ is a constant, v_1^* is the critical hole free volume needed for a molecule to jump into a new position, and V_{FH}^+ is the hole free volume per molecule. In order to obtain better agreement with experimental data, Macedo and Litovitz [56] multiplied the right-hand side of Eq. (13.6.8) by $\exp(-E^*/kT)$, where E^* is the energy needed by a molecule to overcome attractive forces holding it to its neighbors. For a component in a binary mixture of polymer 2 and solvent 1, this implies that the self-diffusion coefficient is as follows:

$$D_1 = D_0 \exp\left(-\frac{E}{RT}\right) \exp\left(-\frac{\gamma[w_1 \hat{V}_1^0(0) + w_2 \xi \hat{V}_2^0(0)]}{\hat{V}_{FH}}\right) \tag{13.6.9}$$

where

E = molar energy needed to overcome attractive forces
ξ = ratio of critical volume of solvent to that of polymer segment
$\hat{V}_1^0(0)$ = occupied volume per gram of solvent
$\hat{V}_2^0(0)$ = occupied volume per gram of polymer
$\hat{V}_{FH}$ = hole free volume per gram of mixture
γ, D_0 = constant

Finally, following Bearman [57], Duda et al. [58] have shown that when the self-diffusion coefficient of the solvent greatly exceeds that of the polymer (this is generally the case) and provided that $w_1 < 1$, the binary diffusion coefficient D could be related to the self-diffusion coefficient as follows:

$$D = \frac{\rho_2 \hat{V}_2 \rho_1 D_1}{RT}\left(\frac{\partial \mu_1}{\partial \rho_1}\right)_{T,P} \tag{13.6.10}$$

where ρ_1 and ρ_2 are the mass densities of the two components, $\hat{V}_2$ is the partial specific volume of the polymer, and μ_1 is the chemical potential of the solvent (see Chapt. 9). The term μ_1 is given by the Flory–Huggins theory as follows:

$$\mu_1 = \mu_1^0 + RT[\ln(1 - \phi_2) + \chi_1\phi_2^2 + \phi_2] \tag{9.3.30}$$

where $1/m$ has been neglected in comparison with unity. Combining the previous three equations gives the desired result [59]:

$$D = D_0\left[\exp\left(-\frac{E}{RT}\right)\right](1-\phi_1)^2$$

$$\times (1 - 2\chi_1\phi_1)\exp\left(\frac{-\gamma[w_1\hat{V}_1^0(0) + w_2\xi\hat{V}_2^0(0)]}{\hat{V}_{FH}}\right) \quad (13.6.11)$$

where the solvent volume fraction is given by

$$\phi_1 = \frac{w_1\hat{V}_1^0}{w_1\hat{V}_1^0 + w_2\hat{V}_2^0} \quad (13.6.12)$$

in which $\hat{V}_1^0$ and $\hat{V}_2^0$ are the specific volumes of the pure components at the temperature of interest.

By examining Eqs. (13.6.5)–(13.6.7) and (13.6.11) and (13.6.12), we find that the variation of D with temperature and composition is known once the following 12 quantities are known: D_0, E, χ_1, $\hat{V}_1^0(0)$, $\hat{V}_2^0(0)$, ξ, K_{11}/γ, K_{12}/γ, $K_{21} - T_{g1}$, $K_{22} - T_{g2}$, $\hat{V}_1^0$ and $\hat{V}_2^0$. Of these, $\hat{V}_1^0(0)$ and $\hat{V}_2^0(0)$ can be obtained using a group contribution method [60], whereas $\hat{V}_1^0$ and $\hat{V}_2^0$ are obtained from pure-component density data. Also, K_{12}/γ and $K_{22} - T_{g2}$ are estimated from polymer viscosity data because the polymer viscosity η_2 can usually be represented in the form

$$\ln\eta_2 = \ln A_2 + \frac{\gamma\hat{V}_2^0(0)/K_{12}}{K_{22} + T - T_{g2}} \quad (13.6.13)$$

where A_2 is a constant. K_{11}/γ and $K_{21} - T_{g1}$ are obtained from solvent viscosity data using an identical procedure. The interaction parameter χ_1 is determined using techniques discussed in Chapter 9. This leaves three unknowns $-D_0$, E, and ξ. These are evaluated with the help of at least three data points known at two or more temperatures.

This process has been illustrated by Duda et al. for a variety of polymers [59] and Figure 13.12 shows the predictive abilities of their theory for the toluene–polystyrene system. Parameter values used are listed in Table 13.1. The results obtained are excellent. These authors also note that the diffusion coefficient given by Eq. (13.6.11) is insensitive to polymer molecular weight, and there is, therefore, no influence of polydispersity. Furthermore, for semi-crystalline polymers above the glass transition temperature, the polymer may be considered to be made up of two phases—one of which has a zero diffusivity [51]. Thus, if the volume fraction of the crystalline phase is ϕ, the effective diffusivity of the polymer is $D\phi$. Finally, Kulkarni and Mashelkar [63] have proposed an altered free-volume-state model that seeks to provide a unified

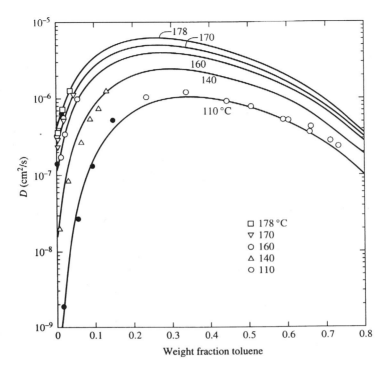

FIGURE 13.12 Test of predictive capabilities of free-volume theory using data for the toluene-polystyrene system. Only data points represented by solid symbols were used to obtain free-volume parameters. (From Duda, J. L., J. S. Vrentas, S. T. Ju, and H. T. Liu: "Prediction of Diffusion Coefficients for Polymer-Solvent Systems," *AIChE J.*, vol. 28, pp. 279–285. Reproduced with the permission of the American Institute of Chemical Engineers Copyright © 1982 AIChE. All rights reserved.)

framework for diffusion in polymer solutions, polymeric gels, blends of polymer melts, and structured solid polymers.

13.7 GAS DIFFUSION IN GLASSY POLYMERS

The diffusion behavior of simple gases (i.e., gases above the critical temperature) in glassy polymers is often quite different from the behavior of the same gases in the same polymers but above the polymer glass transition temperature [64]. In particular, gas solubility shows negative deviations from Henry's law, and the dissolution process is much more exothermic in glassy polymers. These and other

TABLE 13.1 Parameters of Free-Volume Theory for the Toluene (1)–Polystyrene (2) System

Parameter	Value
$\hat{V}_1^0(0)$	$0.917 \text{ cm}^3\text{g}$
$\hat{V}_2^0(0)$	$0.850 \text{ cm}^3/\text{g}$
K_{11}/γ	$2.21 \times 10^{-3} \text{ cm}^3/\text{g K}$
K_{12}/γ	$5.82 \times 10^{-4} \text{ cm}^3/\text{g K}$
$K_{21} - T_{g1}$	-103 K
$K_{22} - T_{g2}$	-327 K
χ_1	0.40
ξ	0.53
D_0	$6.15 \times 10^{-2} \text{ cm}^2/\text{sec}$
E	$5.26 \times 10^3 \text{ cal/g mol}$
$\hat{V}_1^0$ versus T	Data in Ref. 61
$\hat{V}_2^0$ versus T	Data in Ref. 62

Source: Ref. 59.

anomalies have been reviewed by several authors [64–66], and they can be reconciled with the help of the dual-sorption theory [65].

The basic premise of the dual-sorption theory is that a glassy polymer contains *microvoids*, or holes created during the process of cooling the polymer below its glass transition temperature. Therefore, gas molecules can either dissolve in the bulk polymer or go into these holes. Consequently, glassy polymers have a high gas solubility due to the availability of dual-sorption modes. It is found that gas dissolution in the matrix follows Henry's law, whereas that in the holes displays a Langmuir-type dependence. Thus, at equilibrium, the total gas concentration in the polymer can be separated into a normal sorption term c_D and a hole contribution c_H [65]:

$$c = c_D + c_H \tag{13.7.1}$$

where

$$c_D = k_D p \tag{13.7.2}$$

which is similar to Eq. (13.3.13), with k_D being Henry's law constant. The term c_H in Eq. (13.7.1) is given as follows:

$$c_H = \frac{c_H' b p}{1 + bp} \tag{13.7.3}$$

in which c_H is called the *hole-saturation constant* and b is called the *hole-affinity constant*.

From Eq. (13.7.3), we note that when $bp \ll 1$, we have

$$c = (k_D + c'_H b)p \qquad (13.7.4)$$

and when $bp \gg 1$, we have

$$c = k_D p + c'_H \qquad (13.7.5)$$

so that a plot of c versus p at constant temperature consists of straight-line sections, at both low and high pressures, connected by a nonlinear region. This is shown in Figure 13.13 using the solubility data of methane in glassy polystyrene [67].

In view of the foregoing, k_D can be obtained from the slope of the sorption isotherm at high pressures. A knowledge of k_D then allows us to calculate c_H from c, and this leads to an evaluation of the other two model parameters by recasting Eq. (13.7.3) as follows:

$$\frac{p}{c_H} = \frac{1}{c'_H b} + \frac{p}{c'_H} \qquad (13.7.6)$$

so that c'_H and b can be determined from the slope and intercept of the straight-line plot of p/c_H versus p.

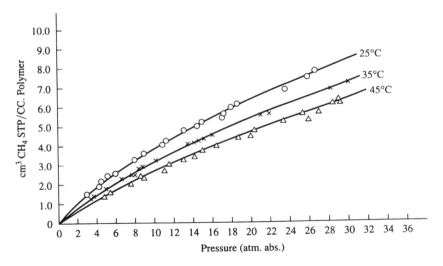

FIGURE 13.13 Solubility of methane in oriented polystyrene. (Reprinted from *J. Membrane Sci.*, vol. 1, Vieth, W. R., J. M. Howell, and J. H. Hsieh: "Dual Sorption Theory," pp. 177–220, 1976 with kind permission from Elsevier Science—NL. Sara Burgerhartstraat 25, 1055 KV Amsterdam, The Netherlands.)

As far as the rate of penetrant transport through glassy polymers is concerned, it is assumed that material contained in the microvoids or holes is completely immobilized and does not contribute to the diffusive flux. It is further assumed that diffusion of the mobile species follows Fick's first law in terms of the concentration gradient of c_D and a constant diffusivity. Thus, the flux in the x direction is

$$J = -D \frac{\partial c_D}{\partial x} \qquad (13.7.7)$$

which when coupled with a mass balance leads to an analogous form of Eq. (13.2.11):

$$\frac{\partial}{\partial t}(c_D + c_H) = D \frac{\partial^2 c_D}{\partial x^2} \qquad (13.7.8)$$

If we assume that there is local equilibrium between the dissolved (mobile) species and the immobilized species, we can eliminate p between Eqs. (13.7.2) and (13.7.3) to obtain a relation between c_H and c_D:

$$c_H = \frac{(c'_H b/k_D)c_D}{1 + (bc_D/k_D)} \qquad (13.7.9)$$

which can be substituted into Eq. (13.7.8) to give a nonlinear partial differential equation involving c_D as the only dependent variable:

$$\left(1 + \frac{c'_H(b/k_D)}{[1 + (b/k_D)c_D]^2}\right) \frac{\partial c_D}{\partial t} = D \frac{\partial^2 c_D}{\partial x^2} \qquad (13.7.10)$$

There is no analytical solution to this equation. However, the nonlinearities disappear at both sufficiently low and sufficiently high pressures (see Example 13.7). Thus, we can obtain the diffusion coefficient D by conducting transient experiments under these conditions or by carrying out steady-state experiments. For the general case, though, Eq. (13.7.10) has to be solved numerically, and this is likely to be the situation in most cases of practical interest.

Example 13.7: If we use Eq. (13.3.19) to determine the diffusion coefficient at low gas pressures in the presence of a dual-sorption mechanism, will the measured value equal the quantity D that appears in Eq. (13.7.10)?

Solution: At very low pressures, c_H is equal to $c'_H bp$, which, in combination with Eq. (13.7.2), means that

$$c_H = \frac{c'_H b}{k_D} c_D$$

When this result is introduced into Eq. (13.7.8), we find that

$$\frac{\partial c_D}{\partial t} = \frac{D}{1 + (c'_H b/k_D)} \frac{\partial^2 c_D}{\partial x^2}$$

and it is clear that the measured diffusion coefficient will be less than D and will equal $D/(1 + c'_H b/k_D)$.

Equation (13.7.10) was solved numerically by Subramanian et al. to predict the increase in mass of a membrane of thickness $2L$, initially free of solute, and exposed to a gas pressure such that the concentration at $x = \pm L$ instantly reached a constant value c_1 [68]. Computations in terms of M_t/M_∞ [see also Eq. (13.3.19)] versus $\sqrt{t}$ were compared with experimental data available in the literature on the absorption of water vapor into a polyimide film. Excellent agreement was found between the two sets of numbers when independently measured values of b, c'_H, and k_D were used in Eq. (13.7.10). The results are shown in Figure 13.14.

In closing this section, it should be noted that the dual-sorption theory has relevance to gas separation by glassy polymers, dyeing of textile fibers, design of pressurized plastic beverage containers, and other applications [65].

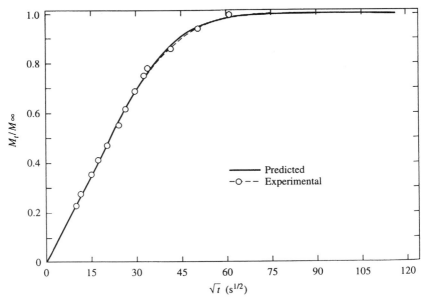

FIGURE 13.14 Fractional absorption of water vapor in Kapton® polyimide at 30°C: membrane thickness $\delta = 5 \times 10^{-3}$ cm (2.0 mils); pressure $p = 9.55$ cm Hg. (From Ref. 68). From J. Polym. Sci. Polym. Phys. Ed., vol. 27, Subramanian, S., J. C. Heydweiller, and S. A. Stern: Dual-mode sorption kinetics of gases in glassy polymers, Copyright © 1989 by John Wiley & Sons, Inc. Reprinted by permission of John Wiley & Sons, Inc.

13.8 ORGANIC VAPOR DIFFUSION IN GLASSY POLYMERS: CASE II DIFFUSION

Organic vapors often interact with polymers and cause them to swell. When sorption and diffusion take place below the polymer glass transition temperature, the rigidity of the polymer chains implies that a significant amount of time is needed before even the surface concentration of the penetrant reaches its equilibrium value. In other words, there are time-dependent effects that result from the polymer relaxation time being comparable to the time scale of diffusion. When polymer relaxation effects dominate the process of vapor transport, the diffusion process becomes quite non-Fickian. This is seen most clearly in a sorption experiment. Based on Eq. (13.3.19), we would expect the initial weight gain $M(t)$ of a glassy polymer sheet exposed to an organic vapor to follow the expression

$$M(t) = bt^n \tag{13.8.1}$$

where n is equal to 0.5, t is time, and b is a constant. Instead, we find that n equals unity. This extreme of anomalous behavior has been called *case II diffusion* [69], as opposed to the expected situation with n equal to 0.5, which is termed *case I diffusion*.

Another aspect of case II diffusion is that there is a sharp boundary separating the inner glassy core of the polymer from the outer solvent-swollen rubbery shell [64,69]. Furthermore, as diffusion proceeds inward, the boundary between the swollen gel and the glassy core moves at constant velocity; this has been determined with optical experiments [70]. Thomas and Windle, for example, studied the weight gain of sheets of polymethyl methacrylate (see Fig. 13.15) suspended in a bath of methanol at 24°C [70,71]. They also observed the movement of the methanol front by coloring the methanol with iodine. Some of their results are shown in Figure 13.16, and these reveal a linear weight gain and a linear front velocity. Note that front penetration is represented two ways in Figure 13.16: (1) by the distance l from the surface of the swollen polymer to the advancing front and (2) by the distance d from the original position of the specimen surface to the front (see Fig. 13.15). Additional optical densitometer experiments demonstrated the existence of step concentration profiles with negligible concentration gradients behind the advancing fronts [70]. Birefringence measurements also indicated that swollen material deformed mechanically by stretching in a direction normal to the front.

Although a large number of theories have been advanced to explain case II diffusion (see articles cited in the literature [64,72], the most successful one appears to be the model of Thomas and Windle [72,73]. Here it is assumed that the volume fraction ϕ of penetrant in the polymer depends on time and only one spatial direction x; that is, $\phi = \phi(x, t)$. The driving force for diffusion is the

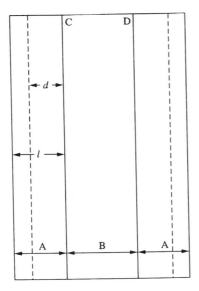

FIGURE 13.15 Diagram defining the two front penetration parameters l and d. (A), swollen region; (B), glassy region; (– – –), original position of surfaces; (C, D), penetrating fronts. (Reprinted from *Polymer*, vol. 19, Thomas N., and A. H. Windle: "Transport of Methanol in Poly(Methyl Methacrylate)," pp. 255–265, Copyright 1978, with kind permission from Elsevier Science Ltd., The Boulevard, Langford Lane, Kidlington OX5 1GB, UK.)

difference in concentration $(\phi_e - \phi)$ across the rubbery–glassy interface, where ϕ_e is the equilibrium value of ϕ. This concentration difference results in an osmotic pressure difference π, which is resisted by the polymer chains on the glassy side of the interface. With time, solute molecules penetrate into the glassy portion, polymer chains relax, and the osmotic pressure difference reduces consequent to a reduction in the concentration difference. Writing an equation for this mechanical equilibrium gives a relation for the time rate of change of ϕ at a fixed position.

As shown in Section 8.3 of Chapter 8, the osmotic pressure of an ideal polymer solution having solvent mole fraction x_1 is $-(RT/v_1) \ln x_1$, where v_1 is the molar volume of the solvent. Consequently, the osmotic pressure difference between two solutions having solvent volume fractions ϕ and ϕ_e is given as follows:

$$\pi = \frac{RT}{v_1} \ln\left(\frac{\phi_e}{\phi}\right) \qquad\qquad (13.8.2)$$

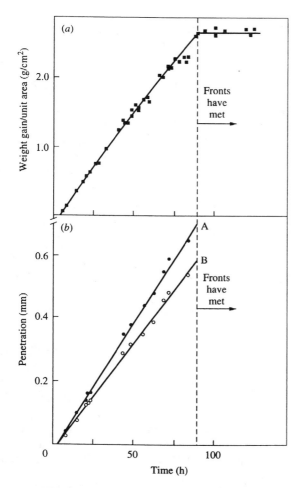

FIGURE 13.16 Data at 24°C for 1-mm (nominal) sheet (actual average thickness 1.18 mm). (a) Weight gain/unit area versus time. (b) Penetration as represented by both l (A) and d (B) versus time. (Reprinted from *Polymer*, vol. 19, Thomas N., and A. H. Windle: "Transport of Methanol in Poly(Methyl Methacrylate)," pp. 255–265, Copyright 1978, with kind permission from Elsevier Science Ltd., The Boulevard, Langford Lane, Kidlington OX5 1GB, UK.)

We would expect that the time rate of change of ϕ would be directly proportional to π but inversely proportional to the viscous resistance of the polymer chains. Thus, we have the following:

$$\frac{\partial \phi}{\partial t} = \frac{\pi}{\eta} \qquad (13.8.3)$$

Because polymer chains deform perpendicular to surfaces of constant ϕ, the relevant viscosity η in Eq. (13.8.3) is the extensional viscosity (see Chap. 14), which can be taken to be independent of the deformation rate at low rates of deformation. The extensional viscosity will, however, depend strongly on ϕ, especially because the polymer goes from the glassy to rubbery state in a very narrow region. If one assumes that

$$\eta = \eta_0 \exp(-m\phi) \qquad (13.8.4)$$

where η_0 and m are constants, then a combination of the previous three equations yields

$$\frac{\partial \phi}{\partial t} = \frac{RT}{v_1 \eta_0} \exp(m\phi) \ln\left(\frac{\phi_e}{\phi}\right) \qquad (13.8.5)$$

which can be rearranged to yield

$$t = \frac{v_1 \eta_0}{RT} \int_{\phi_0}^{\phi_e} \frac{\exp(-m\phi)}{\ln(\phi_e/\phi)} \, d\phi \qquad (13.8.6)$$

and integrated numerically. Computations done in this manner do a fair job of explaining data for the change in surface concentration of polystyrene exposed to *n*-iodohexane vapor [74].

To obtain the concentration profile $\phi(x, t)$ throughout the polymer, we can numerically solve Eq. (13.2.11) with Eq. (13.8.5) as a boundary condition and with a diffusion coefficient that depends on ϕ [75]. Results predict the correct front velocity and concentration profiles. One of the major applications of this work is to polymer debonding and dissolution in photolithography [76]. It should be noted that a possible consequence of the conversion of a glassy polymer to the rubbery state due to sorption of low-molecular-weight solutes is crystallization of the swollen polymer. This complicated situation, which involves coupled diffusion, swelling, and crystallization, has been analyzed in the literature. The reader is referred to the comprehensive work of Kolospiros et al. for details [77].

13.9 POLYMER–POLYMER DIFFUSION

The diffusion coefficient relevant for diffusion in polymer melts is the self-diffusion coefficient [78]. It influences the kinetics of mass-transfer-controlled bulk polymerization reactions, and in injection molding it determines the extent of healing of weld lines, which result from the merging of two streams of the same polymer (see Chap. 15). If polymer molecules in a polymer melt moved in an unhindered manner, we would expect that the self-diffusion coefficient would be given by

$$D = \frac{kT}{f} \tag{13.4.6}$$

with f replaced by $n\zeta$, where n is the number of monomers in the polymer and ζ is the friction experienced by each monomer (see Chap. 14). Because n equals M/M_0, the ratio of polymer to monomer molecular weight, we expect the following relation [78]:

$$D = \left(\frac{kTM}{\zeta}\right)M^{-1} \tag{13.9.1}$$

This inverse relationship between the diffusion coefficient and the molecular weight is, however, not observed, because at large values of M, polymer molecules are thoroughly entangled with each other [79,80]. Consequently, the diffusion coefficient reduces much more rapidly with increasing molecular weight. To predict this behavior, we need a model of how polymer molecules move within an entangled polymer melt. The most successful model, to date, appears to be the reptation model of de Gennes [81,82].

The reptation model (see the very readable review by Osaki and Doi [83] assumes that a given polymer chain faces a number of fixed obstacles in its quest to move within the network of chains. As shown in Figure 13.17, the polymer molecule of interest (represented by a solid line) looks upon the other macromolecules (denoted here by circles) as uncrossable constraints. Consequently, it can move easily along the chain axis but almost not at all in a direction perpendicular to itself. It, therefore, acts as though it is confined within a tube (the dashed region in Figure 13.17). All it can do is slither like a snake (i.e., "reptate") along the tube axis, and this is how it diffuses from one position in the melt to another. Within the confines of the tube, though, its behavior is like that of a freely jointed chain.

To make further progress, we need to relate the distance moved by a molecule in a specified amount of time under the influence of a concentration gradient in a single direction. This can be done as shown in Figure 13.18 [39]. If molecules move a distance δ at one time and take time τ to do so, then, during this time interval, the plane at x can be reached by molecules lying anywhere between $x - \delta$ and $x + \delta$. Because a molecule can move either in the positive x

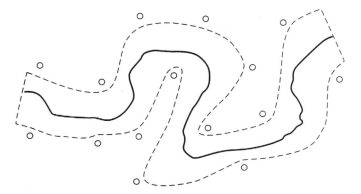

FIGURE 13.17   Reptation in a polymer melt.

direction or in the negative x direction, the net flux in the positive x direction across the plane at x is given by

$$\text{Flux} = \frac{1}{2\tau}\left[\delta c_A\left(x - \frac{\delta}{2}\right) - \delta c_A\left(x + \frac{\delta}{2}\right)\right] = \frac{\delta^2}{2\tau}\frac{dc_A}{dx} \qquad (13.9.2)$$

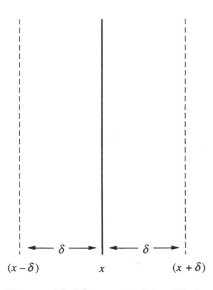

$(x - \delta)$ x $(x + \delta)$

FIGURE 13.18   Model of the diffusion process.

Comparing Eqs. (13.9.2) and (13.2.1) gives us

$$D = \frac{\delta^2}{2\tau} \tag{13.9.3}$$

Returning to the problem of polymer reptation, we find, using Eq. (13.9.3), that the time needed for the chain to diffuse out of the constraining tube is given by

$$\tau = \frac{L^2}{2D} \tag{13.9.4}$$

in which L is the tube length, which also equals the contour length of the polymer.

Viewed from afar, though, the polymer molecule as a whole translates a distance R during the same time interval (see Fig. 13.19) [78]. Thus, the ordinary self-diffusion coefficient of the polymer D_e [using Eq. (13.9.3)] is given by

$$D_e = \frac{R^2}{2\tau} \tag{13.9.5}$$

which, in view of Eq. (13.9.4), becomes

$$D_e = D\frac{R^2}{L^2} \tag{13.9.6}$$

Now, from Section 10.2 of Chapter 10, $R^2 = nl^2$ and $L^2 = n^2l^2$, giving

$$D_e = \frac{D}{n} \tag{13.9.7}$$

Because $D \propto M^{-1}$ [see Eq. (13.9.1)] and $n \propto M$, we have the following:

$$D_e \propto M^{-2} \tag{13.9.8}$$

This simple relationship has been amply verified [84] and attests to the accuracy of the basic physics of the reptation model [21,22,84]. Figure 13.20 shows the data of Klein on the variation of the self-diffusion coefficient of polyethylene with molecular weight [84]; it is found that $D_e \propto M^{2\pm0.1}$. Techni-

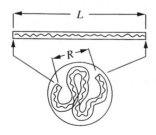

FIGURE 13.19 The tube curvilinear length L is much larger than the end-to-end distance R. (From Ref. 78.)

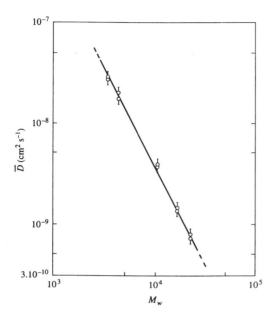

FIGURE 13.20 The variation of $\bar{D}$ with M_w for deuterated polyethylene diffusing in a protonated polyethylene matrix, at a diffusion temperature of $176.0 \pm 0.3°C$. Each point represents a separate experiment and is the mean of 10 separate profiles. The lengths of experimental runs (t_D) vary by a factor of ~ 3 within each pair of experiments for a given DPE fraction. The least-squares best fit to the data is the relation $\bar{D} = 0.2\, M_w^{-2.0\pm0.1}$ and is shown. (Reprinted with permission from *Nature*, vol. 271, Klein, J.: "Evidence for Reptation in an Entangled Polymer Melt," Copyright 1978 Macmillan Magazines Limited.)

ques of measurement, available data, and potential applications have been reviewed by Tirrell [85]. Graessley has used the Doi–Edwards theory [86] of polymer viscoelasticity to obtain, in terms of easily measurable quantities, an expression for the constant of proportionality in Eq. (13.9.8) [87]. This expression predicts, to within a factor of 1.3, the correct magnitude of the data presented in Figure 13.20. Finally, because we still expect the product of the diffusion coefficient and the zero shear viscosity to be essentially constant, the temperature dependence of the diffusion coefficient should be of the Arrhenius form.

13.10 CONCLUSION

In this chapter, we have examined the fundamental equations of mass transfer in binary mixtures, and these involve the mutual diffusion coefficient. We have also presented the basics of the standard techniques used to measure the diffusion coefficient in polymeric systems. Finally, we have looked at the various theories that explain the dependence of the diffusion coefficient on variables such as concentration, temperature, and molecular weight. We have, however, quite intentionally not covered the solution of the mass transport equations when applied to situations of technological interest. Doing this would have made the chapter unmanageably long without contributing significantly to our understanding of polymer physics. In addition, excellent books already exist that do exactly this [13,88]. We have also omitted a detailed discussion of composite materials that typically employ thermosets as the matrix material. Fiber-reinforced plastics are increasingly being used as structural materials in the construction industry; a major barrier to their further widespread use is an incomplete understanding of the influence of environmental effects [89], especially moisture diffusion, on the mechanical properties and durability of such composites [90]. This chapter ought to provide adequate background for someone interested in reading further on diffusion in composites.

REFERENCES

1. Merten, U. (ed.), *Desalination by Reverse Osmosis*, MIT. Press, Cambridge, MA, 1966.
2. Weller, S., and W. A. Steiner, Engineering Aspects of Separation of Gases, *Chem. Eng. Prog.*, 46, 585–590, 1950.
3. Bornside, D. E., C. W. Macosko, and L. E. Scriven, On the Modeling of Spin Coating, *J. Imaging Technol.*, 13, 122–130, 1987.
4. Biesenberger, J. A. (ed.), *Devolatilization of Polymers*, Hanser, Munich, 1983.
5. Chien, Y.W., *Novel Drug Delivery Systems: Fundamentals, Developmental Concepts, Biomedical Assessments*, Marcel Dekker, New York, 1982.
6. Zaffaroni, A., Controlled Delivery Therapy Is Here, *Chemtech*, 6, 756–761, 1976.
7. Brown, H. R., A. C. M. Yang, T. P. Russell, W. Volksen, and E. J. Kramer, Diffusion and Self-Adhesion of the Polyimide PMDA–ODA, *Polymer*, 29, 1807–1811, 1988.
8. Vieth, W. R., *Diffusion in and Through Polymers*, Hanser, Munich, 1991.
9. Baker, R. W., and J. G. Wijmans, Membrane Separation of Organic Vapors from Gas Streams, in *Polymeric Gas Separation Membranes*, D. R. Paul and Y. P. Yampol'skii (eds.), CRC Press, Boca Raton, FL, 1994, 353–397.
10. Cornell, J. W., and B. R. Phillips, Identification of Barre Sources in Circular Knits, *Textile Res. J.*, 49, 105–117, 1979.
11. Welty, J. R., C. E. Wicks, R. E. Wilson, and G. Rorrer, *Fundamentals of Momentum, Heat and Mass Transfer*, 4th ed., Wiley, New York, 2001.
12. Tyrrell, H. J. V., and K. R. Harris, *Diffusion in Liquids*, Butterworths, London, 1984.
13. Bird, R. B., W. E. Stewart, and E. N. Lightfoot, *Transport Phenomena*, 2nd ed.,Wiley, New York, 2002.

14. Northrop, J. H., and M. L. Anson, A Method for the Determination of Diffusion Constants and the Calculation of the Radius and Weight of the Hemoglobin Molecule, *J. Gen. Physiol.*, 12, 543–554, 1929.

15. Wickramasinghe, S. R., D. V. Boger, H. R. C. Pratt, and G. W. Stevens, Diffusion in Elastic Fluids, *Chem. Eng. Sci.*, 46, 641–650, 1991.

16. Boltzmann, L., Zur Integration der Diffusionsgleichung bei Variablen Diffusionscoefficienten, *Ann. Phys. Chem.*, 53, 959–964, 1894.

17. Cussler, E. L., *Multicomponent Diffusion*, Elsevier, Amsterdam, 1976.

18. Crank, J., and G. S. Park, Methods of Measurement, in *Diffusion in Polymers*, J. Crank and G. S. Park (eds.), Academic Press, London, 1968.

19. Dunlop, P. J., B. J. Steel, and J. E. Lane, Experimental Methods for Studying Diffusion in Liquids, Gases, and Solids, in *Physical Methods of Chemistry*, A. Weissberger and B. W. Rossiter (eds.), Wiley, New York, 1972, Vol. I.

20. Klein, J., and B. J. Briscoe, Diffusion of Large Molecules in Polymers, A Measuring Technique Based on Microdensitometry in the Infrared, *Polymer*, 17, 481–484, 1976.

21. Mills, P. J., P. F. Green, C. J. Palmstrom, J. W. Mayer, and E. J. Kramer, Analysis of Diffusion in Polymers by Forward Recoil Spectrometry, *Appl. Phys. Lett.*, 45, 957–959, 1984.

22. Green, P. F., C. J. Palmstrom, J. W. Mayer, and E. J. Kramer, Marker Displacement Measurements of Polymer–Polymer Interdiffusion, *Macromolecules*, 18, 501–507, 1985.

23. Anderson, J. S., and K. Saddington, The Use of Radioactive Isotopes in the Study of the Diffusion of Ions in Solution, *J. Chem. Soc.*, s381–s386, 1949.

24. Robinson, R. A., and R. H. Stokes, *Electrolyte Solutions*, 2nd ed., Butterworths, London, 1959.

25. Crank, J., *The Mathematics of Diffusion*, 2nd ed., Clarendon, Oxford, 1975.

26. Daynes, H. A., The Process of Diffusion Through a Rubber Membrane, *Proc. Roy. Soc. London*, A97, 286–307, 1920.

27. Guo, C. J., D. DeKee, and B. Harrison, Diffusion of Organic Solvents in Rubber Membranes Measured via a New Permeation Cell, *J. Appl. Polym. Sci.*, 56, 823–829, 1995.

28. Stern, S. A., and H. L. Frisch, The Selective Permeation of Gases Through Polymers, *Ann. Rev. Mater. Sci.*, 11, 523–550, 1981.

29. Stannett, V., and M. Szwarc, The Permeability of Polymer Films to Gases—A Simple Relationship, *J. Polym. Sci.*, 16, 89–91, 1955.

30. Frisch, H. L., Factorization of the Activation Energies of Permeation and Diffusion of Gases in Polymers, *J. Polym. Sci.*, B1, 581–586, 1963.

31. Frisch, H. L., The Time Lag in Diffusion, *J. Phys. Chem.*, 61, 93–95, 1957.

32. Frisch, H. L., The Time Lag in Diffusion II, *J. Phys.Chem.*, 62, 401–404, 1958.

33. Meares, P., Transient Permeation of Organic Vapors Through Polymer Membranes, *J. Appl. Polym. Sci.*, 9, 917–932, 1965.

34. Fujita, H., Organic Vapors Above the Glass Transition Temperature, in *Diffusion in Polymers*, J. Crank and G. S. Park (eds.), Academic Press, London, 1968.

35. Duda, J. L., G. K. Kimmerly, W. L. Sigelko, and J. S. Vrentas, Sorption Apparatus for Diffusion Studies with Molten Polymers, *Ind. Eng. Chem. Fundam.*, 12, 133–136, 1973.

36. Duda, J. L., and J. S. Vrentas, Mathematical Analysis of Sorption Experiments, *AIChE J.*, 17, 464–469, 1971.

37. Shah, A. P., R. K. Gupta, H. V. S. GangaRao, and C. E. Powell, Moisture Diffusion through Vinyl Ester Nanocomposites Made with Montmorillonite Clay, *Polym. Eng. Sci.*, 42, 1852–1863, 2002.

38. Saleem, M., A.-F. A. Asfour, D. DeKee, and B. Harrison, Diffusion of Organic Penetrants Through Low Density Polyethylene (LDPE) Films: Effect of Size and Shape of the Penetrant Molecules, *J. Appl. Polym. Sci.*, 37, 617–625, 1989.

39. Einstein, A., *Investigations on the Theory of the Brownian Movement*, Dover, New York, 1956.

40. Reid, R. C., J. M. Prausnitz, and B. E. Poling, *The Properties of Gases and Liquids*, 4th ed., McGraw-Hill, New York, 1977.

41. Cussler, E. L., *Diffusion*, Cambridge University Press, Cambridge, 1984.

42. Vrentas, J. S., and J. L. Duda, Molecular Diffusion in Polymer Solutions, *AIChE J.*, 25, 1–24, 1979.

43. Prilutski, G., R. K. Gupta, T. Sridhar, and M. E. Ryan, Model Viscoelastic Liquids, *J. Non-Newtonian Fluid Mech.*, 12, 233–241, 1983.

44. Chhabra, R. P., P. H. J. Uhlherr, and D. V. Boger, The Influence of Fluid Elasticity on the Drag Coefficient for Creeping Flow Around a Sphere, *J. Non-Newtonian Fluid Mech.*, 6, 187–199, 1980.

45. Kulkarni, M. G., and R. A. Mashelkar, Diffusion Effects in Initiator Decomposition in Highly Viscous and Macromolecular Solutions, *AIChE J.*, 27, 716–724, 1981.

46. Hiss, T. G., and E. L. Cussler, Diffusion in High Viscosity Liquids, *AIChE J.*, 19, 698–703, 1973.

47. Yamakawa, H., and G. Tanaka, Excluded Volume Effects in Linear Polymer Chains: A Hierarchy of Differential Equations, *J. Chem. Phys.*, 47, p. 3991, 1967.

48. Gee, G., Some Thermodynamic Properties of High Polymers, and Their Molecular Interpretation, *Q. Rev. Chem. Soc. London*, 1, 265–298, 1947.

49. Stannett, V., Simple Gases, in *Diffusion in Polymers*, J. Crank and G. S. Park (eds.), Academic Press, London, 1968.

50. Van Krevelen, D. W., *Properties of Polymers*, Elsevier, Amsterdam, 1972.

51. Astarita, G., Heat and Mass Transfer in Solid Polymeric Systems, *Adv. Transp. Processes*, 5, 339–351, 1989.

52. Vrentas, J. S., and J. L. Duda, Diffusion in Polymer–Solvent Systems, I. Reexamination of the Free-Volume Theory, *J. Polym. Sci. Polym. Phys. Ed.*, 15, 403–416, 1977.

53. Vrentas, J. S., and J. L. Duda, Diffusion in Polymer–Solvent Systems, II. A Predictive Theory for the Dependence of Diffusion Coefficients on Temperature, Concentration, and Molecular Weight, *J. Polym. Sci. Polym. Phys. Ed.*, 15, 417–439, 1977.

54. Vrentas, J. S., and J. L. Duda, Solvent and Temperature Effects on Diffusion in Polymer–Solvent Systems, *J. Appl. Polym. Sci.*, 21, 1715–1728, 1977.

55. Cohen, M. H., and D. Turnbull, Molecular Transport in Liquids and Glasses, *J. Chem. Phys.*, 31, 1164–1169, 1959.

56. Macedo, P. B., and T. A. Litovitz, On the Relative Roles of Free Volume and Activation Energy in the Viscosity of Liquids, *J. Chem. Phys.*, 42, 245–256, 1965.

57. Bearman, R. J., On the Molecular Basis of Some Current Theories of Diffusion, *J. Phys. Chem.*, 65, 1961–1968, 1961.

58. Duda, J. L., Y. C. Ni, and J. S. Vrentas, An Equation Relating Self-Diffusion and Mutual Diffusion Coefficients in Polymer-Solvent Systems, *Macromolecules*, 12, 459–462, 1979.

59. Duda, J. L., J. S. Vrentas, S. T. Ju, and H. T. Liu, Prediction of Diffusion Coefficients for Polymer–Solvent Systems, *AIChE J.*, 28, 279–285, 1982.

60. Haward, R. N., Occupied Volume of Liquids and Polymers, *J. Macromol. Sci. Macromol. Chem.*, C4, 191–242, 1970.

61. Timmermans, J., *Physiochemical Constants of Pure and Organic Compounds*, I, Elsevier, Amsterdam, 1950.

62. Fox, T. G., and S. Loshaek, Influence of Molecular Weight and Degree of Cross-linking on the Specific Volume and Glass Temperature of Polymers, *J. Polym. Sci.*, 15, 371–390, 1955.

63. Kulkarni, M. G., and R. A. Mashelkar, A Unified Approach to Transport Phenomena in Polymeric Media—I, *Chem. Eng. Sci.*, 38, 925–939, 1983.

64. Hopfenberg, H. B., and V. Stannett, The Diffusion and Sorption of Gases and Vapours in Glassy Polymers, in *The Physics of Glassy Polymers*, R. N. Haward (ed.), Wiley, New York, 1973.

65. Vieth, W. R., J. M. Howell, and J. H. Hsieh, Dual Sorption Theory, *J. Membr. Sci.*, 1, 177–220, 1976.

66. Stannett, V., The Transport of Gases in Synthetic Polymeric Membranes—An Historic Perspective, *J. Membr. Sci.*, 3, 97–115, 1978.

67. Vieth, W. R., C. S. Frangoulis, and J. A. Rionda, Jr., Kinetics of Sorption of Methane in Glassy Polystyrene, *J. Colloid Interf. Sci.*, 22, 454, 1966.

68. Subramanian, S., J. C. Heydweiller, and S. A. Stern, Dual-Mode Sorption Kinetics of Gases in Glassy Polymers, *J. Polym. Sci. Polym. Phys.*, 27, 1209–1220, 1989.

69. Alfrey, T., E. F. Gurnee, and W. G. Lloyd, Diffusion in Glassy Polymers, *J. Polym. Sci.*, C12, 249–261, 1966.

70. Thomas, N., and A. H. Windle, Transport of Methanol in Poly(Methyl Methacrylate), *Polymer*, 19, 255–265, 1978.

71. Thomas, N. L., and A. H. Windle, Diffusion Mechanics of the System PMMA–Methanol, *Polymer*, 22, 627–639, 1981.

72. Thomas, N. L., and A. H. Windle, A Deformation Model for Case II Diffusion, *Polymer*, 21, 613–619, 1980.

73. Thomas, N. L., and A. H. Windle, A Theory of Case II Diffusion, *Polymer*, 23, 529–542, 1982.

74. Hui, C. Y., K. C. Wu, R. C. Lasky, and E. J. Kramer, Case II Diffusion in Polymers. I. Transient Swelling, *J. Appl. Phys.*, 61, 5129–5136, 1987.

75. Hui, C. Y., K. C. Wu, R. C. Lasky, and E. J. Kramer, Case II Diffusion in Polymers. II. Steady-State Front Motion, *J. Appl. Phys.*, 61, 5137–5149, 1987.

76. Lasky, R. C., T. P. Gall, and E. J. Kramer, Case II Diffusion, in *Principles of Electronic Packaging*, D. P. Seraphim, R. Lasky, and C. Y. Li (eds.), McGraw-Hill, New York, 1989.

77. Kolospiros, N. S., G. Astarita, and M. E. Paulaitis, Coupled Diffusion and Morphological Change in Solid Polymers, *Chem. Eng. Sci.*, 48, 23–40, 1993.

78. Marrucci, G., Molecular Modeling of Flows of Concentrated Polymers, *Adv. Transp. Processes*, 5, 1–36, 1989.

79. Porter, R. S., and J. F. Johnson, The Entanglement Concept in Polymer Systems, *Chem. Rev.*, 66, 1–27, 1966.

80. Graessley, W. W., The Entanglement Concept in Polymer Rheology, *Adv. Polym. Sci.*, 16, 1–179, 1974.

81. de Gennes, P. G., Reptation of a Polymer Chain in the Presence of Fixed Obstacles, *J. Chem. Phys.*, 55, 572–579, 1971.

82. de Gennes, P. G., *Scaling Concepts in Polymer Physics*, Cornell University Press, Ithaca, NY, 1979.

83. Osaki, K., and M. Doi, Nonlinear Viscoelasticity of Concentrated Polymer Systems, *Polym. Eng. Rev.*, 4, 35–72, 1984.

84. Klein, J., Evidence for Reptation in an Entangled Polymer Melt, *Nature*, 271, 143–145, 1978.

85. Tirrell, M., Polymer Self-Diffusion in Entangled Systems, *Rubber Chem. Tech.*, 57, 523–556, 1984.

86. Doi, M., and S. F. Edwards, *The Theory of Polymer Dynamics*, Clarendon, Oxford, 1986.

87. Graessley, W. W., Some Phenomenological Consequences of the Doi–Edwards Theory of Viscoelasticity, *J. Polym. Sci. Polym. Phys. Ed.*, 18, 27–34, 1980.

88. Griskey, R.G., *Polymer Process Engineering*, Chapman and Hall, New York, 1995.

89. Springer, G.S. (ed.), *Environmental Effects on Composite Materials*, Technomic Publishing Co., Lancaster, PA., 1981, Vols. I and II.

90. Kajorncheappunngam, S., *The Effects of Environmental Aging on the Durability of Glass-Epoxy Composites*, Ph.D. thesis, West Virginia University, Morgantown, 1999.

91. Tsay, C.-S., and A. J. McHugh, A Technique for Rapid Measurement of Diffusion Coefficients, *Ind. Eng. Chem. Res.*, 31, 449–452, 1992.

92. Oship, K. A., *High Temperature Permeation of CO_2, O_2, CH_4, and CH_3OH in Polymide Films*, M.S. thesis, State University of New York at Buffalo, Buffalo, 1991.

93. Yano, K., A. Usuki, A. Okada, T. Kurauchi, and O. Kamigaito, Synthesis and Properties of Polyimide–Clay Hybrid, *J. Polym. Sci., A: Polym. Chem.*, 31, 2493–2498, 1993.

94. Arnould, D., and R. L. Laurence, Size Effects on Solvent Diffusion in Polymers, *Ind. Eng. Chem. Res.*, 31, 218–228, 1992.

PROBLEMS

13.1. Ten grams of polystyrene of 300,000 molecular weight is mixed with 90 g of benzene. If the density of the polymer is $1.05 \, \text{g/cm}^3$ and that of the solvent is $0.9 \, \text{g/cm}^3$, determine the polymer concentration in (a) g/cm^3 (b) g mol/cm^3. Also calculate the polymer mass fraction and the polymer mole fraction.

13.2. Consider the steady-state, one-dimensional diffusion of a vapor through a polymer membrane of thickness L. If the molar concentration of the solute at $x = 0$ is c_1 and that at $x = L$ is c_2 (see Fig. 13.4), determine the molar flux in terms of known quantities. It is known that the diffusion coefficient varies with concentration as

$$D = D_0(1 + \alpha c_A)$$

where D_0 and α are constants.

13.3. If the mutual diffusion coefficient depends on concentration, show that the appropriate form of Fick's second law for one-dimensional diffusion is as follows:

$$\frac{\partial c_A}{\partial t} = \frac{\partial}{\partial z}\left(D_{AB}(c)\, \frac{\partial c_A}{\partial x}\right)$$

Also show that the use of a new independent variable $\eta = zt^a$, where a is a constant, converts this equation into an ordinary differential equation. What numerical value of a does one need to use?

13.4. If the experiment described in Example 13.2 is allowed to run for a total of 4 hr, what will be the fractional change in the concentration driving force $(c_2 - c_1)$ over the course of the experiment? Assume that $V_1 - V_2 = 20 \text{ cm}^3$.

13.5. Fill in the missing steps in going from Eq. (13.3.7) to Eq. (13.3.8).

13.6. In Example 13.3, how many moles of the diffusing species are transferred across unit area of the plane at $z = 0$ in time t?

13.7. Apply the results of Problem 13.3 to a free-diffusion experiment in which $D_{AB} = D_{AB}(c)$ and $c_1(0) = 0$ and show that, at any instant t,

$$D_{AB}(c) = -\frac{1}{2t}\left(\frac{\partial z}{\partial c}\right)\int_0^c z\, dc$$

[Hint: You will need to use the fact that at $z = \infty$ or $\eta = \infty$, both c and $dc/d\eta$ are zero.)

13.8. If we assume that the refractive index n of a solution is linearly proportional to the concentration, show that Eq. (13.3.8) becomes

$$n - n_2(0) = \frac{n_1(0) - n_2(0)}{2}(1 - \text{erf } \xi) \tag{i}$$

where z is now measured positive downward.

Tsay and McHugh have shown that fringes of varying intensity can be obtained if a collimated light beam of wavelength λ was passed through the free-diffusion cell of thickness ω [91]. The refractive index n_m corresponding to the minimum light intensity is

$$n_m - n_2(0) = \frac{m\lambda}{\omega} \tag{ii}$$

where m is an integer. Thus, if we measure the variation with time of the location z_m of a specific fringe of constant m value (say, unity), ξ will be constant because n will remain unchanged at value n_1 (i.e., n_m evaluated at $m = 1$).

Tsay and McHugh have shown that for the system dimethyl formamide (DMF) in water, containing 0.635 volume fraction water, z_1^2 equaled $6.766 \times 10^{-5}t$ cm^2 [91]. If $n_1(0) - n_2(0)$ is 0.005, λ is 632.8 nm, and ω is 4 mm, what is the value of the mutual diffusion coefficient? [Hint: Equate the right-hand side of Eq. (i) to the right-hand side of Eq. (ii) with $m = 1$.]

13.9. Use the Stokes–Einstein equation to predict the diffusivity of a chlorine ion at 25°C in a sugar solution of 3.7 P viscosity. Assume that the radius of the ion can be taken to equal 1.8×10^{-8} cm [15].

13.10. Wickramasinghe et al. used the open-ended capillary to independently measure the diffusivity of the Cl$^-$ ion of Problem 13.9 [15]. In this experiment, $Dt/4L^2$ was approximately 0.002. Estimate the time scale of the experiment if L was 2 cm.

13.11. Oship measured the steady-state rate of permeation of CO_2 gas at 155°C through a glassy polyimide film [92]. The membrane area was 7.197 cm^2 and the thickness was 0.0025 cm. When the pressure difference across the membrane was 80 cm Hg, the steady-state molar flow rate was 2.055×10^{-9} g mol/sec. Determine the permeability in barrers where 1 barrer $= 10^{-10}$ cm^3 (STP)/(cm sec cm of Hg).

13.12. Show that if all the clay platelets in Figure 13.6 are uniformly distributed, are aligned parallel to each other, and are square in shape, with thickness h and edge L, then the permeability P of the nanocomposite, in a direction perpendicular to the faces, is given by [93]

$$\frac{P}{P_p} = \frac{1}{1 + (L/2h)\phi}$$

in which P_p is the permeability of the unfilled polymer and ϕ is the volume fraction of filler.

13.13. Calculate the diffusivity of CO_2 in polypropylene at 224°C if the corresponding value at 188°C is 4.25×10^{-5} cm^2/sec [88]. The activation energy is 3 kcal/mol.

13.14. Arnould and Laurence found that a straight-line plot was obtained when the logarithm of the self-diffusion coefficient of acetone in polymethyl methacrylate was plotted against $(K_{22} + T - T_{g2})^{-1}$ at temperatures close to the T_g of the polymer and at small values of the mass fraction of acetone [94]. Show that this behavior is consistent with the free-volume theory in general and with Eq. (13.6.9) in particular. What is the physical significance, if any, of the slope and intercept of such a plot?

13.15. Use the data given in Figure 13.13 to obtain the parameters k_D, C'_H, and b for the methane–polystyrene system at 25°C.

13.16. If we use Eq. (13.3.19) to determine the diffusion coefficient at high gas pressures in the presence of a dual-sorption mechanism, will the measured value equal the quantity D that appears in Eq. (13.7.10)?

13.17. Show that the solution to Eq. (13.3.7) subject to $c(0) = c_s$ and $c(\infty) = 0$, is as follows,

$$\frac{c(z, t)}{c_s} = 1 - \text{erf}\left(\frac{z}{2\sqrt{D_{AB}t}}\right)$$

and thus determine the time required for one polymer melt to interdiffuse into another polymer melt such that c/c_s equals 0.48 at a distance of 1 mm from the interface. Let D_{AB} be 10^{-10} cm^2/sec.

14

Flow Behavior of Polymeric Fluids

14.1 INTRODUCTION

In order to use polymers (whether available in the form of pellets or as melt from a polymerization reactor), the material has to be converted into useful shapes such as fibers, films, or molded articles. This is done using unit operations such as fiber spinning and injection molding, which are analyzed in detail in Chapter 15. Here, we simply mention that flow is an integral part of any shaping operation, and, very frequently, it is useful to know quantities such as the pressure drop needed to pump a polymeric fluid at a specified flow rate through a channel of a given geometry. The answer is easy to obtain if we are working with low-molecular-weight liquids that behave in a Newtonian manner; all we need, by way of material properties, is information about the temperature dependence of the shear viscosity. If the process is isothermal, the shear viscosity is a constant and it can be measured in any one of several ways. Polymer melts and solutions, however, have a steady shear viscosity that depends on the shear rate. Therefore, it is a material function rather than a material constant. For polymeric fluids the typical shape of the steady shear viscosity curve as a function of shear rate is shown in Figure 14.1. At steady state, the viscosity is constant at low shear rates. It usually decreases with increasing shear rate and often becomes constant again at high shear rates. There is, therefore, a lower Newtonian region characterized by the zero shear rate viscosity η_0 and an upper Newtonian region characterized by an

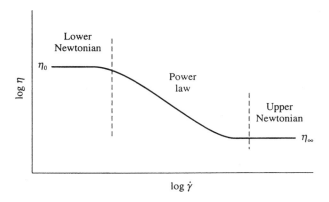

FIGURE 14.1   Qualitative behavior of the steady shear viscosity of polymeric fluids.

infinite shear viscosity η_∞. Between these two regions, the viscosity versus shear rate behavior can usually be represented as a straight line on logarithmic coordinates—this is the *power-law region*.

In the foregoing, we have been careful to append the word "steady" to the shear viscosity. Unlike with Newtonian liquids, the shear stress takes some time to reach a steady value upon inception of shear flow at a constant shear rate. This is sketched in Figure 14.2, which shows that the shear stress can also overshoot the steady-state value. Polymeric fluids are therefore non-Newtonian in the sense that the shear viscosity depends on both shear rate and time. Obtaining the pressure drop corresponding to a given flow rate is, consequently, a slightly more complicated process.

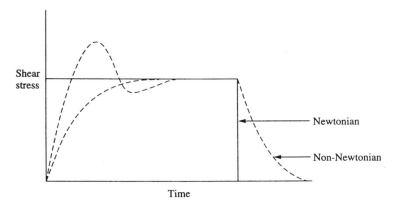

FIGURE 14.2   Start-up and shutdown of shearing at constant shear rate.

A shear-thinning viscosity is not the only non-Newtonian feature of the behavior of polymeric fluids; several other unusual phenomena are observed. If, in the situation depicted in Figure 14.2, the shear rate is suddenly reduced to zero after the attainment of a steady state, low- and high-molecular-weight liquids again behave differently. The stress in the Newtonian fluid goes to zero instantly, but it takes some time to disappear in the polymer. The time scale over which this stress relaxation occurs is known as the *relaxation time* and is denoted by the symbol θ. Additionally, if a small-amplitude sinusoidal strain is imposed on the polymer, the resulting stress is neither in phase with the strain nor out of phase with the strain: There is an out-of-phase component representing energy dissipation and an in-phase component representing energy storage (see Sect. 12.4). Both stress relaxation and the phase difference in dynamic experiments are elastic effects; we say that the polymers are both viscous and elastic (i.e., viscoelastic). In time-dependent flow, the relative extent of these two effects depends on the value of the dimensionless group known as the *Deborah number* (De) and defined as follows:

$$\text{De} = \frac{\theta}{T} \tag{14.1.1}$$

where T is the characteristic time constant for the process of interest. For low values of De, the polymer response is essentially liquidlike (viscous), whereas for high values, it is solidlike (elastic). A further manifestation of viscoelasticity is the swelling of a jet of polymer on emerging from a "die" or capillary. This is shown in Figure 14.3. *Die swell*, or *jet swell*, can be such that D_j/D easily exceeds 2; the corresponding Newtonian value is 1.13. This is true at very low flow rates. At high flow rates, die swell reduces but unstable behavior called *melt fracture* can occur. The jet can become wavy or the surface can become grossly distorted, as sketched in Figure 14.4; the extent of distortion is also influenced by the geometry of the capillary, its surface character, and the properties of the polymer. Note that melt fracture is never observed with Newtonian liquids.

The phenomena just described are interesting to observe and explain. A quantitative description of them is, however, essential for developing models of

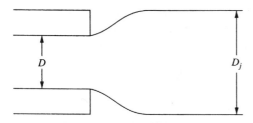

FIGURE 14.3 The die-swell phenomenon.

FIGURE 14.4   Extrudate melt fracture.

various polymer processing operations. Some of these models are discussed in Chapter 15, and these are based on the conservation principles of mass, momentum, and energy, together with appropriate constitutive equations and boundary conditions. They are useful for process optimization and for determining the effect of the various material, geometrical, and processing variables on the properties of the polymeric product. The models also allow us to relate performance variables to machine variables. They are also useful for predicting the onset of flow instabilities. In this chapter, though, we describe methods of measuring the stress response of polymeric fluids in well-characterized flow situations, present the associated methods of data analysis, and give typical results. This naturally leads to a discussion of theories available to explain the observed behavior in terms of material microstructure and to methods of mathematically representing the stress–deformation relations or constitutive behavior. This is the realm of rheology, wherein we examine both polymer solutions and polymer melts. Note that, at a less fundamental level, these measurements can be employed for product characterization and quality-control purposes. A succinct treatment may be found in Ref. [1].

14.2   VISCOMETRIC FLOWS

The flow field that is generated in most standard instruments used to measure rheological or flow properties is a particular kind of shear flow called *viscometric flow*. All of the motion in a viscometric flow, whether in Cartesian or curvilinear coordinates, is along one coordinate direction (say, x_1 in Fig. 14.5), the *velocity varies* along a second coordinate direction (say, x_2), and the third direction is neutral. An illustration of this, shown in Figure 14.5, is the shearing of a liquid between two parallel plates due to the motion of one plate relative to the other. The velocity gradient or shear rate, $\dot{\gamma}$, then is dv_1/dx_2, where v_1 is the only nonzero component of the velocity vector. If the shear rate is independent of position, the flow field is homogeneous and the components of the extra stress [see Eq. (10.5.6) of Chap. 10] are also independent of position. This is intuitively obvious (the extra stress depends on the rate of shear strain), and because the shear rate is the same everywhere, so must be the extra stress. This fact is used to

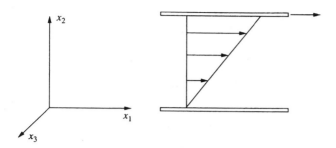

FIGURE 14.5 Viscometric flow in rectangular Cartesian coordinates.

great advantage in several *viscometers* or *rheometers*, as such instruments are called.

As discussed in Section 10.5, there are, in general, only six independent stress components. For a Newtonian liquid, these can be evaluated using Newton's law of viscosity,

$$T_{ij} = \tau_{ij} = \eta \left(\frac{\partial v_i}{\partial x_j} + \frac{\partial v_j}{\partial x_i} \right) \tag{14.2.1}$$

provided that $i \neq j$ and

$$T_{ii} = -p + \tau_{ii} = -p + 2\eta \frac{\partial v_i}{\partial x_i} \tag{14.2.2}$$

in which both i and j can be 1, 2, or 3. For the situation depicted in Figure 14.5, therefore, only τ_{12} is nonzero for a Newtonian liquid. For polymeric fluids, we can use symmetry arguments and show that τ_{13} and τ_{23} are still zero [2], but the normal stresses τ_{ii} can be nonzero. Furthermore, if the τ_{ii} exist, they are even functions of the shear rate. The shear stress, however, is an odd function. Now, note that due to incompressibility, p is indeterminate and we cannot obtain τ_{ii} from a measurement of T_{ii}. However, p can be eliminated if we take stress differences. We can, therefore, define the first and second normal stress differences as follows:

$$N_1(\dot{\gamma}) = T_{11} - T_{22} = \tau_{11} - \tau_{22} \tag{14.2.3}$$
$$N_2(\dot{\gamma}) = T_{22} - T_{33} = \tau_{22} - \tau_{33} \tag{14.2.4}$$

and these must depend uniquely on the shear rate $\dot{\gamma}$, because each of the extra-stress components is a unique function of $\dot{\gamma}$.

The existence of a positive first normal stress difference during shear flow can be used to explain die swell. If the fluid being sheared between parallel plates in Figure 14.5 were to emerge into the atmosphere, T_{11} would obviously equal

$-p_a$, where p_a is atmospheric pressure and is a compressive stress. A positive first normal stress difference would then imply that T_{22} is negative (compressive) and greater than p_a in magnitude. In other words, the upper plate pushes down on the liquid being sheared and the liquid pushes up on the plate with a stress that exceeds p_a in magnitude. Because only atmospheric pressure acts on the outside of the upper plate, it has to be held down by an externally applied force to prevent the fluid from pushing the two plates apart. When the fluid emerges into the atmosphere, there is no plate present to push down on it, and it, therefore, expands and we observe die swell.

Because the shear stress is an odd function of the shear rate and the normal stress differences are even functions, it is customary to define the viscosity function and the first and second normal stress coefficients as follows:

$$\eta(\dot{\gamma}) = \frac{\tau_{12}}{\dot{\gamma}} \tag{14.2.5}$$

$$\Psi_1(\dot{\gamma}) = \frac{N_1}{\dot{\gamma}^2} \tag{14.2.6}$$

$$\Psi_2(\dot{\gamma}) = \frac{N_2}{\dot{\gamma}^2} \tag{14.2.7}$$

which tend to attain constant values η_0, Ψ_{10}, and Ψ_{20} as the shear rate tends to zero. Next, we examine two of the most popular methods of experimentally determining some or all of these quantities.

14.3   CONE-AND-PLATE VISCOMETER

In this instrument, the liquid sample is placed in the gap between a truncated cone and a coaxial disk, as shown in Figure 14.6a. The cone is truncated so that there is no physical contact between the two members. The disk radius R is typically a couple of centimeters, whereas the cone angle α is usually a few degrees. Either of the two members can be rotated or oscillated, and we measure the torque M needed to keep the other member stationary. We also measure the downward force F needed to hold the apex of the truncated cone at the center of the disk. From these measurements, we can determine the three material functions defined by Eqs. (14.2.5)–(14.2.7). Note that F equals zero for a Newtonian liquid.

This flow is a viscometric flow when viewed in a spherical coordinate system, and there is only one nonzero component of the velocity. This component is v_ϕ, which varies with both r and θ; the streamlines are closed circles. If we rotate the plate at an angular velocity Ω, the linear velocity on the plate surface at any radial position is Ωr. On the cone surface at the same radial position, however, the velocity is zero. If the cone angle is small, we can assume that v_ϕ

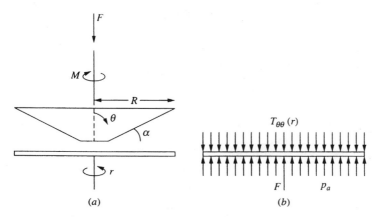

FIGURE 14.6 (a) A cone and plate viscometer; (b) force balance on the plate.

varies linearly across the gap between the cone and the plate. The shear rate $\dot{\gamma}$ at any value of r is then given as follows:

$$\dot{\gamma} = \frac{\Omega r - 0}{r\alpha} = \frac{\Omega}{\alpha} \qquad (14.3.1)$$

in which α is measured in radians and Ω in radians per second. Thus, the shear rate is independent of position within the gap. As a consequence, the stress components resulting from fluid deformation do not depend on position either. Also, because this is a viscometric flow, the only nonzero stress components are the shear stress $\tau_{\phi\theta}$ and the normal stresses $\tau_{\phi\phi}$, $\tau_{\theta\theta}$, and τ_{rr}. Note that, due to symmetry, all derivatives with respect to ϕ are zero. Also note that, by a similar line of reasoning, the stresses will be independent of position even when the plate is oscillated or given a step strain.

If we integrate the shear stress over the cone surface, we can get an expression for the torque M as follows:

$$M = \int_0^R r\tau_{\phi\theta} 2\pi r \, dr = \frac{2\pi R^3 \tau_{\phi\theta}}{3} \qquad (14.3.2)$$

so that

$$\tau_{\phi\theta} = \frac{3M}{2\pi R^3} \qquad (14.3.3)$$

and the viscosity is

$$\eta = \frac{3M\alpha}{2\pi R^3 \Omega} \qquad (14.3.4)$$

In order to obtain the normal stress functions, we need to solve the equations of motion in spherical coordinates [3]. An examination of the θ component of this equation shows $\partial p / \partial \theta = 0$. Thus, p depends on r alone, because derivatives with respect to ϕ are zero. Further, because most polymer fluids are fairly viscous, we can neglect inertia and, as a result, the r component of the equation of motion yields the following [3]:

$$\frac{dp}{dr} = \frac{2}{r}\tau_{rr} - \frac{\tau_{\theta\theta} + \tau_{\phi\phi}}{r} \tag{14.3.5}$$

Integrating with respect to r from R to r gives

$$p(r) = p(R) + (2\tau_{rr} - \tau_{\theta\theta} + \tau_{\phi\phi})\ln\left(\frac{r}{R}\right) \tag{14.3.6}$$

At $r = R$, T_{rr} equals $-p_a$, where p_a is atmospheric pressure. Thus, from the definition of the extra stress, we have

$$-p(R) + \tau_{rr} = -p_a \tag{14.3.7}$$

To make further progress, we examine the equilibrium of the plate and balance forces in the θ direction (Fig. 14.6b). The result is

$$F + \pi R^2 p_a + \int_0^R [-p(r) + \tau_{\theta\theta}]2\pi r \, dr = 0$$

or

$$F + \pi R^2 p_a + \pi R^2 \tau_{\theta\theta} = \int_0^R p(r)2\pi r \, dr \tag{14.3.8}$$

Introducing Eqs. (14.3.6) and (14.3.7) into Eq. (14.3.8), integrating by parts, and simplifying the result gives the first normal stress difference:

$$N_1 = \tau_{\phi\phi} - \tau_{\theta\theta} = \frac{2F}{\pi R^2} \tag{14.3.9}$$

Clearly, the first normal stress difference in shear is a positive quantity.

Finally, if we were to use a pressure transducer to measure $T_{\theta\theta}$ on the plate surface, we would have

$$
\begin{aligned}
T_{\theta\theta} &= -p(r) + \tau_{\theta\theta} \\
&= -p_a - \tau_{rr} + \tau_{\theta\theta} - (2\tau_{rr} - \tau_{\theta\theta} - \tau_{\phi\phi})\ln\left(\frac{r}{R}\right) \\
&= -p_a + N_2 + (N_1 + 2N_2)\ln\left(\frac{r}{R}\right)
\end{aligned}
\tag{14.3.10}
$$

Knowing N_1, we can get N_2 from Eq. (14.3.10). This measurement is not easy to make, but the general consensus is that N_2 is negative and about 0.1–0.25 times the magnitude of N_1.

Shown in Figure 14.7 are data for the shear viscosity of a low-density polyethylene sample (called IUPAC A) as a function of the shear rate over a range of temperatures [4]. These data were collected as part of an international study that involved several investigators and numerous instruments in different laboratories. For this polymer sample, $\bar{M}_n$ is 2×10^4 and $\bar{M}_w$ exceeds 10^6. It is seen that the shear rates attained are low enough that, at each temperature, the zero shear rate viscosity can be identified easily; these η_0 values are noted in Figure 14.7 itself. Clearly, the viscosity values increase with decreasing temperature. If we plot η_0 as a function of the reciprocal of the absolute temperature, we get a straight line, showing that the Arrhenius relation is obeyed; that is,

$$\eta_0 \propto \exp\left(\frac{E}{RT}\right) \qquad (14.3.11)$$

where the activation energy E equals $13.6\,\text{kcal/mol}$ in the present case. The activation energy typically increases as the polymer chain stiffness increases.

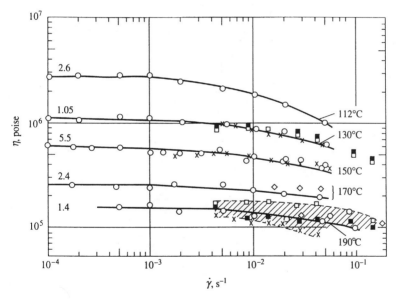

FIGURE 14.7 Determination of zero shear viscosity η_0; sample A. I: □ WRG (25-mm diameter, cone angle $\alpha = 4°$); ■ sample A stabilized, temperature shift of 130°C data using $E = 11.7\,\text{kcal/mol}$ [4]. II: × Kepes (26.15-mm diameter. $\alpha = 21°4'$). IV: ○ WRG modified (72-mm diameter, $\alpha = 4°$). (From Ref. 4.) (Reprinted with permission from Meissner, J.: "Basic Parameters, Melt Rheology, Processing and End-Use Properties of Three Similar Low Density Polyethylene Samples," *Pure Appl. Chem.*, vol. 42, pp. 553–612, 1975.)

Note, though, that close to the glass transition temperature, the WLF equation [see also Eqs. (12.5.4) and (13.6.13) of Chapters 12 and 13, respectively]

$$\log\left(\frac{\eta_0(T)}{\eta_0(T_g)}\right) = -\frac{17.44(T - T_g)}{51.6 + (T - T_g)} \tag{14.3.12}$$

may be a more appropriate equation to use. This is because at temperatures between T_g and $T_g + 100°C$, the viscosity is strongly influenced by increases in free volume; at higher temperatures, we essentially have an activated jump process.

   In addition to temperature, the zero-shear viscosity is also influenced by the pressure, especially at high pressures and especially close to the glass transition temperature. The pressure dependence is again of the Arrhenius type,

$$\eta_0 \propto \exp(Bp) \tag{14.3.13}$$

and is the result of the tendency of the free volume to decrease on the application of a large hydrostatic pressure.

   The first normal stress difference in shear, N_1, on the IUPAC A sample and measured at 130°C is displayed on logarithmic coordinates in Figure 14.8. If we compare this figure to Figure 14.7, we determine that N_1 is comparable to the shear stress at low shear rates but significantly exceeds this quantity at higher shear rates. This happens because of the stronger dependence of N_1 on $\dot\gamma$ compared with the dependence of $\tau_{\phi\theta}$ on $\dot\gamma$. Indeed, straight lines typically

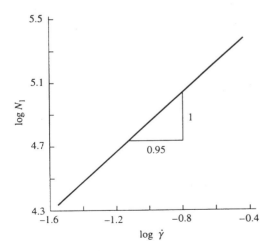

FIGURE 14.8   First normal stress difference in shear for IUPAC A LDPE at 130°C. (From Ref. 4.)

result when these two functions are plotted in terms of the shear rate on logarithmic coordinates. Although the slope of the N_1 plot usually lies between 1 and 2, the maximum value of the slope of the shear stress plot is unity.

Although all of the data discussed so far have been melt data, the shear behavior of polymer solutions is similar to that of polymer melts, except that the stress levels are lower. The same viscometers are used for both fluids, and the cone-and-plate viscometer has become a standard laboratory tool. Indeed, it is the instrument of choice for making simultaneous measurements of the viscosity and the first normal stress difference. This is because data analysis involves few assumptions. There are limitations, though; we can have viscous-heating effects if the viscosity of the fluid is high but, more importantly, centrifugal forces and elastic instabilities cause the sample to be expelled from the gap at high rates of revolution. This limits the instrument to shear rates less than about $100 \ \text{sec}^{-1}$. Difficulties also arise during measurements on filled polymers if the size of the particulates is comparable to the gap size. In such a situation, we use two parallel plates with a large gap. This situation can also be analysed [5], but the analysis is significantly more involved. Also, N_1 and N_2 cannot be obtained separately.

Example 14.1: If $\ln N_1$ is plotted versus $\ln \tau_{\phi\theta}$, a straight line of slope 2 frequently results. Furthermore, the same plot is obtained irrespective of the temperature of measurement. If data are obtained with the help of a cone-and-plate viscometer, how would M be related to F for this relationship to hold?

Solution: We know that $\ln N_1 = 2 \ln \tau_{\phi\theta} + \ln c$, where c is a constant. Consequently,

$$\frac{N_1}{\tau_{\phi\theta}^2} = c$$

and using Eqs. (14.3.3) and (14.3.9) gives

$$\frac{2F}{\pi R^2} = c \left(\frac{3M}{2\pi R^3} \right)^2$$

or

$$F \propto M^2$$

The cone-and-plate viscometer is also routinely used to make time-dependent measurements. If Ω or, alternatively, $\dot{\gamma}$ is not constant but is some specified function of time, the measured quantities M and F also depend on time. Data analysis, however, remains unchanged and the shear stress and N_1 are again given by Eq. (14.3.3) and Eq. (14.3.9), respectively, and are functions of time. Note that today's viscometers can be operated not just at specified values of the shear rate but also at specified values of the shear stress; the shear rate then becomes the dependent variable. Other popular rotational viscometers are the parallel-plate viscometer and the Couette viscometer. These are described in Problems 14.4 and 14.5, respectively.

14.4 THE CAPILLARY VISCOMETER

The need to measure fluid properties at shear rates higher than those accessible with rotational viscometers arises because deformation rates can easily reach 10^5–10^6 sec^{-1} in polymer processing operations. To attain these high shear rates, we use flow through capillaries or slits and calculate the viscometric functions from a knowledge of the pressure drop-versus-flow rate relationship.

Consider the steady flow through a horizontal capillary of circular cross section, as shown in Figure 14.9. In most of the capillary in a region of length L, away from the exit or inlet, the pressure gradient is constant and there is fully developed flow. Thus, $\Delta p/L$ equals dp/dz. In the entry and exit regions, though, there are velocity rearrangements and the actual pressure drops exceed those calculated based on fully developed flow. To neglect these extra pressure losses, we need to use capillaries for which L/D exceeds 50, where D is the capillary diameter and its value is usually greater than about 0.025 cm. For polymer melts, though, short capillaries are used because of thermal degradation problems and because long capillaries require very high pressures. However, methods exist that correct for entrance effects, which are the larger of the two losses [6]. Of course,

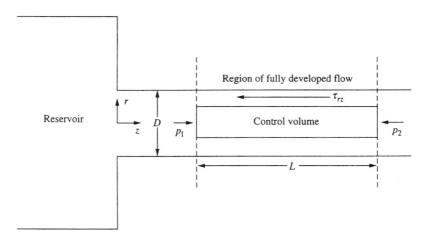

FIGURE 14.9   Schematic diagram of a capillary viscometer.

to obtain consistent results, the ratio of the reservoir diameter to the capillary diameter should also be large—a value of 12–18 is safe.

The fully developed flow region is again a region of viscometric flow because only v_z is nonzero and it varies only with r. Thus, the shear rate $\dot{\gamma}$ is dv_z/dr. However, the deformation is not homogeneous because the shear rate varies from zero at the capillary axis to a maximum at the tube wall. To obtain the viscosity, we need to be able to measure the shear stress and the shear rate at the same location. Although the shear stress can be obtained at any radial position from a macroscopic force balance, the shear rate is most easily calculated at the tube wall. Note that in the analysis that follows, $\partial/\partial z$ is taken to be zero for the extra stresses due to fully developed flow and $\partial/\partial\theta$ is zero due to symmetry. From a macroscopic force balance on the control volume shown in Figure 14.9, we have

$$\pi r^2 (p_1 - p_2) - 2\pi r L \tau_{rz} = 0 \tag{14.4.1}$$

or

$$\tau_{rz} = -\frac{\Delta p r}{2L} \tag{14.4.2}$$

from which we can calculate the wall shear stress, $\tau_{rz}(R)$. The shear stress, as shown, is a positive quantity because Δp is a negative number.

To obtain the shear rate at the wall, we note that the volumetric flow rate Q is given by

$$Q = \int_0^R 2\pi r v_z \, dr \tag{14.4.3}$$

Integrating by parts and knowing that r equals zero at the centerline and v_z equals zero at the tube wall yields the following:

$$Q = -\pi \int_0^R r^2 \dot{\gamma}(r) \, dr \tag{14.4.4}$$

Changing the independent variable from r to τ_{rz} through the use of Eq. (14.4.2) and noting that τ_{rz} depends uniquely on $\dot{\gamma}$ and vice versa gives

$$Q = \frac{8\pi L^3}{(\Delta p)^3} \int_0^{\tau_w} \tau_{rz}^2 \dot{\gamma}(\tau_{rz}) \, d\tau_{rz} \tag{14.4.5}$$

where

$$\tau_w = \tau_{rz}(R) = -\frac{\Delta p R}{2L} \tag{14.4.6}$$

Eliminating Δp in Eq. (14.4.5) through the use of Eq. (14.4.6) and rearranging the result, we have

$$\frac{Q\tau_w^3}{\pi R^3} = -\int_0^{\tau_w} \tau_{rz}^2 \dot{\gamma}(\tau_{rz})\, d\tau_{rz} \tag{14.4.7}$$

Using Leibnitz' rule to differentiate Eq. (14.4.7) with respect to τ_w gives

$$\frac{3\tau_w^2 Q}{\pi R^3} + \frac{\tau_w^3}{\pi R^3}\frac{dQ}{d\tau_w} = -\tau_w^2 \dot{\gamma}(\tau_w) \tag{14.4.8}$$

so that

$$-\dot{\gamma}(\tau_w) = \frac{4Q}{\pi R^3}\left(\frac{3}{4} + \frac{1}{4}\frac{d\ln Q}{d\ln \tau_w}\right) \tag{14.4.9}$$

To measure the viscosity as a function of shear rate, the flow rate is determined as a function of the pressure drop. This can be converted to flow rate as a function of the wall shear stress using Eq. (14.4.6). Therefore, $\ln Q$ can be plotted versus $\ln \tau_w$. At any given value of τ_w, this plot yields the slope $d\ln Q/d\ln \tau_w$, which, with the help of Eq. (14.4.9), gives $\dot{\gamma}(\tau_w)$. Dividing τ_w by this value of $\dot{\gamma}_w$ gives $\eta(\dot{\gamma}_w)$. By repeating this process, we can get η at other values of $\dot{\gamma}_w$. Because it does not matter where the viscosity is measured so long as the shear stress and shear rate are determined at the same location, we can drop the subscript w, resulting in the term $\eta(\dot{\gamma})$.

For a Newtonian liquid the fully developed velocity profile is parabolic and the wall shear rate is as follows (see Problem 14.6):

$$-\dot{\gamma}_w = \frac{4Q}{\pi R^3} \tag{14.4.10}$$

Because $d\ln Q/d\ln \tau_w$ is unity, the result in Eq. (14.4.10) is consistent with Eq. (14.4.9), which can be rewritten as

$$-\dot{\gamma}_w = \dot{\gamma}_{app}\left[\frac{3}{4} + \frac{1}{4}\frac{d\ln Q}{d\ln \tau_w}\right] \tag{14.4.11}$$

in which $\dot{\gamma}_{app}$ is the apparent shear rate, assuming Newtonian liquid behavior, and the quantity in brackets is called the *Rabinowitsch correction factor*.

Figure 14.10 shows capillary viscometry data at 150°C for the IUPAC A LDPE considered earlier in Figure 14.7. It is apparent that these data blend nicely with the data generated using a cone-and-plate viscometer. At shear rates greater than about 1 sec^{-1}, the viscosity-versus-shear rate behavior is linear. Consequently, one can say that the shear stress τ (where subscripts have been dropped for convenience) is given by

$$\tau = K\dot{\gamma}^n \tag{14.4.12}$$

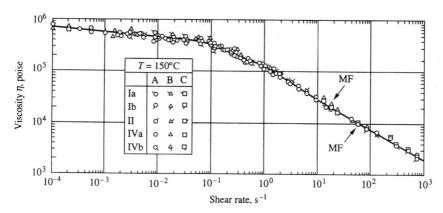

FIGURE 14.10 Viscosity functions for samples A, B, and C at 150°C. Ia, IVa: Weissenberg rheogoniometer (Ia measures at 130°C and shifts the data); II: Kepes rotational rheometer (cone-and-plate); Ib, IVb: capillary viscometers. MF denotes the onset of melt fracture. (From Ref. 4.) (Reprinted with permission from Meissner, J.: "Basic Parameters, Melt Rheology, Processing and End-Use Properties of Three Similar Low Density Polyethylene Samples," *Pure Appl. Chem.*, vol. 42, pp. 553–612, 1975.)

in which K and n are constants. Equation (14.4.12) is known as the *power-law equation*, K is called the *consistency index*, and n is called the *power-law index*. Usually, $n \leq 1$.

Equation (14.4.12) obviously cannot represent the viscosity over the entire range of shear rates, because it suggests the following:

$$\eta = \frac{K}{\dot{\gamma}^{1-n}} \qquad (14.4.13)$$

which predicts a meaningless increase in the viscosity as the shear rate is decreased. A popular alternative model that does not have this shortcoming is the Carreau model [7],

$$\frac{\eta - \eta_\infty}{\eta_0 - \eta_\infty} = [1 + (\lambda_c \dot{\gamma})^2]^{(n-1)/2} \qquad (14.4.14)$$

which has four constants: η_0, η_∞, λ_c, and n. Although this model can accommodate a limiting shear viscosity at both high and low shear rates, the flexibility comes at the expense of greater complexity. Note that the parameter λ_c determines the point of onset of shear thinning; the onset of shear thinning is found to move to larger values of the shear rate as the temperature of measurement is increased (see Fig. 14.7).

Example 14.2: The (unstable) melt fracture behavior illustrated in Figure 14.4 is found to occur in polymer melts when the wall shear stress is about 10^6 dyn/cm^2. If we were to extrude IUPAC A LDPE through a long capillary 0.025 cm diameter at 150°C, at what flow rate might we expect to observe melt fracture?

Solution: By trial and error, we find that the shear rate at which $\eta\dot{\gamma}$ equals 10^6 dyn/cm^2 in Figure 14.10 is about 100 sec^{-1} (as indicated in the figure, melt fracture actually occurs at slightly lower shear rates).

   If we represent this polymer as a power-law fluid, the shear rate at the wall is given by the following (see Problem 14.7):

$$-\dot{\gamma}_w = \frac{4Q}{\pi R^3}\left(\frac{3n+1}{4n}\right)$$

so that Q is known once n is known. From Figure 14.10, however, n equals 0.42. Thus, we have

$$Q = \frac{(-\dot{\gamma}_w)\pi R^3}{(3n+1)/n}$$

$$= \frac{100 \times \pi \times (0.025/2)^3}{(3 \times 0.42 + 1)/(0.42)}$$

$$= 1.14 \times 10^{-4} \text{ cc/s}$$

   Capillary viscometers come in two basic designs, which differ in the method of melt extrusion. In the pressure-driven instrument, gas pressure is used to pump the liquid out of a reservoir and into the capillary, and we measure the volumetric flow rate corresponding to the applied Δp. In a plunger-driven instrument, as the name implies, the flow rate is set and the corresponding Δp is measured. Shear rates of 10^5–10^6 sec^{-1} can easily be achieved with these instruments, with the upper limit on the shear rate being decided by the maximum speed of the plunger in a plunger-driven instrument and the maximum pressure in a pressure-driven instrument. Of course, it must be ensured that viscous dissipation does not lead to nonisothermal conditions; this computation is easy to carry out [8]. The lower limit of operation of a plunger-type viscometer is set by the amount of friction between the plunger and the barrel. As the plunger speed decreases, this friction can become a significant part of the total Δp and make the viscosity measurements unreliable. More details about capillary viscometry are available in the article by Kestin et al. [9]. The theory for flow through slits is similar to that for circular tubes [10]. An advantage of slit viscometry, as shown by Lodge, is that the technique can also be used to measure normal stress differences [10].

14.5 EXTENSIONAL VISCOMETERS

Besides viscometric flow, the other major category of flow that can be generated in the laboratory is extensional flow. In mathematical terms, extensional flow may be represented in a rectangular Cartesian coordinate system x_i by the set of equations for the three components of the velocity vector

$$v_1 = \dot{\varepsilon}_1 x_1 \tag{14.5.1}$$

$$v_2 = \dot{\varepsilon}_2 x_2 \tag{14.5.2}$$

$$v_3 = \dot{\varepsilon}_3 x_3 \tag{14.5.3}$$

which also define the stretch rates $\dot{\varepsilon}_i$. In uniaxial extension at constant stretch rate, $\dot{\varepsilon}_1 = \dot{\varepsilon}$ and $\dot{\varepsilon}_2 = \dot{\varepsilon}_3 = -\dot{\varepsilon}/2$. Interest in this mode of deformation stems from the fact that industrially important polymer processing operations such as fiber spinning are, essentially, examples of uniaxial extension.

In physical terms, the distance between material planes that are perpendicular to the flow direction increases in extensional flow. This is illustrated for uniaxial extension in Figure 14.11a. A material plane is a surface that always contains the same material points or particles. In viscometric flow, the fluid on any surface for which the coordinate x_2 is a constant forms a material plane (see Figure 14.11b). Here, each material plane moves as a rigid body and there is

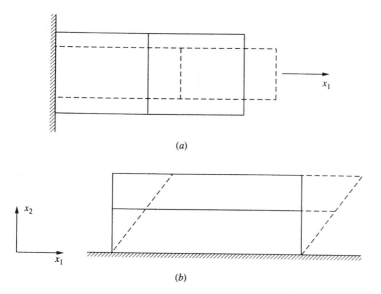

(a)

(b)

FIGURE 14.11 (a) Increase in distance between material planes in a uniaxial extensional flow; (b) Sliding of material planes in viscometric flow. (From Ref. 11.)

relative sliding between neighboring material planes. In extensional flow, as opposed to shear flow, polymer molecules tend to uncoil and ultimately there can even be stretching of chemical bonds, which results in chain scission. Therefore, stresses in the flow direction can reach fairly large values.

Although it is somewhat difficult to visualize how a rod of polymer might be stretched in the manner of Figure 14.11a, this can be done for both polymer melts [11] and polymer solutions [12]. For polymer melts, a cylindrical sample is immersed in an oil bath with one end attached to a force transducer and the other end moved outward so that the stretch rate is maintained constant. Similarly, for polymer solutions, we merely place the liquid sample between two coaxial disks, one of which is stationary and connected to a microbalance, and the other disk moves outward, generating the stretching. The filament diameter thins progressively but remains independent of position. Simultaneously, the filament length l increases exponentially as follows:

$$\ln\left(\frac{l}{l_0}\right) = \dot{\varepsilon}t \tag{14.5.4}$$

which follows directly from Eq. (14.5.1).

In this flow field, there is no shear deformation, and the total stress tensor as well as the extra stress tensor are diagonal. As a consequence, there are only three nonzero stress components, but, due to fluid incompressibility, we can measure only two stress differences. Further, in uniaxial extension, the two directions that are perpendicular to the stretching direction are identical, so there is only one measurable material function: the net tensile stress σ_E, which is the difference $T_{11} - T_{22}$ or $\tau_{11} - \tau_{22}$.

For constant-stretch-rate homogeneous deformation, which begins from rest, a tensile stress growth coefficient is defined as

$$\eta_E^+(t, \dot{\varepsilon}) = \frac{\sigma_E}{\dot{\varepsilon}} \tag{14.5.5}$$

which has the dimensions of viscosity. The limiting value of η_E^+ as time tends to infinity is termed the *tensile, elongational,* or *extensional viscosity,* η_E. In general, η_E is a function of the stretch rate, although in the limit of vanishingly low stretch rates, we have the following [5]:

$$\lim_{\dot{\varepsilon} \to 0}\left(\frac{\eta_E}{\eta_0}\right) = 3 \tag{14.5.6}$$

where η_0 is zero-shear rate viscosity.

Laun and Munstedt have obtained tensile stress growth data on the IUPAC A LDPE sample at 150°C [13,14], and these are shown in Figure 14.12. At a given stretch rate, the stress increases monotonically and, ultimately, tends to a limiting value even for stretch rates as high as 10 sec^{-1}. The extensional viscosity, calculated using these steady-state data, is displayed in Figure 14.13

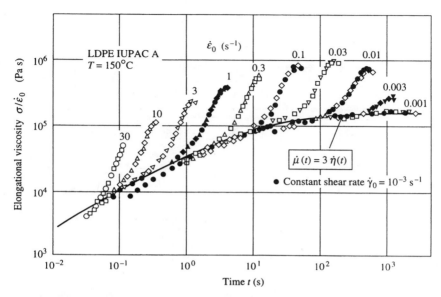

FIGURE 14.12 Time-dependent elongational viscosity at different stretching rates. (From Ref. 13.) Reprinted from Rheol. Acta, vol. 18, Munstedt, H., and H. M. Lann: Elongational behavior of an LDPE MeH: II. Transient behavior in constant stretching rate and tensile creep experiments. Comparison with shear data. Temperature dependence of the elongational properties, 492–504, 1979. Used by permission of Steinkopft Publishers, Darmstadt (FRG).

as a function of the stretch rate, and at low stretch rates, Eq. (14.5.6) is found to hold. As the stretch rate is increased, the extensional viscosity increases, goes through a maximum, and finally decreases to a value even below that of the zero-shear value. The behavior of polymer solutions is qualitatively similar to that of polymer melts, except that the extensional viscosity of polymer solutions can exceed the corresponding shear viscosity by a far wider margin than does the extensional viscosity of polymer melts [15].

Example 14.3: How is the extensional viscosity of a Newtonian fluid related to the shear viscosity?

Solution: Using Eqs. (14.2.2), (14.5.1), and (14.5.2) gives

$$T_{11} - T_{22} = 2\eta \left(\frac{\partial v_1}{\partial x_1} - \frac{\partial v_2}{\partial x_2} \right)$$
$$= 2\eta(\dot{\varepsilon}_1 - \dot{\varepsilon}_2)$$

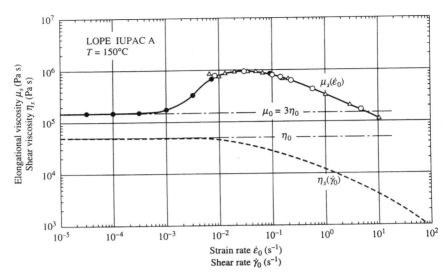

FIGURE 14.13 Steady-state viscosity in shear and elongation as a function of deformation rate. (From Ref. 14.) Reprinted from Rheol. Acta, vol. 17. Laun, H. M., and H. Munstedt: Elongational behavior of an LDPE melt: I. Strain rate and stress dependence of viscosity and recoverable strain in steady state. Comparison with shear data. Influence of interfacial tension, 415–425, 1978. Used by permission of Steinkopff Publishers, Darmstadt (FRG).

For uniaxial extension, however, $\dot{\varepsilon}_1 = -2\dot{\varepsilon}_2 = \dot{\varepsilon}$. Therefore $\sigma_E = 2\eta[\dot{\varepsilon} + (\dot{\varepsilon}/2)] = 3\eta\dot{\varepsilon}$ and $\eta_E = 3\eta$.

Even though commercial instruments are available for making extensional viscosity measurements on polymer melts, this is not a routine measurement. The stretch-rate range of these extensional viscometers is such that the maximum stretch rate that can be achieved is of the order of 1 sec^{-1}; in polymer processing operations, a stretch rate of 100 sec^{-1} is commonplace. Also, not every polymer stretches uniformly, and, even when it does, steady-state stress levels are not always attained. For all of these reasons, extensional viscometry is an area of current research. Additional details regarding extensional and other viscometers may be found in the book by Dealy [16].

14.6 BOLTZMANN SUPERPOSITION PRINCIPLE

Experimental data obtained in viscometric flow or extensional flow are obviously useful for predicting material behavior in a flow field that is predominantly shear or extension. Thus, the shear viscosity function is needed to compute the pressure drop for flow through a channel, whereas the extensional viscosity function can be employed to calculate the force of peel adhesion in a pressure-sensitive adhesive. In addition, the data reveal material behavior and can be used for material characterization and product comparison. However, it is logical to ask whether it is necessary to conduct a new experiment every time we want information in a different flow field or if we want transient data in the same flow field. The answer, we hope, is that behavior in one flow situation can be predicted using data generated in a different flow situation. In other words, a stress constitutive equation for non-Newtonian fluids is sought. This equation would involve a limited number of constants or functions that would be specific to the material being examined and whose numerical values could be determined by conducting one or two experiments in idealized flow fields. In this section, we show how this goal can be partially achieved.

As seen earlier in Problem 12.3 of Chapter 12 and more recently in Figure 14.2, the stress in a polymeric fluid does not decay to zero once deformation is halted. Instead, stress relaxation occurs. Thus, upon imposition of a shear strain γ, we expect the shear stress τ to be

$$\tau(t) = G(t)\gamma \tag{14.6.1}$$

in which $G(t)$ is a modulus function that decays with time.

If a series of strains $\gamma_1, \gamma_2, \ldots$ is imposed on the material at times $t_1, t_2, \ldots$ in the past, the stress at the present time t is a combination of the stresses resulting from each of these strains. If this combination can be taken to be a linear combination (which is the thesis of the Boltzmann superposition principle), we can write the following expression:

$$\tau(t) = G(t - t_1)\gamma_1 + G(t - t_2)\gamma_2 + \cdots \tag{14.6.2}$$

where $t - t_i$ is the time elapsed since the imposition of strain γ_i. Converting the sum in Eq. (14.6.2) into an integral gives

$$\tau(t) = \int_{-\infty}^{t} G(t - s)\, d\gamma = \int_{-\infty}^{t} G(t - s)\dot{\gamma}\, ds \tag{14.6.3}$$

wherein s is a past time and $\dot{\gamma}$ is the rate of deformation. The function $G(t - s)$ is a stress-relaxation modulus and Eq. (14.6.3) is the mathematical form of the Boltzmann superposition principle. This equation can be generalized to three dimensions by rewriting it as follows:

$$\tau_{ij}(t) = \int_{-\infty}^{t} G(t - s)\dot{\gamma}_{ij}\, ds \tag{14.6.4}$$

with the deformation rate components $\dot{\gamma}_{ij}$ being $(\partial v_i/\partial x_j + \partial v_j/\partial x_i)$.

Example 14.4: Is a shear-thinning viscosity consistent with the predictions of the Boltzmann superposition principle?

Solution: If the shear rate is held constant, the shear viscosity according to Eq. (14.6.3) is given by

$$\eta = \frac{\tau(t)}{\dot\gamma} = \int_{-\infty}^{t} G(t-s)\,ds$$

If we allow $t-s$ to equal t' we have

$$\eta = \int_{0}^{\infty} G(t')\,dt' = \text{constant}$$

and, not surprisingly, we find that a linear superposition of stresses does not allow for nonlinear effects such as shear thinning. Shear thinning is a nonlinear effect because the shear stress is less than doubled when we double the shear rate. Thus, the viscosity calculated in this example is the zero-shear viscosity.

The Boltzmann superposition principle is the embodiment of the theory of linear viscoelasticity, and it is valid for both steady and transient deformations, provided that the extent of deformation is low. A specific form of the stress–relaxation modulus may be obtained by permitting the stress response in a polymer to be made up of an elastic contribution and a viscous contribution. Thus, if we again use the Maxwell element encountered previously in Example 12.2 and Figure 12.4 of Chapter 12, the total strain γ in the spring and dashpot combination is (at any time) a sum of the individual strains; that is,

$$\gamma = \gamma_s + \gamma_d \tag{14.6.5}$$

where the subscripts s and d denote the spring and dashpot, respectively.

Because the applied stress τ equals the stress in both the spring and the dashpot, Eq. (14.6.5) implies (see Example 12.1) the following:

$$\frac{d\gamma}{dt} = \frac{d\gamma_s}{dt} + \frac{d\gamma_d}{dt} = \frac{1}{G}\frac{d\tau}{dt} + \frac{\tau}{\eta} \tag{14.6.6}$$

If we assume that $\tau \to 0$ as $t \to -\infty$, the solution of this first-order, linear, nonhomogeneous differential equation is given by

$$\tau(t) = G \int_{-\infty}^{t} e^{-(t-s)/\theta}\left(\frac{d\gamma}{ds}\right) ds \tag{14.6.7}$$

in which s is a dummy variable of integration and θ equals η/G, where G is the constant spring modulus and not the stress relaxation modulus of spring-and-dashpot combination.

Comparing Eqs. (14.6.3) and (14.6.7) yields

$$G(t-s) = G\exp\left(-\frac{t-s}{\theta}\right) \tag{14.6.8}$$

which, when introduced into Eq. (14.6.1), reveals that

$$\frac{\tau(t)}{\gamma} = Ge^{-t/\theta} \tag{14.6.9}$$

and it is seen that the stress decays over a time scale of the order of θ. Consequently, θ is called a *relaxation time*. Although a single relaxation time can be expected to fit data on monodisperse polymers, most polymer samples are polydisperse. It is for this reason that we modify the mechanical analog to make it consist of N Maxwell elements in parallel; each spring has a modulus G_i and each dashpot has a damping constant η_i. By repeating the analysis presented here, we find that

$$G(t-s) = \sum_{i=1}^{N} G_i \exp\left(-\frac{t-s}{\theta_i}\right) \tag{14.6.10}$$

in which $\theta_i = \eta_i/G_i$. The constants G_i and θ_i are usually obtained from the results of dynamic mechanical experiments rather than by conducting a stress–relaxation experiment.

14.7 DYNAMIC MECHANICAL PROPERTIES

As discussed in Section 12.4 of Chapter 12, dynamic mechanical testing is conducted by subjecting a polymer sample to a sinusoidal strain of amplitude γ_0 and frequency ω. Because the strain amplitude is usually of infinitesimal magnitude, we can legitimately apply the Boltzmann superposition principle to this flow situation. Before doing this, though, we change the independent variable in Eq. (14.6.3) from s to t', where $t' = t - s$. Thus, we have

$$\tau(t) = \int_0^\infty G(t')\dot\gamma(t-t')\,dt' \tag{14.7.1}$$

in which $\dot\gamma$ is set to be $\omega\gamma_0 \cos[\omega(t-t')]$. As a result, we have

$$\frac{\tau(t)}{\gamma_0} = \omega\cos(\omega t)\int_0^\infty G(t')\cos(\omega t')\,dt' + \omega\sin(\omega t)\int_0^\infty G(t')\sin(\omega t')\,dt' \tag{14.7.2}$$

which when compared to Eq. (12.4.4) yields the following expressions for the storage and loss moduli, respectively:

$$G'(\omega) = \omega\int_0^\infty G(t')\sin(\omega t')\,dt' \tag{14.7.3}$$

$$G''(\omega) = \omega\int_0^\infty G(t')\cos(\omega t')\,dt' \tag{14.7.4}$$

Example 14.5: How do the storage modulus and loss modulus vary with frequency when $\omega \to 0$?

Solution: As $\omega \to 0$, $\sin(\omega t') \to \omega t'$ and $\cos(\omega t') = 1$, with the result that

$$\lim_{\omega \to 0} G'(\omega) = [\int_0^\infty t' G(t')\, dt']\omega^2 \tag{14.7.5}$$

$$\lim_{\omega \to 0} G''(\omega) = [\int_0^\infty G(t')\, dt']\omega = \eta_0 \omega \tag{14.7.6}$$

where the second equality in Eq. (14.7.6) follows from the result of Example 14.4.

Equations (14.7.5) and (14.7.6) are valid for all materials at low enough frequencies. However, we may not always observe this behavior due to the inability to accurately measure small stresses at very low frequencies. The limiting behavior at very high frequencies can also be obtained (see Problem 14.17). The final expressions are as follows:

$$\lim_{\omega \to \infty} G'(\omega) = G(0) \tag{14.7.7}$$

$$\lim_{\omega \to \infty} G''(\omega) = 0 \tag{14.7.8}$$

Although a knowledge of the stress–relaxation modulus allows us to calculate G' and G'' via Eqs. (14.7.3) and (14.7.4), in practice we find that the experimental measurement of $G'(\omega)$ or $G''(\omega)$ (using a cone-and-plate visco-meter, for example) is much more accurate than the measurement of $G(t')$. Consequently, we measure $G'(\omega)$ and $G''(\omega)$ and use these data to compute $G(t')$. The computed value of $G(t')$ can, in turn, be used to calculate any other linear viscoelastic function through the use of Eq. (14.7.1).

In order to accomplish these objectives, we need methods of interrelating the various linear viscoelastic functions. The general techniques of obtaining one function from another have been discussed by Ferry [17]. In the present case, Baumgaertel and Winter have proposed a particularly simple method [18]. If we introduce Eq. (14.6.10) into Eqs. (14.7.3) and (14.7.4) and carry out the integrations, we get (see also Example 12.2).

$$G'(\omega) = \sum_{i=1}^N \frac{G_i(\omega\theta_i)^2}{1 + (\omega\theta_i)^2} \tag{14.7.9}$$

$$G''(\omega) = \sum_{i=1}^N \frac{G_i(\omega\theta_i)}{1 + (\omega\theta_i)^2} \tag{14.7.10}$$

The terms G_i and θ_i are found by simply fitting Eqs. (14.7.9) and (14.7.10) to measured $G'(\omega)$ and $G''(\omega)$ data using a nonlinear least-squares procedure [18]. In doing so, the choice of N is crucial; a small value of N can lead to errors, whereas a large value of N cannot be justified given the normal errors associated with measuring G' and G''. The set of relaxation times θ_i and moduli G_i is called

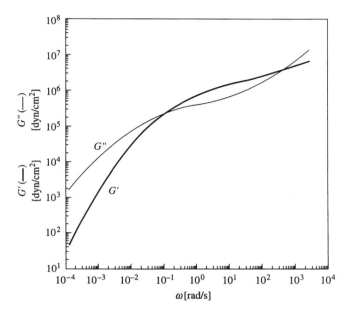

FIGURE 14.14 Master curve of storage and loss moduli of polystyrene.

a discrete-relaxation time spectrum, and we typically choose between one and two relaxation modes per decade of frequency. Figure 14.14 shows a master curve of G' and G'' values in a temperature range of 130–250°C on an injection-molding grade sample of polystyrene; all of the data have been combined by means of a horizontal shift using time–temperature superposition with a 150°C reference temperature according to the procedure of Section 12.5. The stress–relaxation modulus (calculated in the manner of Baumgaertel and Winter) is displayed in Figure 14.15 along with actual stress–relaxation data at 150°C; the agreement could not be better.

The utility of dynamic data appears to go beyond the theoretical applications considered in this section. We find, for example, that the modulus of the complex viscosity η^*, defined as

$$|\eta^*| = [(\eta')^2 + (\eta'')^2]^{1/2} \tag{14.7.11}$$

where $\eta' = G''/\omega$ and $\eta'' = G'/\omega$, when plotted versus frequency, often superposes with the steady shear viscosity as a function of the shear rate [19]. This is known as the Cox–Merz rule, and it provides information about a nonlinear property from a measurement of a linear property. Note that η' is generally called the *dynamic viscosity*.

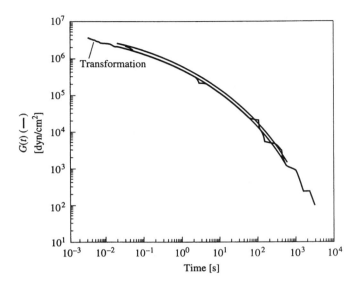

FIGURE 14.15   Stress relaxation modulus of polystyrene at 150°C.

Another useful empiricism, Laun's rule, relates the first normal stress coefficient to dynamic data [20]:

$$\psi_1 = \frac{2\eta''(\omega)}{\omega}\left[1 + \left(\frac{\eta''}{\eta'}\right)^{0.7}\right]_{\omega=\dot{\gamma}} \tag{14.7.12}$$

It should be noted that relating the various material functions to properties such as the molecular weight can only be done with the help of molecular theories. These theories are examined in some detail in the remainder of this chapter.

14.8 THEORIES OF SHEAR VISCOSITY

The shear viscosity of polymer melts depends primarily on the molecular weight, the temperature, and the imposed shear rate; for polymer solutions, the concentration and nature of solvent are additional variables. In one of the earliest theories, [21,22], the shear viscosity was calculated by determining the amount of energy dissipated due to fluid friction in a steady laminar shearing flow at a constant shear rate $\dot{\gamma}$. For this flow situation, depicted in Figure 14.16, the energy dissipated per unit time, per unit volume, P, is given as follows [23]:

$$P = \eta\left(\frac{dv_1}{dx_2}\right)^2 \tag{14.8.1}$$

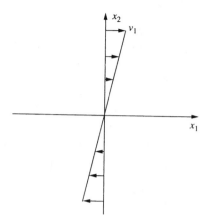

FIGURE 14.16 Polymer molecule in steady laminar shearing flow.

Because (dv_1/dx_2) is $\dot{\gamma}$, the viscosity is known if P can be computed. P, in turn, is calculated as the dot product, or scalar product, of the drag force with the velocity vector.

If, in this flow field, we consider the end-to-end vector of the typical polymer molecule, the two ends move at different velocities in the x_1 direction due to the presence of a velocity gradient. The analysis is simplified if we bring the center of mass of the molecule to rest by imposing the average velocity in the negative x_1 direction. In this case, as is clear from Figure 14.16, the molecule rotates in the clockwise direction. At steady state, the torque due to the imposed shear flow balances the opposing torque due to fluid friction and there is no angular acceleration. The polymer molecule, as represented by the end-to-end vector, rotates at a constant angular velocity ω. The linear velocity at any radial position r is ωr. The drag force F acting on any segment of polymer (say, a monomer unit) is as follows:

$$\mathbf{F} = \zeta \mathbf{v} \qquad (14.8.2)$$

where ζ is the friction coefficient per monomer unit, and Eq. (14.8.2) is similar in concept to Stokes' law for drag on spheres [3].

The work done per unit time against fluid friction on a single polymer segment is given as

$$\mathbf{F} \cdot \mathbf{v} = \zeta |\mathbf{v}|^2 = \zeta (\omega r)^2 \qquad (14.8.3)$$

600 **Chapter 14**

For all of the n segments taken together,

$$\mathbf{F} \cdot \mathbf{v} = \sum_1^N \zeta(\omega r_i)^2 = n\zeta\omega^2\langle r_i^2\rangle \tag{14.8.4}$$

where $\langle r_i^2\rangle$ is the two-dimensional analog of the radius of gyration $\langle s^2\rangle$ encountered earlier. It can be shown that $\langle r_i^2\rangle$ equals $\frac{2}{3}\langle s^2\rangle$ [24], and because $\langle s^2\rangle$ is given by $nl^2/6$ for a freely jointed chain [see Eq. (10.2.17)], we have

$$\mathbf{F} \cdot \mathbf{v} = \frac{n^2 l^2 \zeta \omega^2}{9} \tag{14.8.5}$$

in which the angular velocity, in general, is given by

$$\omega = \frac{1}{2}\left(\frac{\partial v_2}{\partial x_1} - \frac{\partial v_1}{\partial x_2}\right) \tag{14.8.6}$$

which reduces to $\omega = -\dot{\gamma}/2$ for the situation of Figure 14.16. Thus, we have

$$\mathbf{F} \cdot \mathbf{v} = \frac{n^2 l^2 \zeta \dot{\gamma}^2}{36} \tag{14.8.7}$$

To obtain the energy dissipated per unit volume, we need to know the number of polymer molecules per unit volume. For melts, this is $\rho N_A/M$ and for solutions it is cN_A/M, where ρ is density, c is the concentration in mass units, N_A is Avogadro's number, and M is the molecular weight (assuming monodisperse polymer). Multiplying the right-hand side of Eq. (14.8.7) by either $\rho N_A/M$ or cN_A/M and dividing by $\dot{\gamma}^2$ gives the viscosity [recall Eq. (14.8.1)]:

$$\eta = \frac{\rho N_A}{M}\frac{n^2 l^2 \zeta}{36} = \frac{cN_A}{M}\frac{n^2 l^2 \zeta}{36} \tag{14.8.8}$$

Finally, noting that M equals nM_0, where M_0 is the monomer molecular weight, we have

$$\eta = \frac{\rho N_A \zeta}{36M_0}\left(\frac{nl^2}{M}\right)M = \frac{cN_A}{36M_0}\zeta\left(\frac{nl^2}{M}\right)M \tag{14.8.9}$$

and η is predicted to be proportional to M or cM because nl^2/M is not a function of M. Although the dependence on temperature can enter through ζ, it is obvious that Eq. (14.8.9) cannot predict shear thinning, which is almost always observed at high enough shear rates.

When experimental data for the zero-shear viscosity of polymer melts and solutions are examined in light of Eq. (14.8.9), it is found that melts exhibit the expected proportionality with molecular weight (at least for low molecular weights), whereas solutions show a much stronger dependence of the viscosity on molecular weight. The unexpected behavior of solutions apparently results from a dependence of ζ on M and c. When data are corrected to give viscosities at a constant value of ζ, the behavior of polymer solutions also agrees with Eq. (14.8.9); this is shown in Figure 14.17 for both polymer melts and polymer solutions [25]. More extensive data are available in earlier reviews, which also discuss the influence of temperature, chain branching, polydispersity, and solvent viscosity [26,27].

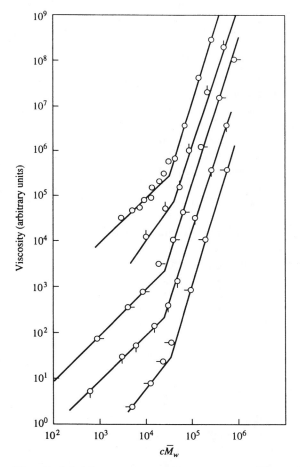

FIGURE 14.17 Viscosity versus the product $c\overline{M}_w$ for polystyrenes at concentrations between 25% and 100%: $\bigcirc$, undiluted at 217°C; $\bigcirc$, 0.55 g/mL in n-butyl benzene at 25°C; $\bigcirc$–, 0.415 g/mL in di-octyl phthalate at 30°C; $\bigcirc$ 0.310 g/mL in di-octyl phthalate at 30°C; and –$\bigcirc$, 0.255 g/mL in n-butyl benzene at 25°C. Data at the various concentrations have been shifted vertically to avoid overlap. (From Ref. 25.) Reprinted with permission from Graessley, W. W.: The entanglement concept in polymer rheology, Adv. Polym. Sci., vol. 16, 1–179, 1974. Copyright 1974 Springer-Verlag GmbH & Co, kG.

An examination of Figure 14.17 reveals that Eq. (14.8.9) is obeyed only up to a critical value of cM. Beyond this critical value, the slope of the viscosity-versus-cM curve increases to about 3.4 on logarithmic coordinates. For undiluted polystyrene the change in slope occurs at a molecular weight of about 32,000 [25], whereas for solutions, the critical molecular weight increases with decreasing concentration. To within the accuracy of experimental data, we find the following:

$$(M_c)_{sol} = \left(\frac{\rho}{c}\right)M_c \tag{14.8.10}$$

where M_c is the value for the molten polymer and is about 300–600 monomer units [28]. This behavior can be explained by saying that when M exceeds M_c, polymer chains begin to become entangled with each other and the entanglement density increases with increasing molecular weight. Entanglements may be visualized as temporary cross-links whose effect is to prevent chain crossability and to vastly increase the fluid friction beyond the value assumed in the Debye–Bueche theory; the net result is a rapid increase in viscosity with increasing molecular weight. In other words, Eq. (14.8.9) is valid only when polymer molecules act independently of each other. When molecules interact, we can no longer say that the energy dissipated by $\rho N_A/M$ molecules equals $\rho N_A/M$ times the energy dissipated by a single molecule acting alone.

Graessley [29] used the concept of intermolecular entanglements to successfully and quantitatively explain the reduction of shear viscosity with increasing shear rate. Now, we talk about the drag force at an entanglement junction and add together the contribution to energy dissipation from all the entanglements in a unit volume of polymer. The result, which is straightforward, is as follows [29]:

$$\eta = \frac{N}{2}\zeta \sum_i J_i(x_2)_i^2 \tag{14.8.11}$$

where N is the number of polymer chains per unit volume, ζ is the friction coefficient at an entanglement junction, and J_i is the number of entanglements between the chain at the origin and the ith chain.

Changes in viscosity with changing shear rate are attributed to shear-induced changes in the entanglement density. In other words, J_i is a function of $\dot{\gamma}$, and its value tends to a constant at low shear rates. It is assumed that a certain amount of time, of the order of the relaxation time, is needed for a polymer chain to both disengage itself from other chains and to get entangled with them. It is further assumed that a polymer chain can entangle itself only with those chains whose centers lie within a sphere of specified radius. Thus, a reduction in entanglement density comes about in a sheared fluid because some polymer molecules, which would otherwise be entangled with the chain under consideration, pass through this sphere of influence with a residence time that is less than the fluid relaxation time. A summation over all of the entanglements formed by the motion of nearby molecules allows us to evaluate the right side of Eq. (14.8.11) and leads to the following [29]:

$$\frac{\eta}{\eta_0} = \frac{2}{\pi}\left(\cot^{-1}\alpha + \frac{\alpha(1 - \alpha^2)}{(1 + \alpha^2)^2}\right) \tag{14.8.12}$$

in which $\alpha = (\eta/\eta_0)[(1/2)\dot{\gamma}\theta_0]$, where θ_0 is the zero-shear-rate relaxation time.

This equation suggests that a unique curve should result if the steady-shear viscosity data of entangling polymers is plotted as η/η_0 versus $\dot{\gamma}\theta_0$. This is, indeed, found to be the case for solutions of narrow-molecular-weight-distribution polystyrene dissolved in n-butyl benzene; these data are displayed in Figure 14.18 [30]. Note that the entanglement theory has also been extended to polydisperse polymers [30].

To properly understand why the zero-shear viscosity of an entangled polymer melt or solution increases so rapidly with increasing molecular weight

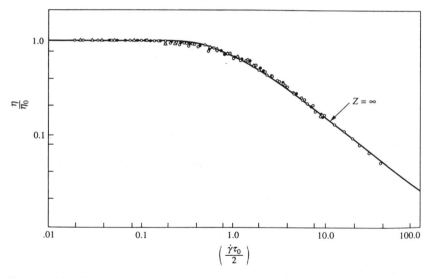

FIGURE 14.18 Master curve composed of data on solutions of narrow-distribution polystyrene in n-butyl benzene. Molecular weights range from 160,000 to 2,400,000, concentrations range from 0.20 to 0.55 g/cm³, and distribution breadths M_w/M_n range from 1.06 to 1.15. The data were shifted parallel to the shear-rate axis to achieve the best fit with the solid line calculated. (Reprinted with permission from W. W. Graessley, Viscosity of Entangling Polydisperse Polymers, J. Chem. Phys., vol. 47, pp. 1942–1953, 1967. Copyright 1967 American Institute of Physics.)

(see Figure 14.17), we need to use a model such as the reptation model, [31,32], which was introduced in Chapter 13.

Section 13.9 of Chapter 13 has shown that, for one-dimensional diffusion,

$$D = \frac{\langle \bar{x}^2 \rangle}{2t} \qquad (14.8.13)$$

where D is the diffusion coefficient and x is the distance traveled by a typical molecule in time t. Applying this equation to the reptating polymer molecule inside the tube shown in Figure 13.15 and letting x be the contour length nl gives the time θ_d taken by the molecule to entirely free itself of the constraining tube. Thus, θ_d is the time needed for the molecule to wriggle out of the tube or to disengage itself from the tube. This is the largest relaxation time that we would observe in a mechanical measurement because it is related to the motion of the entire molecule. It is often called the *terminal relaxation time*. Therefore, we have

$$\theta_d = \frac{(nl)^2}{2D} \qquad (14.8.14)$$

Replacing D with an equation similar to the Stokes–Einstein equation [see Eq. (13.4.6)],

$$D = \frac{kT}{n\zeta} \qquad (14.8.15)$$

gives us [28,33]

$$\theta_d = \frac{\zeta l^2 n^3}{2kT} \qquad (14.8.16)$$

which shows that θ_d is proportional to M^3 because n equals M/M_0.

The zero-shear viscosity η_0 can be obtained from linear viscoelasticity by replacing θ in Eq. (14.6.8) with θ_d and employing the result of Example 14.4:

$$\eta_0 = \int_0^\infty Ge^{-t'/\theta_d} \, dt' = G\theta_d \qquad (14.8.17)$$

or

$$\eta_0 = \left(\frac{G\zeta l^2}{2kTM_0^3} \right) M^3 \qquad (14.8.18)$$

which is the desired result. Even though the 3.4 power dependence is not obtained, the reptation model provides a remarkably consistent interrelationship between the various viscoelastic functions that has been confirmed with observations on linear polymer liquids [34]. As might be expected, the basic picture presented here has been modified by a large number of authors [33,34]. The resulting articles, however, are too numerous to be summarized here.

In closing this section, we summarize the observed viscosity behavior of polymer melts and polymer solutions. The zero-shear-rate viscosity of low-molecular-weight polymers increases linearly with molecular weight; this observation can be explained by the Debye–Bueche theory. Also, the behavior of polymer solutions becomes identical to that of polymer melts if data are plotted against the product of molecular weight with polymer concentration. With increasing molecular weight, polymer–polymer entanglements occur, and the zero-shear viscosity of melts increases at the 3.4–3.6 power of the molecular weight. It has been shown experimentally that this relationship holds for polydisperse polymers as well if the weight-average molecular weight is employed for data representation [35]. This nonlinear increase of viscosity can be understood with the help of the reptation model. For linear, entangled polymers (both monodisperse and polydisperse) the viscosity at a fixed molecular weight and concentration decreases with increasing shear rate, with polydisperse systems showing the onset of shear thinning at a lower shear rate compared to the monodisperse system. As seen earlier in this section, the reduction of viscosity with shear rate can be modeled by Graessley's entanglement theory. Finally, it is found that all of the data for monodisperse samples, including those at different constant temperatures, can be made to superpose if we plot the ratio of the shear viscosity to the zero-shear viscosity as a function of the ratio of the shear rate to the shear rate at which the viscosity has fallen to 80% of the zero-shear value. Additional details may be found in Ref. [35], which also discusses the effect of structural variables such as chain branching.

14.9 CONSTITUTIVE BEHAVIOR OF DILUTE POLYMER SOLUTIONS

When the concentration of polymer molecules in solution is sufficiently low (as shown in Sect. 8.6 of Chap. 8, $[\eta]c < 1$ is the usual criterion), the molecules are isolated from each other and solution behavior can be predicted from a knowledge of the behavior of a single polymer molecule. Because linear polymer molecules act like springs and also possess a finite mass, it is common practice to model an individual polymer molecule as a series of $N + 1$ spheres, each of mass m, connected by N massless springs. The polymer solution then is a noninteracting suspension of these stringy entities in a Newtonian liquid. In the absence of flow, the equilibrium probability that one end of a given polymer molecule is located at a specified distance from the other end is given by Eq. (10.2.12). Under the influence of flow, the strings can uncoil and stretch and also become oriented. If we could calculate the new probability distribution, we could again determine the tension in each spring and, therefore, the contribution to the stress resulting from the deformation of an individual polymer molecule. Multiplying this

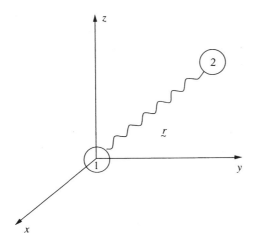

FIGURE 14.19 A linear elastic dumbbell.

contribution by the total number of molecules would yield the stress. We, therefore, have a two-part problem: Suspension micromechanics gives the instantaneous structure and this structure is translated into a stress. This process can become fairly elaborate [36], depending on the value of N and the nature of the elasticity and extensibility of the springs. Here (see Figure 14.19), we consider the simplest possible situation, that of an infinitely extensible, linear elastic dumbbell [37]. The dumbbell consists of two equal masses connected by a spring whose force-versus-extension behavior is given by Eq. (10.3.7). Such a study is useful for pedagogical reasons. In addition, the results have a practical utility because they describe the flow behavior of constant-viscosity elastic liquids called *Boger fluids* [38,39].

14.9.1 Elastic Dumbell Model

Consider a polymer solution containing c polymer molecules per unit volume. (Note the change in definition.) As a consequence, a cube of volume c^{-1} contains exactly one molecule or one dumbbell. For this situation, the polymer contribution to the stress tensor (see Sect. 10.5 of Chap. 10) can be calculated using the method of Bird and co-workers [36,37]. Referring to Figure 14.20, if the end-to-end vector of the dumbbell is $\mathbf{r}$, its projection along $\mathbf{n}$, the unit normal to the arbitrary plane contained within the cube, is $\mathbf{r} \cdot \mathbf{n}$. The probability that this dumbbell cuts the arbitrary plane is $\mathbf{r} \cdot \mathbf{n}$ divided by the length of the cube edge [i.e., $(\mathbf{r} \cdot \mathbf{n})c^{1/3}$]. Further, if the probability that a dumbbell has an end-to-end vector in the range $\mathbf{r}$ and $\mathbf{r} + d\mathbf{r}$, is $p(\mathbf{r})\, d\mathbf{r}$, then the probability that such a

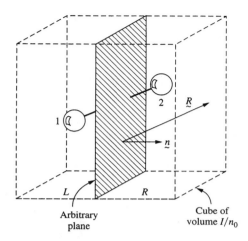

FIGURE 14.20 Calculation of the stress contributed by a dumbbell. (From Ref. 37.) Reprinted with permission from Bird, R. B., H. R. Warner, Jr., and D. C. Evans: Kinetic theory and rheology of dumbell suspensions with Brownian motion, Atd. Polym. Sci. vol. 8, pp. 1–90. 1971. Copyright 1971 Springer-Verlag GmbH & Co. kG.

dumbbell cuts an arbitrary plane having a unit normal $\mathbf{n}$ is $(\mathbf{r} \cdot \mathbf{n})c^{1/3}p(\mathbf{r})\,d\mathbf{r}$. If the force in the dumbbell spring is $\mathbf{F}$ (and has to be aligned with $\mathbf{r}$), the ensemble average of this force is $\langle \mathbf{F} \rangle$ given by

$$\langle \mathbf{F} \rangle = \int \mathbf{F}c^{1/3}(\mathbf{r} \cdot \mathbf{n})p(\mathbf{r})\,d\mathbf{r} \tag{14.9.1}$$

Because the cross-sectional area of the arbitrary plane is $c^{-2/3}$, the average stress vector, $\mathbf{f}$, acting on the plane is $\langle \mathbf{F} \rangle / c^{-2/3}$; that is,

$$\mathbf{f} = c\left[\int (\mathbf{F} \cdot \mathbf{r})p(\mathbf{r})\,d\mathbf{r} \right] \cdot \mathbf{n} \tag{14.9.2}$$

in which the "dyad" $(\mathbf{Fr})$ is the 3×3 matrix obtained by the matrix multiplication of the 3×1 column vector $\mathbf{F}$ with the 1×3 row vector $\mathbf{r}$.

For elastic dumbbells, $\mathbf{F}$ is given by Eq. (10.3.7); that is,

$$\mathbf{F} = K\mathbf{r} \tag{14.9.3}$$

where the constant K equals $3kT/nl^2$. Thus, Eq. (14.9.2) becomes

$$\mathbf{f} = cK[\int (\mathbf{rr})p(\mathbf{r})\,d\mathbf{r}] \cdot \mathbf{n} \tag{14.9.4}$$

Finally, from the definition of the stress tensor $\underset{\sim}{\boldsymbol{\tau}}$ in Eq. (10.5.5), we have

$$\mathbf{f} = \underset{\sim}{\boldsymbol{\tau}} \cdot \mathbf{n} \tag{14.9.5}$$

and by comparison with Eq. (14.9.4),

$$\underset{\sim}{\tau} = cK \int (\mathbf{rr}) p(\mathbf{r}) \, d\mathbf{r} = cK \langle \mathbf{rr} \rangle \tag{14.9.6}$$

where $\langle \mathbf{rr} \rangle$ is the ensemble average of the dyad $\mathbf{rr}$. If $\mathbf{r}$ can be represented as

$$\mathbf{r} = x\mathbf{i} + y\mathbf{j} + z\mathbf{k} \tag{14.9.7}$$

then

$$\mathbf{rr} = \begin{bmatrix} x^2 & xy & xz \\ xy & y^2 & yz \\ xz & yz & z^2 \end{bmatrix} \tag{14.9.8}$$

It can also be shown that the momentum of the beads crossing the plane contributes a term to $\underset{\sim}{\tau}$, but this term simply adds to the isotropic pressure term p. The total stress $\underset{\sim}{T}$ in the polymer solution then is the sum of the solvent stress, which is Newtonian [given by Eqs. (14.2.1) and (14.2.2)], and the polymer stress, which is yet to be calculated but is given formally by Eq. (14.9.6). Thus,

$$\underset{\sim}{T} = -p\underset{\sim}{1} + 2\eta_s \underset{\sim}{D} + cK \langle \mathbf{rr} \rangle \tag{14.9.9}$$

in which $\underset{\sim}{1}$ is the unit matrix having 1's along the diagonal and 0's elsewhere, η_s is the solvent viscosity and $\underset{\sim}{D}$ is called the deformation rate tensor, whose components are given by

$$(\underset{\sim}{D})_{ij} = \frac{1}{2} \left(\frac{\partial v_i}{\partial x_j} + \frac{\partial v_j}{\partial x_i} \right) \tag{14.9.10}$$

Note that when there is no flow, $\underset{\sim}{D}$ is zero and $p(\mathbf{r})$ is given by Eq. (10.2.11). Evaluating the integral in Eq. (14.9.4) gives the following:

$$\langle \mathbf{rr} \rangle = \frac{nl^2}{3} \underset{\sim}{1} \tag{14.9.11}$$

When flow begins from a state of rest, the initial value of $p(\mathbf{r})$ is still given by Eq. (10.2.11), but this changes with time due to flow-induced extension and orientation of the dumbbells. The time-dependent probability distribution can be calculated by analyzing the motion of one bead of the dumbbell with respect to the other.

To calculate $p(\mathbf{r})$ in the presence of flow, consider a dumbbell with one bead anchored to the origin of a rectangular Cartesian coordinate system x, y, z, as shown in Figure 14.19. The other bead is free to move and, at any time, its position is given by a balance of forces that act on it. Although drag due to flow tends to separate the beads and orient the dumbbell, drag is resisted by the spring force and the Brownian motion force. As seen in Chapter 13, Eq. (13.4.1), the

Brownian motion force is equivalent to an osmotic pressure force whose magnitude in one-dimensional flow is $-kT(d \ln c/dx)$; when there is three-dimensional flow, this expression is generalized to $-kT\mathbf{V} \ln c$. Neglecting inertia, the force balance on bead 2 yields

$$\mathbf{F}_{spring} + \mathbf{F}_{osmotic} + \mathbf{F}_{drag} = 0 \qquad (14.9.12)$$

The spring force is given by Eq. (14.9.3), whereas the drag force is

$$\mathbf{F}_{drag} = -\zeta_0(\dot{\mathbf{r}} - \mathbf{v}) \qquad (14.9.13)$$

in which ζ_0 is the friction coefficient given by Stokes' law, $\dot{\mathbf{r}}$ is the velocity of the bead, and $\mathbf{v}$ is the fluid velocity. If the velocity gradient is $\mathbf{V}\mathbf{v}$ (with components $\partial v_i/\partial x_j$), the fluid velocity at $\mathbf{r}$ is $\mathbf{V}\mathbf{v} \cdot \mathbf{r}$ because the velocity at the origin is zero. Introducing the relevant expressions in Eq. (14.9.12) gives

$$-K\mathbf{r} - kT\mathbf{V} \ln c - \zeta_0(\dot{\mathbf{r}} - \mathbf{V}\mathbf{v} \cdot \mathbf{r}) = 0 \qquad (14.9.14)$$

and, solving for $\dot{\mathbf{r}}$, we have

$$\dot{\mathbf{r}} = -\frac{K}{\zeta_0}\mathbf{r} - \frac{kT}{\zeta_0}\mathbf{V} \ln c + \mathbf{V}\mathbf{v} \cdot \mathbf{r} \qquad (14.9.15)$$

Because the dumbbells are neither created nor destroyed, the species conservation equation is as follows:

$$\frac{\partial c}{\partial t} + \mathbf{V} \cdot (\dot{\mathbf{r}}c) = 0 \qquad (14.9.16)$$

Recognizing that it is meaningless to talk about concentration when working with a single bead, we replace c with the probability $p(\mathbf{r})$ and also introduce Eq. (14.9.15) into Eq. (14.9.16) with the following result:

$$-\frac{\partial p}{\partial t} = \mathbf{V} \cdot \left[\left(-\frac{K}{\zeta_0}\mathbf{r} + \mathbf{V}\mathbf{v} \cdot \mathbf{r} \right)p - \frac{kT}{\zeta_0}\mathbf{V}p \right] \qquad (14.9.17)$$

which is called the *diffusion equation*. The solution of this equation for a given flow field (or velocity gradient) yields $p(\mathbf{r}, t)$ which can then be used to evaluate the stress through Eq. (14.9.6).

Example 14.6: Obtain the nonzero components of $\langle \mathbf{rr} \rangle$ for steady laminar shearing flow at a constant shear rate $\dot{\gamma}$.

Solution: If v_x equals $\dot{\gamma}y$ and all other velocity components are zero, then

$$\nabla \mathbf{v} = \begin{bmatrix} 0 & \dot{\gamma} & 0 \\ 0 & 0 & 0 \\ 0 & 0 & 0 \end{bmatrix}$$

$$\nabla \mathbf{v} \cdot \mathbf{r} = \begin{bmatrix} 0 & \dot{\gamma} & 0 \\ 0 & 0 & 0 \\ 0 & 0 & 0 \end{bmatrix} \begin{bmatrix} x \\ y \\ z \end{bmatrix} = \begin{bmatrix} \dot{\gamma}y \\ 0 \\ 0 \end{bmatrix}$$

At steady state, $\partial p/\partial t$ is zero, and the diffusion equation becomes

$$-\frac{K}{\zeta_0}\left(\frac{\partial}{\partial x}(xp) + \frac{\partial}{\partial y}(yp) + \frac{\partial}{\partial z}(zp)\right) + \dot{\gamma}y\,\frac{\partial p}{\partial x} - \frac{kT}{\zeta_0}\left(\frac{\partial^2 p}{\partial x^2} + \frac{\partial^2 p}{\partial y^2} + \frac{\partial^2 p}{\partial z^2}\right) = 0$$

$$(14.9.18)$$

Instead of explicitly solving for $p(r)$, we multiply Eq. (14.9.18) by x^2, xy, y^2, z^2, and so on, in turn, and integrate from $-\infty$ to $+\infty$, assuming that p tends to zero faster than any power of x, y, or z as x, y, and z tend to $\pm\infty$. Also, on physical grounds, $\int_{-\infty}^{\infty} p\,dx\,dy\,dz = 1$.

Multiplying Eq. (14.9.18) by xy and integrating from $-\infty$ to $+\infty$, we find that the very first integral is given as follows [40]:

$$\iiint xy\,\frac{\partial}{\partial x}(xp)\,dx\,dy\,dz = \iint y\left(\int x\,\frac{\partial}{\partial x}(xp)\,dx\right)dy\,dz$$

$$= \iint y\left[x^2 p\big|_{-\infty}^{\infty} - \int xp\,dx\right]dy\,dz \qquad (14.9.19)$$

$$= -\iiint xyp\,dx\,dy\,dz$$

$$= -\langle xy\rangle$$

from the definition of the ensemble average. Continuing in this manner, it is relatively easy to show [40]

$$-\frac{K}{\zeta_0}(-2\langle xy\rangle) + \dot{\gamma}(-\langle y^2\rangle) = 0 \qquad (14.9.20)$$

Further, if the diffusion equation is multiplied by y^2 and integrated, we get

$$\frac{2K}{\zeta_0}\langle y^2\rangle - \frac{2kT}{\zeta_0} = 0 \qquad (14.9.21)$$

and on multiplying by x^2 and integrating, the result is

$$-2\dot{\gamma}\langle xy \rangle + \frac{2K}{\zeta_0}\langle x^2 \rangle - \frac{2kT}{\zeta_0} = 0 \tag{14.9.22}$$

Solving the previous three equations for the three unknowns, $\langle xy \rangle$, $\langle x^2 \rangle$, and $\langle y^2 \rangle$, gives

$$\langle xy \rangle = \frac{\zeta_0 \dot{\gamma}}{2K}\langle y^2 \rangle \tag{14.9.23}$$

$$\langle x^2 \rangle = \frac{kT}{K} + \frac{\zeta_0 \dot{\gamma}}{K}\langle xy \rangle \tag{14.9.24}$$

$$\langle y^2 \rangle = \frac{kT}{K} = \frac{nl^2}{3} \tag{14.9.25}$$

and we can show that $\langle z^2 \rangle$ equals $\langle y^2 \rangle$.

With the results of Example 14.6 in hand, we can use Eq. (14.9.6) to compute the material functions in a steady laminar shearing flow as follows:

$$\tau_{12} = \tau_{xy} = cK\langle xy \rangle = \frac{c\zeta_0 \dot{\gamma}}{6}nl^2 \tag{14.9.26}$$

$$\tau_{11} - \tau_{12} = \tau_{xx} - \tau_{yy} = N_1 = cK[\langle x^2 \rangle - \langle y^2 \rangle] = \frac{cnl^2\zeta_0^2\dot{\gamma}^2}{6K} \tag{14.9.27}$$

$$\tau_{22} - \tau_{33} = \tau_{yy} - \tau_{zz} = N_2 = 0 \tag{14.9.28}$$

Accordingly, the polymer contribution to the shear viscosity is

$$\frac{\tau_{12}}{\dot{\gamma}} = \frac{c\zeta_0 nl^2}{6} \tag{14.9.29}$$

which is proportional to the polymer molecular weight. Consequently, the solution viscosity is given by

$$\eta_{\text{sol}} = \eta_s + \frac{c\zeta_0 nl^2}{6} \tag{14.9.30}$$

which does not depend on the shear rate. This is somewhat unrealistic because shear thinning is generally observed at high enough shear rates even for unentangled systems. In practical terms, though, a constant solution viscosity, independent of shear rate, will be measured if the second term on the right-hand side of Eq. (14.9.30) is negligible compared to the first term. This is the situation that obtains with constant-viscosity, ideal elastic liquids [38]. Regarding the molecular weight dependence, n is proportional to M, but c (the number of molecules per unit volume) is proportional to M^{-1} at a fixed mass concentration. Thus, the molecular-weight dependence of the polymer contribution to the viscosity depends on ζ_0 alone. From Stokes' law, ζ_0 is proportional to the bead radius, which increases with increasing molecular weight. However, the extent of the increase depends on whether the solvent is good or poor. Typically, we find that $\zeta_0 \propto M^a$, where $0.5 \leq a \leq 1$.

As far as elastic effects are concerned, Eqs. (14.9.27) and (14.9.28) state that N_1 is proportional to $\dot{\gamma}^2$ and N_2 equals zero. Both of these relations are found to be true for dilute polymer solutions in high-viscosity solvents [38]. Typical data for N_1 are shown in Figure 14.21.

Although this process can be repeated to obtain the stress predictions for any other flow field, we find that this laborious procedure is not necessary for the special case of the dumbbell model [41]. On multiplying Eq. (14.9.17) by $(\mathbf{rr})$ and integrating by parts, we get a closed-form differential equation for $\langle \mathbf{rr} \rangle$. Replacing $\langle \mathbf{rr} \rangle$ with $\underset{\sim}{\boldsymbol{\tau}}/cK$ through the use of Eq. (14.9.6) yields the equation for the polymer contribution to the stress,

$$\underset{\sim}{\boldsymbol{\tau}} + \theta \frac{\delta \underset{\sim}{\boldsymbol{\tau}}}{\delta t} = 2\eta \, \mathbf{D} \tag{14.9.31}$$

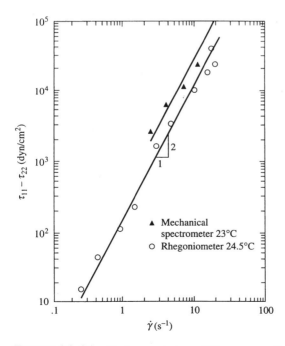

FIGURE 14.21 First normal stress difference as a function of shear rate for a solution of 1000-ppm polyisobutylene in a solvent of polybutene and kerosene. (From Ref. 38.)

in which η is given by the right-hand side of Eq. (14.9.29) and the relaxation time θ equals η/ckT. Equation (14.9.31) is known as the upper-convected Maxwell equation because the form of the equation is similar to Eq. (14.6.6), which results from the use of a Maxwell element. "Upper convected" refers to the appearance of the upper-convected derivative $\delta/\delta t$, defined as follows:

$$\frac{\delta \tau_{ij}}{\delta t} = \frac{\partial \tau_{ij}}{\partial t} + \sum_{m=1}^{3} \left[v_m \frac{\partial \tau_{ij}}{\partial x_m} - \tau_{im} \frac{\partial v_j}{\partial x_m} - \tau_{mj} \frac{\partial v_i}{\partial x_m} \right] \tag{14.9.32}$$

Example 14.7: Obtain an expression for the net tensile stress (neglecting the solvent contribution) in a uniaxial extension experiment according to the upper-convected Maxwell equation.

Solution: It is easy to show [42]

$$\frac{\delta \tau_{11}}{\delta t} = \frac{\partial \tau_{11}}{\partial t} - 2\tau_{11}\dot{\varepsilon} \tag{14.9.33}$$

$$\frac{\delta \tau_{22}}{\delta t} = \frac{\partial \tau_{22}}{\partial t} + \tau_{22}\dot{\varepsilon} \tag{14.9.34}$$

and the equations for τ_{11} and τ_{22} are uncoupled. Introducing Eqs. (14.9.33) and (14.9.34) into Eq. (14.9.31) and noting that the components of $\underset{\sim}{D}$ are given by Eq. (14.9.10), we can integrate Eq. (14.9.31) to yield $\tau_{11} - \tau_{22}$, the extensional stress contributed by the polymer as follows:

$$\tau_{11} - \tau_{12} = \frac{3\eta\dot{\varepsilon}}{(1 - 2\theta\dot{\varepsilon})(1 + \theta\dot{\varepsilon})} - \frac{2\eta\dot{\varepsilon}}{1 - 2\theta\dot{\varepsilon}} \exp\left(-\frac{(1 - 2\theta\dot{\varepsilon})t}{\theta}\right)$$
$$- \frac{\eta\dot{\varepsilon}}{(1 + \theta\dot{\varepsilon})} \exp\left(-\frac{(1 + \theta\dot{\varepsilon})t}{\theta}\right) \tag{14.9.35}$$

which attains a steady value only when $2\theta\dot{\varepsilon} < 1$.

Equation (14.9.31) is the simplest two-parameter equation that correctly predicts the qualitative viscoelastic behavior of polymer solutions. For this reason, it has been extensively used in non-Newtonian fluid mechanics for predicting elastic effects in complex flow fields [43]. Note that it reduces to the Newtonian equation if θ is allowed to become zero. Also, it can be written in an equivalent integral form as follows [43]:

$$\underset{\sim}{\tau}(t) = \frac{\eta}{\theta^2} \int_{-\infty}^{t} \exp\left(-\frac{t - s}{\theta}\right) \underset{\sim}{C}^{-1}(s) \, ds \tag{14.9.36}$$

in which $\underset{\sim}{C}^{-1}$ is the Finger tensor, given earlier by Eq. (10.6.18) of Chapter 10.

14.9.2 Multiple Bead–Spring Models

At the next level of complexity, the polymer molecule can be represented as a necklace of $N + 1$ alternating beads and N Hookean springs. If the presence of the polymer molecule does not disturb the imposed flow field, the result is a "free-draining" molecule, the situation considered by Rouse [44]. At the other extreme, the result can be hydrodynamic interactions, which result in fluid being trapped within the molecule. This result has been accounted for by Zimm [45]. In both cases, we assume that we have a theta solvent so that there are no excluded volume effects. The analysis is identical to the one followed for the dumbbell model, except that now an equation of motion [similar to Eq. (14.9.14)] is written for each bead, which leads to a set of $N + 1$ coupled equations. A closed-form constitutive equation such as Eq. (14.9.31) can be obtained by means of an orthogonal transformation of coordinates, with the new coordinate axes being aligned with the eigenvectors of the matrix describing the coupled set of equations [46]. The final result is as follows:

$$\underset{\sim}{\tau} = \sum_{i=1}^{N} \underset{\sim}{\tau}_i \tag{14.9.37}$$

where

$$\underset{\sim}{\tau}_i + \theta_i \frac{\delta \underset{\sim}{\tau}_i}{\delta t} = 2\eta_i \underset{\sim}{D} \tag{14.9.38}$$

in which η_i equals $\theta_i ckT$. For the Rouse model, the relaxation times θ_i are as follows [25]:

$$\theta_i = \frac{\langle r^2 \rangle n\zeta}{6\pi^2 kTi^2}, \quad i = 1, 2, \ldots \tag{14.9.39}$$

in which $\langle r^2 \rangle$ is the equilibrium mean-square end-to-end distance of the entire molecule, n is the number of monomers, and ζ is the monomeric friction coefficient.

In terms of experimentally measurable quantities, the Rouse relaxation time can be written as [25]

$$\theta_i = \frac{6}{\pi^2} \frac{\eta_0 - \eta_s}{ckTi^2}, \quad i = 1, 2 \tag{14.9.40}$$

Correspondingly, the Zimm relaxation times are as follows [25]:

$$\theta_i = \frac{6}{\pi^2} \frac{\eta_0 - \eta_s}{ckTb_i}, \quad b_i = 1.44, 4.55, 8.60, \ldots \tag{14.9.41}$$

Because both the Rouse and Zimm theories should become more and more accurate as the polymer concentration decreases, they ought to be tested at low

polymer concentrations. Because neither theory predicts a shear-thinning viscosity and because normal stress differences are difficult to measure at low
concentrations, storage modulus data are generally used for experimental validation of the theory. The theoretical expressions for G' and G'' for both models are
the same and identical to those given earlier in Eqs. (14.7.9) and (14.7.10), with
G_i being ckT, except that the θ_i values are different and are given by Eqs.
(14.9.40) and (14.9.41). Note that for polymer solutions, the expression given by
Eq. (14.7.10) is only the polymer contribution $G'' - \omega\eta_s$ to G'' and not the G'' of
the entire solution. Usually, it is more convenient to make G' and $G'' - \omega\eta_s$
dimensionless and use the reduced moduli G'_R and $(G'' - \omega\eta_s)_R$ for comparison
with experimental data. Thus,

$$G'_R = \frac{G'}{ckT} = \sum_{i=1}^{N} \frac{\omega_R^2(\theta_i/\theta_1)^2}{1 + \omega_R^2(\theta_i/\theta_1)^2} \tag{14.9.42}$$

$$(G'' - \omega\eta_s)_R = \frac{G'' - \omega\eta_s}{ckT} = \sum_{i=1}^{N} \frac{\omega_R(\theta_i/\theta_1)}{1 + \omega_R^2(\theta_i/\theta_1)^2} \tag{14.9.43}$$

where the reduced frequency ω_R equals $\omega\theta_1$, with θ_1 being the longest relaxation
time given by either Eq. (14.9.40) or Eq. (14.9.41). For both models, plots of G'_R
and $(G'' - \omega\eta_s)_R$ versus ω_R should have slopes of 2 and 1, respectively, at low
frequencies. At high frequencies, though, the Rouse theory predicts a merger of
the two curves, each with a slope of $1/2$, whereas the Zimm theory says that the
two curves become parallel with a slope of $2/3$. Data of Tam and Tiu on an
aqueous polyacrylamide solution (see Figure 14.22) seem to follow the trend
predicted by the Zimm model, as far as the slopes are concerned [47]. More data
are available in the book by Ferry [17].

Shear thinning can be introduced into the foregoing equations either in a
phenomenological manner [48] or by making the entropic springs nonlinear and
finitely extensible [36]. The resulting expressions have, however, not been tested
extensively.

14.10 CONSTITUTIVE BEHAVIOR OF CONCENTRATED SOLUTIONS AND MELTS

Although the bead–spring models of the previous section can be considered
representative of the behavior of dilute polymer solutions, they cannot be used to
predict the behavior of concentrated solutions or melts due to the formation of
entanglements. Here, we turn instead to network models. The motivation for
using these models is the success of the theory of rubber elasticity in explaining
the stress–strain behavior of a network of polymer molecules linked together by

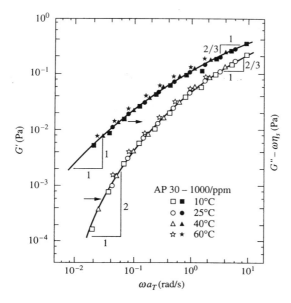

FIGURE 14.22 Master curve of G' and $G'' - \omega\eta_s$ versus ωa_T for 1000 ppm Separan AP30, $\square$, $\blacksquare$: 10°C; $\bigcirc$, $\bullet$, 25°C, $\triangle$, $\blacktriangle$, 40°C, $\stackrel{\star}{\star}$, $\star$, 60°C. (From Ref. 47.) (Reprinted with permission from Tam, K. C., and C. Tiu: "Steady and Dynamic Shear Properties of Aqueous Polymer Solutions," J. Rheol., vol. 33, pp. 257–280, 1989.)

junctions that are permanent in nature. In the case of polymer melts, it is assumed that the entanglements are not permanent but are continually being created and destroyed. This, then, is equivalent to approaching the problem from the other asymptote—that of a solid rather than that of an infinitely dilute solution.

14.10.1 Lodge's Rubberlike Liquid

The assumptions involved in the simplest of network theories have been listed by Lodge [49], and these include the affine deformation assumption. It is also assumed that each segment, which is that part of a polymer molecule contained between two junctions, behaves in a Gaussian manner. This means that the tension in any given segment is proportional to the end-to-end distance of the segment; the constant of proportionality is again $3kT/\langle r^2 \rangle$, with $\langle r^2 \rangle$ being the mean-square end-to-end distance at equilibrium. Thus, the situation is the same as that for a rubber (see Chap. 10), except that the number of junctions changes with time due to flow. At any given time, there is an age distribution of segments. In the absence of flow, of course, the equilibrium distribution function for the network segments at the moment of creation is identical to the equilibrium distribution function for a freely jointed bead–rod chain with no constraints at the end points [7].

Equation (10.7.8) of Chapter 10 has shown that a block of rubber containing N chain segments per unit volume has stress-versus-strain behavior given by

$$\tau_{ij} = NkTC_{ij}^{-1}$$

Similarly, for the rubberlike liquid, we would expect that the contribution to the stress at time t from $v(t, t')\, dt'$ (the number of segments per unit volume created in interval dt' at some past time t') would be

$$v(t, t')kTC_{ij}^{-1}\, dt' \tag{14.10.1}$$

The total stress, however, is the sum of all these contributions:

$$\tau_{ij} = \int_{-\infty}^{t} v(t, t')kTC_{ij}^{-1}\, dt' \tag{14.10.2}$$

in which the function v is still unspecified.

In order to determine v, we assume that chain segments are created at a constant rate equal to $\eta/kT\theta^2$ per unit time per unit volume, with η and θ being unknown constants [7,49]. Thus, during unit time, $\eta/kT\theta^2$ segments are created at time t'. These are assumed to decay at a rate proportional to the number of segments remaining, with the constant of proportionality $1/\theta$. To determine the number of segments $v(t, t')$ still remaining at a later time t, we solve the following equation:

$$\frac{d}{dt}v(t, t') = -\frac{v(t, t')}{\theta} \tag{14.10.3}$$

where $v(t, t')$ is $\eta/(kT\theta^2)$. Clearly, then, we have

$$v(t, t') = \left(\frac{\eta}{kT\theta^2}\right)\exp\left(-\frac{t - t'}{\theta}\right) \tag{14.10.4}$$

Introducing Eq. (14.10.4) into Eq. (14.10.2) gives the equation of the Lodge rubberlike liquid [49]:

$$\tau_{ij} = \frac{\eta}{\theta^2}\int_{-\infty}^{t}\exp[-(t - t')/\theta]C_{ij}^{-1}\, dt' \tag{14.10.5}$$

where the constants η and θ are obtained by comparison with experimental data.

It is remarkable that Eq. (14.10.5) has the same form as Eq. (14.9.36), which was obtained from dilute solution theory. Therefore, it is equivalent in form to Eq. (14.9.31) as well. The parameters in Eq. (14.10.5), though, do not have the same physical significance as the parameters in Eqs. (14.9.31) and (14.9.36). Nonetheless, all of these equations suffer from the same shortcomings, namely that they cannot predict shear thinning. Modern constitutive equations overcome this and other limitations in different ways. A particularly popular modification is the (four-parameter) model of Phan-Thien and Tanner [50].

14.10.2 Other Single-Integral Equations

Rubber elasticity theory can also be used to derive a general class of single-integral equations, as opposed to a particular equation such as that for the Lodge rubberlike liquid. To proceed, consider a cube of rubber, initially of unit edge, in an extensional deformation as shown in Figure 14.23. In the deformed state, the block of rubber has dimensions λ_1, λ_2, and λ_3, which also happen to be the extension ratios.

The work done by force F_1 is

$$dW = F_1 \, d\lambda_1 \tag{14.10.6}$$

so that

$$\tau_{11} = \frac{F_1}{\lambda_2 \lambda_3} = \frac{1}{\lambda_2 \lambda_3} \frac{\partial W}{\partial \lambda_1} = \lambda_1 \frac{\partial W}{\partial \lambda_1} \tag{14.10.7}$$

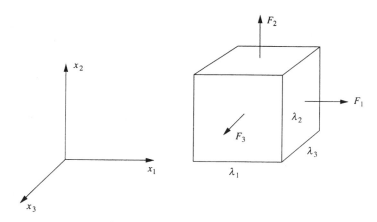

FIGURE 14.23   Extensional deformation of a cube of rubber.

where the last equality follows from volume conservation. Similarly, we have

$$\tau_{22} = \lambda_2 \frac{\partial W}{\partial \lambda_2} \tag{14.10.8}$$

$$\tau_{33} = \lambda_3 \frac{\partial W}{\partial \lambda_3} \tag{14.10.9}$$

where it is clear that W is a function of the stretch ratios.

To make further progress, we recognize that just as the dot product of a vector with itself is independent of the coordinate system used to represent the vector, certain properties of a matrix are also independent of the coordinate system used. There are three such quantities, and they are called the *invariants* of the matrix. For a matrix $\underset{\sim}{A}$ with elements a_{ij}, the invariants are as follows [51]:

$$I_1 = \text{tr } \underset{\sim}{A} = \sum_1^3 a_{ij} \tag{14.10.10}$$

$$I_2 = [(\text{tr } \underset{\sim}{A})^2 - \text{tr } \underset{\sim}{A}^2] \tag{14.10.11}$$

$$I_3 = \text{Determinant of } \underset{\sim}{A} \tag{14.10.12}$$

where tr is an abbreviation for the word "trace."

The Cauchy and Finger tensors for the deformation described in Figure 14.23 are given by

$$\underset{\sim}{C} = \begin{bmatrix} \lambda_1^{-2} & & \\ & \lambda_2^{-2} & \\ & & \lambda_3^{-2} \end{bmatrix} \tag{14.10.13}$$

$$\underset{\sim}{C}^{-1} = \begin{bmatrix} \lambda_1^2 & & \\ & \lambda_2^2 & \\ & & \lambda_3^2 \end{bmatrix} \tag{14.10.14}$$

so that for the Finger tensor

$$I_1 = \lambda_1^2 + \lambda_2^2 + \lambda_3^2 \tag{14.10.15}$$
$$I_2 = \lambda_1^{-2} + \lambda_2^{-2} + \lambda_3^{-2} \tag{14.10.16}$$
$$I_3 = 1 \tag{14.10.17}$$

Because W depends on λ_i, which, in turn, depends on I_1 and I_2 for the Finger tensor, we can conclude the following:

$$
\begin{aligned}
\tau_{11} = \lambda_1 \frac{\partial W}{\partial \lambda_1} &= \lambda_1 \left(\frac{\partial W}{\partial I_1} \frac{\partial I_1}{\partial \lambda_1} + \frac{\partial W}{\partial I_2} \frac{\partial I_2}{\partial \lambda_1} \right) \\
&= \lambda_1 \left(\frac{\partial W}{\partial I_1} 2\lambda_1 + \frac{\partial W}{\partial I_2} (-2\lambda_1^{-3}) \right) \\
&= 2 \frac{\partial W}{\partial I_1} \lambda_1^2 - 2 \frac{\partial W}{\partial I_2} 2\lambda_1^{-2}
\end{aligned}
\tag{14.10.18}
$$

$$
\begin{aligned}
\tau_{22} = \lambda_2 \frac{\partial W}{\partial \lambda_2} &= \lambda_2 \left(\frac{\partial W}{\partial I_1} \frac{\partial I_1}{\partial \lambda_2} + \frac{\partial W}{\partial I_2} \frac{\partial I_2}{\partial \lambda_2} \right) \\
&= \lambda_2 \left(\frac{\partial W}{\partial I_1} 2\lambda_2 + \frac{\partial W}{\partial I_2} (-2\lambda_2^{-3}) \right) \\
&= 2 \frac{\partial W}{\partial I_1} \lambda_2^2 - 2 \frac{\partial W}{\partial I_2} \lambda_2^{-2}
\end{aligned}
\tag{14.10.19}
$$

and similarly for τ_{33}. However, with the help of Eqs. (14.10.13) and (14.10.14), these equations can be written compactly as follows:

$$
\underset{\sim}{\tau} = 2 \frac{\partial W}{\partial I_1} \underset{\sim}{C}^{-1} - 2 \frac{\partial W}{\partial I_2} \underset{\sim}{C}
\tag{14.10.20}
$$

Note that for the theory of rubber elasticity considered in detail in Chapter 10 [see Eq. (10.4.4)], we have

$$
w = \frac{G}{2}(I_1 - 3)
\tag{14.10.21}
$$

where G is the modulus of the rubber. Thus, $\partial W/\partial I_1$ equals $G/2$ and $\partial W/\partial I_2$ equals zero, yielding

$$
\underset{\sim}{\tau} = G\underset{\sim}{C}^{-1}
$$

which is the same as Eq. (10.7.8). Consequently, Eq. (14.10.20) is much more general than Eq. (10.7.8) but contains no molecular information.

To obtain a constitutive equation for viscoelastic liquids using Eq. (14.10.20), we must recognize that, due to stress relaxation, W depends not only on I_1 and I_2 but also on time. If we assume that

$$
W = \int_{-\infty}^{t} m(t - t')U(I_1, I_2) \, dt'
\tag{14.10.22}
$$

in which m and U are as yet unspecified functions, we get the factorable form of the Kaye–BKZ equation [52,53]:

$$\tau = \int_{-\infty}^{t} m(t - t') \left(2 \frac{\partial U}{\partial I_1} \mathbf{C}^{-1} - 2 \frac{\partial U}{\partial I_2} \mathbf{C} \right) dt' \tag{14.10.23}$$

Because m and U are unspecified functions, it it possible to predict most observed rheological phenomena by picking specific functional forms for these quantities [7].

A particular form of Eq. (14.10.23) that deserves special mention is the one resulting from the theory of Doi and Edwards [54] and based on the reptation idea of de Gennes [32]. These authors assumed that upon deformation of the polymer, stress relaxation took place by two separate mechanisms. In the first instance there was a rapid retraction of the chain within the deformed tube (see Figure 13.15 of Chap. 13), and in the second instance, there was slow diffusion of the chain out of the original tube by the process of reptation. Within the confines of the tube, the polymer acts like a Rouse chain. This leads to the following linear stress relaxation modulus [33]:

$$G(t) = G \sum_{i=1,3,\dots}^{\infty} i^{-2} \exp\left(-\frac{ti^2}{\theta_d} \right) \tag{14.10.24}$$

which is close to a single exponential function $\exp(-t/\theta_d)$ over a fairly wide range of times around θ_d, the time for chain disengagement. As shown in Section 14.7, experimentally obtained dynamic moduli can be employed in conjunction with Eqs. (14.7.9) and (14.7.10) to yield the discrete relaxation time spectrum, which, in turn, gives the stress relaxation modulus through the use of Eq. (14.6.10). Reptation model parameters can then be obtained by fitting Eq. (14.10.24) to the stress–relaxation modulus.

By experiment, it is found that polydispersity has a strong effect on the shapes of the G' and G'' functions and, therefore, on the relaxation time spectrum [55]. By theory, it is found that the reptation model can be used to relate the stress–relaxation modulus of polydisperse polymers to the corresponding results for monodisperse polymers using what is essentially a mixing rule that involves the molecular-weight distribution (MWD) [56]. By inverting this mixing rule, it is possible to obtain the molecular-weight distribution from the relaxation time modulus of a polydisperse polymer [57]. This procedure can be tailored to both addition polymers that typically have a log normal distribution and to condensation polymers that have a "most probable" distribution, and it has several advantages: (1) For polymers, such as polypropylene, that dissolve in common solvents only at high temperatures, rheological methods of obtaining the MWD are much less expensive than the use of light scattering or size-exclusion chromatography. (2) For polymers, such as fluoropolymers, that do not dissolve in common solvents, this is the only technique of determining MWD. (3) For all polymers, rheology is much better able to detect the presence of a high-molecular-weight tail or the presence of long-chain branching as compared to chromatography; the largest molecules in a polymer sample often determine how easy or how difficult it is to process the material.

Osaki and Doi have discussed the predictions of the Doi–Edwards theory in nonlinear viscoelasticity and compared these with available experimental data [33]. Although this model has a remarkable degree of internal consistency, agreement with data is not always quantitative. As a consequence, the model has been modified on a number of occasions and is still a subject of research [34,58,59].

14.11   CONCLUSION

This chapter has been a guided tour through the difficult study of rheology. The difficulties stem primarily from the very mathematical nature of the material, especially where constitutive modeling is concerned. Because the practitioner involved in polymer processing requires at least a qualitative understanding of non-Newtonian fluid mechanics, non-Newtonian effects and measurement techniques have been discussed at the beginning of the chapter. In addition, the theories presented are those that have provided the basic underpinning for much of the technical literature that is read by rheologists today. A remarkable feature of these theories is that the upper-convected Maxwell equation (or the Lodge rubberlike liquid) emerges as the simplest viscoelastic constitutive equation irrespective of the approach used. Although this equation has several short-comings, some of which have been noted in the chapter, it has been extensively used to model polymer processing operations; it should also be the first choice of anyone wishing to understand the qualitative features of any new flow situation involving polymeric liquids. The best evidence that this equation has merit is provided by the existence of the Boger fluid, a dilute solution of a high-molecular-weight polymer dissolved in a very viscous solvent. This liquid has a constant viscosity and is highly elastic. In recent years, it has been used as the test fluid for verifying computations done using the Maxwell model. The theory and practice of rheology have, of course, advanced far beyond the limited material contained in this chapter. Current developments may be found in the numerous books on rheology, including those cited in the chapter and several that have been published recently [55,60–65].

REFERENCES

1. Doraiswamy, D., The Rheology of Non-Newtonian Fluids, in *The Engineering Handbook*, R. C. Dorf (ed.), CRC Press, Boca Raton, FL, 1996, pp. 375–386.

2. Coleman, B. D., H. Markovitz, and W. Noll, *Viscometric Flows of Non-Newtonian Fluids*, Springer-Verlag, New York, 1966.

3. Denn, M. M., *Process Fluid Mechanics*, Prentice-Hall, Englewood Cliffs, NJ, 1980.

4. Meissner, J., Basic Parameters, Melt Rheology, Processing and End-Use Properties of Three Similar Low Density Polyethylene Samples, *Pure Appl. Chem.*, 42, 553–612, 1975.

5. Walters, K., *Rheometry*, Chapman & Hall, London, 1975.

6. Bagley, E. B., End Corrections in the Capillary Flow of Polyethylene, *J. Appl. Phys.*, 28, 624–627, 1957.

7. Bird, R. B., R. C. Armstrong, and O. Hassager, *Dynamics of Polymeric Liquids*, 2nd ed., Wiley, New York, 1987, Vol. 1.

8. Brinkman, H. C., Heat Effects in Capillary Flow I, *Appl. Sci. Res.*, A2, 120–124, 1951.

9. Kestin, J., M. Sokolov, and W. Wakeham, Theory of Capillary Viscometers, *Appl. Sci. Res.*, 27, 241–264, 1973.

10. Collyer, A. A., and D. W. Clegg (eds.), *Rheological Measurement*, 2nd ed., Chapman & Hall, London, 1998.

11. Gupta, R. K., Extensional Rheometry of Polymer Melts, in *Adhesive Bonding*, L. H. Lee (ed.), Plenum, New York, 1991, pp. 75–95.

12. Sridhar, T., V. Tirtaatmadja, D. A. Nguyen, and R. K. Gupta, Measurement of Extensional Viscosity of Polymer Solutions, *J. Non-Newtonian Fluid Mech.*, 40, 271–280, 1991.

13. Munstedt, H., and H. M. Laun, Elongational Behavior of an LDPE Melt. II. Transient Behavior in Constant Stretching Rate and Tensile Creep Experiments. Comparison with Shear Data. Temperature Dependence of the Elongational Properties, *Rheol. Acta*, 18, 492–504, 1979.

14. Laun, H. M., and H. Munstedt, Elongational Behavior of an LDPE Melt. I. Strain Rate and Stress Dependence of Viscosity and Recoverable Strain in Steady State. Comparison with Shear Data. Influence of Interfacial Tension, *Rheol. Acta*, 17, 415–425, 1978.

15. Tirtaatmadja, V., and T. Sridhar, A Filament Stretching Device for Measurement of Extensional Viscosity, *J. Rheol.*, 37, 1081–1102, 1993.

16. Dealy, J. M., *Rheometers for Molten Plastics*, Van Nostrand Reinhold, New York, 1982.

17. Ferry, J. D., *Viscoelastic Properties of Polymers*, 3rd ed., Wiley, New York, 1980.

18. Baumgaertel, M., and H. H. Winter, Determination of Discrete Relaxation and Retardation Time Spectra from Dynamic Mechanical Data, *Rheol. Acta*, 28, 511–519, 1989.

19. Cox, W. P., and E. H. Merz, Correlation of Dynamic and Steady Viscosities, *J. Polym. Sci.*, 28, 619–622, 1958.

20. Laun, H. M., Predictions of Elastic Strains of Polymer Melts in Shear and Elongation, *J. Rheol.*, 30, 459–501, 1986.

21. Debye, P., The Intrinsic Viscosity of Polymer Solutions, *J. Chem Phys., 14, 636–639, 1946.*

22. Bueche, F., Viscosity, Self-diffusion, and Allied Effects in Solid Polymers, *J. Chem. Phys., 20, 1959–1964, 1952.*

23. Welty, J. R., C. E. Wicks, and R. E. Wilson, *Fundamentals of Momentum,* Heat, and Mass Transfer, 3rd ed., Wiley, New York, 1984, p. 246.

24. Bueche, F., *Physical Properties of Polymers*, Wiley, New York, 1962.

25. Graessley, W. W., The Entanglement Concept in Polymer Rheology, *Adv. Polym. Sci.*, 16, 1–179, 1974.

26. Porter, R. S., and J. F. Johnson, The Entanglement Concept in Polymer Systems, *Chem. Rev.*, 66, 1–27, 1966.

27. Berry, G. C., and T. G. Fox, The Viscosity of Polymers and Their Concentrated Solutions, *Adv. Polym. Sci.*, 5, 261–357, 1968.

28. Klein, J., The Onset of Entangled Behavior in Semidilute and Concentrated Polymer Solutions, *Macromolecules*, 11, 852–858, 1978.

29. Graessley, W. W., Molecular Entanglement Theory of Flow Behavior in Amorphous Polymers, *J. Chem. Phys.*, 43, 2696–2703, 1965.

30. Graessley, W. W., Viscosity of Entangling Polydisperse Polymers, *J. Chem. Phys., 47, 1942–1953, 1967.*

31. de Gennes, P. G., *Scaling Concepts in Polymer Physics*, Cornell University Press, Ithaca, N.Y., 1979.

32. de Gennes, P. G., Reptation of a Polymer Chain in the Presence of Fixed Obstacles, *J. Chem. Phys.*, 55, 572–579, 1971.

33. Osaki, K., and M. Doi, Nonlinear Elasticity of Concentrated Polymer Systems, *Polym. Eng. Rev.*, 4, 35–72, 1984.

34. Graessley, W. W., Entangled Linear, Branched, and Network Polymer Systems— Molecular Theories, *Adv. Polym. Sci., 47, 67–117, 1982.*

35. Graessley, W. W., Viscoelasticity and Flow in Polymer Melts and Concentrated Solutions, in *Physical Properties of Polymers*, J. E. Mark (ed.), American Chemical Society, Washington, DC, 1984.

36. Bird, R. B., C. F. Curtiss, R. C. Armstrong, and O. Hassager, *Dynamics of Polymeric Liquids*, 2nd ed., vol. 2, Wiley, New York, 1987.

37. Bird, R. B., H. R. Warner, Jr., and D. C. Evans, Kinetic Theory and Rheology of Dumbbell Suspensions with Brownian Motion, *Adv. Polym. Sci.*, 8, 1–90, 1971.

38. Prilutski, G., R. K. Gupta, T. Sridhar, and M. E. Ryan, Model Viscoelastic Liquids, *J. Non-Newtonian Fluid Mech.*, 12, 233–241, 1983.

39. Boger, D. V., A Highly Elastic Constant-Viscosity Fluid, *J. Non-Newtonian Fluid Mech.*, 3, 87–91, 1977/78.

40. Gordon, R. J., *Structured Fluids, Certain Continuum Theories and Their Relation to Molecular Theories of Polymeric Materials*, Ph.D. dissertation, Princeton University, Princeton, NJ, 1970.

41. Marrucci, G., The Free Energy Constitutive Equation for Polymer Solutions from the Dumbbell Model, *Trans. Soc. Rheol.*, 16, 321–330, 1972.

42. Denn, M. M., and G. Marrucci, Stretching of Viscoelastic Liquids, *AIChE J.*, 17, 101–103, 1971.
43. Crochet, M. J., A. R. Davies, and K. Walters, *Numerical Simulation of Non-Newtonian Flow*, Elsevier, Amsterdam, 1984.
44. Rouse, P. E., Jr., A Theory of the Linear Viscoelastic Properties of Dilute Solutions of Coiling Polymers, *J. Chem. Phys.*, 21, 1272–1280, 1953.
45. Zimm, B. H., Dynamics of Polymer Molecules in Dilute Solutions: Viscoelasticity, Flow Birefringence and Dielectric Loss, *J. Chem. Phys.*, 24, 269–278, 1956.
46. Lodge, A. S., and Y. J. Wu, Constitutive Equations for Polymer Solutions: Derived from the Bead/Spring Model of Rouse and Zimm, *Rheol. Acta*, 10, 539–553, 1971.
47. Tam, K. C., and C. Tiu, Steady and Dynamic Shear Properties of Aqueous Polymer Solutions, *J. Rheol.*, 33, 257–280, 1989.
48. Gordon, R. J., and A. E. Everage, Jr., Bead–Spring Model of Dilute Polymer Solutions: Continuum Modifications and an Explicit Constitutive Equation, *J. Appl. Polym. Sci.*, 15, 1903–1909, 1971.
49. Lodge, A. S., Constitutive Equations from Molecular Network Theories for Polymer Solutions, *Rheol. Acta*, 7, 379–392, 1968.
50. Phan-Thien, N., and R. I. Tanner, A New Constitutive Equation Derived from Network Theory, *J. Non-Newtonian Fluid Mech.*, 2, 353–365, 1977.
51. Larson, R. G., *Constitutive Equations for Polymer Melts and Solutions*, Butterworths, Boston, MA, 1988.
52. Kaye, A., College of Aeronautics, Note No. 134, Cranford, UK, 1962.
53. Bernstein, B., A. E. Kearsley, and L. Zapas, A Study of Stress Relaxation with Finite Strain, *Trans. Soc. Rheol.*, 7, 391–410, 1963.
54. Doi, M., and S. F. Edwards, Dynamics of Concentrated Polymer Systems, Parts 1–4, *J. Chem. Soc. Faraday Trans. II*, 74, 1789–1801, 1802–1817, 1818–1832, 1978; 75, 38–54, 1979.
55. Gupta, R. K., *Polymer and Composite Rheology*, 2nd ed., Marcel Dekker, New York, 2000.
56. des Cloizeaux, J., Relaxation of Entangled and Partially Entangled Polymers in Melts, *Macromolecules*, 25, 835–841, 1992.
57. Mead, D. W., Determination of Molecular Weight Distributions of Linear Flexible Polymers from Linear Viscoelastic Material Functions, *J. Rheol.*, 38, 1797–1827, 1994.
58. Marrucci, G., The Doi–Edwards Model Without Independent Alignment, *J. Non-Newtonian Fluid Mech.*, 21, 329–336, 1986.
59. Marrucci, G., and N. Grizzuti, The Doi–Edwards Model in Slow Flows, Predictions on the Weissenberg Effect, *J. Non-Newtonian Fluid Mech.*, 21, 319–328, 1986.
60. Barnes, H. A., J. F. Hutton, and K. Walters, *An Introduction to Rheology*, Elsevier, Amsterdam, 1989.
61. Dealy, J. M., and K. F. Wissbrun, *Melt Rheology and Its Role in Plastics Processing*, Van Nostrand Reinhold, New York, 1990.
62. White, J. L., *Principles of Polymer Engineering Rheology*, Wiley–Interscience, New York, 1990.

63. Collyer, A. A. (ed.), *Techniques in Rheological Measurement*, Chapman & Hall, London, 1993.
64. Macosko, C. W., *Rheology*, VCH, New York, 1994.
65. Morrison, F. A., *Understanding Rheology*, Oxford University Press, New York, 2001.

PROBLEMS

14.1. Sketch the shear stress-versus-shear rate behavior corresponding to the viscosity-versus-shear rate behavior depicted in Figure 14.1. Indicate the lower and upper Newtonian regions clearly on your sketch.

14.2. Fill in the missing steps needed to go from Eq. (14.3.8) to Eq. (14.3.9).

14.3. Calculate the percentage change in the zero-shear viscosity of the LDPE sample shown in Figure 14.7 if the temperature is increased from 150°C to 151°C. What do you conclude?

14.4. If, in Figure 14.6, we replace the cone by a coaxial plate also of radius R, we get the parallel-plate viscometer. If the vertical gap between the two plates is H, where $H \ll R$, show that the shear rate at any radial position (in cylindrical coordinates now) is $\omega r/H$. How is the viscosity of a Newtonian liquid related to the measured torque in this instrument?

14.5. In a couette viscometer, the liquid sample is kept in the annular gap between two concentric cylinders, as shown in Figure P14.5. If one

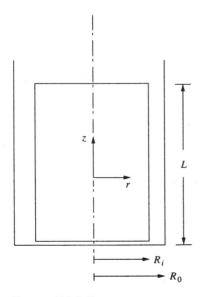

FIGURE P14.5   A couette viscometer.

cylinder is rotated relative to the other at an angular velocity ω, obtain an expression, in terms of $\tau_{\theta r}$, for the torque that has to be applied to the other cylinder to keep it stationary. Assume that $R_0 - R_i$ is small enough that the circumferential velocity varies linearly across the gap. Also assume that the viscometer length L is large enough that the influence of the ends can be ignored. How is the liquid viscosity related to the measured torque? Do not assume Newtonian fluid behavior.

14.6. Let τ_{rz} be $-\eta \, dv_z/dr$ in Eq. (14.4.2) and solve for the velocity profile $v_z(r)$ for the fully developed tube flow of a Newtonian liquid. Integrate the velocity profile over the tube cross section to relate the volumetric flow rate Q to the pressure gradient. Substitute this relationship back into Eq. (14.4.2) and evaluate the result at $r = R$ to obtain Eq. (14.4.10).

14.7. What is the Rabinowitsch correction factor for a power-law fluid?

14.8. The capillary is always positioned vertically in commercial capillary viscometers. Yet, the analysis in Section 14.4 was for a horizontal capillary. Does the theory need to be modified in any way? Justify your answer.

14.9. Capillary viscometer data on a food emulsion are given as follows in terms of the wall shear stress as a function of the volumetric flow rate. Use Eq. (14.4.9) to compute the wall shear rate corresponding to each datum point and thus obtain a graph of the shear viscosity versus the shear rate. The capillary radius is 0.094 cm.

τ_w (dyn/cm^2)	Q (cm^3/sec)
2271	0.0045
2711	0.0067
3196	0.0111
3554	0.0161
3948	0.0224
4200	0.0284

14.10. Given here are shear viscosity data at 20°C for a 500-ppm solution of a high-molecular-weight polyacrylamide in distilled water. Obtain the best-fit parameters for the power-law and Carreau models.

Shear rate (sec^{-1})	Viscosity (P)
0.022	3.46
0.040	3.42
0.075	3.19

0.138	2.89
0.255	2.53
0.471	2.16
0.870	1.76
1.607	1.38
2.970	1.06
5.490	0.79
10.150	0.58
18.740	0.42
34.600	0.30

14.11. Several articles in the technical literature suggest that melt fracture is caused by the violation of the no-slip boundary condition at the tube wall. In other words, at high enough flow rates, $v_z(R)$ equals V_s, which is called a *slip velocity*.

Beginning with Eq. (14.4.3), redo the analysis in Section 14.4 using the slip boundary condition and show that the equivalent form of Eq. (14.4.7) is as follows:

$$\frac{4Q}{\pi R^3} = \frac{4V_s}{R} - \frac{4}{\tau_w^3} \int_0^{\tau_w} \tau_{rz}^2 \dot{\gamma}(\tau_{rz}) \, d\tau_{rz}$$

By partially differentiating this equation with respect to $1/R$ while holding τ_w constant, suggest a method of obtaining V_s with the help of pressure drop-versus-flow rate data.

14.12. A pressure-driven capillary viscometer is used to extrude polyethylene terephthalate (PET) at $300°C$, as shown in Figure P14.12. If PET can be assumed to be a Newtonian liquid with a shear viscosity of $1000\,P$ and density of $1.3\,g/cm^3$, what is the volumetric flow rate through the capillary? You might want to use the results of Problem 14.6. Assume isothermal conditions. If the filament emerging from the capillary at point B is stretched continuously with a net downward force of 7×10^4 dyn, will the volumetric flow rate change? If so, by what percentage?

14.13. Sketch the force-versus-time behavior corresponding to any one of the filament stretching runs in Figure 14.12.

14.14. Use Eq. (14.6.4) to obtain expressions for the first and second normal stress differences during shearing at a constant shear rate. What do you conclude from these expressions?

14.15. What is the relation between the extensional viscosity and the shear viscosity according to the Boltzmann superposition principle?

14.16. Use Eq. (14.6.7) to determine the time dependence of the shear stress when a fluid is sheared at a constant rate $\dot{\gamma}$ beginning from a state of rest at $s = 0$.

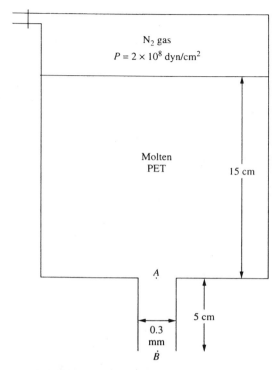

FIGURE P14.12 Extrusion of PET using a pressure-driven capillary viscometer.

14.17. Derive Eq. (14.7.7) by integrating Eq. (14.7.3) by parts and noting that (a) $G(t') \to 0$ as $t' \to \infty$ and (b) dG/dt' is a negative exponential. Similarly, derive Eq. (14.7.8), beginning with Eq. (14.7.4).

14.18. Is the neglect of the solvent contribution to the stress in the solution justified in Example 14.7? Justify your answer by doing some calculations.

14.19. If a polymer solution is sheared at a constant shear rate $\dot{\gamma}$, beginning from a state of rest, how does the shear stress vary with time if we use the upper-convected Maxwell model?

14.20. Equation (14.9.31) gives only the polymer contribution to the stress. What is the equivalent equation that gives the combined polymer and solvent contribution?

14.21. Obtain an expression for the first normal stress difference in shear at steady state according to the upper-convected Maxwell equation. Is there a solvent contribution in this case?

15

Polymer Processing

15.1 INTRODUCTION

Thus far, we have talked about how polymers are synthesized, how they are characterized, and how they behave as solids, melts, or in the form of solutions. Ultimately, however, it is necessary to convert the polymer into useful products. Typically, these may be rods, pipes, films, fibers, or molded articles. Most frequently these materials are made from a single polymer. Increasingly, though, blends, filled polymers, and composite materials are used. It should be noted that even when a single polymer is used, it is rarely a chemically pure material. Almost invariably, it contains additives that act as dyes, plasticizers, antioxidants, and so on. A variety of physical structures can result, depending on the kind of polymer and additives used and also on how these materials are processed. Because the final structure obtained determines the physical properties of the product, the process used has to be chosen with care. For inexpensive, high-volume disposable items such as beverage containers or toys, the most inexpensive process is used. The manufacture of high-value-added sophisticated items such as compact disks or optical lenses, on the other hand, requires a great deal of engineering and an intimate knowledge of the fundamentals of polymer behavior.

In this chapter, we discuss just three of the most common polymer processing operations-extrusion, injection molding, and fiber spinning. Because most polymers are sold in the form of pellets, an extruder is required to melt, homogenize, and pump the thermoplastic material. Although articles such as tubes, rods, and flat sheets can be made by extrusion, an extruder is often coupled with other polymer processing machinery. A knowledge of extrusion is therefore a prerequisite for studying other polymer processing operations. Note that extrusion is a continuous operation. As opposed to this, injection molding is a cyclic operation used to make a very wide variety of low- and high-technology items of everyday use. It is now also being used to fabricate ceramic heat engine components, which have complex shapes and are useful for high-temperature applications. Ceramics such as silicon carbide and silicon nitride are hard, refractory materials, and injection molding is one of the very few processes that can be used for the purpose of mass production. Fiber spinning is studied not only because it is the mainstay of the synthetic textiles industry but also because it is used to make novel fibers out of liquid-crystalline polymers, graphite, glass, and ceramics for use in composite materials. It is also a process where polymer elasticity becomes important due to the extensional nature of the flow field. In contrast, extrusion and injection molding are shear-dominated processes.

Although this is the last chapter of the book, it is really an introduction to the very practical and fascinating topic of manufacturing items made from polymeric materials. The major purpose of this chapter is to describe these three operations and also to show how first principles are used to mathematically simulate these or any other process. Such analyses are a sine qua non for improving product quality, for designing new products, and for process optimization. The material presented here is necessarily simplified, and we have restricted ourselves wherever possible to steady-state and isothermal operations; extensions to more realistic situations are conceptually straightforward, and we have provided citations of the appropriate technical literature. For more details on the process of constructing mathematical models, the reader is directed to the excellent book by Denn [1].

15.2 EXTRUSION

This is the most common polymer processing operation. It is generally used to melt and pump thermoplastic polymers through a die, which gives a desired shape to the extrudate. Although extrusion can be carried out using pressure-driven and plunger-driven devices, it is the screw extruder that is used almost universally in industrial applications. Extruders can contain multiple screws, but we shall initially focus on the single-screw extruder, which consists of a helical screw rotating inside a cylindrical barrel. This is shown schematically in Figure 15.1 [2]. The polymer is generally fed to the extruder in the form of pellets through a hopper, which leads to the channel formed between the screw and the barrel. If

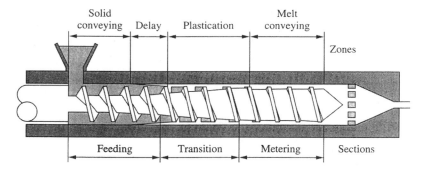

FIGURE 15.1   Schematic representation of a single-screw extruder. (From Ref. 2)

the polymer being processed is hygroscopic, it is usually dried beforehand and the hopper is blanketed by a dry inert gas such as nitrogen. Although the barrel is heated to a temperature above the melting point of the polymer, the region very close to the base of the hopper is often water cooled to prevent polymer from melting in the hopper and forming a solid plug, which would block additional polymer from entering the extruder. The rotation of the screw forces the polymer to move along the channel, and it does so initially as a solid plug, then as a semisolid, and, finally, as a melt. The channel depth is usually fairly large in the solids-conveying zone and it decreases progressively as the polymer melts and ultimately becomes constant in the melt zone. Although it is convenient to think in terms of these three separate zones and the screw geometry often reflects this thinking, it is probably true that the processes of solids conveying, melting, and melt pressurization occur simultaneously. For the purposes of analysis, though, we shall still treat the three zones separately. Of course, if the extruder is fed directly by a polymerization reactor, as happens during the manufacture of synthetic textiles by melt spinning, the solids-conveying and melting zones are absent. Only the melt zone remains; it is also called the *metering zone.*

   The purpose of any mathematical model of steady-state extrusion is to relate quantities such as energy dissipation, the volumetric flow rate, the melting profile, the temperature profile, and the pressure profile to the extruder geometry, to the processing variables such as barrel temperature and screw rpm, and to the material properties of the polymer. In order to accomplish this task, we also need to know the details of the die that is attached to the extruder. Because the process, in general, is nonisothermal and the rheological models are nonlinear, analytical solutions cannot be obtained for realistic cases of interest. Invariably, numerical techniques of solution have to be used. Here, though, we will consider the simplest possible cases with a view toward both elucidating the physics of the problem and illustrating the approach to be taken for problem solving. Analyses for more realistic situations are available in the literature; here, the mathematics is more complicated, but the basic approach is the same. It should be realized, though, that the process of determining the screw geometry to yield desired extruder performance is much more difficult than determining extruder performance for a given geometry.

15.2.1 Screw Geometry

A section of a simplified screw is shown in Figure 15.2 to define the variables that characterize the screw geometry [3]. The notation used is that of Tadmor and Klein [3]. The inside diameter of the barrel is D_b, whereas the screw diameter is D; both of these quantities can range from 1 to 12 in. A typical value of the ratio of the screw length to its diameter is 24. The channel depth is H, and it is clear from Figure 15.1 that both H and D vary with axial position. The radial clearance between the tip of the flights and the inner surface of the barrel is δ_f, and L is the axial distance moved by the screw during one full revolution. The width of the screw flight in the axial direction is b, and the width in a direction perpendicular to the flight is e. Finally, W is the distance between flights measured perpendicular to the flights, and θ, the helix angle, is the angle between the flight and the plane perpendicular to the screw axis. In general, θ, b, and W vary with radial position; nonetheless, we will take them to be constant. We will also assume that δ_f is negligible and that there is no leakage of material over the flights. Also, for purposes of analysis, we will assume that D_b is approximately the same as D.

 To begin the analysis, we recognize that flow occurs because friction at the surface of the barrel makes the plastic material slide down the channel and go toward the extruder exit as the screw is rotated. This motion of a material element, resulting from the relative velocity between the barrel and the screw, can be

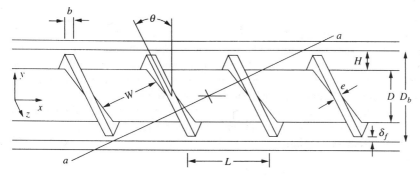

FIGURE 15.2 Line a–a indicates a cut perpendicular to the flight at the barrel surface. (From Ref. 3.)

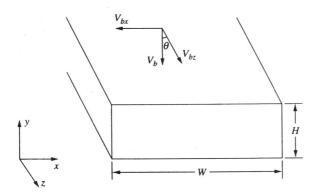

FIGURE 15.3 Polymer flow in the channel between the screw and barrel surface.

studied more easily by allowing the barrel to rotate in a direction opposite to that of screw rotation and holding the screw stationary. Further, if we realize that the curvature of the screw is hardly felt by the polymer, we can consider that the polymer is moving down a long, rectangular cross-sectional channel due to the movement of the upper surface. This is shown in Figure 15.3. In effect, we can unwind the channel and use Cartesian coordinates for the analysis.

15.2.2   Solids-Conveying Zone

This is the region from the point at which material enters the hopper to a point in the extruder channel where melting begins. Although a large number of models have been proposed to determine the flow rate of solids in this region, the accepted analysis is that of Darnell and Mol [4], which is presented here in simplified form.

As polymer pellets move down the channel, they become compacted into a plug that moves at a velocity V_p in the down channel or z direction. In general, there is slip between the plug and both the barrel and screw surfaces. The barrel moves at a velocity V_b, which in magnitude equals πDN, where N is the revolutions per unit time of the screw; this velocity vector makes an angle θ to the down-channel direction. Clearly, the velocity of the barrel relative to the plug is $(\mathbf{V}_b - \mathbf{V}_p)$ and it makes an angle $(\theta + \phi)$ to the z direction; this is the angle at which the barrel appears to move for an observer moving with the plug.

With reference to Figure 15.4, we have

$$\tan \phi = V_{\text{pl}}\left(V_b - \frac{V_{\text{pl}}}{\tan \theta} \right)^{-1} \tag{15.2.1}$$

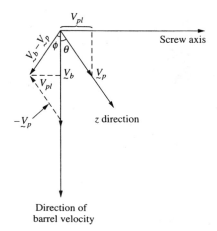

Screw axis

z direction

Direction of
barrel velocity

FIGURE 15.4 Diagram showing the different velocity vectors.

where V_{pl} is the component of the plug velocity along the screw axis. Thus,

$$V_{pl} = V_b \frac{\tan \phi \tan \theta}{\tan \phi + \tan \theta} \qquad (15.2.2)$$

The volumetric flow rate, Q_s, of the solid plug is the product of the axial component of the plug velocity and the area for flow in that direction,

$$Q_s = V_{pl} \pi DH \qquad (15.2.3)$$

which, in combination with Eq. (15.2.2), gives

$$Q_s = (\pi^2 D^2 NH) \frac{\tan \phi \tan \theta}{\tan \phi + \tan \theta} \qquad (15.2.4)$$

where all the quantities except ϕ are known; ϕ is obtained by a simultaneous force and moment balance on a section of the solid plug, as shown in Figure 15.5.

The force that causes the plug to move is the force of friction, F_1, between the barrel and the plug. As mentioned previously, the barrel velocity relative to the plug is in a direction that makes an angle $\theta + \phi$ to the z direction. This, therefore, is the direction of F_1. The magnitude of F_1 is given by

$$F_1 = f_b pW \ dz \qquad (15.2.5)$$

where f_b is the coefficient of friction between the plug and the barrel surface, p is the isotropic pressure within the plug, and dz is the thickness of the plug. A pressure gradient develops across the plug and the force due to this is as follows:

$$F_6 - F_2 = HW \ dp \qquad (15.2.6)$$

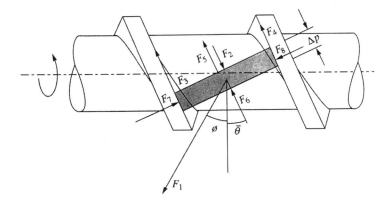

FIGURE 15.5   Forces acting on the solid plug. (From Ref. 3.)

As will be evident later, dp is a positive number, so pressure increases with increasing z.

Normal forces act on the plug at the flights due to the presence of the isotropic pressure. Thus, we have

$$F_8 = pH \ dz \qquad\qquad (15.2.7)$$

The normal force that acts on the other flight is

$$F_7 = pH \ dz + F^* \qquad\qquad (15.2.8)$$

where F^* is a reaction force.

Finally, there are friction forces on the two flights and on the screw surface, as shown in Figure 15.5. Their magnitudes are as follows:

$$F_3 = f_s F_7 \qquad\qquad (15.2.9)$$
$$F_4 = f_s F_8 \qquad\qquad (15.2.10)$$
$$F_5 = f_s pW \ dz \qquad\qquad (15.2.11)$$

where f_s is the coefficient of friction between the plug and the screw surface.

Each of the forces F_1 to F_8 can be resolved into a component parallel to the screw axis and a component perpendicular to the screw axis. These must sum to zero because the plug does not accelerate. The axial force balance takes the following form:

$$F_1 \sin \phi + (F_6 - F_2) \sin \theta - (F_7 - F_8) \cos \theta + (F_3 + F_4 + F_5) \sin \theta = 0$$
$$(15.2.12)$$

Introducing expressions for the various forces into Eq. (15.2.12) and rearranging, we find that

$$F^* = \frac{A_1 p \, dz + A_2 \, dp}{\cos \theta - f_s \sin \theta} \tag{15.2.13}$$

where

$$A_1 = f_b W \sin \phi + 2Hf_s \sin \theta + Wf_s \sin \theta \tag{15.2.14}$$
$$A_2 = HW \sin \theta \tag{15.2.15}$$

Another expression for F^* can be obtained by a moment balance about the axis of the screw,

$$\frac{D}{2} [F_1 \cos \phi - (F_6 - F_2) \cos \theta - (F_7 - F_8) \sin \theta$$
$$- (F_3 + F_4 + F_5) \cos \theta] = 0 \tag{15.2.16}$$

with the result that

$$F^* = \frac{B_1 p \, dz - B_2 \, dp}{\sin \theta + f_s \cos \theta} \tag{15.2.17}$$

where

$$B_1 = f_b W \cos \phi - 2Hf_s \cos \theta - Wf_s \cos \theta \tag{15.2.18}$$
$$B_2 = HW \cos \theta \tag{15.2.19}$$

Equating the two expressions for F^* and rearranging yields

$$\frac{dp}{dz} = -\frac{(A_1 K - B_1)}{(A_2 K + B_2)} p \tag{15.2.20}$$

where

$$K = \frac{\sin \theta + f_s \cos \theta}{\cos \theta - f_2 \sin \theta} \tag{15.2.21}$$

Integrating Eq. (15.2.20) from $z = 0$, near the hopper base, where $p = p_B$ to z gives

$$p = p_B \exp\left(-\frac{(A_1 K - B_1)}{(A_2 K + B_2)} z\right) \tag{15.2.22}$$

and pressure rises exponentially with distance z. If z_b is taken as the length of the solids-conveying zone in the down-channel direction, p at z_b gives the pressure at the end of this zone.

We began this analysis seeking the angle ϕ so that we could calculate the solids-conveying rate from Eq. (15.2.4). Now, ϕ is found to be given implicitly in terms of the pressure rise by Eq. (15.2.22), because it is hidden in the constants A_1 and B_1. Thus, a knowledge of Δp is needed for the determination of Q_s. If Δp is nonzero, we would have to calculate this quantity. In principle, we can work backward from the extruder exit and compute p at z_b using the condition that the mass flow rate has to be the same throughout the extruder. Note from Eq. (15.2.4) that the volumetric flow rate increases linearly with the screw speed. If, on the other hand, the flow rate is known, we can determine ϕ from Eq. (15.2.4) and the pressure rise from Eq. (15.2.22). To calculate the pressure at the end of the solids-conveying region, though, we need the value of p_B, the pressure at the base of the hopper. For this, it is necessary to examine the flow of granular solids in a conical bin. This has been done, and results are available in the literature [5,6].

Example 15.1: What is the maximum possible rate of solids conveying through an extruder?

Solution: The flow rate is maximum when there is no obstruction at the extruder exit (i.e., $\Delta p = 0$) and when there is no friction between the polymer and the screw surface. Under these conditions, Eq. (15.2.12) becomes

$$F_1 \sin \phi - F^* \cos \theta = 0$$

whereas Eq. (15.2.16) takes the form

$$F_1 \cos \phi - F^* \sin \theta = 0$$

with the result that

$$\tan \phi = \cot \theta$$

which when inserted into Eq. (15.2.4) leads to the desired result:

$$Q_{\max} = \pi^2 D^2 N H \sin \theta \cos \theta$$

In closing this section we mention that Chung has observed that the model of Darnell and Mol is strictly valid only up to the point that the polymer begins to melt [7]. Because the barrel temperature is kept above the melting point of the polymer, a layer of liquid forms fairly quickly and coats the solid plug. As a consequence, Eqs. (15.2.5) and (15.2.9)–(15.2.11) have to be modified and the forces F_1, F_3, F_4, and F_5 calculated using the shear stress in the molten polymer film. A result of this modification is that ϕ becomes a function of the screw revolutions per minute (rpm) [7,8]. We also mention that Campbell and Dontula have proposed a new model that does not require us to assume that the screw is stationary [9]. This model appears to give better agreement with experimental data.

15.2.3 Melting Zone

Melting of the polymer occurs due to energy transfer from the heated barrel and also due to viscous dissipation within the polymer itself. This melting does not happen instantly but takes place over a significant part of the screw length. The purpose of any analysis of the melting process is to predict the fraction of polymer that is melted at any down-channel location and to relate this quantity to material, geometrical, and operating variables.

Maddock [10] and Tadmor and Klein [3] studied the melting process by "carcass analysis": They extruded colored polymer and stopped the extruder periodically. By cooling the polymer and extracting the screw, they could track the progress of melting and also determine the sequence of events that ultimately resulted in a homogeneous melt. They found that a thin liquid film was formed between the solid bed of the polymer and the barrel surface. This is shown in Figure 15.6. Because of the relative motion between the barrel and the polymer bed, the molten polymer was continually swept from the thin film in the x direction into a region at the rear of the bed between the flight surface and the bed. Liquid lost in this manner was replaced by freshly melted polymer so that the film thickness δ and the bed thickness both remained constant. As melting proceeded, the solid polymer was transported at a constant velocity V_{sy} to the thin film–solid bed interface and, correspondingly, the bed width X decreased with increasing down-channel distance.

A large number of models, of varying degrees of complexity, exist for calculating X as a function of distance z [3,8,11,12]. In the simplest case, it is assumed that the solid polymer is crystalline with a sharp melting point T_m and a latent heat of fusion λ and that the molten polymer is a Newtonian liquid. It is also assumed that the solid and melt physical properties such as the density, specific

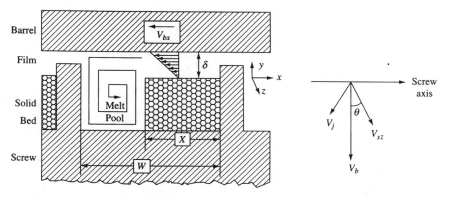

FIGURE 15.6 Melting of a polymer inside the extruder. (From Ref. 2.)

heat, and thermal conductivity remain constant, independent of temperature. The barrel temperature is taken to be T_b and the screw temperature is T_s. Furthermore, the velocity of the solid bed in the z direction, V_{sz}, is taken to be constant. Because there is no accumulation of mass anywhere within the extruder, this velocity is obtained from Eq. (15.2.4) as follows:

$$V_{sz} = \frac{Q_s}{HW} \tag{15.2.23}$$

From the viewpoint of an observer moving with the solid polymer, the barrel appears to move with a velocity $\mathbf{V}_j$ (see Figure 15.6),

$$\mathbf{V}_j = \mathbf{V}_b - \mathbf{V}_{sz} \tag{15.2.24}$$

so that (see Fig. 15.6)

$$|\mathbf{V}_j|^2 = |\mathbf{V}_b|^2 + |\mathbf{V}_{sz}|^2 - 2|\mathbf{V}_b||\mathbf{V}_{sz}|\cos\theta \tag{15.2.25}$$

If we now consider a coordinate system xyz with its origin located at the solid bed–melt interface (see Fig. 15.6) and moving with a velocity V_{sz}, the differential form of the energy balance as applied to the melt film is given by

$$k_m \frac{d^2T}{dy^2} + \eta\left(\frac{dv}{dy}\right)^2 = 0 \tag{15.2.26}$$

where k_m is the melt thermal conductivity, η its viscosity, and $v = |\mathbf{V}_j|$. Here, it is assumed that that temperature varies only in the y direction and conduction is the only mode of heat transfer. Note that the second term in Eq. (15.2.26) results from viscous dissipation; this term equals the constant quantity $\eta(v/\delta)^2$, provided that this is considered to be a flow between two infinite parallel plates.

Integrating Eq. (15.2.26) twice with respect to y and using the conditions $T(0) = T_m$ and $T(\delta) = T_b$ gives the following:

$$\frac{T - T_m}{T_b - T_m} = \frac{\eta v^2}{2k_m(T_b - T_m)\delta}\frac{y}{\delta}\left(1 - \frac{y}{\delta}\right) + \frac{y}{\delta} \tag{15.2.27}$$

The heat flux into the interface from the melt film is

$$q_{y1} = -k_m \frac{dT}{dy}(0) = -\frac{k_m}{\delta}(T_b - T_m) - \frac{\eta v^2}{2\delta} \tag{15.2.28}$$

Within the solid polymer bed, the energy balance is

$$\rho_s c_s V_{sy} \frac{dT}{dy} = k_s \frac{d^2T}{dy^2} \tag{15.2.29}$$

where the various symbols have their usual meanings and the subscript s denotes the solid polymer.

For the purpose of integrating Eq. (15.2.29), it is assumed that the temperature far away from the interface at $y = -\infty$ is the screw temperature T_s. Because $T(0)$ is T_m, then, we have within the solid bed

$$\frac{T - T_s}{T_m - T_s} = \exp\left(\frac{V_{sy}\rho_s c_s}{k_s} y\right) \tag{15.2.30}$$

The heat flux into the interface from the solid bed is

$$q_{y2} = -k_s \frac{dT}{dy}(0) = -\rho_s c_s V_{sy}(T_m - T_s) \tag{15.2.31}$$

The difference in heat fluxes across the interface represents the energy needed to melt the polymer per unit time per unit interface area. Thus, we have

$$|q_{y1}| - |q_{y2}| = V_{sy}\rho_s\lambda \tag{15.2.32}$$

Introducing Eqs. (15.2.28) and (15.2.31) into Eq. (15.2.32) gives

$$V_{sy}\rho_s\lambda = \frac{k_m}{\delta}(T_b - T_m) + \frac{\eta v^2}{2\delta} - \rho_s c_s V_{sy}(T_m - T_s) \tag{15.2.33}$$

which involves two unknowns, δ and V_{sy}.

An additional relation between δ and V_{sy} is needed, and this is obtained by proposing that the mass that enters the melt film in the y direction from the solid bed all leaves with the melt in the x direction. Consequently, we have the following:

$$V_{sy}\rho_s X = \left(\frac{V_{bx}}{2}\right)\rho_m\delta \tag{15.2.34}$$

Although Eq. (15.2.34) does relate δ to V_{sy}, it also involves a new unknown X. Solving for V_{sy} from Eq. (15.2.33) and introducing the result into Eq. (15.2.34) gives

$$\delta = \left(\frac{2k_m(T_b - T_m) + \eta v^2}{V_{bx}\rho_m[c_s(T_m - T_s) + \lambda]}\right)^{1/2} X^{1/2} = c_1 X^{1/2} \tag{15.2.35}$$

where the constant c_1 is defined by Eq. (15.2.35).

The polymer that melts has to come from the bed of the solid polymer whose width X decreases with down-channel distance. This change in width is obtained from a mass balance on the solid polymer as follows:

$$-\rho_s V_{sz} H \frac{dX}{dz} = \left(\frac{V_{bx}}{2}\right)\rho_m\delta \tag{15.2.36}$$

which can be combined with Eq. (15.2.35) to yield

$$-\frac{dX}{dz} = \frac{V_{bx}}{2} \frac{\rho_m}{\rho_s} \frac{c_1 X^{1/2}}{V_{sz}H} = c_2 X^{1/2} \tag{15.2.37}$$

where the constant c_2 is defined by Eq. (15.2.37). Integrating Eq. (15.2.37) using $X(0) = W$ gives

$$\frac{X}{W} = \left(1 - \frac{c_2 z}{2W^{1/2}}\right)^2 \tag{15.2.38}$$

Complete melting will occur when $X = 0$. The total down-channel distance z_T needed for this is

$$z_T = \frac{2W^{1/2}}{c_2} \tag{15.2.39}$$

The foregoing analysis combines experimental observations with fundamentals of transport phenomena to analytically relate X to z. It is obviously quite restrictive. Given the known behavior of polymeric fluids, we can immediately think of a number of modifications, such as making the shear viscosity in Eq. (15.2.26) depend on temperature and shear rate. We can also relax the assumption that $T(-\infty) = T_s$ and assume that the screw is adiabatic [13]. These modifications have all been done, and the results are available in the literature. The modifications bring model predictions closer to experimental observations but also necessitate numerical or iterative calculations. The various melting models have been reviewed by Lindt [14].

Finally, we note that there is usually a region preceding the melting zone wherein a melt film exists but in the absence of a melt pool. This is called the *delay zone*; its length is small, typically one to two screw turns [11]. Empirical correlations exist for estimating the extent of the delay zone [3].

15.2.4 The Melt Zone

The completely molten polymer entering the melt zone is usually pressurized before it leaves the extruder. Pressure builds up because relative motion between the barrel and the screw forces the polymer downstream, but the exit is partially blocked by a shaping die. In modeling this zone, we are interested in relating the volumetric flow rate to the screw rpm and in calculating the pressure rise for a given volumetric flow rate. This is fairly easy to do if we take the polymer to be Newtonian and the melt conveying process to be isothermal. All that we have to do is to solve the Navier–Stokes equation in the z direction for the flow situation shown earlier in Figure 15.3. Whereas the z component of the barrel velocity contributes to the volumetric flow rate out of the extruder, the x component merely causes recirculation of fluid, because there is no leakage over the flights.

Because polymer viscosities are usually very large, the Reynolds numbers are small and we can safely neglect fluid inertia in the analysis that follows. In addition, we can take pressure to be independent of y because the channel depth is very small in comparison to both the width and length. The simplified form of the z component of the Navier–Stokes equation is as follows:

$$\frac{\partial p}{\partial z} = \eta \frac{\partial^2 v_z}{\partial y^2} \tag{15.2.40}$$

where we have assumed fully developed flow in the z direction. As a consequence, $\partial p/\partial z$ is constant.

Integrating Eq. (15.2.40) subject to the conditions $v_z = 0$ at $y = 0$ and $v_z = V_{bz}$ at $y = H$ gives

$$v_z = V_{bz}\frac{y}{H} - \frac{y(H-y)}{2\eta}\frac{\partial p}{\partial z} \tag{15.2.41}$$

The volumetric flow rate is given by

$$Q = W \int_0^H v_z \, dy \tag{15.2.42}$$

Introducing Eq. (15.2.41) into Eq. (15.2.42) gives

$$Q = \frac{V_{bz}HW}{2} - \frac{WH^3}{12\eta}\frac{\Delta p}{L} \tag{15.2.43}$$

where Δp is the total change in pressure over the entire length L of the melt-conveying zone and V_{bz} is given by $\pi DN \cos \theta$.

Example 15.2: In their study, Campbell et al. rotated the barrel and measured the volumetric flow rate Q as a function of the screw rotational speed N in revolutions per second for the extrusion of a Newtonian oil under open-discharge conditions [15]. How should the measured value of Q depend on N? For their extruder, $D = 5.03$ cm, $W = 1.1$ cm, $H = 0.36$ cm, and $\theta = 6.3°$.

Solution: In the absence of a die, Δp is zero and Eq. (15.2.43) predicts that

$$Q = \frac{\pi \times 5.03 \times N \cos(6.3°) \times 0.36 \times 1.1}{2}$$
$$= 3.11N \text{ cm}^3/\text{sec}$$

and the result does not depend on the shear viscosity of the polymer.

The experimental results of Campbell et al., along with the theoretical predictions, are shown in Figure 15.7 [15]. Although the measured output varies

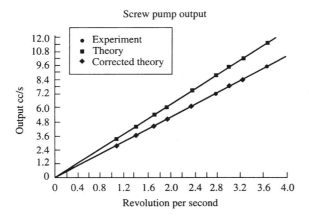

FIGURE 15.7   Theoretical and experimental flow rate for barrel rotation. (From Ref. 15.)

linearly with N as expected, the theory overpredicts the results. This happens because the theory is valid for an infinitely wide channel, whereas there is actually just a finitely wide channel. To correct for the finite width, we multiply the theoretical output by the following correction factor [15]:

$$F_d = \frac{16W}{\pi^3 H} \sum \frac{1}{i^3} \tanh\left(\frac{i\pi H}{2W}\right)$$

                                                                             (15.2.44)

which, for conditions of Example 15.2, has a value of 0.82. Excellent agreement is found with data when this correction is applied, and the line marked "corrected theory" passes through all the data points in Figure 15.7. Interestingly enough, this perfect agreement between theory and practice is not obtained when the extruder is operated with the barrel stationary and the screw moving. Campbell et al. ascribe this to the small contribution made to the flow by the helical flights [15]. They have proposed a theory that takes this contribution into account and eliminates the mismatch [15].

The flow rate calculated in Example 15.2 is the maximum possible flow rate through the extruder. In general, the extruder is equipped with a die whose shape depends on whether the end product will be rods, tubes, fibers, or flat films. In this case, the Δp in Eq. (15.2.43) is nonzero and positive. To obtain the volumetric flow rate now, we need an independent equation for the pressure rise; this is developed by considering the flow of the polymer through the die. For Newtonian liquids, we can easily show that the volumetric flow rate through the die has to be proportional to the pressure drop across the die. For the fully developed flow through a rod die (a tube), for example (see Chap. 14), we have

$$Q = \frac{\pi R^4}{8\eta} \frac{\Delta p}{l}$$ (15.2.45)

where R is the tube radius and l is the tube length. This equation allows us to calculate the pressure at the extruder–die interface because the pressure at the die exit is atmospheric.

If you read the previous section again, you will realize that the change in pressure across the melting zone remains an unknown quantity. Because the melt pool geometry is similar to the screw channel geometry, the pressure rise is calculated by again solving the Navier–Stokes equation.

15.2.5 Overall Extruder Simulation

In order to predict extruder performance, we have to solve the equations describing the three different extruder zones in sequence. This is because information generated in one zone is an input for a different zone. Even then, the solution is iterative [16]. We begin by assuming the mass flow rate and calculating the pressure change across each zone for a specified rpm. If the calculated pressure at the die exit, obtained by adding up the individual pressure drops, differs from atmospheric, we guess the flow rate again and repeat the computations until agreement is obtained. Typical pressure profiles for the extrusion of polyethylene through a 4.5-cm diameter extruder are shown in Figure 15.8. It can sometimes happen, due to the nonlinear nature of the equations involved, that there is more than one admissible solution. This implies the existence of multiple steady states, which can be the cause of surges in throughput as the extruder cycles between different steady states.

In any proper extruder simulation, we also must take into account the nonisothermal nature of the process, because a significant amount of heat is generated due to viscous dissipation. In addition, we cannot ignore the shear-thinning behavior of the molten polymer. All this, along with the iterative nature of the calculations, mandates the use of a computer. However, it is possible to obtain reasonable answers to fairly realistic steady-state problems using just a personal computer [2]. Using computer models of the type described by Vincelette et al. [2], it is possible to try to optimize extruder performance with respect to energy consumption or temperature uniformity or any other criterion of interest. Note again, though, that in this brief treatment of single-screw extrusion, we have not considered phenomena such as leakage flow over the flights and the variation of the coefficient of friction between the polymer and the extruder surface. We have also not accounted for complexities arising from features such as extruder venting, a variable channel depth, and the change in parameters (such

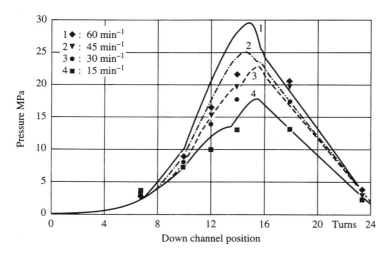

FIGURE 15.8 Predicted and experimental pressure profiles $N = 1$ (1), 0.8 (2), 0.6 (3), 0.4 (4), 0.2 (5) min^{-1}. (From Ref. 2)

as the helix angle) with axial position. For a discussion of all these effects, the reader is referred to advanced texts on the subject [3,6,17].

15.2.6 Extruders as Mixers

A single-screw extruder is an excellent device for melting and pumping polymers, but it is not an efficient mixer due to the nature of the flow patterns inside the extruder channels. During polymer processing, however, we often need to do the following:

1. Blend in additives such as pigments, stabilizers, and flame retardants
2. Add fillers such as carbon black, mica or calcium carbonate
3. Disperse nanofillers such as montmorillonite and carbon nanotubes
4. Mix elastomers such as ABS or EPDM that act as toughening agents
5. Put in reinforcements like short glass fibers
6. Make polymeric alloys using two miscible plastics
7. Carry out reactive extrusion for the synthesis of copolymers.

For all of these purposes, it is generally desirable that the dispersed phase leaving the extruder be in the form of primary particles (not agglomerates), if solid, and that it have molecular dimensions, if liquid. Also, it is necessary that the concentration of the dispersed phase be uniform throughout the mixture; this is a particularly challenging proposition when the process has to be run in a continuous manner and the dispersed phase is only a small volume fraction of the total mixture.

A simple method of providing mixing in a single-screw extruder, especially for liquid–liquid systems, is by introducing mixing pins that are a series of obstacles protruding from the screw surface; these force the polymer stream to divide and recombine around the pins. Alternately, one may add mixing devices to the end of the extruder. These devices may be motionless, or they may have moving parts such as in the case of kneading gears. Motionless mixers or static mixers consist of blades or obstructions that are placed lengthwise in a tube, forming open but intersecting channels [18]. When the extruder forces liquid to flow through these channels, the melt has to go around the obstacles and recombine periodically, and this ultimately leads to a very homogeneous mixture. A recent innovation in static mixer technology has been the development of the extensional flow mixer [19]. Here, material is made to flow through a series of converging and diverging regions of increasing intensity. This again results in a fine and well-dispersed morphology, but at the expense of a higher pressure drop. Note that a single-screw extruder in combination with a static mixer is well suited for the manufacture of polymer alloys or blends and for making color concentrates or master batches.

An immense amount of plastic, particularly nylon, polycarbonate, and polyester, is compounded with short glass fibers. The addition of up to 40 wt% glass to the polymer significantly increases the heat distortion temperature and allows the compounded product to be used for under-the-hood automotive applications. The process of compounding polymers with glass fibers is typically carried out with the help of *twin-screw extruders*; multiple strands of the glass-filled resin are extruded into a water bath and then cut, often under water, to give pellets that are ready for injection molding into finished products. As the name implies, a twin-screw extruder (TSE) has two screws which are generally parallel to each other, but, unlike single-screw extruders, the screw diameter here does not vary with axial position. If the screws turn in opposite directions, we have a counterrotating TSE, whereas if the two rotate in the same direction, we have a corotating TSE. Shown in Figure 15.9 is the top view of a corotating machine [20]; this is a fully intermeshing extruder because the two (identical) screws are situated as close to each other as possible. Note that the counterrotating geometry features screws having leads of opposite hands. A beneficial consequence of intermeshing screws is that, during extruder operation, each screw wipes the entire surface of the other screw. The polymer can, therefore, not remain inside the extruder for long periods of time, and thermal degradation is prevented. The residence time distribution is narrow, and the residence time can be made as short as 5–10 sec [21]. Flushing the extruder between different product grades is therefore quick and easy.

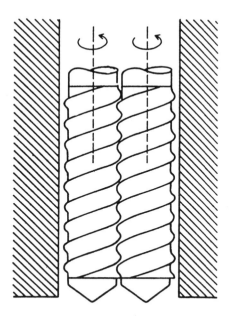

FIGURE 15.9 Cut-away view of a self-wiping, corotating twin-screw extruder. (From Ref. 20.)

If the screw cross section is circular and the screws are intermeshing, the barrel cross section has to be a figure 8, and this is shown in Figure 15.10. Here, each screw has three tips (i.e., it is trilobal), although newer extruders are bilobal. The intermeshing nature of the screws dictates the channel geometry, and, if we section a screw parallel to its axis, the channel shape that results is the one presented in Figure 15.11. The channel shape changes as we change the number of tips (parallel channels), and this affects the average shear rate in the channel and the mixing characteristics of the extruder. Each screw of a corotating TSE is assembled by sliding a number of modular elements on to a shaft in a desired sequence and then locking the elements in place. Two of the basic elements that make up any screw are *screw bushings* (also called *conveying elements*) and *kneading disks*. When conveying elements are employed, the screws look like those shown in Figure 15.9. If we unwind the channels of trilobal conveying elements, we get five parallel channels as shown in Figure 15.12. As the screws turn, the polymer is conveyed from one screw to the other and back, and the flow is very much a drag flow, quite like that in a single-screw extruder.

The reason that a TSE is considered to be superior to a single screw extruder is due to the mixing provided by the presence of kneading disks. A

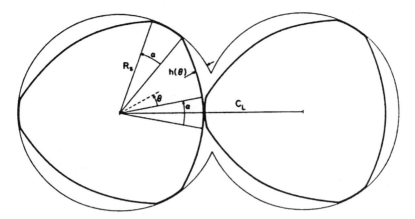

FIGURE 15.10 Cross-sectional view of a three-tip screw in a self-wiping, corotating twin-screw extruder. (From Ref. 20.)

collection of kneading disks is known as a *kneading block* (KB), and a typical KB is shown in Figure 15.13 [22]. A KB consists of a number of disks having the same cross section (or number of lobes) as the conveying elements but stuck together at different stagger angles. The result may be looked upon as a conveying element having a helix angle of 90°. Material is sheared and severely squeezed as it goes through the KB, and the extent of deformation can be controlled by varying the number of disks, their width, and the stagger angle. Indeed, by changing the nature and location of the different elements, we can alter the "severity" of the screw. This is what makes a corotating twin-screw extruder

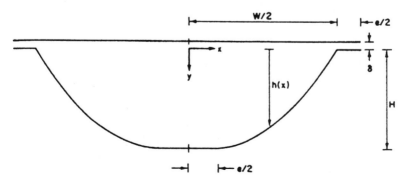

FIGURE 15.11 Channel configuration corresponding to Figures 15.9 and 15.10. (From Ref. 20.)

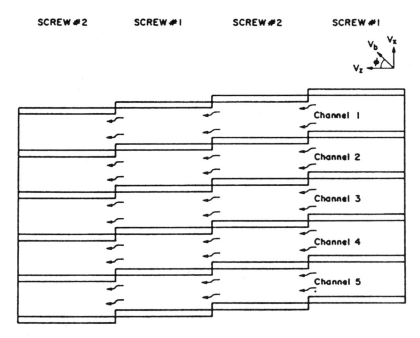

FIGURE 15.12 Unwound channels of three-tip screw elements. Arrows show the motion of the fluid as it is transferred from one screw to the other in the intermeshing region. (From Ref. 20.)

so versatile. Note that TSEs allow for multiple feeding ports at different locations on the barrel so that glass, for example, may be introduced after the polymer has melted and also for the application of vacuum for purposes of devolatalization. Also, a TSE is typically "starve-fed," and there is no relation between the screw rpm and the extruder throughput. In other words, we do not use a hopper to feed the extruder. Instead, a single-screw device feeds solid polymer to the TSE at any desired rate, and as a consequence, the conveying channels of the TSE are only partially filled with polymer; the degree of fill is typically 25–50%, and it changes as the screw pitch changes. There is, therefore, no pressurization of melt in the screw bushings. A consequence of this is a decoupling of the different parts of the extruder, and what happens in one portion of the extruder does not instantly affect what happens in another portion of the extruder. Through the KBs, though, the conveying capacity is small, material backs up, and in the region before the KB the degree of fill reaches 100%. Here, the polymer gets pressurized, and it is this increase in pressure that forces the polymer to go through the kneading disks. For additional details, we refer the reader to excellent books on the topic [6,21,23].

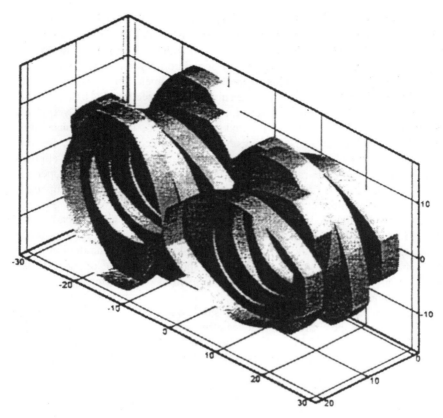

FIGURE 15.13 Isometric view of a bilobal kneading block. (From Ref. 22.)

15.3 INJECTION MOLDING

One of the conceptually simplest methods of fabricating a plastic component, though complex in geometry, is to make a mold or cavity that is identical in shape and size to the article of interest and to fill it with a molten polymer, which then solidifies and yields the desired product. This is the essence of the process of injection molding, and machines are now available that can mass produce items ranging in weight from a fraction of an ounce to several pounds and do so with little or no human intervention. As a consequence, production costs are low, but start-up costs can be high due to the high costs of both the injection-molding machine and the molds themselves. The process is versatile, though, and can be used to mold thermoplastics as well as thermosets. In addition, fillers can be added to make high-strength composite materials and foaming agents can be added to reduce the density of the molded article. By simultaneous or sequential injection of two polymers into the same cavity, it is possible to make a part with a foamed core and a dense skin. During injection molding (as shown in Fig. 15.14), we can also inject an inert gas such as nitrogen into the mold so that it channels through the less viscous sections of the molten polymer. This results in weight reduction and also allows us to produce curved, hollow sections [24]. Some of the polymers that are commonly used for injection molding are polyethylene, polypropylene, and polystyrene for making containers, toys, and housewares. Polyesters are used for gears, bearings, electrical connectors, switches, sockets, appliance housings, and handles. Nylons are used for high-temperature applications such as automobile radiator header tanks and, for anticorrosion properties, acetals are preferred for making items such as gears, bearings, pump impellers, faucets, and pipe fittings. Other moldable polymers that are frequently encountered are polymethyl methacrylate for lenses and light covers, and polycarbonates and ABS for appliance housings and automobile parts.

As shown in schematic form in Figure 15.14, an injection-molding machine is essentially a screw extruder attached to a mold. The action of the extruder results in a pool of molten polymer directly in front of the screw tip, and this causes a buildup in pressure that forces the screw to retract backward. Once a predetermined amount of polymer (the shot size) has been collected, screw rotation stops and the entire screw moves forward like a plunger pushing material into the mold. This type of machine is therefore known as a *reciprocating-screw injection-molding machine*. Once the polymer has solidified, the mold is opened,

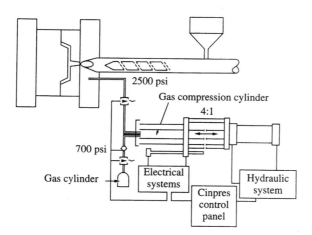

FIGURE 15.14 In the CINPRES I process, a controlled volume of inert gas (oxygen-free N_2) is injected through the nozzle into the center of the still-molten polymer. (From Ref. 24.)

the part is removed, and the cycle of operations is repeated. Typical cycle times range from a few seconds to a minute. Injection-molding machines are normally described in terms of the screw diameter, the maximum shot size in ounces, and the force in tons with which the mold is clamped to the injection unit of the machine.

A mold is typically composed of two parts, called the *cavity* and the *core*. The cavity gives the molding its external form, whereas the core gives it the internal form. This is seen in Figure 15.15, which shows a mold used to make a plastic tumbler. It is clear that an empty space having the shape and size of the tumbler is formed when the cavity and core are clamped together. Most molds designed for long service life are made from alloy steels and can cost several tens of thousands of dollars. To consistently make moldings having the correct dimensions, it is necessary that the mold material be wear resistant and corrosion resistant and not distort during thermal cycling; chrome and nickel plating are common. Details of mold design and of the mechanical aspects of opening and closing molds and ejecting solidified parts are available in the book by Pye [25]. Note that most molds are water-cooled. Also, a mold frequently has multiple cavities.

The melt that collects near the reciprocating screw leaves the injection unit through a nozzle that is essentially a tapered tube, which is often independently heated [26] and may also contain a screen pack. The simplest way to connect the nozzle to the mold shown in Figure 15.15 is through the use of a sprue bush, which is another tapered passage of circular cross section, as shown in Figure 15.16. For multicavity molds, however, we need a runner system to join the sprue to the gate or entry point of each cavity. Because we want all the cavities to fill up at the same time, the runners have to be balanced; one possible arrangement is

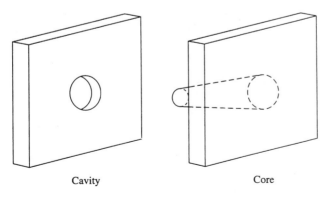

Cavity   Core

FIGURE 15.15 A simple mold.

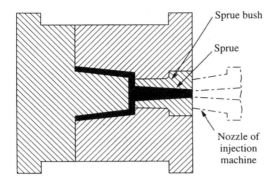

FIGURE 15.16   Feed system for single-impression mold. (From Pye, R. G. W.: *Injection Mould Design*, 4th ed., Longman, London, 1989. Reprinted by permission of Addison Wesley Longman Ltd.)

shown in Figure 15.17. The runner system should be such that the cavities fill up rapidly with a minimum amount of pressure drop, and this suggests the use of a large cross-sectional area. At the same time, though, we want the material in the runner to solidify quickly after mold filling, and this is possible if the runner cross section is small. It can, however, not be so small that the runner freezes before the mold is full, because this results in a useless molding called a *short shot*. For these reasons and the fact that the polymer in the runner has to be removed with the molding and recycled, the actual runner length and diameter are a compromise meant to satisfy conflicting requirements. Similarly, the gate diameter has to be small for ease of runner removal but large enough that the high shear rates and viscous heating in the gate region do not result in thermal and mechanical degradation of the melt.

If we monitor the pressure at the gate as a function of time during commercial injection molding, we typically get the result shown in Figure 15.18, so that the overall process can be divided into three distinct stages. In the first stage, the mold fills up with polymer and there is a moderate increase in pressure. Once the mold is full, the second stage begins and pressure rises drastically in order that additional material be packed into the mold to compensate for the shrinkage caused by the slightly higher density of the solid polymer relative to the melt. Finally, during the cooling stage, the gate freezes and there is a progressive reduction in pressure. As Spencer and Gilmore explain, if flow into the mold did not take place during the packing stage, the formation of a solid shell at the mold walls would prevent the shrinkage of the molten polymer inside [27]. Stresses would then arise, leading to the collapse of the shell and the formation of sink marks. If the shell were thick enough to resist collapse, a

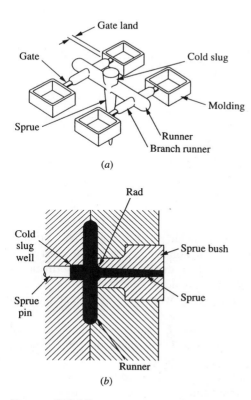

(a)

(b)

FIGURE 15.17 Feed system: (a) typical shot consisting of moldings with sprue, runner, and gates attached; (b) section through feed portion of mold. (From Pye, R. G. W.: *Injection Mould Design*, 4th ed., Longman, London, 1989. Reprinted by permission of Addison Wesley Longman Ltd.)

vacuum bubble would form and, in any case, some of the thermal stresses would remain in the molded article. A consequence of the frozen-in stress would be that dimensional changes could occur on raising the temperature of the molding.

Even though the presence of the packing stage eliminates thermal stresses, this is a mixed blessing. This is because polymer chain extension and orientation occur along with flow during the packing part of the cycle. Unlike that during the filling stage, this orientation is unable to relax due to the increased melt viscosity and attendant large relaxation times resulting from cooling. The orientation therefore remains frozen-in and leads to anisotropic material properties. In addition, dimensional changes again take place on heating, and the material tends to craze and crack much more easily. The packing time is therefore picked

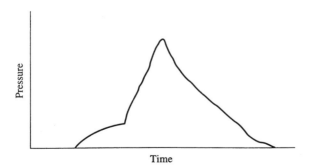

FIGURE 15.18   Typical gate pressure profile during injection molding.

to minimize stresses resulting from both quenching and polymer chain orientation. A further level of complexity arises if the polymer is crystallizable. Because
the rate and nature of crystallization depend on the thermal and deformational
history, different morphologies can be obtained, depending on how the molding
cycle is run, and this can endow the molding with totally different physical
properties.

The major machine variables in injection molding include the melt
temperature, mold temperature, injection speed, gate pressure, the packing
time, and the cooling time. The interplay of these variables determines the
pressure, temperature, and velocity profiles within the mold and the position and
shape of the advancing front during mold filling. These field variables, in turn,
determine the structure that is ultimately witnessed in the molded part. Any
mathematical model of injection molding is therefore directed at calculating the
values of the primary microscopic variables as a function of time during the
molding cycle. This is done by examining the three stages of the injection
molding cycle separately. Given the complexity of the process, this is not an easy
job. Nonetheless, we shall try to illustrate the procedure using a simple
rectangular mold geometry.

15.3.1   Mold Filling

White and Dee carried out flow visualization studies for the injection molding of
polyethylene and polystyrene melts into an end-gated rectangular mold [28].
Experiments were conducted under isothermal conditions and also for situations
where the mold temperature was below the polymer glass transition temperature
or the melting point, as appropriate. The apparatus used was a modified capillary
rheometer in which, instead of a capillary die, a combined-nozzle mold assembly
was attached to the barrel, as shown in Figure 15.19. The mold could be heated to

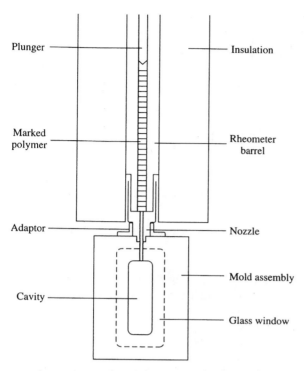

Plunger

Marked
polymer

Adaptor

Cavity

Insulation

Rheometer
barrel

Nozzle

Mold assembly

Glass window

FIGURE 15.19 Injection-molding apparatus. (From Ref. 28.)

any desired temperature and it had a glass window that permitted observation of
the flow patterns within. Results for the flow-front progression and the streamline
shape are displayed in Figure 15.20 for isothermal experiments conducted at
200°C. Here, the fluid was injected slowly and there was no influence of the
polymer type. In each case, there was radial flow at the gate and, once the corners
were filled, the front shape became almost flat and this front moved forward and
filled the mold. The corresponding results for injection into a cold mold at 80°C
are shown in Figure 15.21 [28]. Now, the front is much more curved and there is
pronounced outward flow toward the mold wall. In addition, because of the
increase in polymer viscosity resulting from cooling, stagnant regions develop in
corners and near the mold wall.

To a good approximation, most of the polymer that flows into the mold can
be considered to be flowing in almost fully developed flow between two parallel
plates. For this situation, the melt at the midplane moves at a velocity that is
greater than the average velocity. At the front, however, the fluid not only has to

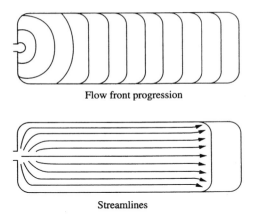

Flow front progression

Streamlines

FIGURE 15.20 Mold filling under isothermal conditions. (From Ref. 28.)

slow down but also has to curve toward the mold wall, as shown in Figure 15.22. This phenomenon is called *fountain flow*, and it happens because there is no flow at the mold wall and the only way to fill the region near the wall is by fluid flowing to the wall from the central core. This has been demonstrated quite strikingly by Schmidt, who used colored markers to show that the first tracer to enter the cavity gets deposited near the gate, whereas the last tracer shows up on the mold wall at the far end of the molded plaque [29]. A consequence of fountain flow is that fluid elements near the front get stretched during the journey to the

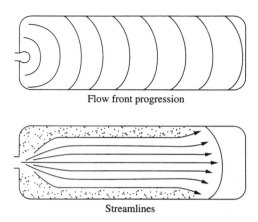

Flow front progression

Streamlines

FIGURE 15.21 Filling a cold mold. (From Ref. 28.)

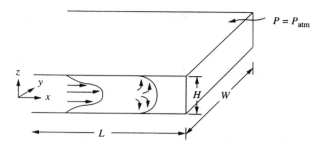

FIGURE 15.22 Fountain flow during mold filling.

mold wall and freeze in an extended conformation with the chain axis aligned with the overall flow direction.

The results of White and Dee also revealed that if fluid injection was rapid, a jet emerged from the gate, struck the mold wall at the far end, and piled up upon itself, as shown in Figure 15.23 [30]. Later, a front entered the gate and filled the mold. The advancing front, however, could not entirely absorb the piled-up material within itself, and the solidified molding showed evidence of jetting in the form of *weld lines*. Note that whenever flow splits around an insert in the mold, the two fronts meet later and the place where they meet shows up as a weld line. A weld line usually represents a region of weakness and is undesirable. One way to eliminate jetting is to place an obstruction directly in front of the gate or to mold against a wall. Oda et al. have found that if the die swell were large enough, the jet thickness would equal the mold thickness and the polymer would contact the mold walls; the mold walls would then act as a barrier and jetting would not occur [30].

In order to simulate mold filling, we can normally neglect inertia and body force terms in the equations of motion due to the very viscous nature of polymer

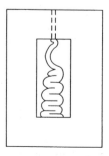

FIGURE 15.23 Jetting in injection-mold filling. (From Ref. 30)

melts. Using the rectangular Cartesian coordinate system shown in Figure 15.22, we have

$$\frac{\partial p}{\partial x} = \frac{\partial \tau_{zx}}{\partial z} \tag{15.3.1}$$

$$\frac{\partial p}{\partial y} = \frac{\partial \tau_{zy}}{\partial z} \tag{15.3.2}$$

and there is no pressure variation in the thickness direction because the mold thickness H is at least an order of magnitude smaller than the mold length L or mold width W. For the same reason, we can neglect the derivatives of the stresses in the x and y directions compared to the derivative in the z direction.

Integrating Eqs. (15.3.1) and (15.3.2) from $z = 0$ to any other value of z and assuming symmetry gives

$$\tau_{zx} = \frac{\partial p}{\partial x} z = \eta \frac{\partial v_x}{\partial z} \tag{15.3.3}$$

$$\tau_{zy} = \frac{\partial p}{\partial y} z = \eta \frac{\partial v_y}{\partial z} \tag{15.3.4}$$

where the viscosity itself may depend on temperature and the rate of deformation, even though we have represented the shear stress in terms of a product of a shear viscosity and a shear rate. Solving for the velocity components by integrating from the mold wall to z gives

$$v_x = \frac{\partial p}{\partial x} \int_{-H/2}^{z} \frac{\xi}{\eta} \, d\xi = \frac{\partial p}{\partial x} I \tag{15.3.5}$$

$$v_y = \frac{\partial p}{\partial x} \int_{-H/2}^{z} \frac{\xi}{\eta} \, d\xi = \frac{\partial p}{\partial y} I \tag{15.3.6}$$

where the integral in these equations has been denoted by I. If we introduce these expressions for the velocity into the incompressible continuity equation

$$\frac{\partial v_x}{\partial x} + \frac{\partial v_y}{\partial y} = 0 \tag{15.3.7}$$

we get

$$I \left(\frac{\partial^2 p}{\partial x^2} + \frac{\partial^2 p}{\partial y^2} \right) + \frac{\partial p}{\partial x} \frac{\partial I}{\partial x} + \frac{\partial p}{\partial y} \frac{\partial I}{\partial y} = 0 \tag{15.3.8}$$

If the flow is almost fully developed, the shear rate as well as the temperature and, consequently, the viscosity varies strongly in the thickness direction but weakly in the x and y directions. The x and y derivatives of I can, therefore, be neglected and we have

$$\frac{\partial^2 p}{\partial x^2} + \frac{\partial^2 p}{\partial y^2} = 0 \tag{15.3.9}$$

This equation was derived independently by White [31] and by Kamal and co-workers [32,33]. Now, we define a stream function ψ as follows [33]:

$$v_x = -\frac{\partial \psi}{\partial y} IG(t) \tag{15.3.10}$$

$$v_y = \frac{\partial \psi}{\partial x} IG(t) \tag{15.3.11}$$

so that the continuity equation is identically satisfied. Here, the function $G(t)$ accounts for the time variation of pressure. Differentiating Eq. (15.3.10) with respect to y and Eq. (15.3.11) with respect to x and subtracting one equation from the other gives

$$\frac{\partial v_x}{\partial y} - \frac{\partial v_y}{\partial x} = -IG(t)\left(\frac{\partial^2 \psi}{\partial y^2} + \frac{\partial^2 \psi}{\partial x^2}\right) \tag{15.3.12}$$

Using Eqs. (15.3.5) and (15.3.6), however, we find that the left-hand side of Eq. (15.3.12) equals zero. Consequently, we have

$$\frac{\partial^2 \psi}{\partial x^2} + \frac{\partial^2 \psi}{\partial y^2} = 0 \tag{15.3.13}$$

and both the pressure and the stream function obey Laplace's equation. Based on potential flow theory, we know that lines of constant pressure are orthogonal to the streamlines. This fact allows us to obtain the stream function once the isobars have been computed; it is not necessary to determine the function $G(t)$. The pressure boundary conditions that we use are that the pressure at the gate is specified and that the advancing front is a line of constant pressure. Because molds are vented, this constant pressure is atmospheric. In addition, because $v_y = 0$ at the mold walls, Eq. (15.3.6) implies that $\partial p/\partial y = 0$ there. Because Eq. (15.3.9) and the associated boundary conditions do not involve the rheological properties of the polymer, the pressure variation, the front shape, and the streamlines are independent of the fluid rheology. The velocity profiles and, consequently, the mold-filling time, however, involve the integral I and must, as a result, depend on the rheology. These conclusions have been verified experimentally [33]. Note that fluid elasticity has been conspicuously absent from the previous development. This was the case because the flow was essentially a viscometric flow. Indeed, pressure profiles computed using viscoelastic rheological models are the same as those obtained with inelastic fluid models in this situation [34].

Because the temperature variation in the x and y directions has been neglected, the differential energy balance including viscous dissipation becomes

$$\rho c \frac{\partial T}{\partial t} = k \frac{\partial^2 T}{\partial z^2} + \eta \left[\left(\frac{\partial v_x}{\partial z} \right)^2 + \left(\frac{\partial v_y}{\partial z} \right)^2 \right] \tag{15.3.14}$$

The boundary conditions applicable to this equation are as follows: The mold temperature is T_M at all times and the melt temperature at the gate is T_0. Also, there is symmetry about $z = 0$. The spatial velocity derivatives appearing in Eq. (15.3.14) can be eliminated [35] with the help of Eqs. (15.3.3) and (15.3.4) so that

$$\rho c \frac{\partial T}{\partial t} = k \frac{\partial^2 T}{\partial z^2} + \frac{z^2}{\eta} \left[\left(\frac{\partial p}{\partial x} \right)^2 + \left(\frac{\partial p}{\partial y} \right)^2 \right] \tag{15.3.15}$$

The simplest realistic representation for the viscosity is a generalization of the power-law model,

$$\eta = A \exp \left(\frac{\Delta E}{RT} \right) \mathrm{II}_D^N \tag{15.3.16}$$

where II_D is the second invariant of the rate of deformation tensor and A and N are constants. For the case at hand,

$$\mathrm{II}_D = \left(\frac{\partial v_x}{\partial z} \right)^2 + \left(\frac{\partial v_y}{\partial z} \right)^2 = \frac{z^2}{\eta^2} \left[\left(\frac{\partial p}{\partial x} \right)^2 + \left(\frac{\partial p}{\partial y} \right)^2 \right] \tag{15.3.17}$$

Substituting Eq. (15.3.17) into Eq. (15.3.16) gives the following [35]:

$$\eta = \left(A \exp \left(\frac{\Delta E}{RT} \right) \left\{ z^2 \left[\left(\frac{\partial p}{\partial x} \right)^2 + \left(\frac{\partial p}{\partial y} \right)^2 \right] \right\}^N \right)^{1/(2N+1)} \tag{15.3.18}$$

and the viscosity is known once the temperature and the pressure distribution are known.

The computations are begun by assuming an initial value of the average injection velocity and a constant initial temperature [35]. This allows us to locate the position x_F of the melt front a time instant Δt later. At time Δt, Eq. (15.3.9) is solved to obtain the pressure variation so that we can find the temperature variation from Eqs. (15.3.15) and (15.3.18). Similarly, the velocity components are obtained from Eqs. (15.3.5), (15.3.6), and (15.3.18). Integrating the x component of the velocity over the mold cross section yields the volumetric flow rate Q into the mold. The quotient Q/WH is the new front velocity;

multiplying this quantity by Δt and adding to $x_F(\Delta t)$ gives $x_F(2\Delta t)$. The previous calculations are repeated many times until the mold fills. In this manner, we can obtain the mold-filling time as well as the pressure and temperature within the filled mold; these form inputs to any model of the packing stage.

The procedure just described is very simple and does not yield information about the velocity and temperature profile rearrangement near the front due to fountain flow. To obtain this information, we must allow for a nonzero value of v_z and impose stress boundary conditions on the free surface. This has been done by Kamal et al., who used a Marker and Cell computational scheme to solve the problem [36]. For more complex mold shapes, we must resort to finite-difference or finite-element methods, for which several commercial computer programs are now available [37]. These programs are useful for balancing runners, helping decide gate locations, avoiding thermal degradation due to high shear rates, predicting weld-line formation due to the meeting of flow fronts, and eliminating the possibility of short shots. In addition, we can explore trends with varying (1) injection temperature, (2) pressure, and (3) other machine variables.

Example 15.3: How long would it take to fill the rectangular mold shown in Figure 15.22 with an isothermal Newtonian liquid if the pressure at the gate is held fixed at value p_0? Assume that v_x is the only nonzero velocity component.

Solution: Using Eq. (15.3.3) gives

$$\frac{\partial v_x}{\partial z} = \frac{1}{\eta} \frac{\partial p}{\partial x} z$$

Integrating with respect to z and using the boundary condition of zero velocity at $z = H/2$ gives

$$v_x = \frac{1}{2\eta} \frac{\partial p}{\partial x} \left(z^2 - \frac{H^2}{4} \right)$$

The instantaneous volumetric flow rate Q is given by

$$Q = 2 \int_0^{H/2} v_x W \, dz = -\frac{W}{\eta} \frac{\partial p}{\partial x} \frac{H^3}{12}$$

If the advancing front is located at $x = L^*$ at time t, $\partial p/\partial x$ equals $-p_0/L^*$. Also, from a mass balance, Q is equal to $WH(dL^*/dt)$. Substituting these expressions for Q and $\partial p/\partial x$ into the above equation gives

$$WH \frac{dL^*}{dt} = \frac{W H^3}{\eta} \frac{p_0}{12 L^*}$$

or

$$L^* \, dL^* = \frac{H^2}{12\eta} p_0 \, dt$$

Integrating from time zero to t, we have

$$\frac{L^{*2}}{2} = \frac{H^2}{12\eta} p_0 t$$

The mold fills up when $L^* = L$. From the above equation, this time is $6\eta L^2 / H^2 p_0$.

15.3.2 Mold Packing and Cooling

Once the mold is filled, there is no more free-surface flow, but melt still enters the cavity to compensate for shrinkage caused by cooling, which results in an increased density. Because the change in density is the major phenomenon taking place during packing, the assumption of incompressibility is obviously invalid, and we need an equation relating polymer density to pressure and temperature. Kamal et al. assume the following [32]:

$$(p + w)(V - V_0) = R_c T \tag{15.3.19}$$

where V is the specific volume or the reciprocal of the density ρ and w, V_0, and R_c are constants. The appropriate form of the continuity equation is now

$$\frac{\partial \rho}{\partial t} + \frac{\partial}{\partial x}(\rho v_x) = 0 \tag{15.3.20}$$

where we have assumed that flow is essentially in the x direction alone. Polymer compressibility does not alter the momentum balance, which again leads to

$$v_x = \frac{\partial p}{\partial x} I \tag{15.3.5}$$

Because both I and ρ will be independent of x, the previous two equations imply that

$$\frac{\partial \rho}{\partial t} + I\rho \frac{\partial^2 p}{\partial x^2} = 0 \tag{15.3.21}$$

If necessary, we can eliminate ρ between Eqs. (15.3.19) and (15.3.21) and obtain a single partial differential equation relating pressure to position, time, and temperature [32]. This has to be solved simultaneously with the energy balance,

$$\rho c \frac{\partial T}{\partial t} = k \frac{\partial^2 T}{\partial z^2} + \frac{z^2}{\eta} \left(\frac{\partial p}{\partial x} \right)^2 \tag{15.3.22}$$

wherein it is generally not necessary to account for the temperature and pressure dependence of the density, specific heat, and thermal conductivity.

Because we know the temperature and pressure in the mold at the end of the filling stage and the time dependence of pressure at the gate, we can numerically solve Eqs. (15.3.19), (15.3.21), and (15.3.22) to obtain temperature as a function of z and t and pressure as a function of x and t. Through the use of Eq. (15.3.3), this information then allows us to calculate the shear-stress distribution in the molding at the end of packing.

After filling and packing are complete, cooling continues, but in the absence of flow. Now, we merely need to solve the heat-conduction equation,

$$\rho c \, \frac{\partial T}{\partial t} = k \, \frac{\partial^2 T}{\partial z^2} \tag{15.3.23}$$

subject to the temperature distribution existing at the end of the packing stage. If the polymer is crystallizable, Eq. (15.3.23) will also involve the heat of crystallization. Because Eq. (15.3.23) is linear, an analytical series solution is possible [32].

The major effect of polymer cooling is that it retards stress relaxation and, as mentioned previously in the chapter, some of the stress remains frozen-in even after the molding has completely solidified. This stress relaxation cannot be predicted using an inelastic constitutive equation (why?); the simplest equation that we can use for the purpose is the upper-convected Maxwell model. In the absence of flow, the use of this model yields

$$\tau_{zx}(t, x) = \tau_{zx}(0, x) \exp\left(-\int_0^t \frac{dt'}{\lambda_0}\right) \tag{15.3.24}$$

where λ_0 is the temperature-dependent zero-shear relaxation time and temperature shift factors may be used to relate $\lambda_0(T)$ to the relaxation time at a reference temperature T_R [38]. The simultaneous solution of Eqs. (15.3.23) and (15.3.24) gives an estimate of the orientation stresses remaining unrelaxed in the injection-molded part. Even though we have considered only shear stresses (because they are the ones that influence mold filling), the deformation of a viscoelastic fluid results in unequal normal stresses as well. Because the thermomechanical history of mold filling is known, the magnitude of the normal stresses may be computed with the help of an appropriate rheological model. Alternately, we can take a short-cut and empirically relate the first normal stress difference in shear N_1 to the shear stress. The result is as follows [39,40]:

$$N_1 = A\tau_{zx}^{\beta} \tag{15.3.25}$$

where A and β are constants specific to the polymer used. These normal stresses also contribute to the residual stresses in the molded part.

15.3.3 Molding Microstructure

The simplified injection-molding model described in the previous sections gives the temperature, pressure, velocity, and stress distribution at any time during the molding cycle. Of greatest interest to the end user, however, are the mechanical properties of the molded article, and these depend on the microstructure. As explained in the previous chapters, properties such as the modulus, yield strength, microhardness, shrinkage, and fracture behavior depend on the amount, size, and nature of crystals and also on the orientation of polymer chains in the amorphous and crystalline regions of the material. Because the microstructure varies with position within the molding, these properties also vary with position. The goal of the current research is to relate the thermomechanical history of the resin during processing to the development of microstructure.

At present, there is no simple way of predicting the orientation of polymer chains in the crystalline regions. In the amorphous regions, however, we can relate chain orientation to fluid stresses through the use of optical birefringence. The latter quantity is defined as the difference Δn between the principal refractive indices parallel and perpendicular to the stretch direction for a uniaxially oriented polymer. For shearing flows, using the stress-optic law gives the following [41]:

$$\Delta n = n_1 - n_2 = C(N_1^2 + 4\tau_{12}^2)^{1/2} \qquad (15.3.26)$$

where C is the stress-optical coefficient of the material. Because the right-hand side of Eq. (15.3.26) is known from the injection-molding models, the birefringence can be calculated. The value Δn is related to the amorphous orientation function, f_{am}, as follows:

$$f_{am} = \frac{\Delta n}{\Delta_{am}^{\circ}} \qquad (15.3.27)$$

where Δ_{am}° is the intrinsic birefringence of the amorphous phase.

To determine the extent of crystallinity in the solidified molding, an equation is necessary for the rate of crystallization in terms of temperature so that we can follow the development of crystallinity during the cooling stage. For use with injection molding, Lafleur and Kamal recommend the equation of Nakamura et al. [38], which is based on the Avrami equation (see Chap. 11). Their experimental results show good agreement between the experimental and calculated crystallinity values for the injection molding of high-density polyethylene [42].

The calculated distributions of crystallinity and orientation functions can be used in conjunction with models of the type developed by Seferis and Samuels [43] to predict quantities such as the Young's modulus. Conceptually, this closes the loop among polymer processing, polymer structure, and polymer properties.

15.4 FIBER SPINNING

A significant fraction of the total worldwide polymer usage is devoted to applications related to clothing, carpeting, and furnishing, and this demand is the basis of the synthetic fiber industry. The process of fiber manufacture involves the extrusion of a polymeric fluid in the form of long, slender filaments, which are solidified and wound up on bobbins. If a polymer solution is used, solvent is removed by evaporation with the help of a hot gas (dry process) or by means of a coagulation bath (wet process). Fibers of cellulose acetate and polyvinyl acetate are made by the former technique, whereas polyacrylonitrile and cellulose fibers are made by the latter technique. If a polymer melt (such as a nylon, polyester, or a polyolefin) is used, solidification occurs due to cooling, and this process of melt spinning is therefore the simplest one to consider. Even so, the journey that a polymer molecule takes in going from the extruder to the finished fiber has so many twists and turns that it is necessary to analyze separate parts of the process rather than the entire process itself. Here, we will examine melt spinning in as much detail as possible. Practical details of all three processes are available in the books by Walczak [44] and Ziabicki [45].

A schematic diagram of the melt-spinning process is shown in Figure 15.24. Molten polymer leaving the extruder passes through a gear pump, which provides accurate control of the flow rate. It then goes through a filter pack, which may consist of sand, wire mesh, or steel spheres, and whose purpose is to

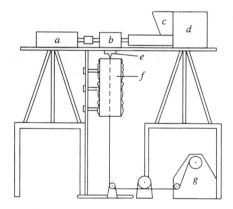

FIGURE 15.24 Schematic diagram of the melt-spinning apparatus: (a) gear pump drive, (b) gear pump, (c) hopper purged with nitrogen gas, (d) extruder, (e) spinneret, (f) insulated isothermal oven, and (g) winder. (From *J. Appl. Polym. Sci.*, vol. 34, Gupta R. K., and K. F. Auyeung: "Crystallization in Polymer Melt Spinning," Copyright © 1987 by John Wiley & Sons, Inc. Reprinted by permission of John Wiley & Sons, Inc.)

homogenize the temperature, degrade any polymer gels, and remove foreign matter that may plug the capillaries used to shape the fibers. The capillaries may be noncircular in shape and are drilled into a spinneret or die. The multiple filaments leaving the spinneret are cooled by cross-flow air in a chimney that may contain several spinnerets. Simultaneously, the filaments are gathered together to form a yarn, which is stretched by the action of rollers, leading to a significant decrease in the fiber cross-sectional area. A liquid called a *spin finish*, which facilitates fiber drawing and also prevents electrostatic charging, is applied to the yarn, which is then wound onto a bobbin called a *doff*.

A spinneret is essentially a stainless-steel disk containing a dozen to a few hundred holes arranged in a regular pattern, as in Figure 15.25a. If circular cross-sectional fibers are to be made, each capillary can be machined as shown in Figure 15.25b. The most critical aspect of a spinneret is the diameter of the capillaries; each diameter has to be the same. This is because diameter variations result in large flow-rate variations. Recall that for a given pressure drop, the flow rate of a Newtonian liquid is proportional to the fourth power of the diameter. Because dirt particles tend to block the capillaries, a very small capillary diameter cannot be used; a value less than 0.01–0.02 in. may not allow continuous operation. The larger the diameter is, the larger is the spinline draw ratio, defined as the ratio of the capillary cross-sectional area to the solidified fiber cross-sectional area. Especially with isothermal operations, we find that increasing the draw ratio above a critical value results in periodic diameter oscillations called *draw resonance* [47]. These and all other dimensional and structural variations are undesirable because they can lead to uneven dye uptake, which shows up as barre or stripes in a colored fabric. Of course, the maximum flow rate through the capillaries is limited by the occurrence of melt fracture, which reveals itself as extrudate distortion. This can be minimized by the use of a small capillary entrance angle α. Letting α range from 60° to 90° instead of using a flat entry (180°) also results in the elimination of recirculating regions in the corners of the

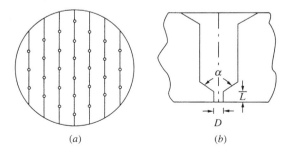

(a) (b)

FIGURE 15.25 (a) Spinneret used for fiber spinning; (b) a simple capillary.

abrupt contraction where polymer might remain for a long time and thermally degrade. The incentive for employing noncircular cross-sectional capillaries is to alter the optical properties of the spun fibers. We find that circular fibers yield shiny fabrics.

The major process variables are volumetric flow rate, extrusion temperature, quench air temperature and velocity, take-up speed, and draw ratio. The influence of some of these variables has already been discussed. For crystallizable polymers, the lower limit of the extrusion temperature is about 20°C above the melting point. Otherwise, the crystallites do not all melt, which makes it difficult to control the structure of fiber. The upper limit is set by thermal degradation. It is generally preferable to operate at a high temperature because the lowered melt viscosity implies ease in processing, but the added heat has to be removed during cooling and solidification. Most published data on commercial operations are on polyethylene terephthalate (PET), and much of the following discussion applies to this polymer. PET fibers are generally made at wind-up speeds of 1000–3000 m/min, although newer machines operate at speeds in excess of 5000 m/min. For this polymer, the residence time on the spinline is usually less than 1 sec, the length of the molten zone is about 1 m, and the maximum stretch rate is of the order of 100 sec^{-1}. At spinning speeds below 3000 m/min, the fiber that results is almost totally amorphous and its tensile properties are not very good. Therefore, it is heated to a temperature between the glass transition temperature and the melting point in a second process step and stretched to induce crystallization. This is a slow process that cannot be done in-line with melt spinning. It is therefore labor-intensive. The crystalline fiber tends to shrink on heating, but this can be prevented by heat-setting or annealing at a high temperature, either at constant length or under tension. Finally, an additional heat treatment may be applied to crimp or texture the yarn and make it bulky in order to improve its feel or "hand."

The manufacture of an acceptable fiber involves a very large number of process steps that are carried out to alter the fiber morphology. As a consequence, it is difficult to predict the effect of changing a single variable in any given process step on the final fiber and its properties. This makes a reading of the textiles literature and interpretation of published results very difficult. Although attempts have been made to relate morphological changes to fundamental variables [48], the state of the art is not, for example, to predict, a priori, the size, shape, and amount of crystals and the orientation of polymer chains within the fiber at the end of any process step. Models that can be formulated and that do exist relate only to the melt-spinning step and only for amorphous fibers. Detailed reviews of single-filament models are available, and these yield the steady-state temperature, velocity, and stress profiles in the molten fiber [49–52]. In addition, they reveal the conditions under which the fiber diameter is likely to be sensitive to changes in the process variables.

In the next section, we develop, from first principles, the equations governing the behavior of a fiber monofilament of circular cross section. We derive solutions to some limiting cases and also discuss solutions obtained by others to the more general set of equations. These results not only enhance our understanding of the melt-spinning process but also have practical utility. Denn has shown that on-line feedback control of fiber spinning is impractical due to the very large number of spinning machines in a synthetic fiber plant [53]. Fiber-spinning models can help to identify regions of operation where small changes in operating variables do not influence fiber dimensions or fiber structure. In addition, it has been observed that there is a good correlation between polymer orientation in the spun fiber (as measured by the optical birefringence) and fiber mechanical properties such as tenacity and modulus [54]. Furthermore, the stress in the fiber at the point of solidification is found to be proportional to the optical birefringence [54]. Because fiber-spinning models can predict stresses, we have a means of predicting fiber mechanical properties, at least for PET fibers.

15.4.1   Single-Filament Model

The basic equations describing the steady-state and time-dependent behavior of a molten spinline made up of a single fiber of circular cross section were first developed by Kase and Matsuo [55]. It was assumed that the filament cross-sectional area and polymer velocity and temperature varied with axial position down the spinline but that there were no radial variations. The model, therefore, is a one-dimensional model that is valid provided that the fiber curvature is small. To derive the equations, we consider the differential control volume shown in Figure 15.26 and carry out a simultaneous mass, momentum, and energy balance.

Because the rate of accumulation of mass within a control volume equals the net rate at which mass enters the control volume, we have

$$\frac{\partial}{\partial t}(\rho A \Delta x) = \rho v A|_x - \rho v A|_{x+\Delta x} \tag{15.4.1}$$

or

$$\frac{\partial A}{\partial t} = -\frac{\partial}{\partial x}(Av) \tag{15.4.2}$$

where A is the filament cross-sectional area.

Similarly, the rate at which the x-momentum accumulates within a control volume equals the net rate at which x momentum enters the control volume due to

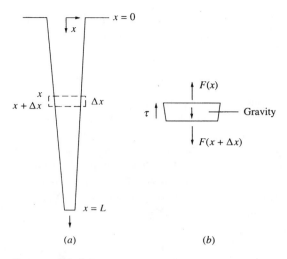

FIGURE 15.26 (a) Control volume used for mass, momentum, and energy balances during fiber spinning; (b) forces acting on the control volume.

polymer flow plus the sum of all x directed external forces acting on the control volume. As a consequence (see Fig. 15.26b), we have

$$\frac{\partial}{\partial t}(\rho A v \Delta x) = \rho v^2 A|_x - \rho v^2 A|_{x+\Delta x} + F_{x+\Delta x} - F_x + \rho A g \Delta x - 2\pi R \tau \Delta x$$

$$(15.4.3)$$

in which F is the tensile (rheological) force within the filament and τ is the shear stress due to air drag. The second-to-last term in Eq. (15.4.3) results from the presence of gravity, and the small contribution due to surface tension has been neglected. Dividing both sides of Eq. (15.4.3) by Δx and taking the limit as $\Delta x \to 0$ gives

$$\rho \frac{\partial}{\partial t}(Av) = -\rho \frac{\partial}{\partial x}(Av^2) + \frac{\partial F}{\partial x} + \rho A g - 2\pi R \tau \qquad (15.4.4)$$

Finally, neglecting radiation, the rate of energy accumulation within a control volume equals the net rate at which energy enters the control volume due to polymer flow minus the rate at which energy leaves due to heat transfer by convection to the cross-flow air. Therefore, we have

$$\frac{\partial}{\partial t}(\rho c A T \Delta x) = \rho c A T v|_x - \rho c A T v|_{x+\Delta x} - 2\pi R h \Delta x (T - T_a) \qquad (15.4.5)$$

where c is the polymer specific heat, h is the heat transfer coefficient, and T_a is the temperature of the cooling air. From Eqs. (15.4.2) and (15.4.5), we have

$$\frac{\partial T}{\partial t} = -v \frac{\partial T}{\partial x} - \frac{2h}{\rho c} \sqrt{\frac{\pi}{A}} (T - T_a) \tag{15.4.6}$$

Before we can obtain solutions to the foregoing model equations, we need to specify the functional forms of F, τ, and h. Whereas F depends on the rheological properties of the polymer, τ and h are obtained by considering the velocity and temperature profiles in the boundary layer adjacent to the fiber surface. Sakiadis has used boundary layer theory to obtain an expression for the drag coefficient, c_f, for the laminar flow of air surrounding a moving, continuous, infinite, circular cylinder of constant diameter D emerging from a slot in a wall and traveling at a constant velocity v [56]. The resulting drag coefficient is defined as

$$c_f = \frac{\tau}{(0.5 \rho_a v^2)} \tag{15.4.7}$$

where ρ_a is the density of air. Equation (15.4.7) is found to underpredict the experimentally measured drag force by a significant amount [52]. In a Reynolds number range of 20–200, experimental data are best represented as follows [57]:

$$c_f = 0.27 \, \mathrm{Re}^{-0.61} \tag{15.4.8}$$

where Re is the Reynolds number, $Dv\rho_a/\eta_a$, and η_a is the viscosity of air. Theoretical analyses for the heat transfer coefficient also suffer from the limitation that diameter attenuation with an attendant increase in the filament velocity cannot be taken into account [58]. Other complicating factors are the presence of cross-flow air and the fact that filaments generally do not remain still but vibrate in a transverse manner. From experiments conducted using a stationary, heated wire, Kase and Matsuo have proposed the following [55]:

$$\frac{hD}{k_a} = 0.42 \mathrm{Re}^{0.334} (1 + K) \tag{15.4.9}$$

where k_a is the thermal conductivity of air and the quantity K, which equals $0.67 \, v_y/v$, arises due to the presence of cross-flow air at a velocity v_y. When h is derived from measurements made on an actual spinline, the result is slightly different; George suggests the use of the following [59]:

$$h = 1.37 \times 10^{-4} (v/A)^{0.259} \left[1 + \left(\frac{8v_y}{v} \right)^2 \right]^{0.167} \tag{15.4.10}$$

in which cgs units have been employed.

15.4.2   Steady-State Behavior of Single Filaments

The single-filament model presented in the previous section consists of a set of coupled, nonlinear, partial differential equations. Although a general solution does not exist, a number of analytical solutions can be derived for simplified forms of the equations under steady-state conditions; these have been reviewed by Petrie [50] among others. If the temperature is constant, the mass and momentum balances become

$$Q = Av = \text{constant} \tag{15.4.11}$$

$$\rho Q \frac{dv}{dx} = \frac{dF}{dx} + \rho Ag - 2\pi R\tau \tag{15.4.12}$$

where Q is the volumetric flow rate. In addition, if the fluid is Newtonian, we have

$$\frac{F}{A} = 3\eta \frac{dv}{dx} \tag{15.4.13}$$

in which η is the shear viscosity of the polymer. If spinning is carried out at low speeds, air drag can be neglected and the last three equations can be combined to yield a single equation in v:

$$3\eta \left(\frac{d^2v}{dx^2} - \frac{1}{v}\left(\frac{dv}{dx}\right)^2 \right) = -\rho g + \rho v \frac{dv}{dx} \tag{15.4.14}$$

In order to solve Eq. (15.4.14), two boundary conditions are needed. Typically, these are

$$v(0) = v_0 \tag{15.4.15}$$

$$v(L) = v_0 D_R \tag{15.4.16}$$

where L is the spinline length and D_R is called the *draw ratio*.

     If, in addition to all the previous assumptions, we also assume that the force contributed by gravity is small, then the solution to Eq. (15.4.14) is as follows [60]:

$$v = c_1 \left(c_2 e^{-c_1 x} - \frac{\rho}{3\eta} \right)^{-1} \tag{15.4.17}$$

and the constants c_1 and c_2 are determined through the use of the two boundary conditions. A further simplification occurs if inertia is also negligible [i.e., if the right-hand side of Eq. (15.4.14) is zero]. In this case, either by using Eq. (15.4.14) or by letting the Reynolds number in the dimensionless version of Eq. (15.4.17) go to zero, we can show that

$$v = v_0 \exp\left(\frac{x \ln D_R}{L}\right) \tag{15.4.18}$$

Combining Eq. (15.4.18) with Eqs. (15.4.11) and (15.4.13) yields

$$A = A_0 \exp\left(-\frac{x \ln D_R}{L}\right) \tag{15.4.19}$$

$$F = \frac{3\eta(\ln D_R)A_0 v_0}{L} \tag{15.4.20}$$

Example 15.4: Calculate the force needed to isothermally draw a single filament of a Newtonian liquid of 1000 P shear viscosity at a draw ratio of 20. The spinline length is 2 m, and the volumetric flow rate is 3.5 cm^3/min.

Solution: If we use Eq. (15.4.20) we can find force as follows:

$$F = \frac{3 \times 1000(\ln 20)(3.5)}{200} = 157.2 \text{ dyn}$$

In attempting to compare experimental data on a Newtonian sugar syrup with these results, we find that although the neglect of inertia, air drag, and surface tension is justified, the neglect of gravity is not [61]. Indeed, gravity can constitute as much as 75% of the total spinline force.

Kase has made use of the simplified Newtonian results to check the validity of the flat-velocity profile assumption (i.e., the assumption that the axial velocity v is independent of radial position r) [62]. Kase assumed that the actual velocity profile was of the form

$$v = v_0 \exp\left(\frac{x \ln D_R}{L}\right)\left(1 + a_2 r^2 + a_4 r^4 + a_6 r^6 + \cdots\right) \tag{15.4.21}$$

and introduced this expression into the Navier–Stokes equations. By comparing terms of the same power in r, he solved for the coefficients a_2, a_4, a_6, and so on. He found that under most conceivable spinning conditions, the higher-order terms in Eq. (15.4.21) were negligible and, for all practical purposes, the velocity profile was flat across the filament.

Turning now to data on the nonisothermal, low-speed spinning of polymer melts such as PET, we find that Eq. (15.4.13) (Newtonian rheology) can often still be used, but that the coefficient 3η on the right-hand side has to be replaced by β, a temperature-dependent quantity whose value can exceed $3\eta(T)$. If β is large enough, as happens with polymer melts, the momentum balance [Eq. (15.4.12)] simplifies to

$$\frac{dF}{dx} = 0 \qquad (15.4.22)$$

so that Eqs. (15.4.11) and (15.4.13) imply the following:

$$\frac{dA}{dx} = -\frac{FA}{\beta Q} \qquad (15.4.23)$$

The steady-state energy balance is obtained from Eq. (15.4.6):

$$\frac{dT}{dx} = -\frac{2h}{\rho c Q}\sqrt{\pi A}(T - T_a) \qquad (15.4.24)$$

Eliminating dx between the two previous equations, separating the variables, and integrating gives

$$\int \frac{\rho c \, dT}{\beta(T - T_a)} = \frac{2\sqrt{\pi}}{F}\int \frac{h}{\sqrt{A}}\, dA \qquad (15.4.25)$$

so that F is known if the limits of integration are taken to correspond to the spinneret and the winder. Once F is determined, Eq. (15.4.25) yields A in terms of T, which can be introduced into Eq. (15.4.24) to give T as a function of x and, consequently, A as a function of x. Kase and Matsuo used this procedure to generate area and temperature profiles for the low-speed spinning of polypropylene melts [54]. Typical results are shown in Figure 15.27; agreement with experimental data is good. These and similar steady-state results can be used to explore the influence of the various operating variables on the filament velocity, temperature, and stress profiles.

At higher spinning speeds, we cannot neglect inertia, gravity, and air drag in the force balance, and all of the terms in Eq. (15.4.12) have to be retained. As a consequence, results have to be obtained numerically. Using this method, George has found very good agreement with data on PET up to spinning speeds as high as 3000 m/min. [59]. Simulations show that using the full Eq. (15.4.12) leads to a significantly higher value of the stress at the freeze point, but the velocity and temperature profiles are only slightly altered. Because the freeze-point stress is the key variable of interest, it is essential to use the full set of equations.

Thus far, we have limited our discussion to the melt spinning of Newtonian liquids. For materials that are more elastic than PET, this restriction needs to be relaxed. The simplest way to do this is by using the upper-convected Maxwell model introduced in Chapter 14. In this case, the equation equivalent to Eq. (15.4.13) is

$$\frac{F}{A} = \tau_{xx} - \tau_{rr} \qquad (15.4.26)$$

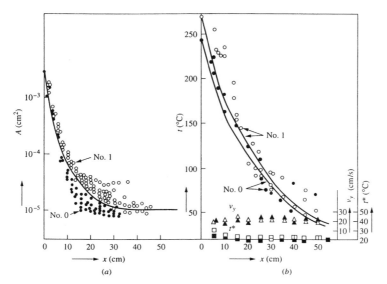

FIGURE 15.27   Filament cross section A, filament temperature t, air speed v_y, and air temperature t^* at different spinneret temperatures t_N: ($\bigcirc$, $\bullet$) experimental; (—) theoretical. (From Ref. 54.) From J. Appl. Polym. Sci., vol. 11, Kase, S., and T. Matsuo: Studies on melt spinning: II. Steady state and transient solutions of fundamental equations compared with experimental results, Copyright © 1967 by John Wiley & Sons, Inc. Reprinted by permission of John Wiley & Sons, Inc.

where the extra stresses τ_{xx} and τ_{rr} in the axial and radial directions, respectively, are given implicitly by

$$\tau_{xx} + \theta\left(v\,\frac{d\tau_{xx}}{dx} - 2\tau_{xx}\,\frac{dv}{dx}\right) = 2\eta\,\frac{dv}{dx} \tag{15.4.27}$$

$$\tau_{rr} + \theta\left(v\,\frac{d\tau_{rr}}{dx} + \tau_{rr}\,\frac{dv}{dx}\right) = -\eta\,\frac{dv}{dx} \tag{15.4.28}$$

in which θ is the relaxation time and η is the shear viscosity; both θ and η are constant if the temperature is constant.

Before proceeding further, it is convenient to make the equations dimensionless by introducing the following dimensionless variables:

$$u = \frac{v}{v_0}, \qquad s = \frac{x}{L}, \qquad T = \frac{\tau_{xx}Q}{Fv_0}$$

$$P = \frac{\tau_{rr}Q}{Fv_0} \tag{15.4.29}$$

With the help of these variables, Eqs. (15.4.26)–(15.4.28) become

$$T - P = u \tag{15.4.30}$$

$$T + \alpha\left(u\frac{dT}{ds} - 2T\frac{du}{ds}\right) = 2\varepsilon\frac{du}{ds} \tag{15.4.31}$$

$$P + \alpha\left(u\frac{dP}{ds} + P\frac{du}{ds}\right) = -\varepsilon\frac{du}{ds} \tag{15.4.32}$$

which involve the two dimensionless groups

$$\alpha = \frac{\theta v_0}{L} \quad \text{and} \quad \varepsilon = \frac{\eta Q}{FL} \tag{15.4.33}$$

These are three equations in the three unknowns (T, P, and u) and they can be combined into a single second-order ordinary differential equation. The final result is as follows [63]:

$$\frac{u}{du/ds} - 3\varepsilon + \alpha u - \frac{\alpha u^2\, d^2u/ds^2}{(du/ds)^2} - 2u\alpha^2\frac{du}{ds} = 0 \tag{15.4.34}$$

which can be numerically solved subject to the boundary conditions of Eqs. (15.4.15) and (15.4.16). Unfortunately, ε is not known because F is not known. Indeed, F is sought as part of the solution. One strategy now is to simply do a parametric mapping: Plot $u(s)$ versus s for various assumed values of α and ε. A comparison with experimental data then involves using the measured force F as a model input. Typical velocity profiles for $D_R = 20$ are shown in Figure 15.28. Here, α and ε combinations have been picked in such a way that $T(0) = 1$. It is seen that velocity profiles become more and more linear as α, or equivalently fluid elasticity, is increased. This is in accord with experimental observations. In addition, it can be shown that, everything else being equal, the viscoelastic model predicts much higher stress levels than the Newtonian model examined earlier [63]. A quantitative comparison between data and simulations is difficult to carry out because no polymer melt is known to behave exactly as a Maxwell liquid. However, agreement can be obtained between data on dilute polymer solutions in very viscous solvents and the predictions of the Oldroyd model B [64]. This fluid model is a linear combination of the Maxwell and Newtonian models.

If we examine the low-speed, nonisothermal manufacture of synthetic fibers of a viscoelastic polymer, then the inclusion of the energy balance in the previous set of equations results in the emergence of one additional dimensionless group, St, the Stanton number. By numerically solving the entire set of equations, Fisher and Denn discovered that the energy balance was only weakly coupled to the

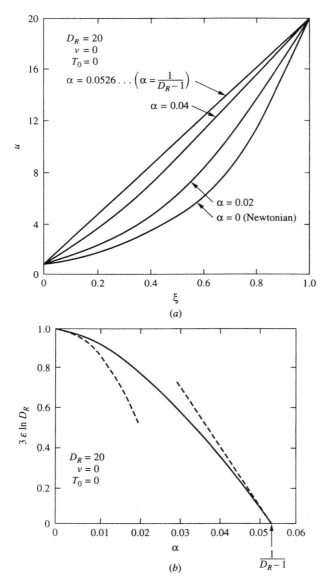

FIGURE 15.28 (a) Dimensionless velocity as a function of distance, $D_R = 20$, $v = 0$, $T_0 = 1$; (b) dimensionless reciprocal force as a function of viscoelastic parameter, $\alpha = 1$, $D_R = 20$, $v = 0$, $T_0 = 1$. (From Denn, M. M., C. J. S. Petrie, and P. Avenas: "Mechanics of Steady Spinning of a Viscoelastic Liquid," *AIChE J.*, vol. 21, pp. 791–799. Reproduced with the permission of the American Institute of Chemical Engineeres Copyright © 1975 AIChE. All rights reserved.)

momentum balance [65]. To a good approximation, the temperature profile could be described by the following:

$$T = T_a + (T_0 - T_a) \exp(-St\ s) \tag{15.4.35}$$

where

$$St = 1.67 \times 10^{-4} \frac{L}{\rho c A_0^{5/6} v_0^{2/3}} \tag{15.4.36}$$

The numerical coefficient in Eq. (15.4.36) arises from the use of cgs units and the form of the heat transfer coefficient proposed by Kase and Matsuo [54,55]. The uncoupling of the energy balance means that the temperature down the spinline is known in advance. Consequently, if the temperature dependence of η and θ is known, Eqs. (15.4.26)–(15.4.28) can again be solved to yield the velocity and stress profiles. Computations show that the effect of cooling is to increase the initial rate of diameter attenuation for viscoelastic fibers [65].

At high spinning speeds, we cannot neglect inertia, air drag, and gravity, so the complete Eq. (15.4.12) has to be used. These effects are accounted for in the computer model of Gagon and Denn [66]. These authors compare their viscoelastic simulations with the data of George [59], but find that due to the relatively inelastic nature of PET, their model offers little improvement over the Newtonian model used by George. Comparisons have not been made with data on more elastic polymers.

15.4.3 Multifilament Spinning and Other Concerns

In order to simulate industrial fiber-spinning operations, we must address the question of multifilament spinning. If fibers are spun as a bundle, they are likely to interact with each other. If we focus on a single fiber in a group of fibers, then the presence of other fibers is likely to affect the temperature and velocity of the air around the given fiber. This would alter both the air drag and heat transfer around the fiber with the result that fiber properties may become asymmetric. In addition, there would be property variations among fibers in the same yarn. Thus, although the basic equations remain unchanged, we need to compute the air velocity and air temperature around each fiber or each row of fibers. This is attempted by some means in two of the published computer models [67,68]. Results obtained are useful in designing spinnerets to minimize interfiber and intrafiber property differences.

One aspect of fiber spinning that we have intentionally not discussed in this chapter is that of time-dependent behavior. This behavior can take different forms: Fibers may break or their diameter may vary in a periodic or random manner. This may be the result of instabilities or the sensitivity to external disturbances. A proper understanding can be gained only by solving the complete time-dependent equations presented earlier. Although much progress has been made in this direction, significant problems remain. For details, the interested reader should consult the reviews available in the literature [47,50–53,69].

Another unresolved problem has to do with predicting structure development during fiber spinning at high speeds. Although a large body of data has been assembled and partly rationalized [70], the information gathered cannot be explained or predicted based on first principles. This remains a challenge for the future.

15.5 CONCLUSION

This chapter has been a sort of capstone. We have shown how it is possible to couple the laws of conservation of mass, momentum, and energy with a knowledge of the fundamentals of polymer behavior to analyze polymer processing operations. Although we illustrated this with three important processes, we could equally well have chosen any other process. (Descriptions and analyses of other processes are available in standard books on polymer processing [18,71–74]). Indeed, these same three processes are likely to change as technology changes. However, the tools for analyzing these changes remain unchanged. The analysis itself can be carried out at different levels of complexity, and the knowledge gained depends on the level of sophistication employed. Nonetheless, information obtained even with the use of rather naive models can often be quite useful. Ultimately, however, numerical computations have to be used if quantitative agreement is desired between model predictions and experimental observations. What is important, though, is that the physics embodied in the models be correct. If the physics is in error, no amount of powerful mathematical techniques can set the results right. We hope that a reading of this book has conveyed this message in a clear manner.

REFERENCES

1. Denn, M. M., *Process Modeling*, Longman, New York, 1986.
2. Vincelette, A. R., C. S. Guerrero, P. J. Carreau, and P. G. Lafleur, A Model for Single-Screw Plasticating Extruders, *Int. Polym. Process.*, 4, 232–241, 1989.
3. Tadmor, Z., and I. Klein, *Engineering Principles of Plasticating Extrusion*, Reinhold, New York, 1970.
4. Darnell, W. H., and E. A. J. Mol, Solids Conveying in Extruders, *Soc. Plast. Eng. J.*, 12(4), 20–29, 1956.
5. Walker, D. M., An Approximate Theory for Pressures and Arching in Hoppers, *Chem. Eng. Sci.*, 21, 975–997, 1966.
6. Rauwendaal, C., *Polymer Extrusion*, 3rd rev. ed., Hanser, Munich, 1994.

7. Chung, C. I., New Ideas about Solids Conveying in Screw Extruders, *Soc. Plast. Eng. J.*, 26(5), 32–44, 1970.

8. Chung, C. I., Plasticating Screw Extrusion of Polymers, *Adv. Transp. Processes*, 6, 105–132, 1989.

9. Campbell, G. A., and N. Dontula, Solids Transport in Extruders, *Int. Polym. Process.*, 10, 30–35, 1995.

10. Maddock, B. H., A Visual Analysis of Flow and Mixing, *Soc. Plast. Eng. J.*, 15, 383–389, 1959.

11. Hiner, H. C., *A Model for the Plasticating Extrusion of Reactive Systems*, M.S. thesis, State University of New York at Buffalo, Buffalo, 1984.

12. Amellal, K., P. G. Lafleur, and B. Arpin, Computer Aided Design of Single-Screw Extruders, in *Polymer Rheology and Processing, A. A. Collyer and L. A. Utracki (eds.), Elsevier Applied Science, London, 1990.*

13. Chung, C. I., A New Theory for Single-Screw Extrusion, *Mod. Plastics*, 45(13), 178–198, 1968.

14. Lindt, J. T., Mathematical Modeling of Melting of Polymers in a Single-Screw Extruder, A Critical Review, *Polym. Eng. Sci.*, 25, 585–588, 1985.

15. Campbell, G. A., P. A. Sweeney, and J. N. Felton, Experimental Investigation of the Drag Flow Assumption in Extruder Analysis, *Polym. Eng. Sci.*, 32, 1765–1770, 1992.

16. Fenner, R. T., Extrusion (Flow in Screw Extruders and Dies), in *Computational Analysis of Polymer Processing*, J. R. A. Pearson and S. M. Richardson (eds.), Applied Science, London, 1983.

17. Stevens, M. J., *Extruder Principles and Operation*, Elsevier Applied Science, London, 1985.

18. Middleman, S., *Fundamentals of Polymer Processing*, McGraw-Hill, New York, 1977.

19. Bourry, D., F. Godbille, R. E. Khayat, A. Luciani, J. Picot, and L. A. Utracki, Extensional Flow of Polymeric Dispersions, *Polym. Eng. Sci.*, 39, 1072–1086, 1999.

20. Denson, C. D., and B. K. Hwang, Jr., The Influence of the Axial Pressure Gradient on Flow Rate for Newtonian Liquids in a Self Wiping, Co-Rotating Twin Screw Extruder, *Polym. Eng. Sci.*, 20, 965–971, 1980.

21. Wildi, R. H., and C. Maier, *Understanding Compounding*, Hanser, Munich, 1998.

22. Bravo, V. L., A. N. Hrymak, and J. D. Wright, Numerical Simulation of Pressure and Velocity Profiles in Kneading Elements of a Co-Rotating Twin Screw Extruder, *Polym. Eng. Sci.*, 40, 525–541, 2000.

23. White, J. L., *Twin Screw Extrusion*, Hanser, Munich, 1991.

24. Theberge, J., IM Alternatives Produce Performance Advantages, *Plastics Eng.*, 27–31, February 1991.

25. Pye, R. G. W., *Injection Mould Design*, 4th ed., Wiley, New York, 1989.

26. Rubin, I. I., *Injection Molding, Theory and Practice*, Wiley–Interscience, New York, 1973.

27. Spencer, R. S., and G. D. Gilmore, Residual Strains in Injection Molded Polystyrene, *Mod. Plastics*, 28, 97–108, 155, December 1950.

28. White, J. L., and H. B. Dee, Flow Visualization for Injection Molding of Polyethylene and Polystyrene Melts, *Polym. Eng. Sci.*, 14, 212–222, 1974.

29. Schmidt, L. R., A Special Mold and Tracer Technique for Studying Shear and Extensional Flows in a Mold Cavity During Injection Molding, *Polym. Eng. Sci.*, 14, 797–800, 1974.

30. Oda, K., J. L. White, and E. S. Clark, Jetting Phenomena in Injection Mold Filling, *Polym. Eng. Sci.*, 16, 585–592, 1976.

31. White, J. L., Fluid Mechanical Analysis of Injection Mold Filling, *Polym. Eng. Sci.*, 15, 44–50, 1975.

32. Kamal, M. R., Y. Kuo, and P. H. Doan, The Injection Molding Behavior of Thermoplastics in Thin Rectangular Cavities, *Polym. Eng. Sci.*, 15, 863–868, 1975.

33. Kuo, Y., and M. R. Kamal, The Fluid Mechanics and Heat Transfer of Injection Mold Filling of Thermoplastic Materials, *AIChE J.*, 22, 661–669, 1976.

34. Isayev, A. I., and C. A. Hieber, Toward a Viscoelastic Modelling of the Injection Molding of Polymers, *Rheol. Acta*, 19, 168–182, 1980.

35. Ryan, M. E., and T. S. Chung, Conformal Mapping Analysis of Injection Mold Filling, *Polym. Eng. Sci.*, 20, 642–651, 1980.

36. Kamal, M. R., E. Chu, P. G. Lafleur, and M. E. Ryan, Computer Simulation of Injection Mold Filling for Viscoelastic Melts with Fountain Flow, *Polym. Eng. Sci.*, 26, 190–196, 1986.

37. Manzione, L. T. (ed.), *Applications of Computer Aided Engineering in Injection Molding*, Hanser, Munich, 1987.

38. Lafleur, P. G., and M. R. Kamal, A Structure-Oriented Computer Simulation of the Injection Molding of Viscoelastic Crystalline Polymers, Part I. Model with Fountain Flow, Packing, Solidification, *Polym. Eng. Sci.*, 26, 92–102, 1986.

39. Oda, K., J. L. White, and E. S. Clark, Correlation of Normal Stresses in Polystyrene Melts and Its Implications, *Polym. Eng. Sci.*, 18, 25–28, 1978.

40. Greener, J., and G. H. Pearson, Orientation Residual Stresses and Birefringence in Injection Molding, *J. Rheol.*, 27, 115–134, 1983.

41. Oda, K., J. L. White, and E. S. Clark, Influence of Melt Deformation History on Orientation in Vitrified Polymers, *Polym. Eng. Sci.*, 18, 53–59, 1978.

42. Kamal, M. R., and P. G. Lafleur, A Structure-Oriented Computer Simulation of the Injection Molding of Viscoelastic Crystalline Polymers, Part II. Model Predictions and Experimental Results, *Polym. Eng. Sci.*, 26, 103–110, 1986.

43. Seferis, J. C., and R. J. Samuels, Coupling of Optical and Mechanical Properties in Crystaline Polymers, *Polym. Eng. Sci.*, 19, 975–994, 1979.

44. Walczak, Z. K., *Formation of Synthetic Fibers*, Gordon and Breach Science, New York, 1977.

45. Ziabicki, A., *Fundamentals of Fibre Formation*, Wiley, London, 1976.

46. Gupta, R. K., and K. F. Auyeung, Crystallization in Polymer Melt Spinning, *J. Appl. Polym. Sci.*, 34, 2469–2484, 1987.

47. Petrie, C. J. S., and M. M. Denn, Instabilities in Polymer Processing, *AIChE J.*, 22, 209–236, 1976.

48. Samuels, R. J., *Structured Polymer Properties*, Wiley, New York, 1974.

49. White, J. L., Dynamics and Structure Development During Melt Spinning of Fibers, *J. Soc. Rheol. Japan*, 4, 137–148, 1976.

50. Petrie, C. J. S., *Elongational Flows*, Pitman, London, 1979.

51. Denn, M. M., Continuous Drawing of Liquids to Form Fibers, *Annu. Rev. Fluid Mech.*, 12, 365–387, 1980.
52. White, J. L., Dynamics, Heat Transfer, and Rheological Aspects of Melt Spinning, A Critical Review, *Polym. Eng. Rev.*, 1, 297–362, 1981.
53. Denn, M. M., Fibre Spinning, in *Computational Analysis of Polymer Processing*, J. R. A. Pearson and S. M. Richardson (eds.), Applied Science, London, 1983.
54. Kase, S., and T. Matsuo, Studies on Melt Spinning, II. Steady State and Transient Solutions of Fundamental Equations Compared with Experimental Results, *J. Appl. Polym. Sci.*, 11, 251–287, 1967.
55. Kase, S., and T. Matsuo, Studies on Melt Spinning, I. Fundamental Equations on the Dynamics of Melt Spinning, *J. Polym. Sci.*, A3, 2541–2554, 1965.
56. Sakiadis, B. C., Boundary-Layer Behavior on Continuous Solid Surfaces, III. The Boundary Layer on a Continuous Cylindrical Surface, *AIChE J.*, 7, 467–472, 1961.
57. Gould, J., and F. S. Smith, Air-Drag on Synthetic-Fibre Textile Monofilaments and Yarns in Axial Flow at Speeds of up to 100 Metres per Second, *J. Textile Inst.*, 72, 38–49, 1980.
58. Bourne, D. E., and D. G. Elliston, Heat Transfer Through the Axially Symmetric Boundary Layer on a Moving Circular Fibre, *Int. J. Heat Mass Transfer*, 13, 583–593, 1970.
59. George, H. H., Model of Steady-State Melt Spinning at Intermediate Take-up Speeds, *Polym. Eng. Sci.*, 22, 292–299, 1982.
60. Matovich, M. A., and J. R. A. Pearson, Spinning a Molten Threadline—Steady State Isothermal Viscous Flows, *Ind. Eng. Chem. Fundam.*, 8, 512–520, 1969.
61. Chang, J. C., and M. M. Denn, An Experimental Study of Isothermal Spinning of a Newtonian and a Viscoelastic Liquid, *J. Non-Newtonian Fluid Mech.*, 5, 369–385, 1979.
62. Kase, S., Studies on Melt Spinning, III. Velocity Field Within the Thread, *J. Appl. Polym. Sci.*, 18, 3267–3278, 1974.
63. Denn, M. M., C. J. S. Petrie, and P. Avenas, Mechanics of Steady Spinning of a Viscoelastic Liquid, *AIChE J.*, 21, 791–799, 1975.
64. Sridhar, T., R. K. Gupta, D. V. Boger, and R. Binnington, Steady Spinning of the Oldroyd Fluid B, II. Experimental Results, *J. Non-Newtonian Fluid Mech.*, 21, 115–126, 1986.
65. Fisher, R. J., and M. M. Denn, Mechanics of Nonisothermal Polymer Melt Spinning, *AIChE J.*, 23, 23–28, 1977.
66. Gagon, D. K., and M. M. Denn, Computer Simulation of Steady Polymer Melt Spinning, *Polym. Eng. Sci.*, 21, 844–853, 1981.
67. Yasuda, H., H. Ishihara, and H. Yanagawa, Computer Simulation of Melt Spinning and its Application to Actual Process, *Sen-I Gakkaishi*, 34, 20–27, 1978.
68. Tung, L. S., R. L. Ballman, W. J. Nunning, and A. E. Everage, Computer Simulation of Commercial Melt Spinning Processes, *Proceedings of the Third Pacific Chemical Engineering Congress*, Seoul, Korea, 1983, pp. 117–123.
69. Denn, M. M., and J. R. A. Pearson, An Overview of the Status of Melt Spinning Instabilities, *Proceedings of the Second World Congress Chemical Engineers*, Montreal, Canada, 1981, pp. 354–356.

70. Ziabicki, A., and H. Kawai (eds.), *High-Speed Fiber Spinning*, Wiley, New York, 1985.
71. Tadmor, Z., and C. G. Gogos, *Principles of Polymer Processing*, Wiley, New York, 1979.
72. Pearson, J. R. A., *Mechanics of Polymer Processing*, Elsevier Applied Science, London, 1985.
73. Agassant, J. F., P. Avenas, J. P. Sergent, and P. J. Carreau, *Polymer Processing*, Hanser, Munich, 1991.
74. Grulke, E. A., *Polymer Process Engineering*, PTR Prentice-Hall, Englewood Cliffs, NJ, 1994.

PROBLEMS

15.1. Calculate the solids-conveying rate as a function of the screw rpm for an extruder for which $D = 5$ cm, $H = 0.75$ cm, $W = 4.25$ cm, and $\theta = 20°$. Assume that there is no pressure rise and that, for the polymer being extruded, $f_s = 0.1$ while f_b has the value 0.3. What would be the flow rate if the screw was frictionless?

15.2. Under what conditions will the solids-conveying rate through an extruder fall to zero? What is the pressure rise across the solids conveying zone in this case?

15.3. How does the maximum possible pressure rise in the metering section of an extruder depend on the screw rpm?

15.4. Relate the volumetric flow rate of a power law fluid to the screw rpm for flow through the metering section of an extruder that is operated at open discharge. How does the power-law index influence the results? Why?

15.5. If the flights are attached to the barrel and the screw (or shaft) is rotated, does the "extruder" actually extrude polymer? How does the volumetric flow rate in the metering section depend on the rate of rotation?

15.6. If the oil used in Example 15.2 has a viscosity of 1.6 P, how does the flow rate vary with screw rpm when a rod die is attached to the extruder? The die radius is 1 mm and the die length is 1 cm. Assume that for the extruder, L equals 20 cm.

15.7. By solving the Navier–Stokes equation in the x direction, obtain an expression for the x component of the velocity in the metering section of an extruder. Eliminate the pressure gradient in the x direction from this expression by noting that there cannot be any net flow in the x direction.

15.8. Use the results of Section 15.2.4 and Problem 15.7 to obtain an expression for the work done (per unit down-channel distance) by the moving barrel on the polymer melt in the metering section of the extruder.

15.9. If the flow rate is kept constant in Example 15.3, how does the pressure at the gate vary with time?

15.10. In reaction injection molding, two liquids that can react with each other are injected simultaneously into the mold. As reaction occurs, the viscosity increases and, for this reason, we want the filling stage to be over before an appreciable amount of reaction has taken place. With reference to Figure 15.22, what should be the minimum value of the constant flow rate Q if the two components are injected in equal concentrations and if the reaction follows second-order kinetics, that is,

$$\frac{dc}{dt} = -kc^2$$

where k is a rate constant, and we do not want c/c_0 to fall below 0.9? Here, c_0 is the initial concentration of either reactant.

15.11. Use Eqs. (15.4.11)–(15.4.13) to actually derive Eq. (15.4.14).

15.12. Make Eq. (15.4.17) dimensionless. Then, let the Reynolds number tend to zero and thereby obtain Eq. (15.4.18).

15.13. For the same pressure drop, by what percentage amount does the flow rate through a capillary change if the capillary radius is increased by 1%?

15.14. The equivalent form of Eq. (15.4.13) for the power-law fluid is

$$\frac{F}{A} = \eta_1 \left(\frac{dv}{dx}\right)^n$$

where η_1 is the extensional viscosity and n is the power-law index. Obtain the equivalent form of Eq. (15.4.18). Do you recover Eq. (15.4.18) if $n = 1$ and $\eta_1 = 3\eta$?

15.15. If the spinneret diameter in Example 15.4 is 0.7 mm, what is the residence time of a fluid element on the spinline?

Subject Index